AGRISCIENCE
Fundamentals & Applications

AGRISCIENCE
Fundamentals & Applications

ELMER L. COOPER

NOTICE TO THE READER

Publisher does not warrant or guarantee any of the products described herein or perform any independent analysis in connection with any of the product information contained herein. Publisher does not assume, and expressly disclaims, any obligation to obtain and include information other than that provided to it by the manufacturer.

The reader is expressly warned to consider and adopt all safety precautions that might be indicated by the activities described herein and to avoid all potential hazards. By following the instructions contained herein, the reader willingly assumes all risks in connection with such instructions.

The publisher makes no representations or warranties of any kind, including but not limited to, the warranties of fitness for particular purpose or merchantability, nor are any such representations implied with respect to the material set forth herein, and the publisher takes no responsibility with respect to such material. The publisher shall not be liable for any special, consequential or exemplary damages resulting, in whole or in part, from the readers' use of, or reliance upon, this material.

Cover Photo Credits: Background photo copyright Grant Heilman Photography.
Small photos top, left, and right courtesy of Cornell University Photography.
Small center photo courtesy of Agway, Inc.

Delmar Staff
Associate Editor: Joan M. Gill
Managing Editor: Gerry East
Project Editor: Carol Micheli
Production Supervisor: Karen Seebald
Design Coordinator: Susan Mathews

For information address Delmar Publishers Inc.
3 Columbia Circle Drive, PO Box 15-015
Albany, NY 12212-5015

20 19 18 17 16 15 14 13 12

Printed in the United States of America
Published simultaneously in Canada by Nelson Canada,
a division of The Thomson Corporation

Library of Congress Cataloging-in-Publication Data

Cooper Elmer L.
Agriscience: fundamentals & applications / Elmer L. Cooper.
p. cm.
Summary: An agriscience textbook exploring such topics as environmental technology, plant sciences, integrated pest management, interior and exterior plantscape, animal sciences, food science, and agribusiness.
ISBN 0-8273-3394-3.—ISBN 0-8273-3395-1 (Instructor's guide)
1. Agriculture. [1. Agriculture.] I. Title.
S495.C78 1990
630—dc20 89-23586
CIP
AC

Contents

Preface

Agriculture education has undergone a drastic series of changes over the past decade and it continues to change just as the industry and business of agriculture has experienced such changes. Today's agriculture education demands new approaches in teaching and new directions for learning. *AGRISCIENCE: FUNDAMENTALS AND APPLICATIONS* is a new textbook that addresses these new approaches and directions.

Written in a clear, easy-to-read style, *AGRISCIENCE: FUNDAMENTALS AND APPLICATIONS* integrates basic biological and technological concepts with principles of production agriculture. The student is taken on a journey through all of today's agricultural areas including environmental technology, plant sciences, integrated pest management, plant sciences, interior and exterior plantscape, animal sciences, food science and agribusiness. The text is an introduction to the field of agriscience designed to acquaint the student with all scientific and production concepts that are integrated throughout the text. Urban, suburban, exurban and rural agriscience concerns are addressed.

Ten major sections of this text are further divided into thirty-six units. Each unit contains an Agriscience Career profile, "performance-based objectives," "terms-to-know," and a list of "materials-needed." Chapters are clearly sub-divided and the chapter-end reviews include objective questions, essay questions, and many student activities. Mathematical and verbal skills are emphasized in both the activities and the review questions. Scientific concepts and agricultural projects are addressed in the activities. A comprehensive glossary can be found at the end of the text. The career profiles along with several agriscience career chapters are designed to acquaint the student with the many technical and professional careers associated with agribusiness and agriscience.

AGRISCIENCE: FUNDAMENTALS AND APPLICATIONS is a text designed for use in agriscience courses, agricultural survey courses, introductory agriculture courses, and general agriculture courses. It is an overview that introduces the student to many agriscience concepts, and readies them for a later in-depth treatment of each subject.

This text was prepared with the help of many individuals, associations, and corporations. Without their help and their commitment to agriculture and agriculture education, this text could not be properly written.

REVIEWERS

Ferrell M. Bridwell
Paul M. Dorman High School
Spartanburg, South Carolina

Gerald McDonald
Spring Branch Education Center
Houston, Texas

Marlies Harris
University of California, Davis
Davis, California

Steve McKay
Anderson Valley Agriculture Institute
Booneville, California

Robert A. Sills
Warren Hills Regional High School
Washington, New Jersey

Glenn B. Sims
Windsor High School
Windsor, Illinois

Jim Welles
Cherokee High School
Rogersville, Tennessee

Acknowledgments

The author and publisher wish to express their appreciation to the many individuals, FFA associations, and organizations who have supplied photographs and information necessary for the creation of this text.

A very special thank you to all the folks at the National FFA organization and the USDA photo libraries who provided many of the excellent photographs found in this textbook. Because of their efforts this is a better book. Special thanks also, to:

National FFA Organization
Bill Stagg, Photographer
United States Department of Agriculture
Agway, Inc.
FFA Chapter, Denmark, Wisconsin
Ken Seering, Photographer
The faculty and students of Anderson Valley High School, Boonville, California
Michael T. Gill, Bedford, Texas
U.S. Fish and Wildlife Service
Charles Cadieux, Photographer
Bob Hines, Artist
Lee Emery, Photographer
Earl W. Craven, Photographer
Bureau of Sport Fisheries & Wildlife
E. R. Kalmbach, Photographer
Jack Dermid, Photographer
E. P. Hadden, Photographer
Gale Monson, Photographer
Rogers Vocational Center, Gardendale, Alabama
Roy Holsomback, Photographer
John Deere Company
American Veterinary Medical Association
American Soybean Association
American Fisheries Society
John Scarola, Photographer
American Shetland Pony Club, Peoria, Illinois
Omaha Livestock Market, Inc.
Riburn Industries, Inc.
SHOWCASE Magazine, copyright 1989
Cailoway Nurseries, Hurst, Texas
Christian Children's Fund
Alison Cross, Public Relations Representative
FFA Chapter, Gilbert, Arizona
Joe Granio, Photographer
American Quarter Horse Association, Amarillo, Texas
Kevin Mathias, PhD, College Park, Maryland
The New Pesticide User's Guide, by Bert L. Bohmont
Vocational Agricultural Service, University of Illinois at Urbana-Champaign
From *The Fate of an Insecticide Discharged by Aircraft* (Flint and van den Bosch, 1977, after von Rumker et al., 1974)
Science of Food and Agriculture, CAST
Texas Vocational Instructional Services
From *Introductory Horticulture*, Third Edition, by Edward Reiley and Carroll Shry, copyright 1988 by Delmar Publishers Inc.
From *Ornamental Horticulture*, by Edward Plaster, copyright 1987 by Delmar Publishers Inc.
From *Soil Science and Management*, by Jack Ingels, copyright 1985 by Delmar Publishers Inc.
University of Maryland Cooperative Extension Service
The United Nations Fund for Population Activities
Instructional Materials Service, Texas A&M University
Flint, M. L. and R. van den Bosch, 1981. *Introduction to Integrated Pest Management*. Plenum Press, NY, page 178.
Penn State University, Entomology Department and Agricultural Extension Education Department
From *Maryland Magazine*, Autumn 1988, "Choking Off Life of the Bay" by Johnstone Quinan, copyright The Washington Post.
From *Turfgrass Science and Management*, by Robert Emmons, copyright 1984 by Delmar Publishers Inc.
From *Dairy Farm Management*, by Thomas Quinn, copyright 1980 by Delmar Publishers Inc.
From *Modern Livestock and Poultry Production*, Third Edition, by James Gillespie, copyright 1989 by Delmar Publishers Inc.
From *Agricultural Mechanics: Fundamentals and Applications*, by Elmer Cooper, copyright 1987 by Delmar Publishers Inc.
Chicago Board of Trade
From *Managing our Natural Resources*, by William Camp, copyright 1988 by Delmar Publishers Inc.
Childworld Magazine
USDA, Soil Conservation Service
Denmark High School, Agriculture Education Department, Photo by Jim Jones

Honey Bee Research, Weslaco, Texas
C. W. Hardeman, Scottsboro High School Agriculture Education Department
George Rogers Area Vocational-Technical School
ChemLawn Corporation
Brouwer Equipment, Ltd.
American Rabbit Breeders Association
National Livestock and Meat Board
Stebco Products

All photos in the color insert are courtesy of The National FFA Organization.

The author and publisher gratefully acknowledge the unique expertise provided by the contributing editors in the text. Their work provides a special assurance that the content in many units is at the cutting edge of the respective topic and is particularly appropriate for the level of the text. The contributing authors are:

Robert S. DeLauder, Agriscience Instructor, Damascus, Maryland;
Thomas S. Handwerker, Ph.D., Department of Agriculture, University of Maryland at Princess Anne;
Curtis F. Henry, Business Manager, College of Agriculture, University of Maryland at College Park;
Robert G. Keenan, Agriscience Instructor, Landsdowne High School, Baltimore, Maryland;
J. Kevin Mathias, PhD., Institute of Applied Agriculture, University of Maryland at College Park;
Regina A. Smick, EdD., Academic Advisor and Instructor, College of Agriculture, Virginia Tech, Blacksburg; and
Gail P. Yeiser, Instructor, Institute of Applied Agriculture, University of Maryland at College Park.

Finally, the work of Dollye L. Cooper, wife of the author and technical problem solver was especially helpful in standardizing the manuscripts of contributing authors. Her help in word processing and other details of manuscript preparation was invaluable.

Agriscience in the Information Age

The Science of Living Things

OBJECTIVE To recognize the major sciences that explain the development, existence, and improvement of living things.

Competencies to be Developed

After studying this unit, you will be able to

- ☐ define agriscience.
- ☐ discover agriscience in the world around us.
- ☐ relate agriscience to agriculture, agribusiness, and renewable natural resources.
- ☐ state the major sciences that support agriscience.
- ☐ describe basic and applied sciences that relate to agriscience.

TERMS TO KNOW

Agriscience
Agriculture
Agriculture/agribusiness and renewable natural resources
Agribusiness
Renewable natural resources
Technology
High technology
Aquaculture
Agricultural engineering
Animal science technology
Crop science
Soil science
Biotechnology
Integrated pest management
Organic food
Water resources
Environment
Turf
Biology
Chemistry
Biochemistry
Entomology
Agronomy
Horticulture
Ornamentals
Animal sciences
Agricultural economics
Agricultural education

MATERIALS LIST

writing materials
newspapers and magazines
encyclopedias

Life in the United States and throughout the world is changing every moment of our lives. The space we occupy as well as the people we work and play with may be constant for a brief time. However, these are quick to change with time and circumstances. The things we need to know and the resources we have to use are constantly shifting as the world turns around us.

Humans have the gift of intelligence—the ability to learn and know (Figure 1-1). This permits us to compete successfully with the millions of other creatures that share the earth with us (Figure 1-2). In ages past, humans have not always fared well in this competition. Wild animals had the advantages of speed, strength, numbers, hunting skills, and superior senses over humans. These superior senses of sight, smell, hearing, heat sensing, and reproduction all helped certain animals, plants, and microbes to exercise control over humans in order to meet their own needs.

The cave of the cave dweller, lake of the lake dweller, and cliff of the cliff dweller indicate human reliance on natural surroundings for basic needs (food, clothing, and shelter) (Figure 1-3). Those early homes gave humans some protection from animals and unfavorable weather. However, they were still exposed to the ravishes of disease, the pangs of hunger, the sting of cold, and the oppression of heat.

The world of agriscience has changed the comfort, convenience, and safety of people today. In the United States, we spend only 14% of our wages to feed ourselves (Figure 1-4). And, the agriculture/agribusiness and renewable natural resources of the nation provide materials for clothing and housing at an equally attractive price.

FIGURE 1-1
Humans have the gift of intelligence and the ability to learn and know. *(Courtesy Anderson Valley High School Agriculture Education Department)*

AGRISCIENCE DEFINED

Agriscience is a relatively new term which you are not likely to find in your dictionary. *Agriscience* is the application of scientific principles and new technologies to agriculture. *Agriculture* is defined as the activities concerned with the production of plants and animals, and the related supplies, services, mechanics, products, processing, and marketing. Actually, modern agriculture covers so many activities that a simple definition is not possible. So, the U.S. Department of Education uses the phrase "*agriculture/agribusiness and renewable natural resources*" to refer to the broad range of activities in agriculture. Agriculture generally has some tie-in or tie-back to animals or plants. However, production agriculture,

FIGURE 1-3
Early humans had to rely on natural settings to shield them from danger and the elements.

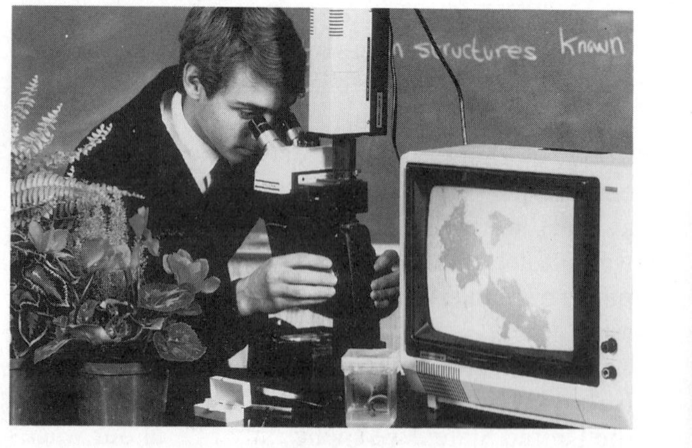

FIGURE 1-2
The gift of intelligence has permitted humans to compete with animals even though most animals are superior to humans in other ways.

FIGURE 1-4
Americans spend only 14% of their income on food. *(Courtesy USDA)*

or farming and ranching, accounts for only one-fifth of the total jobs in agriculture (Figure 1-5). The other four-fifths of the jobs in agriculture are nonfarm and nonranch jobs. *Agribusiness* refers to commercial firms that have developed with or stem out of agriculture (Figure 1-6).

Renewable natural resources are the resources provided by nature that can replace or renew themselves. Examples of such resources are wildlife, trees, and fish. Some occupations in renewable natural resources are game trapper, forester, and waterman (someone who uses boats and specialized equipment to harvest fish, oysters, and other seafood).

Technology is defined as the application of science to an industrial or commercial objective. Hence, the word agriscience was coined to describe the application of high technology to agriculture. *High technology* refers to the use of electronics and ultramodern equipment to perform tasks and control machinery and processes (Figure 1-7). It plays an important role in the industry of agriculture.

Agriscience includes many endeavors. Some of these are aquaculture, agricultural engineering, animal science technology, crop science, soil science, biotechnology, integrated pest management, organic foods, water resources, and environment. *Aquaculture* means the growing and management of living things in water, such as fish. *Agricultural engineering* means the application of mechanical and other engineering principles in agricultural settings. *Animal science technology* refers to the use

FIGURE 1-5
Farming and ranching accounts for approximately one-fifth of the agricultural jobs in the United States. ***(Courtesy Denmark High School Agriculture Education Department)***

FIGURE 1-6
Agribusinesses are important to the life of most communities.

FIGURE 1-7
High technology is used extensively in agriscience. ***(Courtesy USDA)***

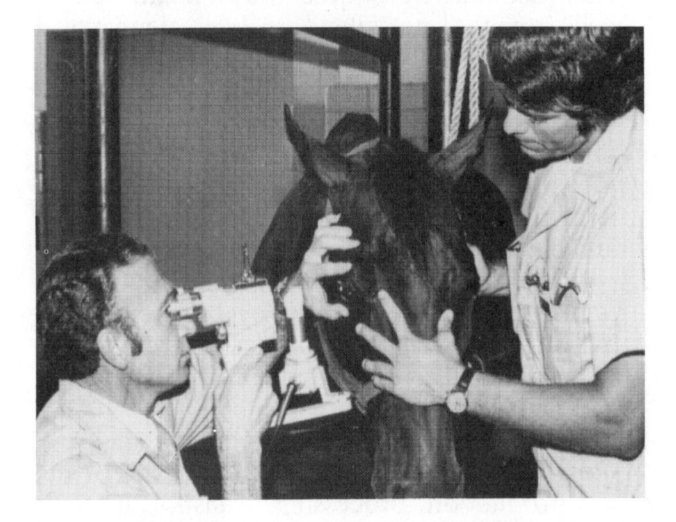

FIGURE 1-8
Veterinarians use animal sciences to keep our pets and production animals healthy. ***(Courtesy of the American Veterinary Medical Association)***

of modern principles and practices for animal growth and management (Figure 1-8). *Crop science* refers to use of modern principles in growing and managing crops. *Soil science* refers to the study of the properties and management of soil to grow plants. *Biotechnology* is a relatively new term referring to the management of the characteristics transmitted from one generation to another (Figure 1-9).

The phrase *integrated pest management* refers to the many different methods used together to control insects, diseases, rodents, and other pests. *Organic food* is a term used for foods that have been grown without the use of chemical pesticides. *Water resources* cover all aspects of water conservation and management. Finally, *environment* refers to the space and mass around us. This generally means air, water, and soil.

AGRI-PROFILE

CAREER AREA: AGRISCIENTIST

Students experience the wonder of the science and the technology of living things. ***(Courtesy of the National FFA Organization)***

In recent years science has played an increasing role in the lives of plants and animals and the people around them. These living bodies include plants ranging in size from microscopic bacteria to the huge redwood and giant sequoia trees. And they include animals from the one-celled amoeba to elephants and whales.

Only recently has science identified the nature of viruses and permitted humans to observe the submicroscopic world in which they exist. The electron microscope, radioactive tracers, computers, electronics, robots, and biotechnology are just a few of the developments that have revolutionized the world of living things. We call this world agriscience. Through agriscience, humans can control their destinies better than at any time in known history.

Agriscience spans many of the major industries of the world today. Some examples are production, processing, transportation, selling, distribution, recreation, environmental management, and professional services. Studies and experiences in a wide array of basic and applied sciences are appropriate preparation for a career in agriscience.

AGRISCIENCE AROUND US

Whether you live in the city, town, or country, you are surrounded by the world of agriscience. Plants use water and nutrients from the soil and release water and oxygen into the air. Animals provide companionship as pets and assistance with work. Both plants and animals are sources of food. Many microscopic plants and animals are silent garbage disposals. They decay the unused plant and animal remains around us. This process returns nutrients to the soil and has many other benefits to our environment and well-being.

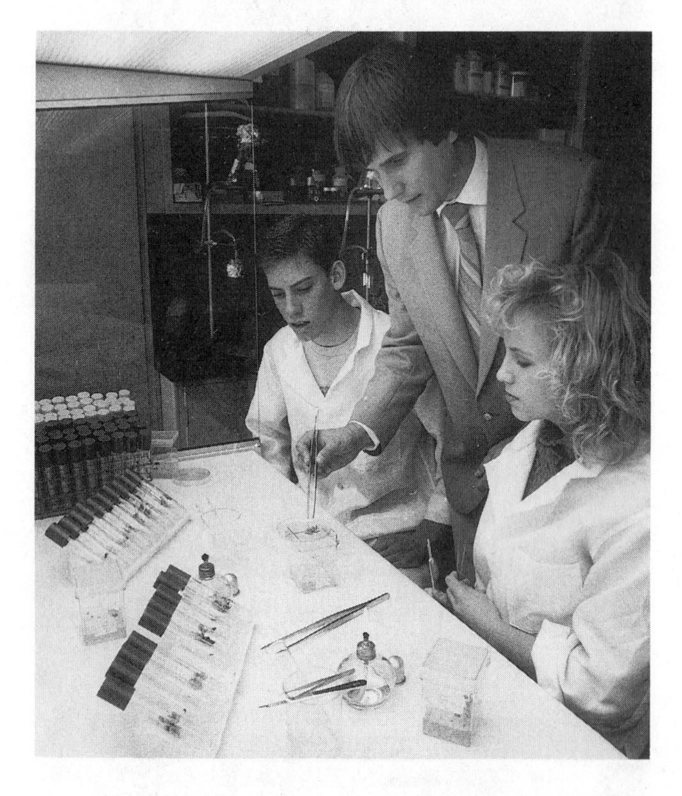

FIGURE 1-9

Genetic engineering and other forms of biotechnology have become one of the most important priorities in research today. ***(Courtesy of Anderson Valley High School Agriculture Education Department)***

Agriscience encompasses the wildlife of our cities and country, and the fish and other life in streams, ponds, lakes, and oceans. Plants are used extensively to decorate homes, businesses, shopping malls, buildings, and grounds. When one crop is used less, another takes its place. This occurs even where land changes from farm use to suburban and urban use.

Corn has long been referred to as king among crops in the United States. Yet, in one state, turf grass is challenging corn as the number one crop. *Turf* is grass used for decorative or soil-holding purposes. This change has occurred as more land has been used for roads, housing, businesses, institutions, recreation, and, in general, nonfarm use (Figure 1-10).

Many of the flowers used by florists in the United States come from Colombia, South America, and other foreign countries. Bulbs come from Holland and meat products are imported from Argentina. Lumber is shipped from the United States to Japan, only to return in the form of plywood and other processed lumber commodities. A drop in the price of sow bellies or an unexpected change in the price of grain futures in Chicago can affect business and investment around the world (Figure 1-11).

The great water-control projects on the Colorado River have permitted the transformation of the American Southwest from a desert to irrigated lands. This is now an area of intensive crop production which has stimulated national population shifts. Water management has transformed the great dust bowl of the American West into the "bread basket" of the world.

Agriscience helps create the magazines, newspapers, and journals of the nation. Similarly, radio and television rely on reporters, wildlife biologists, extension specialists, and others in agriscience. They help create programs about plants, animals, wildlife, market conditions, homemaking, gardening, lawn care, and dozens of related topics.

AGRISCIENCE AND OTHER SCIENCES

Agriscience is really the application of many sciences. Colleges of agriculture and life sciences do research and teach students in these sciences. *Biology* is the basic science of the plant and the animal kingdoms. *Chemistry* is the science that deals with the characteristics of elements or simple substances. It includes the behavior of substances when combined with other substances. *Biochemistry* focuses on chemistry as it applies to living matter. These three are referred to as basic sciences. One must understand the basic sciences to work in the applied sciences.

Applied sciences utilize basic sciences in practical ways. For instance, *entomology* is the science of insect life. A knowledge of biology and chemistry is necessary to understand insects and other animal and plant life.

Agronomy is the science of soils and field crops, while *horticulture* is the science of fruits, vegetables, and ornamentals. *Ornamentals* are plants used for their appearance. Examples are flowers, shrubs, trees, and grass. *Animal sciences* involve animal growth, care, and management. They include veterinary medicine, animal nutrition, and animal production and care.

Agricultural economics addresses the management of agricultural resources, including farms and

FIGURE 1-10
Turf has become one of America's main crops. ***(Courtesy Anderson Valley High School Agriculture Education Dept.)***

FIGURE 1-11
Marketing in agriscience has become big business. ***(Courtesy of the Chicago Board of Trade)***

agribusinesses. Farm policy and international trade are important components of agricultural economics. *Agricultural education* covers teaching and program management in agriculture. Agricultural communications, journalism, extension service, and community development are components of agricultural education. These and other disciplines are part of agriscience.

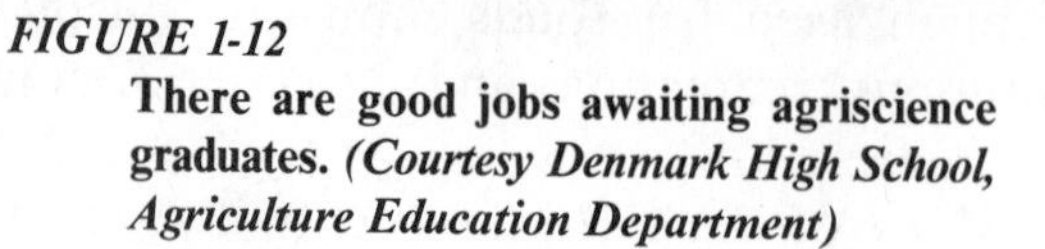

FIGURE 1-12
There are good jobs awaiting agriscience graduates. ***(Courtesy Denmark High School, Agriculture Education Department)***

A Place for You in Agriscience

What about career opportunities in agriscience? In 1988, it was reported that the nation's agricultural colleges were enjoying the strongest demand for graduates in a decade. A U.S. Department of Agriculture (USDA) study group forecasted a national shortage of 4,000 agricultural and life sciences graduates per year. Further, employers are offering higher salaries and more job variety to agriscience college graduates than ever before. Similarly, graduates of high school agricultural, horticultural, or other agriscience programs can obtain good jobs and have rewarding careers (Figure 1-12). These are described in a later unit in this text. By studying agriscience, you open the door to exciting educational programs and careers that bring life and prosperity.

Student Activities

1. Write the "Terms To Know" and their meanings in your notebook.
2. List examples of animals that have better senses than humans. Indicate the sense(s) along with the animal.
3. Write a paragraph or more each on cave, lake, and cliff dwellers, and explain how the type and location of their homes provided protection from 1) animals and 2) unfavorable weather. An encyclopedia would be a good resource for this activity.
4. Ask your teacher to assign you to a small discussion group to talk about the responses to number 3.
5. Place a map of your school community on a bulletin board, with a colored map pin in every location of a farm and/or agribusiness in your school community.
6. Talk to your County Extension Agent or another agricultural leader regarding the importance and role of agriculture/agribusiness and renewable natural resources in your county.
7. Select one of the sciences mentioned in this unit, prepare a written report on the meaning and nature of that science, then read it to the class.

Self-Evaluation

A. MULTIPLE CHOICE

1. Humans have the ability to learn and know. This is known as

   a. achievement.
   b. intelligence.
   c. intuition.
   d. spontaneity.

2. The percentage of an average U.S. worker's pay used for food is

   a. 10%.
   b. 14%.
   c. 50%.
   d. 74%.

3. The best term to describe the application of scientific principles and new technologies to agriculture is

   a. agribusiness.
   b. renewable natural resources.
   c. farming.
   d. agriscience.

4. Harmful insects, rodents, and diseases are all referred to as

   a. animals.
   b. plants.
   c. pests.
   d. parasites.

5. Agriscience encompasses

   a. wildlife and fish.
   b. ornamental plants and trees.
   c. farms and agribusinesses.
   d. all of the above and more.

6. Irrigated lands are generally used for

   a. intensive crop production.
   b. wildlife refuges.
   c. forests.
   d. boating and fishing.

7. An example of a basic science is

   a. agronomy.
   b. aquaculture.
   c. horticulture.
   d. chemistry.

8. An example of an applied science is

   a. animal science.
   b. biochemistry.
   c. biology.
   d. chemistry.

9. One relationship of agriscience with many other sciences is that

   a. agriscience is the application of many other sciences.
   b. agriscience is entirely different from all other sciences.
   c. agriscience is an old term and an old science.
   d. agriscience is a very narrow science and easily defined.

10. The career and job outlook in agriscience is

    a. a strong demand for college graduates.
    b. a shortage of 4,000 trained workers per year.
    c. higher salaries are being offered.
    d. all of the above.

B. MATCHING

______	1. Aquaculture	a. Commercial firms in agriculture
______	2. Renewable resource	b. Electronics and ultramodern equipment
______	3. Agribusiness	c. Growing in water
______	4. Chemistry	d. Basic science of plants and animals
______	5. High technology	e. Can replace itself
______	6. Biology	f. Characteristics of elements
______	7. Organic food	g. Space and mass around us
______	8. Environment	h. Grown without chemical pesticides

C. COMPLETION

1. Integrated pest management refers to the use of many different methods used together to ______ .
2. The transformation of the American Southwest from desert to irrigated lands was made possible, in part, by water-control projects on the ______ river.
3. By studying agriscience, you open the door to exciting educational programs, which should lead to ______ .

UNIT 2 Healthful Places to Live

OBJECTIVE To determine important elements of a desirable environment.

Competencies to Be Developed

After studying this unit, you will be able to

- ☐ survey the variety of living conditions in our society.
- ☐ describe the conditions of desirable living spaces.
- ☐ discuss the influence of climate on our environment.
- ☐ compare the influence of humans, animals, and plants on the environment.
- ☐ examine the problems of an inadequate environment.
- ☐ consider why agriscience is critical to our future.

TERMS TO KNOW

Sewerage system
Safe water
Polluted
Condominium
Townhouse
Famine
Contaminate
Urine
Feces
Parasite
Immune

MATERIALS LIST

paper
pen or pencil
current newspaper

Living conditions in the world vary extensively. In all countries, there are some very desirable places to live (Figure 2-1). Yet even in the highly developed countries, there are vast areas of poverty. How do you explain the differences in living conditions from one place to another? Why do living conditions vary from one community to another? From one neighborhood to another? From one house to another? The wealth and preferences of individuals explain some of the differences. Yet, the environment has much to do with the quality of life. It also has much to do with the way we feel about ourselves and others.

VARIETY IN LIVING CONDITIONS

The United Nations "State of World Population" report stated that, in 1987, the population of the world would reach the 5 billion mark (Figure 2-2). Somewhere, a new child had the distinction of being the 5 billionth human being living on the planet Earth (Figure 2-3). What was the home like where "Baby 5 Billion" was born? Was there adequate food? Was that child warm, but not too warm or too cold? Was there sufficient food? Was the child kept free from serious illness? Did the family have a house or good living space they could call home? Did they have clothing to permit them to live and work outside the home in relative comfort? What was the quality of life of others around the 5 billionth human being? Did that child survive and is that person living a happy life? Positive answers to these questions would indicate a good environment for a person to live. These same questions should be asked for the rest of humankind.

FIGURE 2-1

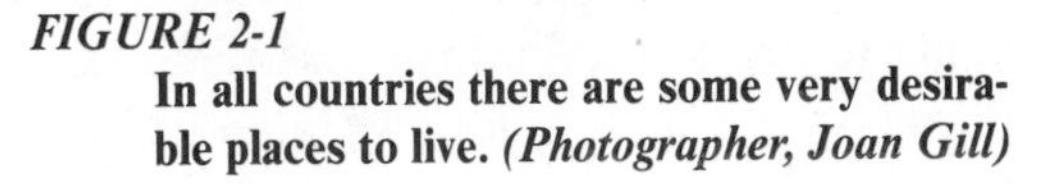

In all countries there are some very desirable places to live. ***(Photographer, Joan Gill)***

FIGURE 2-2
The population of the world exceeds five billion. *(Courtesy USDA)*

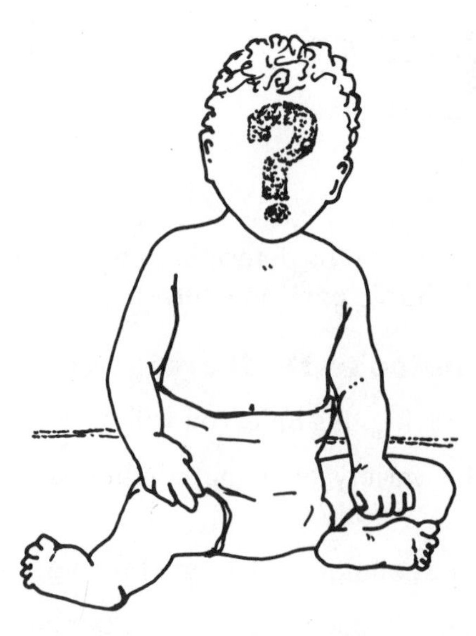

FIGURE 2-3
What kind of life for *Baby Five Billion*?

The Homes We Live In

Homes of the Very Poor

Homes of the most destitute range in size and quality from nothing to a piece of cardboard or a scrap of wood on the ground. Many survive the freezing winters with only a blanket on the warm sidewalk grates of our modern cities. For others, housing may take the form of a grass hut or a shack made of wood, cardboard, or scraps of sheet metal. The outdoors must provide water and washing areas, and serve as a receptacle for human waste. Large families generally share such homes with pets, poultry, or other livestock (Figure 2-4).

In cities, the poor frequently live in old buildings that are in very bad condition and with plumbing that does not work. Drugs, crime, poisonous lead paint, and disease are typical hazards for such people. The steamy streets and sidewalks provide little relief from the oppressive summer heat.

Homes of the Less Fortunate

People with modest sources of income may have homes that are simple, but provide basic protection from the elements. Such homes may be of wood, stone, masonry blocks, sheet metal, brick, or other fairly permanent material. The presence of windows and doors may provide some protection from the elements and the possibility for some privacy.

These people frequently have access to water that is safe to drink, but bathrooms are nonexistent and toilets do not have safe sewerage-disposal systems. A *sewerage system* receives and treats human waste (Figure 2-5). To be regarded as safe, a sewerage system must decompose human waste and release by-products that are free from harmful chemicals and disease-causing organisms. In most countries of the world, people rely on creeks or rivers to supply their drinking water, bathe the family, wash the clothes, and carry away the human waste.

FIGURE 2-4
Animals and people share the same home.

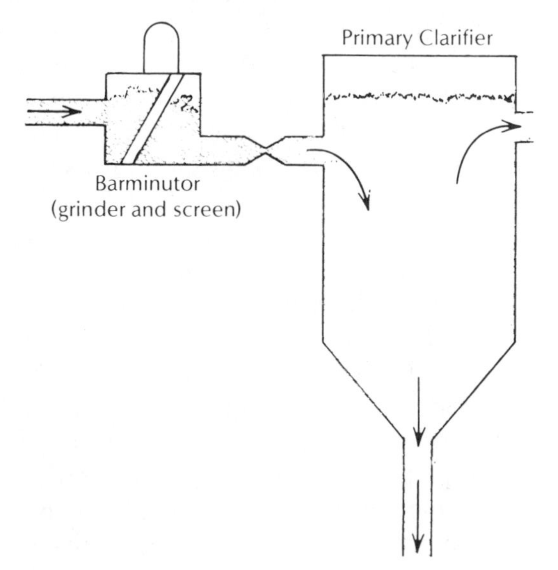

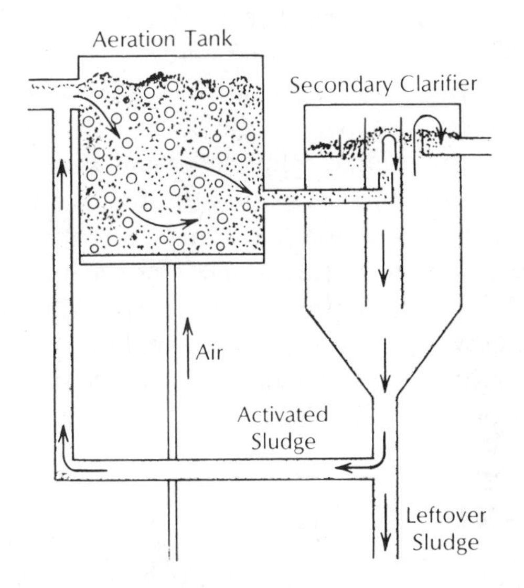

FIGURE 2-5
Safe sewerage disposal is essential to good health. ***(From Camp/Managing Our Natural Resources, copyright 1988 by Delmar Publishers Inc.)***

Housing for the Lower Classes People with low or modest income, in developed countries such as the United States, Canada, and Europe, may live in housing with bathrooms and running water. However, maintenance of the systems may be poor and ignorance of the users may cause conditions that are health hazards. Such living areas may have central heat in cold climates or modern stoves to provide heat for comfort and cooking. People in rural areas as well as cities may have some comforts of home if this is a personal priority.

In the undeveloped countries of Central and South America, Asia, and Africa, the lower classes are fortunate if there is a source of *safe water* (water that is free of harmful chemicals and disease-causing organisms) at the village center (Figure 2-6). Modest and simple running-water facilities are generally the first evidence of community development in many such areas. Even a single faucet with unpolluted water is a major step forward for many communities. *Polluted* means containing harmful chemicals or organisms.

Homes of the Middle and Upper Classes

The middle- and upper-class people of the world can afford and enjoy housing that is clean, safe, and convenient. Construction is generally of wood, masonry, concrete, glass, or steel. Such living spaces are often found as single houses. They are found in both rural and urban areas. In towns, villages, and urban areas, homes may also be in the form of townhouses, condominiums, or apartment buildings. A *condominium* is an apartment building or unit in which the apartments are individually owned. All living space of a single unit is on one floor. A *townhouse* is one of a row of houses connected by common side walls. Each unit is generally two or three stories high, giving the occupants more variety of living space.

FIGURE 2-6
For much of the world's population, even a single community faucet with safe water would be a luxury. ***(Courtesy of Christian Children's Fund, Alison Cross, Public Relations Representative)***

Modern homes are commonplace in the United States. However, this has been true only for a decade or two. During the early centuries of our history, most people living in town enjoyed better housing than those living in the country. However, in the 1940s and later, improved roads and other factors enabled rural residents to afford housing as good as that found in town, and some farmers, ranchers, and other rural people enjoyed running water, bathrooms, central heating, carpeted floors, and time-saving electrical appliances. But it took 20 to 30 years for this standard of living to reach a majority of the rural homes.

AGRI-PROFILE

CAREER AREA: COMMUNITY DEVELOPMENT
Youth activities creating healthful places to live lead to lifetime careers. ***(Courtesy of the National FFA Organization)***

Community developers work in a variety of settings ranging from home and family advising to the legislative halls of the nation. In the United States, the community developer may be the business person who buys the land and develops a plan for roads, open spaces, housing units, businesses, schools, churches, and other uses. Such plans must then be presented at public hearings and approved by various government and environmental agencies.

Others in community development may work with families and agencies to improve existing neighborhoods. Goals may include improvement of nutrition, availability of food, quality of air, purity of water, reduction of pollution, protection from diseases, reduction of accidents, and quality of recreational and health services. Workers in agriscience careers contribute substantially in creating healthful places to live.

Food Until the 1970s, much of the world went to bed hungry. Only a few countries had sufficient food for their people. And all countries had problems with distribution. So, not everyone had appropriate food for proper nutrition. Today, major *famines* (widespread starvation) are still a fact of life. During the latter part of the 1980s, serious famines have occurred in Africa, Asia, the Caribbean, and other parts of the world.

United Nations scientist John Tanner concluded that, in theory, the world could feed itself; but in practice it could not. In 1987, there were 730 million people not getting enough food for an active working life. And, in view of the world's recent population explosion, there were more hungry people than ever before. While some countries enjoy an adequate food supply from their own production and imports, most have many individuals who do not get proper nutrition.

Family Family may well be the dominant force that shapes the environment for most individuals. The family has control of the household activities and sets the priorities of its members. The family has considerable influence over the attractiveness of its surroundings and the warmth of relations among individuals.

For some, the family chooses the neighborhood and community where they live. A wise choice, however, is based on having the knowledge of better opportunities and the necessary resources to move to a better environment. For most of the world's population, the community where individuals are born is the community where they are raised and spend their lives.

Neighborhood and Community Neighborhood and community have substantial influence on the environment in which we live. Some communities have tree-lined country roads with attractive fields, pastures, or woodlands to provide variety in the landscape (Figure 2-7). Other communities may have the advantage of attractive homes, businesses, or community centers. Urban areas may boast high-rise buildings for work and residence. These provide beautiful vistas of city lights or harbor scenes of commerce and recreation.

Neighborhoods and villages are parts of larger communities. These are influenced greatly by the families who live in the immediate area. If families work together toward common goals, they can shape the character, education, religious activities,

FIGURE 2-7
A tranquil community *(Photographer, Michael Gill)*

FIGURE 2-8
Topography is an important factor impacting on the environment. *(Courtesy USDA)*

social outlets, employment opportunities, and other broad aspects of their environment.

Climate and Topography Climate and topography are also important factors affecting our environment. Average annual temperatures are very high near the equator. Yet people living near ocean waters, even in tropical areas, enjoy a moderate climate with cool breezes most of the time. Inland, the inhabitants are likely to experience hot, humid weather with high rates of rainfall. The high rainfall, in turn, stimulates heavy plant growth, resulting in jungle conditions. Similarly, sea-level elevations may create balmy 80° temperatures, while a short trip to the top of a nearby mountain may reveal snow on its peak (Figure 2-8).

Northern areas, such as Alaska, may border on the Arctic Circle and have long frigid winters. Yet those same latitudes enjoy summers suitable for short-season crops. People inhabit most areas of the earth, so the climate and topography where they find themselves create environmental conditions that influence their quality of life.

INFLUENCE OF HUMANS, ANIMALS, AND PLANTS

Humans have the option to favorably influence the environment. Or we can work against the forces of nature to reduce the beauty, healthfulness, and safety of our surroundings. Humans and animals have body processes that use nutrients from the food we eat. We also have processes to remove waste products, poisons, and disease organisms from food, drink, and the air. However, the body is limited in its capacity to remove the poisons and organisms. Therefore, to remain healthy, humans and animals must limit their exposure to disease organisms and poisons.

Contamination by Humans and Animals A major problem for humans and animals is to avoid *contaminating* (to add material that will change the purity or usefulness of a substance) food and water with secretions from their own bodies. *Urine* and *feces* are liquid and solid body wastes, respectively. Therefore, they are serious contaminants of food and water. Similarly, however, certain diseases can be spread by body contact or by breathing contaminated air. An example would be breathing air expelled by a sneeze of a person with a cold virus or sore throat (Figure 2-9).

There are serious animal diseases spread to other animals through contact with their own body wastes. If animals have plenty of living space, this generally does not cause serious problems. But, as with humans, when animals are concentrated, health hazards increase. Fortunately, most diseases are spread among a given species of animal and not from one species to another. For instance, most diseases of dogs do not spread to cats. Similarly, most diseases of animals do not infect humans. However, there are some animal disorders that cause human sickness. Internal parasites (an organism that lives on another organism with no benefit to the host) and brucellosis are examples of animal disease organisms which may be transferred from animals to humans and create enormous health problems.

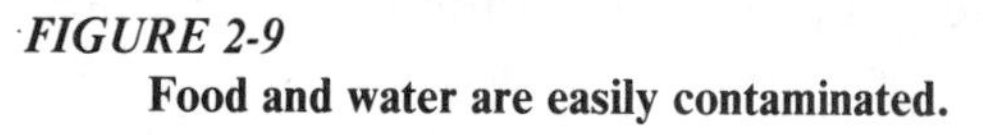

FIGURE 2-9
Food and water are easily contaminated.

FIGURE 2-10
Cockroaches are serious household pests that may contaminate food or beverages. *(Courtesy USDA)*

Contamination By Insects Insects impact heavily on our environment. The cockroach is an unwelcome guest in many households of the world (Figure 2-10). Some cockroaches feed on human waste and then on the food of humans. They transmit disease from waste material to food and water. In poor housing conditions, they can move from household to household. In doing so, they leave disease organisms and possible illness in their wake.

Plants are not immune from diseases and insect pests, nor damage from animals and humans. *Immune* means not harmed by. A plant relies on the nutrients and water it can extract from the soil and air. It cannot move, like animals and people, to sources of food and water. Therefore, it must depend on its tremendous capacity to reproduce in order to survive as a species. Reproduction is the plant's main function in life. However, plants accommodate both man and animals as they struggle to reproduce themselves. In doing so, they become an indispensable part of the environment.

Contamination By Chemicals Certain chemicals are serious threats to our environment (Figure 2-11). Oil spills and industrial chemical discharges have caused serious problems in rivers and streams. Chemical pesticides continue to threaten wildlife, fish, shellfish, beneficial insects, microscopic organisms, plants, animals, and humans. Over 25 years ago, American biologist Rachel Carson shocked the world with her book, *Silent Spring*. She presented convincing evidence of environmental damage being done by pesticides.

In 1972, DDT was banned in the United States because of its damaging effects on the environment. This insecticide had been used to control mosquitoes, which carried the dreaded malaria organism. DDT was also a very effective chemical used against flies and it enjoyed widespread use in homes, on farms and ranches, and wherever flies were a problem. Yet, because of its damaging effect on desirable organisms, it had to be discontinued and safer substitutes were found.

DDT was consumed by birds who ate insects killed by the insecticide. The chemical caused bird eggshells to be so thin that the eggs broke before the baby birds were born. The result was the disappearance in large areas of some species of birds.

It was reported that over 10,000 different pesticides are registered for use in the states that surround one of our major coastal bays. Needless to

FIGURE 2-11
Chemicals are needed in our modern society but threaten our health if misused or abused.

FIGURE 2-12
Plants are necessary to conserve our soil.

say, careful management and control of so many toxic materials is absolutely essential. It will require the utmost care to avoid unacceptable damage to our environment.

PROBLEMS WITH AN INADEQUATE ENVIRONMENT

As we ponder the life of "baby 5 billion," mentioned earlier in this unit, we must wonder if we are doing our part to preserve and enhance the environment. Plants, animals, insects, soil, water, and air must be kept in reasonable balance or all will suffer. Excessive plant growth can infringe upon the space for humans and animals. Yet excessive populations of humans and animals can consume plant species until it is unable to adequately reproduce itself. Too many animals can compete excessively with humans for food, water, and space. Some species of insects are regarded as harmful by people because they feed upon desirable crops or bother humans or livestock. However, many species of insects are beneficial to plants, animals, or humans.

Humans and animals tend to consume or remove plants, which hold soil in place and prevent erosion from wind and water (Figure 2-12). For instance, during the 1960s, most of the forests of China were cut and not replanted. Rapid and alarming soil erosion followed. Today, the government places a high priority on reforestation. Soil is needed to hold nutrients until plants need them. Similarly, we need the soil to filter and store clean water for plant growth and human and animal consumption. Plants take water from the soil and release water and oxygen to the air, which benefits humans and animals.

The number of human beings in the world is growing at the rate of 150 every minute; 220,000 a day; 80 million a year. At this rate, the earth's population will reach 6 billion by the year 2000; 7 billion by 2010; and 8 billion by 2022. Can the earth sustain such population growth? Will humans find enough to eat? Will we destroy our environment and, in doing so, destroy the system which supports life itself? Will we survive the competition of such population growth, but sacrifice our quality of life? Might we, in fact, improve our quality of life by using our intelligence to improve our environment?

AGRISCIENCE IN OUR GROWING WORLD

The keys to a prosperous future, indeed the bottom line for survival of the world's population, can be found in agriscience. Agriscience is the science of food production, processing, and distribution. It is the system that supplies fiber for building materials, rope, silk, wool, and cotton. It provides the grasses and ornamental trees and shrubs that beautify our landscapes, protect the soil, filter out dust and sound, and supply oxygen to the air.

Agriscience accounts for 20% of the jobs in the United States. It is the mechanism that permits the United States and other developed countries of the world to enjoy high standards of living. It is the system that nondeveloped countries are using in their efforts to feed and clothe their bulging populations. They look to agriscience for the necessary technology to enter the 21st century on a par with other nations. As we look to the 21st century, we must look to agriscience to maintain and improve our quality of life.

Student Activities

1. Write the "Terms To Know" and their meanings in your notebook.
2. Develop a bulletin board that illustrates the components of our environment.
3. Collect newspaper articles that describe environmental problems in your community.
4. Prepare a two- or three-page paper describing a good environment in which to live. Include factors such as home, community, air, water, cleanliness, wildlife, plants, and animals.
5. Ask your teacher to invite a public health official to your class to discuss health problems in the community and how they could be reduced by improving the environment.
6. Draw a chart that illustrates some relationships among plants, animals, trees, soil, water, air, and people.
7. Write all of the unfamiliar terms in this unit and look up their definitions in a dictionary.

Self-Evaluation

A. MULTIPLE CHOICE

1. In 1987, the world's population was

   a. 2 million.
   b. 200 million.
   c. 3 billion.
   d. 5 billion.

2. The material used for the most simple homes of the very poor is

   a. brick.
   b. cardboard.
   c. masonry block.
   d. wood.

3. To be regarded as safe, a sewerage system must

   a. be connected to a city system.
   b. be constructed from concrete block.
   c. decompose human waste.
   d. discharge into a stream or river.

4. Safe water is

   a. any water pumped from wells.
   b. water collected from a roof.
   c. free of harmful chemicals and organisms.
   d. water taken from free-flowing rivers.

5. Apartments on one level in large buildings and owned by the residents are called

   a. condominiums.
   b. single houses.
   c. townhouses.
   d. villas.

6. In 1987, the United Nations regarded the world's food capability as

a. it could feed itself in theory, but not do so in practice.
b. it could probably never keep up with population growth.
c. food supplies outpaced demand, and reduced production was recommended.
d. widespread famine could not be helped by better distribution.

7. Contaminants of food and water include

a. registered pesticides.
b. contact by cockroaches.
c. feces and urine.
d. all of the above.

8. While plants cannot move to food and water, they survive because of

a. their capacity to reproduce.
b. their ability to survive without food and water.
c. their roots, which extract water from any material.
d. parasites that convert water to nutrients.

9. The population of the world is projected to increase to

a. 6 billion by the year 2000.
b. 7 billion by the year 2010.
c. 8 billion by the year 2022.
d. All of the above.

10. Agriscience in the United States

a. accounts for 20% of the jobs.
b. is likely to diminish in importance.
c. reduces our standard of living.
d. is being replaced by biotechnology.

B. MATCHING

______	**1.** Neighborhood	**a.** Release oxygen into the air
______	**2.** 730 million	**b.** A priority in China
______	**3.** Starvation	**c.** Registered pesticides in one bay area
______	**4.** Parasite	**d.** Part of a community
______	**5.** Immune	**e.** Not harmed by
______	**6.** DDT	**f.** Number too hungry for an active work life
______	**7.** 10,000	**g.** Famine
______	**8.** Cockroach	**h.** Lives on another organism
______	**9.** Reforestation	**i.** Banned insecticide
______	**10.** Plants	**j.** Feeds on human waste and food

Human Efforts to Improve the Environment

OBJECTIVE To explore efforts made to improve the environment.

Competencies to Be Developed

After studying this unit, you will be able to

- ☐ compare world population patterns and trends.
- ☐ identify significant historical developments in agriscience.
- ☐ state practices used to increase productivity in agriscience.
- ☐ identify important research achievements in agriscience.
- ☐ describe future research priorities in agriscience.

TERMS TO KNOW

Ratio
Reaper
Combine
Moldboard plow
Cotton gin
Corn picker
Barbed wire
Milking machine
Tractor
Improvement by selection
Selective breeding
Genetics
Heredity
Genes
Legume
Tofu
Katahdin
BelRus
Aerosol
Beltsville Small White
Green Revolution
Feedstuff
Mastitis
Coccidiosis
Impatiens
Hybrid
Deficiency
Laser
USDA
Ice-minus
Bacteria
X-Gal
Microbe

MATERIALS LIST

paper
pencil or pen
encyclopedias
agriscience magazines

The human race can survive without the conveniences of industry and science. This was demonstrated after the fall of mighty nations, empires, and civilizations. Assyria, Egypt, Greece, Rome, China, and Mexico all had famous civilizations which later fell, with successive generations reverting to primitive methods. In each case, there is evidence of advanced culture, language, math, and engineering. Despite the loss of advanced knowledge, however, the people of these nations were able to survive. The possibility of survival was related to their ability to feed themselves. This was achieved by using satisfactory agricultural practices or by bartering goods for food.

Today, agriscience sustains human existence and determines the general status of nations. For if a few can produce the food and fiber to support the population, then many are freed to do other things. Since the first colonists, agriculture has been a major force in America. Today, the United States is a major world supplier of food. It is also a major supplier of fiber for clothing and trees for lumber, posts, piling, paper, and wood products. The use of ornamental plants and acreage devoted to recreation was probably never greater in the history of our country.

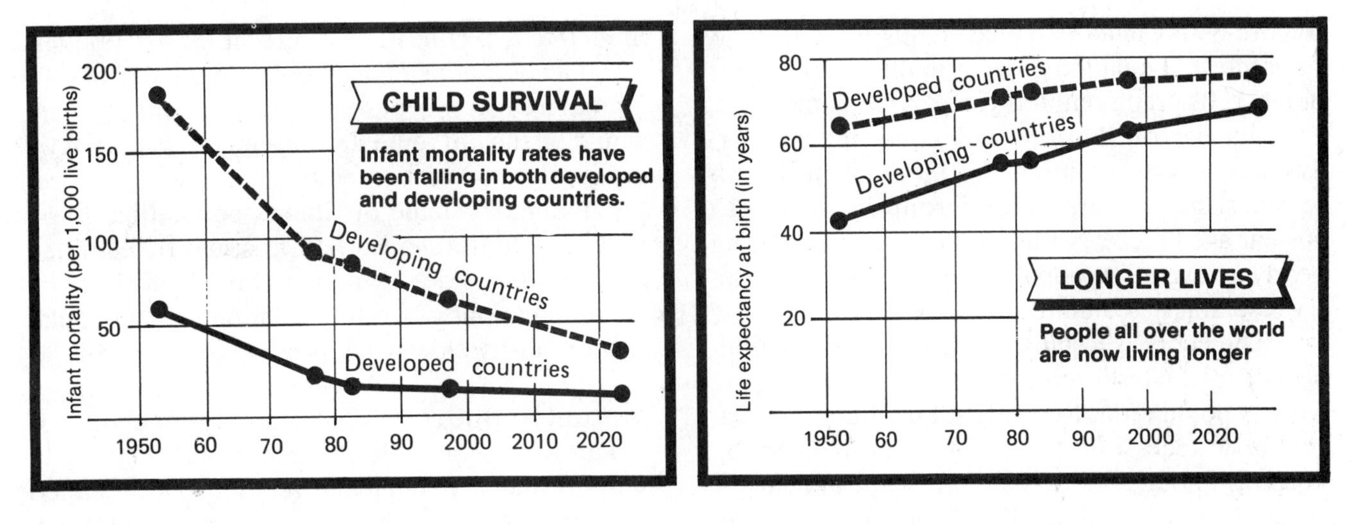

FIGURE 3-1

Worldwide child survival and life expectancy *(Courtesy United Nations Fund for Population Activities)*

A NEW AGE In 1987, the United Nations organization reported that more children than ever before were surviving to adulthood. It also indicated that adults were living longer (Figure 3-1). Together these trends mean more population growth and more pressure on the environment. Advancements in medical science and services have made good health and longer lives a reality, but only for those who can afford good nutrition and modern health services. Similarly, through agriscience we have made substantial gains in providing food, fiber, and shelter for the world (Figure 3-2). At the same time, the environment has stayed reasonably clean, considering the impact of bulging populations.

Changing Population Patterns

In the past, persons under the age of 25 constituted the world's largest population group. This occurred because children were valued for the help they provided in making the family living. Further, the young were the backbone of a nation's labor force and provided the manpower for the nations' armies. In most countries, the young respected their elders and provided for the needs of the elderly within the family.

Today, we are in a new age, based on population patterns. Honduras has the traditional pattern, with its largest number of citizens in the 0- to 5-year-old age group. Then the number per age group drops off to the smallest number occurring in the over 80-years-of-age group. When the Honduras population groups are displayed by sex in a bar graph, the graph takes the shape of a pyramid

#1.
Labor required to produce wheat, corn, and cotton (in hours)

	1800	1935-39	1955-59	1980-84
Wheat (100 bu.)	373	67	17	7
Corn (100 bu.)	344	108	20	3
Cotton (1 bale)	601	209	74	5

#2.
Yields per acre of wheat, corn, and cotton

	1800	1940	1960	1985-86
Wheat (bu.)	15	15	20	34
Corn (bu.)	25	29	55	118
Cotton (lb.)	154	253	446	630

FIGURE 3-2

Changes in agriscience productivity *(Courtesy "Agriculture Yesterday & Today," USDA)*

(Figure 3-3). Canada's pattern is slightly different. Its greatest population group is in the 15- to 20-year bracket. Its graph reminds you of a Christmas tree, with its narrow bottom and cone appearance. Sweden, a country known for its excellent health services, has its largest age group in the 35- to 40-year age bracket, with all other age groups being nearly as large. The population graph for that country resembles a column.

The People's Republic of China has over 1 billion people. This is approximately one-fifth of the world's population. Recently, China has been reasonably successful at feeding its population by keeping about 70% of its work force on farms. In contrast, only 2- to 3% of the work force in the United States is necessary to operate the nation's farms.

A Bold Experiment In the mid 1970s, China implemented a policy whereby each couple was limited to one child. They called it the 4-2-1 policy. This means that extended families consisted of four grandparents, two parents, and one child. What would be the outcome if such a policy were strictly enforced for several generations? You would expect the pyramidal shape of China's population graph to change to the shape of a Christmas tree, and, in time, to an upside-down pyramid! What would be the implications of feeding a nation with a population of mostly elderly people?

Mutual Support The working population of a nation must provide the goods and services to support the very young and very old. This is called mutual support. Each person, during his lifetime, passes through the stages of receiving, giving, and receiving support. The population patterns discussed previously provide indications of the ratio or proportion of working-age people to the very young and very old.

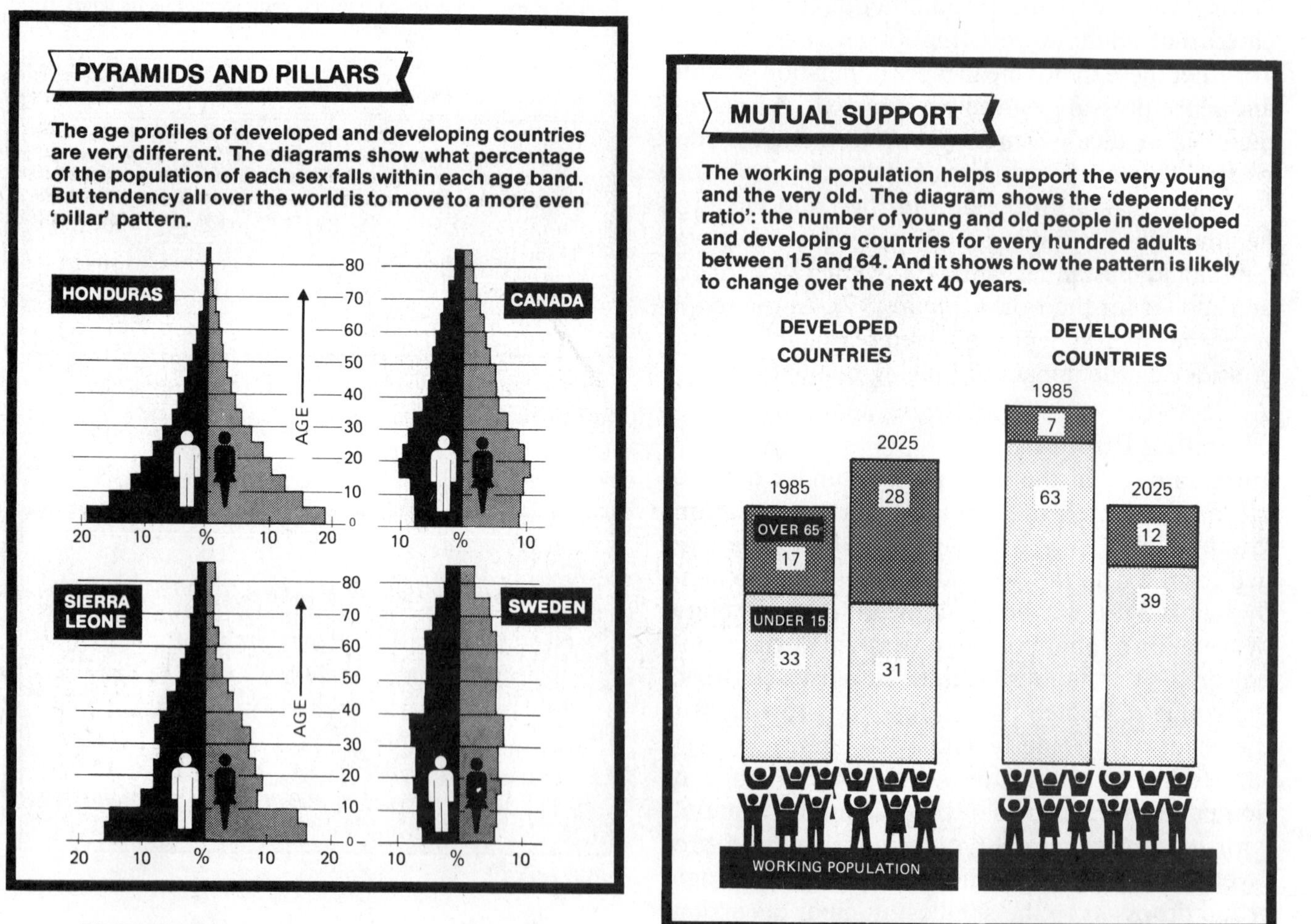

FIGURE 3-3
Age profiles and population patterns for developing and developed countries *(Courtesy United Nations Fund for Population Activities)*

FIGURE 3-4
Dependency ratios for developing and developed countries *(Courtesy United Nations Fund for Population Activities)*

Ratio means proportion. It is expressed as one value to another, such as 1 to 1, 1 to 2, 1 to 10, 10 to 1, or 5 to 2. Ratio may also include three values, such as 10-6-4, indicating the proportion of major ingredients in a bag of fertilizer. Countries that are highly developed tend to have a lower ratio of dependent young and old people to working people. Developing countries, on the other hand, now tend to have higher ratios (Figure 3-4). But, by the year 2025, it is projected that the ratio of working-age people to the very young and very old will be lower in developing countries than in developed countries.

One possible explanation for such a shift is that the current high population of the very young in developing countries will become the broad band of working-age citizens of the future. This will probably be accompanied by reduced birth rates and fewer young people, which will create a narrow base or the Christmas tree shape in the population graph.

FIGURE 3-5

In the 1800s, there were many inventions that helped revolutionize agriculture in the United States and a few other countries. ***(Courtesy Gilbert High School Agriculture Department; owner Warren Helm)***

IMPACT OF AGRISCIENCE

History records little progress in agriculture for thousands of years. Then, starting in the early 1800s, the use of iron spurred inventions that revolutionized agriculture in the United States, British Isles, and northern Europe (Figure 3-5). However, for most of the world, the progress has been much slower.

There is evidence of humans, thousands of years ago, using forked pieces of tree limbs to loosen soil for planting. Yet similar devices are in use today in some parts of the world (Figure 3-6). In contrast, agriscience in the United States enables the average farmer to produce enough food and fiber for 112 people. However, large farm operators produce enough food and fiber for 200 or more people (Figure 3-7).

Progress Through Engineering

During colonial times, in what was to become the United States, 90% of the people were farmers. The technology was so limited that most people had to raise their own food or hunt or fish to survive. They could not produce or catch enough food to offer for sale, so most people had to be farmers, fishermen, hunters, or trappers in order to survive.

Mechanization through inventive engineering was an important factor in America's agricultural development. The change from 90% to about 3% of the population being farmers evolved over a 200-year period. Machines helped make this possible. The old saying that "necessity is the mother of invention" suggests the relationship between an inventor's problem and the use of previously ac-

FIGURE 3-6

Even today, there are people using pieces of tree limbs for plows in some countries, suggesting that agricultural progress has been painfully slow worldwide.

FIGURE 3-7
The large farm operator in the United States today produces enough food and fiber for over 200 people. *(Courtesy Denmark High School Agriculture Education Department)*

quired skills to solve that problem. The solution is frequently a new device or machine.

American Inventors America provided the inventors of many of the world's most important agricultural machines (Figure 3-8). In 1834, Cyrus McCormick invented the *reaper,* a machine to cut small grain. Later, a threshing device was added to the reaper and the new machine was called a combine. The *combine* cut and threshed the grain in the field. Today, one modern combine operator can cut and thresh as much grain in one day as 100 persons could cut and bundle in the 1830s.

Thomas Jefferson's invention of an iron plow to replace the wooden plow of the time was of great signficance. Later, in 1837, a blacksmith named John Deere experienced the frustration of prairie soil sticking to the cast-iron plows of the time. It became apparent that Jefferson's invention would not work in the rich prairie soils of the Midwest. Through numerous attempts at shaping and polishing a piece of steel cut from a saw blade, the steel *moldboard plow* evolved. That plow permitted plowing of the rich prairie soils of the great American West and launched the beginning of the John Deere Company.

In 1793, Eli Whitney invented the cotton gin. The *cotton gin* removes the cotton seed from cotton fiber. This paved the way for an expanded cotton and textile industry. In 1850, Edmund W. Quincy invented the mechanical *corn picker,* which removes ears of corn from the stalks. During the same era, Joseph Glidden developed *barbed wire,* with sharp points to discourage livestock from touching fences. This effective fencing permitted establishment of ranches with definite boundaries. In 1878, Anna Baldwin invented a *milking machine* to replace hand milking. And, in 1904, Benjamin Holt invented the *tractor,* which became the source of power for belt-driven machines as well as for pulling.

Formation of Machinery Companies Many of the early inventors worked alone or with one or two partners. They were all workers in the area of agricultural mechanics and, as such, in agriscience. By the early 1900s, the inventors or other enterprising people had formed companies to produce agricultural machinery or process agricultural products. This made invention a continuing process. Successive invention was used to improve earlier

FIGURE 3-8
America provided the inventors for many of the world's most important agricultural machines. *(Courtesy Denmark High School Agriculture Education Department)*

inventions and develop new equipment and supplies to meet the needs of a changing agriculture.

The development of mechanical cotton pickers and corn harvesters greatly expanded the output per farm. Significant expansion of American agriculture also resulted from the development of irrigation technology. Since the end of World War II, the mechanization of American agriculture has moved at a breath-taking pace (Figure 3-9).

Mechanizing Undeveloped Countries

In the undeveloped countries of the world, many engineers, teachers, and technicians have sought simple, tough, reliable machines to improve agriculture. In such countries, America's highly developed, complex, computerized, and expensive machinery does not work for long. Most countries do not have people trained for the variety of agricultural mechanics jobs that are needed to support America's agriculture.

A machine with rubber tires is useless if a tire is damaged and repair services are not available. Similarly, failure of an electronic device may cause a $50,000 piece of machinery to become junk in the hands of an unskilled person in a country without appropriate repair facilities. This is the case in most undeveloped countries in Central and South America, Asia, and Africa. For the undeveloped nations of the world, other aspects of agriscience must become the vehicles for advancing agricultural productivity.

Improving Plant and Animal Performance

Humans have improved on nature's support of plant and animal growth since they discovered that loosening the soil and planting seeds could result in new plants. Even prior to that discovery, they probably aided growth by keeping animals away from plants until fruit or other plant parts edible to humans were yielded.

FIGURE 3-9
Since World War II, the mechanization of American agriculture has moved at a breath-taking pace. ***(Courtesy Denmark High School Agriculture Education Department)***

Improvement by Selection

History documents the domestication of the dog, horse, sheep, goat, ox, and other animals thousands of years ago. Improvement by selection soon followed. *Improve-*

AGRI-PROFILE

CAREER AREA: ENVIRONMENTAL MANAGEMENT
Environmental management requires skills in observation, analysis, and interpretation. ***(Courtesy of the National FFA Organization)***

Management of the environment requires the attention of consumers as well as professionals. However, specialists in air and water quality, soils, wildlife, fire control, automotive emissions, and factory emissions all help maintain a clean environment against tremendous population pressures in many localities. Helicopter, airplane, and satellite crews gather important data for scientific analysis to help monitor the quality of our environment.

Individuals in environmental careers may work indoors or outdoors; in urban or rural settings; in boats, planes, factories, laboratories or parks; and in positions ranging from laborer to professional. Environmental concerns are high on global agendas today as nations attempt to head off global hunger and pollution.

ment by selection means picking the best plants or animals for producing the next generation (Figure 3-10).

As people bought, sold, bartered, and traded, they were able to get animals that had desirable characteristics, such as speed, gentleness, strength, color, size, and milk production. By obtaining animals with characteristics they preferred, the offspring of those animals would tend to imitate the characteristics of the parents. By accident, the owner was practicing *selective breeding,* or the selection of parents to get desirable characteristics in the offspring.

The chariot armies of the Egyptians and Romans, the might of the Chinese emperors, the speed of the invading barbarians into northern Europe, the strength of mounts carrying knights into battle, and the evasive Arabians of the desert provide convincing testimony to man's early success at breeding horses for specific purposes.

Improvement by Genetics

An Austrian monk named Gregor Johann Mendel is credited for discovering the effect of *genetics* (the biology of heredity) on plant characteristics. *Heredity* is the transmission of characteristics from an organism to its offspring through the genes of reproductive cells. *Genes* are the components of cells that determine the individual characteristics of living things. Mendel experimented with garden peas. He observed that there was a definite pattern in the way different characteristics were passed down from one generation to another.

In 1866, Mendel published a scientific paper reporting the results of his experiments. He had discovered that certain characteristics occurred in pairs; for example, short and tall in pea plants. Further, he observed that one characteristic seemed to be dominant over the other. If tall was the dominant characteristic, then a short plant crossed with a tall plant produced a tall plant. Similarly, any plant that was short did not possess a gene for tallness. Two crossed short plants always had short plants as offspring. This happened because there were no tall characteristics in either parent to dominate the characteristic of the offspring.

Mendel's work provides an excellent example of the power of the written word. His discoveries and conclusions would have been lost if they were not recorded. The usefulness of his discoveries was not recognized until long after his death. In 1900, other scientists reviewed his writings and built upon the knowledge he had reported. Today biologists credit his work as being the foundation for the scientific study of heredity. Principles of heredity apply to animals as well as plants. More information on the principles of heredity can be found in the units of the sections entitled Plant Sciences and Animal Sciences.

FIGURE 3-10
Improvement by selection has been an important process in agricultural development. ***(Courtesy Agway, Inc.)***

Improving Life through Agriscience Research

Unlocking the Secrets of the Soybean

Americans have long appreciated the extensive research on the peanut done by the American scientist George Washington Carver. Carver is credited with finding over 300 uses for the peanut. These include food for humans, feed for livestock, cooking fats and oils, cosmetics, wallboard, plastics, paints, and explosives.

Less known are the secrets of the soybean. The Chinese have known for centuries that the soybean is a versatile plant with many uses. Calling it the "yellow jewel," the Chinese are said to have

FIGURE 3-11

The soy bean is the world's most important source of vegetable oil and provides the basic materials for hundreds of products. ***(Courtesy of the American Soybean Association)***

grown the soybean 3,000 years ago. The strong flavor of the soybean itself is not appealing, but the bean is a legume and very nutritious. A *legume* is a plant that hosts nitrogen-fixing bacteria. These bacteria convert nitrogen from the air to a form that can be used by plants. Legume plants are excellent sources of protein for humans and animals.

A Chinese scholar is believed to have first made tofu from soybeans in 164 B.C. *Tofu* is a popular Chinese food made by boiling and crushing soybeans, coagulating the resulting soy milk, and pressing the curds into desired shapes. Today, tofu is a major food in the diet of China's population of 1 billion people. It provides a reasonably healthful diet. Tofu can be fermented; marinated; smoked; steamed; deep-fried; sliced; shredded; made into candy; or shaped into loaves, cakes, or noodles.

Soy oil is the world's most plentiful vegetable oil. It is first extracted from the soybean, and the material that is left is processed into a protein-rich livestock feed known as soybean oil meal. The components of the soybean are used for hundreds of items. These range from dozens of food products, to lubricants, paper, chalk, paint, and plastics (Figure 3-11). As early as 1940, Henry Ford evidently shocked journalists with an unusual demonstration. He slammed an ax into a Ford automobile-trunk door made from highly-resilient plastic. The new plastic was made from soybeans.

Baked Potatoes from the Northeast

Many improvements in our way of life can be traced to agriscience research. For instance, the U.S. Department of Agriculture developed many pest-resistant varieties of potatoes. A case in point is the work with the *Katahdin,* a popular potato variety of the 1930s. From the Katahdin, scientists developed the *BelRus,* a superior baking variety bred to grow well in the Northeast.

The Common Aerosol

Prior to World War II, death from malaria was commonplace in the tropics. The deaths of American soldiers from malaria triggered intensive research on the control of mosquitoes, the carrier of the malaria-causing organism. Development of the "bug bomb" resulted. Our present-day *aerosol* (a can with contents under pressure) resulted from that early research (Figure 3-12).

Turkey for the Small Family

In your grandparents' time, Thanksgiving was probably observed by having all the relatives in to consume the typical 30 lb. turkey. As families became smaller and more scattered, the need for such large birds decreased. But even people with small families liked turkey. The 30 lb. bird was too much, so the problem was to develop a breed of turkey that weighed only 8 to 12 lb. at maturity (Figure 3-13). The solution was the *Beltsville Small White* turkey, named after the Beltsville Agricultural Research Center in Maryland, where the breed was developed.

FIGURE 3-12

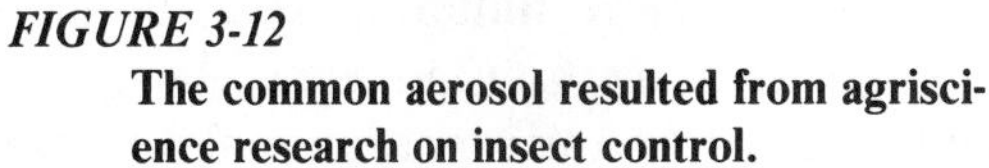

The common aerosol resulted from agriscience research on insect control.

FIGURE 3-13
Since development of the Beltsville Small White breed of turkeys, it is practical for even small families to enjoy whole turkeys. ***(Courtesy USDA)***

The Green Revolution During the 1950s, starvation was rampant in many countries of the world. A major question was: could the world's agriculture sustain the new population growth? The solution was partly in the development of new, higher-yeilding, disease- and insect-resistant varieties of small grain for developing countries. The result — the *Green Revolution,* a process whereby many countries became self-sufficient in food production by utilizing improved varieties and practices.

Cultivated Blueberries Wild blueberries were enjoyed in earlier times when people had time to pick the tiny berries growing in the wild. But labor costs were too high to harvest such berries for sale. The solution—development of high-quality, large-fruited blueberry varieties from the wild. This started the new and valuable cultivated-blueberry industry of today.

Nutritional Values Until rather recently, feeding of animals and human nutrition were based on poor methods of feed and food analysis. The problem—how can one recommend what to feed or what to eat if the content of food for humans and crops for livestock cannot be accurately determined? The solution—develop detergent chemical methods for determining the nutritional value of *feedstuff* (any edible material used for animals). The procedures are now used widely throughout the world in both human and animal nutrition.

Biological Attractants The use of chemical pesticides provided short-term solutions to many insect-control problems. However, it soon became apparent that chemicals have disadvantages and other means of control must be found. A partial solution—discover chemicals that insects produce and give off to attract their mates (Figure 3-14). These were, in time, produced in the laboratory. Laboratory production of these chemicals permitted mass trapping of insects to survey insect populations for integrated pest-management programs.

Examples of contributions made by agriscience could go on and on. Some of these will be presented in subsequent units of the text.

Recent Breakthroughs in Agriscience

Mastitis Reduced The mastitis organism has always been a serious problem for dairy farmers. *Mastitis* is an infection of the milk-secreting glands of cattle, goats, and other milk-producing animals. The resulting loss of milk production adds millions of dollars yearly to the cost of milk in the United States. Recent research developed abraded plastic

FIGURE 3-14
Attractants are man-made chemicals which imitate odors of females ready for mating. They are valuable products for attracting and killing insect pests without using toxic chemicals. ***(Courtesy USDA)***

loops for insertion into cattle udders. The procedure resulted in a 75% reduction in clinical mastitis. The reduction in infections resulted in increased milk production, averaging nearly 4 lb. of milk per cow per day.

Human Nutrition Recent studies in human nutrition have demonstrated the benefits of decreasing fat in the diet and increasing the proportion of fat from vegetable sources. This practice reduces high blood pressure. Much of the progress in human nutrition has grown out of research on animals and plants by agriscientists.

Fire-Ant Control Fire ants infest 230 million acres in the southern areas of the United States. Their presence in the warmer climates of the world is a constant threat to the well-being of humans and livestock. A new synthetic control for fire ants increases the ratio of nonproductive drone ants to worker ants. This ratio change gradually weakens the colony and causes it to die.

New Hope for Coccidiosis Control *Coccidiosis* is a disease that costs poultry growers nearly $300 million a year in the United States alone. Recently, the U.S. Department of Agriculture and research efforts by private industry genetically engineered a parasite constituent which stimulates birds to develop immunity to coccidiosis. Hopefully this will be the first step in the process of developing an effective vaccine against this persistent pest.

Exotic Flowers Horticulturists, gardeners, and hobbyists will be delighted with the new varieties of *Impatiens* (a popular, easy to grow, summer-flowering plant). Plant explorers introduced exotic new germ plasm and plant breeders developed a new technique called ovule-culture to develop hybrids and new kinds of Impatiens. A *hybrid* is the offspring of a plant or animal derived from the crossing of two different species or varieties.

Satellites and Nitrogen-Gas Lasers Nutrient deficiencies in growing corn and soybean crops are not easy to detect from the ground. *Deficiency* means something less available than needed for optimum growth. Yet new technology now permits the monitoring of deficiencies of iron, nitrogen, potassium, and other nutrients using nitrogen-gas *lasers* (a device used to determine wavelengths given off by the plants) from satellites. These wavelengths indicate the level of various nutrients in the plant (Figure 3-15).

Sugar Beet and Rice Hybrids The development of new varieties is a technique that has been used in agriscience for several decades to improve plant performance. A recent breakthrough has provided a sugar-beet hybrid with a high ratio of taproot weight to leaf weight. The hybrid yields about 15% more sugar per acre than previous varieties. On the other side of the world, Chinese agronomists recently developed hybrid rice. Hybrid rice is capable of yielding up to 40% more rice per acre than traditional varieties.

A study of USDA (an acronym for the U.S. Department of Agriculture) publications reveals great numbers of improved varieties, new products, and superior processes discovered or developed through agriscience research.

Importance of Being Blue During the mid and late 1980s, biotechnology became an important word in the scientific world. Biotech or biotechnology companies sprang up overnight and large agribusiness corporations spent millions of dollars on research in biotechnology. Federal and state governments had to scramble to debate and consider new laws and policies regarding biotechnical research to assure public safety.

Of great concern was the fear that scientists would modify the genetic makeup of organisms, which could damage our environment. There was also the concern that these organisms would multiply and spread out of control. Such organisms would have the potential for catastrophic damage.

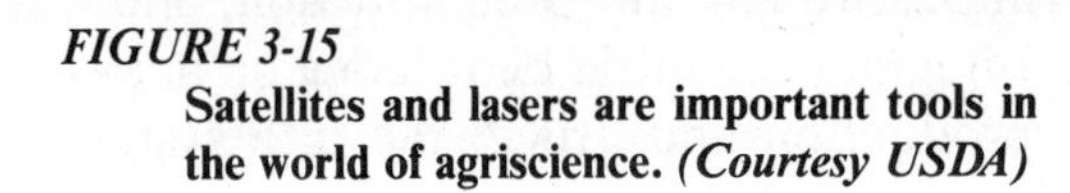

FIGURE 3-15
Satellites and lasers are important tools in the world of agriscience. ***(Courtesy USDA)***

In 1988, California scientists made the first outdoor tests of ice-minus. *Ice-minus* is bacteria that is genetically altered to retard frost formation on plant leaves. *Bacteria* are single-celled plants. Also in 1988, genetically altered bacteria were reportedly injected into elm trees in an effort to control Dutch Elm disease. Further, bacteria were genetically modified so researchers could readily detect their presence in the environment. This bacteria turns a brilliant shade of blue in the presence of a compound called *X-Gal* (Figure 3-16). It is hoped that other such procedures will be developed to increase the ease of tracing genetically changed *microbes* (any living organism that requires the aid of a microscope to be seen) in the future.

FIGURE 3-16
Ice-minus is genetically engineered bacteria whose presence can be detected because it turns a brilliant shade of blue when mixed with a material called X-Gal. ***(Courtesy USDA)***

AGRISCIENCE AND THE FUTURE

Agriscience will become more important in the future. As the world's population increases, it will require a more sophisticated agriscience industry to provide the food, clothing, building materials, ornamental plants, recreation areas, and open-space needs for the world's billions. Americans will have to work more in the international arena, as more countries become highly competitive in agriscience. Research and development will continue to play a dominant role as they lead the way in agriscience expansion in the future.

The USDA Agricultural Research Service has developed a strategic plan to guide the research efforts of the future. That strategic plan identifies and explains the main problems that confront the food and agriscience industry. It charts the minimum course of action that will provide the research needed for solutions. The six objectives of the strategic plan are to:

1. manage and conserve the nation's soil and water resources for a stable and productive agriculture;
2. maintain and increase the productivity and quality of crop plants;
3. increase the productivity of animals and the quality of animal products;
4. improve the delivery system and conversion of raw agricultural commodities into food and useful products for domestic consumption and export;
5. promote optimum human health and performance through improved nutrition; and
6. Integrate scientific knowledge on agricultural production and processing into systems that optimize resource management and facilitate the transfer of technology to end-users.

In 1988, a new and special research objective was developed to meet recent threats to the stability of American farms and other aspects of agriscience. American farm surpluses, low farm product prices, increased farmer reliance on chemical pesticides, lower government subsidies, and growing concerns about contamination of drinking water and food were cited as reasons for the new initiative. The new objective is to conduct research to find ways to reduce the cost of production while maintaining production efficiency. The cost of fertilizers, chemicals, machinery, supplies, and equipment is believed to be too high for the farmer to operate at a profit in the future. The new thrust is called "low input agricultural research and extension."

Student Activities

1. Write the "Terms To Know" and their meanings in your notebook.
2. Look up three prominent American inventors and describe the events that led to the invention that made them famous.
3. Make a model of one of the machines that strongly influenced agriscience development.
4. Make a collage depicting some important discoveries, inventions, and developments in agriscience.
5. Assume that a bar graph depicting China's population for 1975 was pyramid-shaped like that of Honduras. Make a bar graph to represent what the population pattern will look like after two generations of 4-2-1 families.

Self-Evaluation

A. MULTIPLE CHOICE

1. Historically, the largest age group of a typical nation was

   a. under 25.
   b. 26 to 50.
   c. 51 to 75.
   d. over 75.

2. Developing countries tend to have a high ratio of

   a. working-age people to the very young and very old.
   b. tractors to farm workers.
   c. nonfarm workers to farm workers.
   d. elderly to working-age people.

3. Agricultural development would be described as

   a. rapid until about 1850.
   b. very slow in most countries.
   c. very rapid in the United States in the last 50 years.
   d. rapid in the past, but decreasing today.

4. The inventor of the iron plow was

   a. Cyrus McCormick.
   b. John Deere.
   c. Joseph Glidden.
   d. Thomas Jefferson.

5. The combine is a combination of the reaper and a

   a. corn picker.
   b. cotton gin.
   c. cotton picker.
   d. threshing device.

6. Gregor Johann Mendel is remembered for his experimentation with

   a. birds.
   b. horses.
   c. insects.
   d. peas.

7. Yellow jewel is the name given by the Chinese to

   a. a special type of horse.
   b. a very young emperor.
   c. garden peas.
   d. soybeans.

8. Katahdin and BelRus are varieties of

a. bean sprouts.
b. celery.
c. potatoes.
d. soybeans.

9. Beltsville Small White is a

a. breed of rabbit.
b. breed of turkey.
c. type of building.
d. variety of soybeans.

10. The great advance in world food production in the 1960s was called the

a. biological attractants.
b. Green Revolution.
c. Greening of America.
d. Great Leap Forward.

B. MATCHING

_______	**1.** Aerosol	**a.**	Sense nutrient deficiencies
_______	**2.** Barbed wire	**b.**	Turns blue in X-Gal
_______	**3.** Coccidiosis	**c.**	"Bug bomb"
_______	**4.** Cotton gin	**d.**	Nitrogen fixation
_______	**5.** Genes	**e.**	Disease of poultry
_______	**6.** Ice-minus	**f.**	Sharp points to discourage livestock
_______	**7.** Impatiens	**g.**	Colorful flower
_______	**8.** Laser	**h.**	Eli Whitney
_______	**9.** Legume	**i.**	Heredity
_______	**10.** Low-input agriculture	**j.**	Curd from soybeans
_______	**11.** Mastitis	**k.**	John Deere
_______	**12.** Milking machine	**l.**	New research objective
_______	**13.** Steel moldboard plow	**m.**	Cyrus McCormick
_______	**14.** Reaper	**n.**	Infection of milk-secreting glands
_______	**15.** Tofu	**o.**	Anna Baldwin

The 21st Century and You

Career Options in Agriscience

OBJECTIVE To survey the variety of career opportunities in agriscience, observe how they are classified, and consider how you can prepare for careers in agriscience.

Competencies to Be Developed

After studying this unit, you will be able to

- ☐ define agriscience and its major divisions.
- ☐ describe the opportunities for careers in agriscience.
- ☐ compare the scope of job opportunities in farm and off-farm agriscience jobs.
- ☐ list activities in middle school, high school, and thereafter to help prepare for agriscience careers.
- ☐ identify resource people for obtaining career assistance in agriscience.

TERMS TO KNOW

Production agriscience
Agriscience processing, products, and distribution
Horticulture
Forestry
Agriscience supplies and services
Agriscience mechanics
Profession
Agriscience professions

MATERIALS LIST

paper
pencil or pen
bulletin board materials
agriscience magazines and pictures
Occupational Outlook Handbook or other agriscience career references

Life is possible without many of our modern conveniences, but not without food. Basic to life is an adequate supply of suitable food and other products of the soil, air, and water. This includes food for nourishment, fiber for clothing, and trees for lumber. Less obvious are alcohols for fuel and solvents, oils for home and industry, and oxygen for life itself. The industry that provides these vital basic commodities is agriscience. American agriscience is the world's largest commercial industry, with assets of nearly $1 trillion (Figure 4-1).

DEFINITION Agriscience is a term that includes all jobs relating in some way to plants, animals, and renewable natural resources. Such jobs occur indoors and outdoors. They include people in banking and finance; radio, television, and satellite communications; engineering and design; construction and maintenance; research and education; and environmental protection. All are in the field of agriscience if their products or services are related to plants, animals, and other renewable natural resources.

PLENTY OF OPPORTUNITIES Approximately 21 million people are employed in agriscience careers. About 400,000 people are needed each year to fill positions in this field. Of those vacancies, only 100,000 are currently being filled by people trained in agriscience (Figure 4-2). That means there are many opportunities for you. Only 20% of the careers in agriscience require college degrees, so you can use what you learn today in your current job, on

HOW MUCH IS ONE TRILLION DOLLARS?

You can count $1 trillion ($1,000,000,000,000) by using the following procedure:

One bill every second
Sixty bills per minute
Thirty-six hundred bills per hour
Eighty-six thousand per day
Thirty-one million five hundred thirty-six thousand per year
And continue counting for thirty-one thousand seven hundred and ten years!

FIGURE 4-1 **The agriscience industry in the United States has assets of nearly one trillion dollars.**

Attractive Employment Outlook for Agriscience
400,000 people are needed each year as replacements
100,000 trained people available
300,000 openings each year for additional people trained in agriscience

FIGURE 4-2
The employment outlook is good for people trained in agriscience.

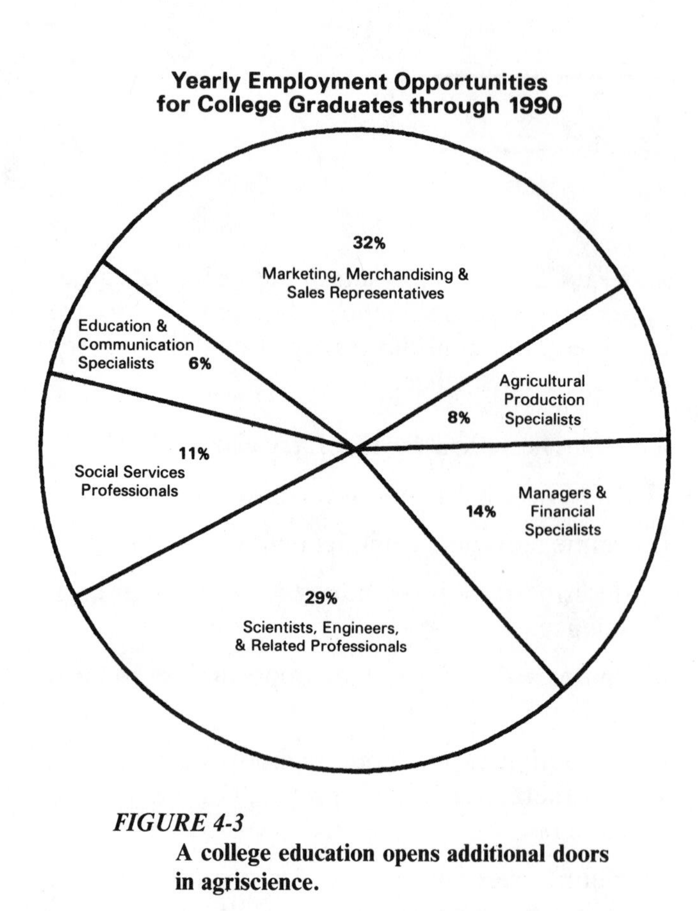

FIGURE 4-3
A college education opens additional doors in agriscience.

your farm, or in agriscience classes to go directly to fulltime employment. However, if you choose to pursue a college degree in agriscience, many additional career opportunities will be open to you (Figure 4-3).

Careers That Help Others

Helping others is an extra bonus with a career in agriscience. Of the jobs in processing, marketing, production, natural resources, mechanics, banking, education, writing, and other areas, many are people-oriented jobs. This means you have the extra benefit of being a product or process specialist and receive the special appreciation of others.

Careers That Satisfy

You might ask, "How can an agriscience career benefit me?" "What's in it for me?" Of course, there's the money. In agriscience, salaries vary tremendously from job to job. Generally, the better qualified individuals will be able to earn more. Agriscience industries employ one-fifth of all workers in the United States. And, there are job openings for skilled individuals at various levels of expertise. You can be a sharp, well-paid agriscience mechanic right out of high school, or work for an advanced degree and be an agriscience engineer. You can do what you decide is best for you.

THE WHEEL OF FORTUNE

Agriscience is like a wheel with a large hub. The hub of that wheel is production agriscience, or farming and ranching. The rest of the wheel is the nonfarm and nonranch activities in agriscience. Since so many opportunities for rewarding careers exist in that wheel, it may be called a wheel of fortune.

Production Agriscience

Production agriscience is farming and ranching. It involves the growing and marketing of field crops and livestock. Careers in this area account for one-fifth of all jobs in agriscience. In 1988, the average U.S. farmer produced enough food and fiber for at least 112 people (Figure 4-4). Large farm operators produced enough to feed over 200.

Most other agriscience careers are involved with goods or services that flow toward or away from production agriscience. Workers in those careers permit American farmers to supply goods so efficiently that American consumers spend only 14% of their income on food. This is the lowest percentage in the world. Out of five workers in agriscience, four have jobs that are not on farms. The nonfarm agriscience jobs may be in rural, suburban, or urban settings. The agriscience wheel consists of the hub, which contains one-fifth of the workers in production on farms, and the rest of the wheel, which contains the four-fifths of the nonproduction-type jobs that are off the farm (Figure 4-5).

Agriscience Processing, Products, and Distribution

Spin the wheel of fortune! What comes up for you? Agriscience processing, products, and distribution (Figure 4-6)!

FIGURE 4-4
The average United States farmer produces enough food and fiber for 112 people.

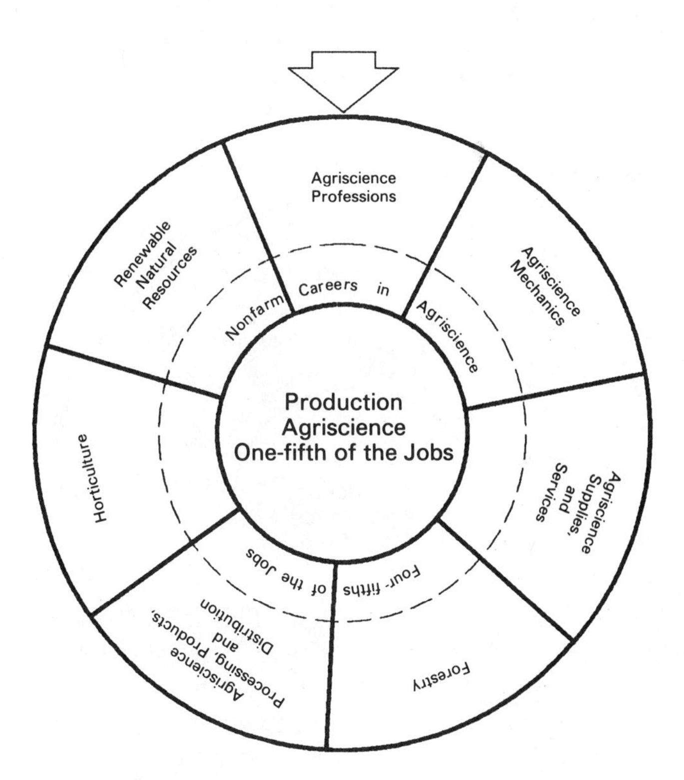

FIGURE 4-5
Agriscience may be compared to a wheel of fortune.

Agriscience processing, products, and distribution are those parts of the industry that haul, grade, process, package, and market commodities from production sources. Pick any item of food, clothing, or other commodity. Trace it back to its source. Except for metals and stone, most objects can be traced back to a farm or ranch, forest, greenhouse, or other agriscience production facility. If you consider a deluxe hamburger, you can trace the beef, mayonnaise, tomato, lettuce, pickle, catsup, mustard, relish, bun, and sesame seeds back to farms where they were produced (Figure 4-7). The same is true of many ingredients in soda, coffee, chocolate, or any other beverage you choose.

Check the label in your coat. Is it made of cotton, nylon, polyester, leather, vinyl, rubber, or wool? Each can be traced to a farm, ranch, or plantation. Your search may take you to a Maryland farm, a California ranch, a Colombian coffee plantation, or a trapper's lodge.

People with careers and jobs in agriscience processing, products, and distribution make it all possible. From hauling to selling, processing to mer-

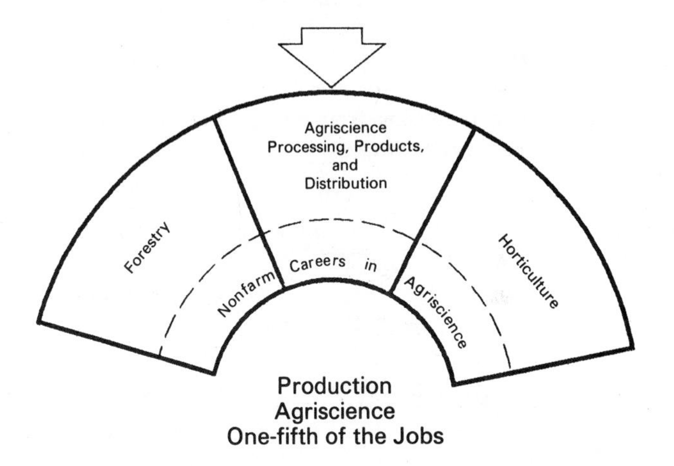

Ag Establishment Inspector
Butcher
Cattle Buyer
Christmas Tree Grader
Cotton Grader
Farm Stand Operator
Federal Grain Inspector
Food & Drug Inspector
Food Processing Supervisor
Fruit & Vegetable Grader
Fruit Distributor
Fruit Press Operator
Flower Grader
Grain Broker
Grain Buyer
Grain Elevator Operator
Hog Buyer
Livestock Commission Agent
Livestock Yard Supervisor
Meat Inspector
Meatcutter
Milk Plant Supervisor
Produce Buyer
Produce Commission Agent
Quality Control Supervisor
Tobacco Buyer
Weights & Measures Official
Winery Supervisor
Wool Buyer

FIGURE 4-6
Spin the wheel of fortune! What comes up for you? Agriscience processing, products, and distribution!

chandising, inspection, and research—the commodity moves from its source to consumption. The USDA reports that the producer's share of the food dollar is as low as 15.4 cents for cereal and bakery products. The rest is for handling, processing, and distribution.

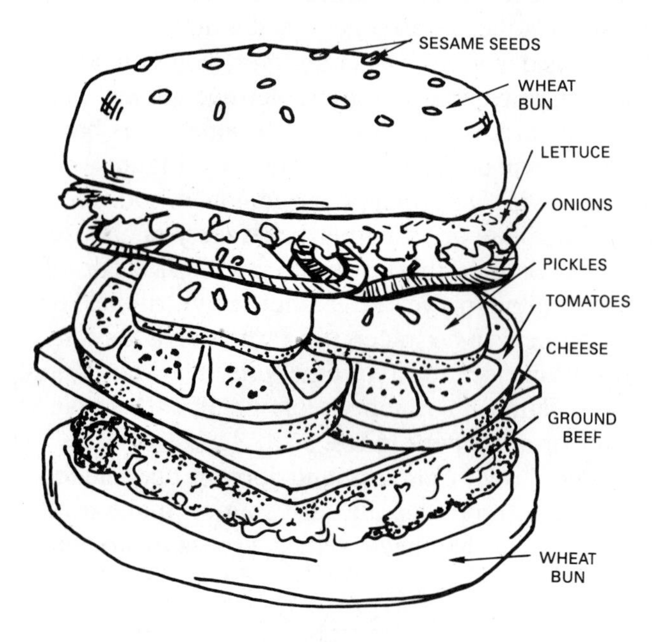

FIGURE 4-7
The components of a deluxe cheeseburger may have come from several states or even different countries.

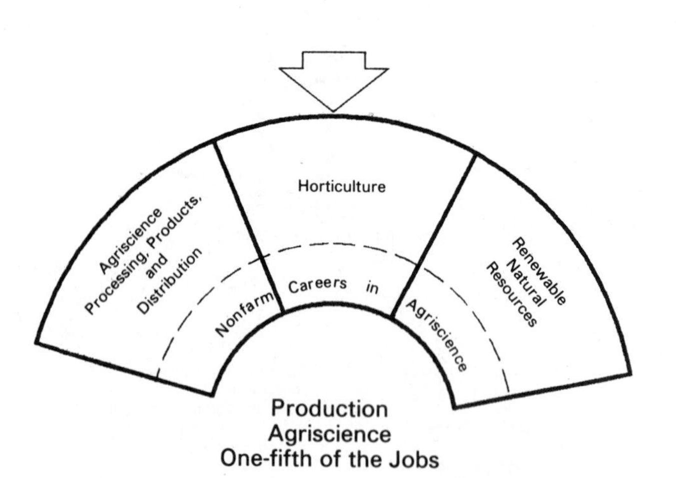

FIGURE 4-8
Spin the wheel of fortune! What comes up for you? Horticulture!

Horticulture Spin the wheel of fortune! What comes up for you? Horticulture (Figures 4-8 & 4-9)!

Horticulture includes producing, processing, and marketing fruits, vegetables, and ornamental plants (turf grass, flowers, shrubs and trees grown and used for their beauty). Horticultural production could be thought of as farming. But, since it is generally done on small plots, the production of horticultural crops is classified with horticulture rather than farming. Horticultural commodities are high-labor and high-income commodities.

FIGURE 4-9
Horticulture provides many opportunities in urban as well as rural areas.

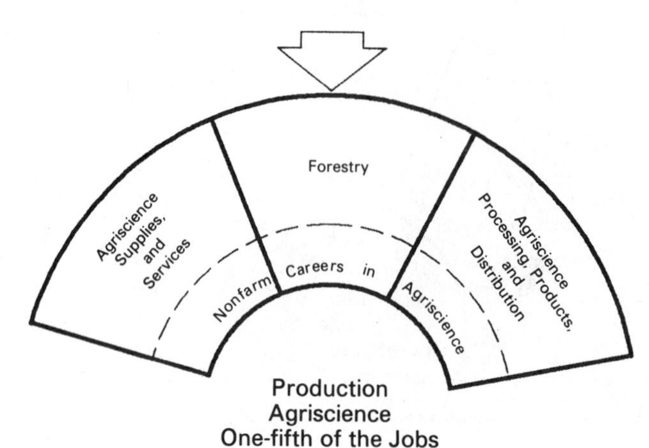

FIGURE 4-10
Spin the wheel of fortune! What comes up for you? Forestry!

The landscape designer, golf course superintendent, greenhouse supplier, greenhouse manager, flower wholesaler, floral market analyst, florist, strawberry grower, vegetable retailer—all are horticulturists. Recent data show more than 110,000 people employed by the floral industry alone.

Forestry Spin the wheel of fortune! What comes up for you? Forestry (Figure 4-10)!

Forestry is the industry that grows, manages, and harvests trees for lumber, poles, posts, panels, pulpwood, and many other commodities. Americans have huge appetites for wood products.

Careers in forestry range from growing tree seedlings to marketing wood products. Many jobs in forestry are outdoors and require the use of large machines to cut trees, drag logs, and load trucks (Figure 4-11). Other jobs are service-oriented, such as the state or district forester whose job is to give advice and administer governmental programs. Many find enjoyable careers in forestry research, teaching, wood technology, and marketing.

Renewable Natural Resources Spin the wheel of fortune! What comes up for you? Renewable natural resources (Figures 4-12 & 4-13)!

Renewable natural resources involve the management of wetlands, rangelands, water, fish, and wildlife. All require people with an appreciation for nature and scientific knowledge of plants and animals. This area of agriscience is attractive to those who enjoy working in parks, on game preserves, or with landowners to preserve and enhance natural habitat, plants, and wildlife. Water quality and soil conservation are state and regional concerns of high priority. New career opportunities in natural resource management are resulting from new efforts to save our rivers and bays.

FIGURE 4-11
The forestry industry provides many opportunities for work outdoors. ***(From Camp/Managing Our Natural Resources, copyright 1988 by Delmar Publishers Inc.)***

Agriscience Supplies and Services Spin the wheel of fortune! What comes up for you? Agriscience supplies and services (Figures 4-14 & 4-15)!

Agriscience supplies and services are businesses that sell supplies to agencies that provide services for people in agriscience. Examples of supplies are seed, feed, fertilizer, lawn equipment, farm

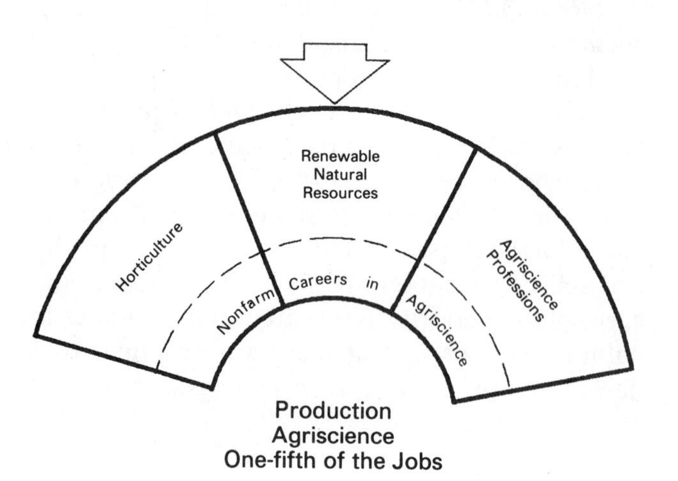

Animal Behaviorist
Animal Ecologist
Animal Taxonomist
Environmental Conservation Officer
Environmentalist
Fire Warden
Forest Fire Fighter/Warden
Forest Ranger
Game Farm Supervisor
Game Warden
Ground Water Geologist
Park Ranger
Range Conservationist
Resource Manager
Soil Conservationist
Trapper
Water Resources Manager
Wildlife Manager

FIGURE 4-12
Spin the wheel of fortune! What comes up for you? Renewable natural resources!

FIGURE 4-13
Effective resources management is critical for maintaining a balanced environment. ***(Courtesy of the United States Department of Agriculture)***

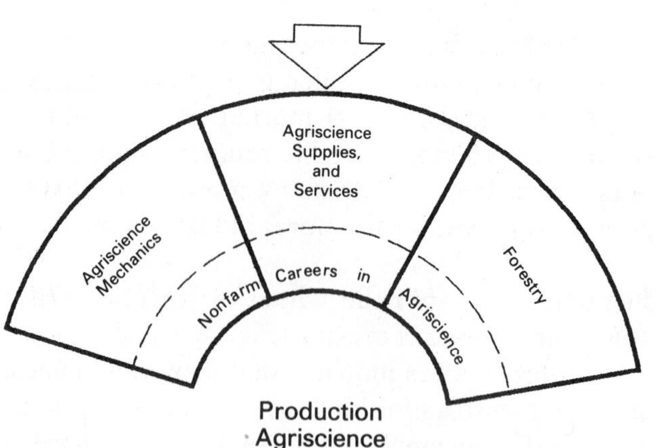

Aerial Crop Duster
Ag Aviator
Ag Chemical Dealer
Ag Equipment Dealer
Animal Groomer
Animal Health Products Distributor
Animal Inspector
Animal Keeper
Animal Trainer
Artificial Breeding Distributor
Artificial Breeding Technician
Artificial Inseminator
Biostatician
Chemical Applicator
Chemical Distributor
Computer Analyst
Computer Operator
Computer Programmer
Computer Salesperson
Custom Operator
Dairy Management Specialist
Dog Groomer
Farm Appraiser
Farm Auctioneer
Farrier
Feed Mill Operator
Feed Ration Developer & Analyst
Fertilizer Plant Supervisor
Fiber Technologist
Field Inspector
Field Sales Representative, Agricultural Equipment
Field Sales Representative, Animal Health Products
Field Sales Representative, Crop Chemicals, Machinery
Harness Maker
Harvest Contractor
Horse Trainer
Insect & Disease Inspector
Kennel Operator
Lab Technician
Meteorological Analyst
Pest Control Technician
Pet Shop Operator
Poultry Field Service Technician
Poultry Hatchery Manager
Poultry Inseminator
Sales Manager
Salesperson
Service Technician
Sheep Shearer

FIGURE 4-14
Spin the wheel of fortune! What comes up for you? Agriscience supplies and services!

FIGURE 4-15
Agriscience supplies and services provide the vital materials and services to keep a trillion dollar industry moving. ***(Courtesy Denmark High School Agriculture Education Department)***

machinery, hardware, pesticides, and building supplies. These businesses are operated by owners, managers, mill operators, truck drivers, sales personnel, bookkeepers, field representatives, clerks, and others.

People in these jobs provide the supplies for agriculture. However, there are many in agriscience who seldom handle the commodities themselves. Instead they provide a service. Those who provide legal assistance, write agricultural publications, advise agriculturists on money matters, or provide advice on crops, livestock, pest control, or soil fertility are working in service occupations. Such jobs are for those who are more people-oriented than commodity-oriented.

Agriscience Mechanics Spin the wheel of fortune! What comes up for you? Agriscience mechanics (Figures 4-16 & 4-17)!

Are you fascinated by tools and equipment? Are you challenged by something that does not work? Are you creative and like to build things? If so, a career in agriscience mechanics may be for you. *Agriscience mechanics* is the design, operation, maintenance, service, selling, and use of power units, machinery, equipment, structures, and utilities in agriscience.

Agriscience mechanics includes the use of hand and power tools, woodworking, metalwork-

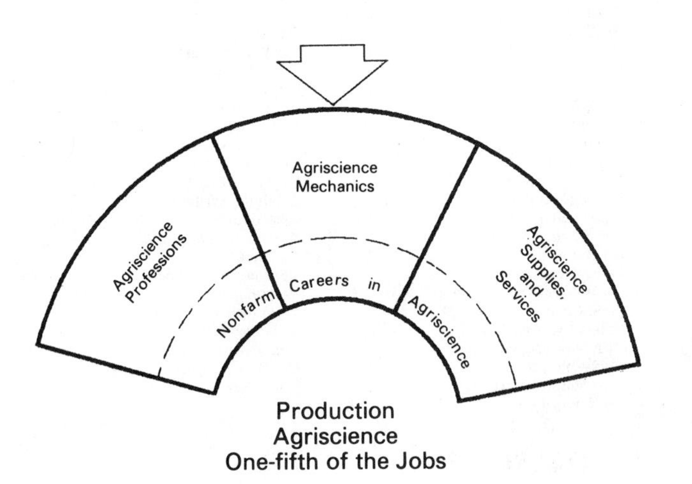

Ag Construction Engineer	Irrigation Engineer
Ag Electrician	Land Surveyor
Ag Equipment Designer	Machinist
Ag Plumber	Parts Manager
Ag Safety Engineer	Research Engineer
Diesel Mechanic	Safety Inspector
Equipment Operator	Soil Engineer
Hydraulic Engineer	Welder

FIGURE 4-16
Spin the wheel of fortune! What comes up for you? Agriscience mechanics!

AGRI-PROFILE

CAREER AREA: AGRISCIENCE TECHNICIAN/ **PROFESSIONAL**

Whether you prefer to be a technician or a professional, agriscience offers a broad array of career possibilities. ***(Courtesy of the National FFA Organization)***

A recent assessment of career opportunities in agriscience by the United States Department of Agriculture revealed a bright outlook for college graduates. Other studies indicate a need for workers trained in agriscience for technical-level jobs.

The agriscience technician may be broadly trained in plant and animal sciences and employable in many fields. For the sharp individual with good work habits and a broad background in plant and animal sciences, the choices are extensive. Add agriscience mechanics skills to the package and the individual has access to dozens of career options. Figure V4 illustrates some of them. Can you add others?

Education for agriscience careers should begin at the high school level or earlier. If one plans to work at the technician level, an early start will be especially helpful. The technician is expected to have first-hand experience and detailed knowledge of procedures and techniques. Such knowledge comes with experience in shops, laboratories, farms, greenhouses, fisheries, and on-the-job training situations, as well as in the classroom. Technicians generally have some special training beyond high school level, while professionals are required to obtain degrees at the bachelors, masters, or doctorate levels.

FIGURE 4-17
Careers in agriscience mechanics are varied and challenging. ***(Courtesy Denmark High School Agriculture Education Department)***

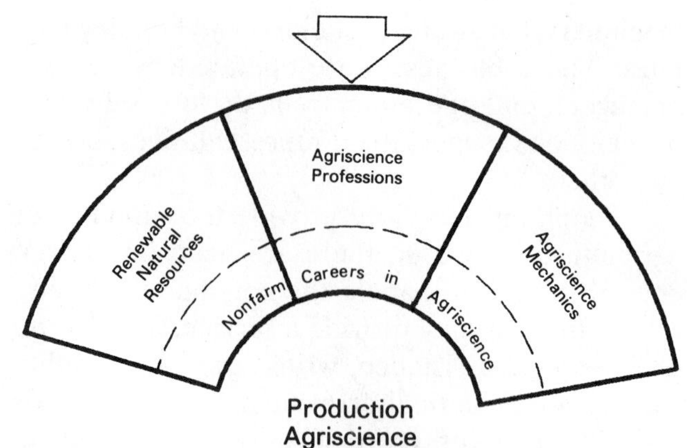

Ag Accountant
Ag Advertising Executive
Ag Association Executive
Ag Consultant
Ag Corporation Executive
Ag Economist
Ag Educator
Ag Extension Agent
Ag Extension Specialist
Ag Journalist
Ag Lawyer
Ag Loan Officer
Ag Market Analyst
Ag Mechanics Teacher
Ag News Director
Agriculture Attaché
Agronomist
Animal Cytologist
Animal Geneticist
Animal Nutritionist
Animal Physiologist
Animal Scientist
Agriculturist
Avian Veterinarian
Bacteriologist
Biochemist
Bioengineer
Biophysicist
Botanist
Computer Specialist
Credit Analyst
Dairy Nutrition Specialist
Dendrologist
Electronic Editor
Embryologist
Environmental Educator
Entomologist
Equine Dentist
4-H Youth Assistant
Farm Appraiser
Farm Broadcaster
Farm Investment Manager
Food Chemist
Foreign Affairs Official
Graphic Designer
Herpetologist
Horticulture Instructor
Hydrologist
Ichthyologist
Information Director
International Specialist
Invertebrate Zoologist
Land Bank Branch Manager
Limnologist
Magazine Writer
Mammalogist
Marine Biologist
Marketing Analyst
Media Buyer
Microbiologist
Mycobiologist
Nematologist
Organic Chemist
Ornithologist
Ova Transplant Specialist
Paleobiologist
Parasitologist
Pharmaceutical Chemist
Photographer
Plant Cytologist
Plant Ecologist
Plant Geneticist
Plant Nutritionist
Plant Pathologist
Plant Taxonomist
Pomologist
Poultry Scientist
Public Relations Manager
Publicist
Publisher
Reproductive Physiologist
Rural Sociologist
Satellite Technician
Scientific Artist
Scientific Writer
Silviculturist
Software Reviewer
Soil Scientist
Vertebrate Zoologist
Veterinarian
Veterinary Pathologist
Virologist
Viticulturist
Vocational Agriculture Instructor/
FFA Advisor

FIGURE 4-18
Spin the wheel of fortune! What comes up for you? Agriscience professions!

ing, welding, electricity, plumbing, tractor and machinery mechanics, hydraulics, terracing, drainage, painting, and construction. Choose your level—indoors or out. Choose your role—employee, employer, or professional.

Agriscience Professions

Spin the wheel of fortune! What comes up for you? Agriscience professions (Figures 4-18 & 4-19)!

The word *profession* means an occupation requiring an education, especially law, medicine, teaching, or the ministry. *Agriscience professions* are those professional jobs that deal with agriscience situations. They cut across all divisions in the wheel of fortune.

Consider the agriscience teacher and the Cooperative Extension agent. Both must have a bachelor's or master's degree to be qualified. They may teach subjects in several or all divisions of agriscience.

Consider the veterinarian, agriscience attorney, research scientist, geneticist, or engineer—all these professions require advanced degrees and high levels of training. If you can meet the standard, then a career in an agriscience profession may be for you.

Computers in Agriscience

The use of computers is extensive in agribusiness. This means there are many opportunities to combine computer skills with agriscience settings. These range from farms and ranches to horticulture and mechanics, from business and industry to research.

PREPARING FOR AN AGRISCIENCE CAREER

Career education is an important part of public education today.

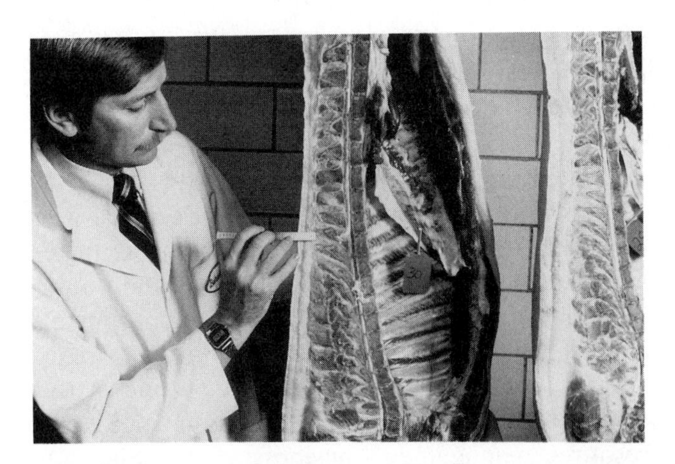

FIGURE 4-19

Professional workers are becoming more important than ever as agriscience enters the information age *(Courtesy Agway, Inc.)*

Since the early 1970s, the career education movement has spread from occupational education programs to the general curriculum. Many school systems now emphasize career education from kindergarten through adulthood.

As you consider a career in agriscience it is important to consider how to meet the requirements to get started in that career. Some young people have an early start on careers in agriscience. They may have grown up on a farm or ranch. Or their parents may have worked in one of the other areas of agriscience, such as horticulture, resource management, business, or teaching. Or they might have obtained jobs or worked for friends or neighbors who had agriscience businesses. The following are some suggestions for preparing for a career in agriscience.

While in Middle School

- □ Make science projects with plants, animals, soil, water, energy, ecology, conservation, and wildlife.
- □ Research and report on the above.
- □ Join 4-H or Scouts and choose agricultural projects and merit badges.
- □ Volunteer to work on lawn, garden, greenhouse, farm, or conservation projects.
- □ Study career education.

While in High School

- □ Enroll in agriscience classes, including plant science, animal science, agriscience mechanics, agribusiness, and farm management.
- □ Enroll in college-preparatory courses in English, math, and science.
- □ Join the FFA (Future Farmers of America) and participate in leadership, citizenship, and agriscience activities.
- □ Develop a broad, supervised, occupational experience program in agriscience.
- □ Acquire hands-on, skill-development experiences.

After High School

- □ Obtain an agricultural job and plan ways to get additional training while on the job.
- □ Enter a community college and take courses that will transfer to the college of agriculture and life sciences at your state university.
- □ Enter a two-year program in technical agriculture.
- □ Enter a college of agriculture and life sciences and obtain a bachelor's degree (B.S.), master's degree (M.S.), and/or doctorate (Ph.D.).

You can obtain information on careers, schools, and colleges from many sources. The following suggestions may be helpful to you.

High School Agriscience Teachers

- □ Agriscience mechanics
- □ General agriscience
- □ Animal sciences
- □ Horticulture

Cooperative Extension Service

- □ Listed in your phone book under county or city government

State Department of Education

- □ Specialist in agriculture/agribusiness and renewable natural resources
- □ FFA State Executive Secretary

Community Colleges and Other Post-secondary Institutions

- □ Occupational Dean, Community College
- □ Director or Dean, Institute or Technical School
  Typical programs include: agriscience business management, farm production and management, ornamental horticulture, water resources, and wildlife.
- □ Dean, College of Agriculture and Life Sciences
  Typical programs include: agricultural education, agricultural engineering, agricultural and resource economics, agronomy (crops and soils), animal sciences, food science, forestry, horticulture, natural resources management, and poultry science.
- □ Agricultural Education Coordinator, Department of Agricultural and Extension Education, Land Grant University

As new technologies and job opportunities emerge, so will the need for well-trained and educated new people. Agriscience is a diverse field with job opportunities available at all levels. Pick your area of interest, determine the level at which you wish to operate, obtain the appropriate education for the job, and follow through with a rewarding career into the 21st century!

Student Activities

1. Write the "Terms To Know" and their meanings in your notebook.
2. Using paper and pencil, calculate the amount of time needed to count the dollar value of assets in agriscience in the United States, as suggested in Figure 4-1.
3. Develop a bulletin board that illustrates the "Wheel of Fortune," with its listing of the broad categories of jobs or divisions in agriscience.
4. Develop a collage that illustrates the many jobs in agriscience.
5. Write the names of the divisions of agriscience such as "Production Agriscience," "Agriscience Processing, Products, and Distribution," "Horticulture," etc., and list five (5) jobs under each that look interesting to you.
6. Pick a job from the lists you developed for number 5 and write a one-page description of the job. Include the following sections in your paper:
   - **a.** job title;
   - **b.** education/training required to get the job and advance in the field;
   - **c.** working conditions on the job;
   - **d.** advantages/benefits of the job;
   - **e.** disadvantages of the job;
   - **f.** salaries of beginning and advanced workers in the field; and
   - **g.** aspects of the job that you like.
7. Using the job you researched for number 6, or another job or career area, list the things you should do during and after high school to prepare for a career in that area.
8. Determine the name and address of an appropriate official of a school or college and request information from them about educational opportunities for you in their institution.
9. Develop a list of agriscience jobs in which computers are used.

Self-Evaluation

A. MULTIPLE CHOICE

1. The industry that provides commodities that are basic to life is

   a. aerospace.　　c. biotechnology.
   b. agriscience.　　d. transportation.

2. The number of workers in agriscience in the United States is approximately

   a. 21 million.　　c. 100,000.
   b. 100 million.　　d. 400,000.

3. The percentage of total jobs in agriscience that require a college education to enter the job is

   a. 10%.
   b. 20%.
   c. 41%.
   d. 60%.

4. The products and services that are provided in most of the areas in the agriscience "Wheel of Fortune" seem to flow to or originate from

   a. agriscience processing, products, and distribution.
   b. agriscience professions.
   c. horticulture.
   d. production agriculture.

5. The management of wetlands comes under the area of

   a. agriscience processing, products, and distribution.
   b. agriscience professions.
   c. horticulture.
   d. renewable natural resources.

6. Of all agriscience jobs, the percentage that are *not* on farms or ranches is

   a. 20%.
   b. 40%.
   c. 60%.
   d. 80%.

7. On the average, a producer's share of each dollar spent on food in the United States is about

   a. 11 cents.
   b. 15 cents.
   c. 25 cents.
   d. 75 cents.

8. The number of floral industry workers in the United States is about

   a. 110,000.
   b. 220,000.
   c. 500,000.
   d. 1,000,000.

9. A student may join 4-H or scout groups to learn agriscience concepts as early as

   a. college.
   b. high school.
   c. middle school.
   d. none of the above.

10. Agriscience classes in high school would logically include extensive instruction in

    a. plants, animals, and agribusiness.
    b. plants, animals, and social sciences.
    c. plants, mechanics, and higher math.
    d. food, fiber, and physics.

B. MATCHING

______	**1.** Production	**a.**	Teacher or veterinarian
______	**2.** Processing and distribution	**b.**	Hydraulics
______	**3.** Horticulture	**c.**	Seed, feed, or lawn supply
______	**4.** Forestry	**d.**	Farming or ranching
______	**5.** Natural resources	**e.**	Lumber
______	**6.** Supplies and services	**f.**	Ornamentals
______	**7.** Mechanics	**g.**	Grading and packaging
______	**8.** Professions	**h.**	Wildlife

Supervised Agricultural Experience in Agriscience

OBJECTIVE To learn the rationale for and plan a supervised agricultural experience program.

Competencies to Be Developed

After studying this unit, you will be able to

- ☐ define supervised agricultural experience program (SAEP) terms.
- ☐ determine the place and purposes of SAEPs in agriscience programs.
- ☐ determine the types of supervised agricultural experience activities.
- ☐ explore the opportunities for supervised agricultural experience programs.
- ☐ set personal goals for an SAEP.
- ☐ plan your personal SAEP.

TERMS TO KNOW

Simulate
Real-world experience
On-the-job training
Supervised Agricultural Experience Program (SAEP)
Supervised
Agricultural
Experience
Program
FFA
Project
Enterprise
Production or Productive Project
Improvement Project
Supplementary Agriscience Skills
Resources Inventory
Resumé

MATERIALS LIST

paper
pencil or pen
bulletin board materials
Student Interest Survey form (Figure 5-7)
Resources Inventory form (Figure 5-8)
Selecting a Supervised Agricultural Experience Program form (Figure 5-9)
Experience Inventory form (Figure 5-13)
Placement Agreement (Figure 5-14)
Improvement Project Plan and Summary (Figure 5-15)
Supplementary Agriscience Skills Plan and Record (Figure 5-16)

Agriscience programs in various schools teach basic principles and practices in plant and animal sciences, resources management, business management, agriscience mechanics, landscape design, leadership, and personal development. Such programs emphasize reading and math skill development and the application of scientific principles.

Classrooms are excellent places to learn fundamentals and theory through reading, study, discussion, and planning. School laboratories, such as greenhouses, agricultural mechanics shops, school land demonstration plots, farms, and animal production facilities, help provide experiences that simulate real-world experiences. *Simulate* means to look or act like. *Real-world experience* means conducting the activity in the daily routine of our society. Simulation is an excellent way to learn. It imitates the real world and may provide an ideal setting for the activity.

However, classroom experiences and simulation lack the thoroughness of real-world experiences. Also, the personal relationships, such as employer–employee, supervisor–subordinate, owner–worker, sales person–customer, owner–government, owner–community, fellow workers, and employee competition, are rarely present in simulated activities. Therefore, a program is needed for the student to obtain real-world experiences and on-the-job training if agriscience education is to lead to a successful career. *On-the-job training* means experience obtained while working in an actual job setting. In agriscience, the method used for students to obtain real-world experiences is referred to as the *supervised agricultural experience program* (*SAEP*).

SUPERVISED AGRICULTURAL EXPERIENCE PROGRAM DEFINED

A supervised agricultural experience program consists of all supervised agriscience experiences learned outside of the regularly scheduled classroom or laboratory. *Supervised* means to be looked after and directed. *Agricultural* means business, employment, or trade in agriculture, agribusiness, or renewable natural resources. *Experience* means anything or everything observed, done, or lived through. *Program* means the total plans, activities, experiences, and records of the supervised occupational activities.

PURPOSE OF SAEPS

Supervised agricultural experience programs provide opportunities for learning by doing. They provide the means for you to learn with the help of your teacher, parents, employer, and other adults experienced in the area of your interest. Student SAEPs are an essential part of effective agriscience programs.

Some important purposes and benefits of supervised agricultural experience programs are:

1. provide educational and practical experiences in a specialized area of agriscience;
2. provide the opportunity to become established in an agriscience occupation;
3. provide opportunities for earning while learning;
4. create opportunities for earning after graduation;
5. develop interests in additional areas of agriscience;
6. develop valuable work skills such as
   a. appreciate importance of honest work,
   b. improve personal habits,
   c. develop superior work habits,
   d. establish good relationships with others,
   e. keep effective records,
   f. fill out useful reports,
   g. follow instructions and regulations,
   h. contribute to your occupation, and
   i. contribute to your family, community, and nation;
7. permit individualization of instruction;
8. permit recognition for individual achievement; and
9. become established in an agriscience business.

SAEP AND THE TOTAL AGRISCIENCE PROGRAM

High school agriscience programs may be comprehensive and provide students with agricultural experience programs, leadership development programs, and laboratory experiences, along with classroom instruction. Leadership skills are developed through the FFA program, as discussed in Unit 6. These components are integrated so they complement each other (Figure 5-1). This integration should provide the most effective program for the student and make the best use of teacher time.

SAEP and Classroom Instruction

The SAEP is planned as part of the classroom instruction. Students use this instruction to learn how to plan an SAEP, decide what types are possible, choose activities for their own SAEP, and make appropriate arrangements with parents, teacher, and employer. The student conducts the SAEP under the supervision of the teacher, who provides instruction on the necessary topics. The teacher also arranges for small group and individual instruction. This permits the student to use the classroom and laboratory to solve problems encountered with the SAEP.

SAEP and the FFA

The SAEP also overlaps with the FFA (an intracurricular youth organization for students enrolled in occupational agriscience programs). It has many activities that encourage students to do more in their SAEPs and provides contests that more fully develop useful skills for SAEPs. Further, the FFA provides awards and other recognition for achievements in SAEPs. These frequently lead to trips and other travel experiences which greatly enrich the student's education.

FFA and Classroom Instruction

Finally, FFA is merged with classroom and laboratory instruction in a number of ways. The teacher has instructional units on the FFA in the classroom. Further, instruction is provided in public speaking, parliamentary procedure, and other leadership skills. The classroom setting can become the place where FFA teams polish their skills in preparation for upcoming contest or award activities. Therefore, the students who obtain the most benefit from their agriscience program are

A COMPREHENSIVE AGRISCIENCE PROGRAM

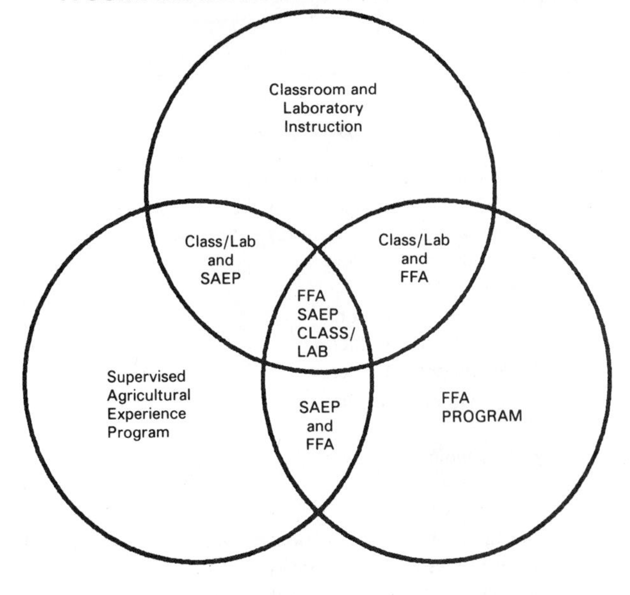

FIGURE 5-1
A comprehensive agriscience program should provide the student appropriate classroom/ laboratory instruction, supervised agricultural experiences, and personal development through FFA. ***(Courtesy Anderson Valley High School Agriculture Education Department)***

those who take advantage of the opportunities provided through SAEPs and the FFA.

TYPES OF SAEP PROGRAMS

Supervised agricultural experiences grow out of planned programs (Figure 5-2). The word *project* is used to describe a series of activities related to a single objective or enterprise, such as raising rabbits, building a porch, or improving wildlife habitat. The term *enterprise* generally refers to a type of animal or plant. Examples are dairy, beef, or rabbits, and corn, hay, turf, or poinsettia.

FIGURE 5-2
Effective planning results in meaningful supervised agricultural experiences. ***(Courtesy Anderson Valley High School Agriculture Education Department)***

Production or Productive Enterprises—Area A

Students usually have an area of major focus and effort in

SUPERVISED AGRICULTURAL EXPERIENCE PROGRAMS					
AREA A				AREA B	AREA C
Production or Productive Enterprises					
Crop and Animal Enterprise (ownership)	Agribusiness Enterprise (ownership)	Farm Placement (for wages and experience)	Agribusiness Placement (for wages and experience)	Improvement Projects	Supplementary Agriscience Skills

FIGURE 5-3

The three areas of a complete SAEP are production or productive enterprises, improvement projects, and supplementary agriscience practices.

their SAEPs. This area generally consumes most of student's time on a day-to-day basis. It should be an ongoing effort which is challenging, leads to learning agriscience concepts on a continuing basis, and generates income.

Income from production or productive enterprises may be profit from raising and selling animal or plant commodities that the student owns. Or it may be income received from owning and operating a business, like a landscaping business. It may also be income from wages received from working on a farm or in some other agribusiness. This is referred to as the production or productive component of the SAEP program, Area A (Figure 5-3). A *production or productive project* is an enterprise conducted for wages or profit (Figure 5-4).

PRODUCTION/PRODUCTIVE ENTERPRISES

A. A production enterprise is a crop, livestock, or agribusiness venture which the student has a degree of *ownership*. The student may own all or part of the enterprise.
   1. Types of Crop Enterprises
      a. Corn Production
      b. Soybean Production
      c. Small Grain Production
      d. Greenhouse Production
      e. Nursery Production
      f. Hay Production
      g. Vegetable Production
      h. Fruit Production
      i. Forestry Production
      j. Christmas Tree Production
   2. Types of Livestock Enterprises
      a. Commercial Cow-Calf Production
      b. Registered Breeding Stock Production
      c. Market Beef Production
      d. Dairy Production
      e. Feeder Pig Production
      f. Market Swine Production
      g. Sheep Production
      h. Poultry Production
      i. Horse Production
      j. Rabbit Production
   3. Types of Agribusiness Enterprises
      a. Lawn Service
      b. Custom Farm Work
      c. Trapping and Pelt Sales
      d. Hunting Guide Service
      e. Tree Service
      f. Farm and Garden Supply Service
      g. Artificial Insemination Business
      h. Animal Care and Boarding
      i. Winery
      j. Fishing and Crabbing for Sales
      k. Custom

FIGURE 5-4

Agriscience students have many production projects from which to choose.

Home Improvement Projects—Area B Enterprises conducted for wages or profit provide great incentive to work and learn. However, students can obtain additional experiences by conducting projects that improve the beauty, convenience, safety, value, or efficiency of the home or surroundings. Such projects are called *improvement projects,* Area B (Figure 5-3). They are conducted for the learning experience and improvement of the surroundings. No wages or profit is received from improvement projects. All students, regardless of home or family conditions, can arrange to conduct improvement projects (Figure 5-5).

Supplementary Agriscience Skills—Area C The broadest and most flexible area of the SAEP is supplementary agriscience skills. *Supplementary agriscience*

EXAMPLES OF IMPROVEMENT PROJECTS

A. Soil Improvement Programs
1. Liming
2. Fertilizing
3. Drainage
4. Erosion control
5. Plow under green manure
6. Introduce a cropping system
7. Soil sampling

B. Building Improvement Programs
1. Painting
2. Window repair
3. Roof repair
4. Foundation repair
5. Floor repair
6. Siding repair
7. Door repair
8. Electric wiring
9. Water systems installation
10. Heating
11. Lighting protection
12. Feeding floor construction
13. Remodeling
14. Home sewage system installation

C. Fence Improvement Programs
1. Construction of new fence
2. Fence replacement and repair
3. Construction of flood gates
4. Construction and repair of gates

D. Homestead Improvement Programs
1. Plan and set out a windbreak
2. Seed or reseed lawn
3. Plant shrubs and trees
4. Clean up homestead

E. Orchard, Small Fruits and Vegetables Improvement Programs
1. Plan and set out a fruit tree orchard
2. Plan and set out a small fruit garden
3. Plan and grow a home vegetable garden
4. Renovate an old orchard

F. Weed Control Programs
1. Spray major weed areas on farm
2. Mow or spray weeds in fence rows
3. Pull weeds in corn
4. Clip weeds in permanent pastures

G. Insect and Pest Control
1. Rat control
2. Corn borer control
3. Japanese beetle control
4. Livestock parasite control

H. Farm Management Programs
1. Keep farm accounts
2. Plan farm safety program
3. Inventory farm equipment
4. Keep checking account records

I. Agricultural Shop Programs

J. Agricultural Machine Repair and Reconditioning Programs

K. Livestock Improvement Programs

L. Crop Improvement Programs

M. Landscaping Improvement Programs
1. *General Clean-up Program*
   a. Remove dead trees or shrubs
   b. Remove and discard dead branches
   c. Remove unsightly junk, trash, and woodpiles
   d. Provide a specific storage area for all lawn and horticulture equipment
   e. Repair or remove broken lawn furniture
   f. Improve the grade if necessary
   g. Remove, replace or repair fences, sidewalks, step railings or porches
   h. Transplant trees, shrubs or flowers
   i. Make or improve the driveway
   j. Pick up nails and broken glass
   k. Remove unsightly rocks
2. *Grounds Maintenance Practices*
   a. Mow the lawn
   b. Trim or prune trees and shrubs
   c. Fertilize the lawn, trees, shrubs and flowers
   d. Apply herbicides
   e. Apply insecticides
   f. Water the lawn, trees, shrubs and flowers
   g. Repair trees
   h. Brace trees
   i. Prevent sunburn of plants
   j. Prevent insect damage and disease
   k. Set up, adjust and move a sprinkler
   l. Edge the lawn
   m. Stake trees
   n. Set up rain gauge
   o. Record precipitation from rain gauge
   p. Mulch trees, shrubs and flowers
   q. Rake the lawn
3. *Improvement and Beautification Activities*
   a. Plant new trees and shrubs
   b. Draw a landscape plan
   c. Seed, plug or sod lawn
   d. Plant flowers
   e. Relocate and replant trees and shrubs
   f. Renovate existing lawn
   g. Build a patio
   h. Make a window box
   i. Build a trellis
   j. Plant a windbreak

FIGURE 5-5
Improvement projects are available for agriscience students regardless of the home situation.

EXAMPLES OF SUPPLEMENTARY SKILLS

A. AGRIBUSINESS
1. Operate cash register
2. Display merchandise
3. Keep inventory records
4. Write sales tickets
5. Deliver products
6. Set up machinery
7. Assemble equipment
8. Greet customers
9. Close sales
10. Answer telephones
11. Order merchandise
12. Operate adding machine
13. Wrap meat
14. Cut carcass into wholesale cuts
15. Cut wholesale cuts into retail cuts
16. Compute sales tax

B. AGRICULTURE MECHANICS
1. Arc weld metals
2. Oxy-acetylene weld metals
3. Cut detals with oxy-acetylene
4. Operate farm machinery
5. Operate wood power tools
6. Operate metal power and hand tools
7. Recondition and sharpen tools
8. Service air cleaner
9. Service electric motor
10. Store machinery
11. Install rings on pistons
12. Grind valves
13. Wire electrical convenience outlet
14. Change transmission fluid
15. Fasten sheet metal with rivets
16. Repair flat tire
17. Calibrate field sprayer
18. Pour concrete
19. Lay reinforcement steel
20. Use farm level

C. CORN
1. Select seed
2. Plant seed
3. Prepare seedbed
4. Calibrate corn planter
5. Adjust planter for depth
6. Apply dry fertilizer
7. Apply anhydrous ammonia
8. Conduct corn variety test
9. Check harvest losses
10. Apply herbicides and insecticides
11. Operate combine
12. Operate corn picker
13. Identify weeds
14. Identify insects
15. Cultivate corn
16. Dry corn artificially

D. FORAGES
1. Innoculate legume seeds
2. Rotate pastures
3. Greenchop forages
4. Bale hay
5. Renovate permanent pastures
6. Combine grasses and legumes
7. Graze pastures properly
8. Rotate pastures

E. HORTICULTURE
1. Plant vegetable garden
2. Prepare garden plot
3. Plant fruit trees
4. Bud-graft
5. Cleft-graft
6. Whip-graft
7. Prepare growing medium
8. Sterilize soil
9. Pot plants
10. Water greenhouse plants
11. Root cuttings
12. Harvest crop
13. Process vegetables
14. Store produce
15. Fertilize plants
16. Determine plant diseases
17. Treat deficiency symptoms
18. Force blooming

F. SMALL GRAINS
1. Calibrate grain drill
2. Plant small grains
3. Broadcast fertilizer
4. Select adapted varieties
5. Clean seeds
6. Harvest small grains

G. SOIL MANAGEMENT
1. Test soil for lime requirements
2. Lime soils
3. Seed grass waterway
4. Rotate crops
5. Plant windbreak
6. Test for fertilizer
7. Fertilize soils
8. Lay drainage tile
9. Terrage fields
10. Farm on contour
11. Aerate soil
12. Control erosion

H. SOYBEANS
1. Test for germination
2. Innoculate seed
3. Take soil sample
4. Control weeds
5. Treat seed for storage
6. Market beans
7. Harvest soybeans
8. Check harvest losses

FIGURE 5-6
There are hundreds of supplementary skills which are useful for employment in agriscience.

skills, Area C, are agriscience skills or practices that are learned anywhere except in the instructional settings at school, production enterprises, or improvement projects in the SAEP. Students are encouraged to obtain supplementary agriscience skills by working on the home farm, by doing volunteer work, by helping neighbors, doing odd jobs, or conducting leisure-time activities (Figure 5-6).

EXAMPLES OF SUPPLEMENTARY SKILLS *(cont'd.)*

I. BEEF CATTLE

1. Assist cow in calving
2. Dehorn
3. Castrate bull calves
4. Disinfect navel of baby calves
5. Select herd sire
6. Select replacement heifers
7. Creep feed calves
8. Cull poor producers
9. Ear tag
10. Tatoo
11. Remove warts
12. Drench
13. Treat for bloat
14. Treat for external parasites
15. Select feeders
16. Trim hooves
17. Vaccinate for blackleg
18. Vaccinate for IBR
19. Vaccinate for brucellosis
20. Formulate a balanced ration
21. Ring bull
22. Fit animal for show or fair
23. Analyze production records
24. Palpate to determine pregnancy

J. DAIRY CATTLE

1. Select replacement heifers
2. Dry cows
3. Control external parasites
4. Assist newborn calves in getting colostrum
5. Vaccinate heifers for brucellosis
6. Test cows for T.B.
7. Cull low producers
8. Dehorn
9. Production test cows
10. Participate in DHIA
11. Treat mastitis
12. Operate milking machines
13. Clean facilities after milking
14. Artificially inseminate cows
15. Prevent milk fever
16. Formulate balanced ration
17. Wash udder with disinfectant
18. Detect abnormal milk with strip-cup
19. Feed according to production
20. Clean and sterilize utensils

K. POULTRY

1. Clean and disinfect brooder
2. Cull poor producers
3. Select pullets
4. Sex baby chick or poults
5. Prevent cannibalism

K. POULTRY *(cont'd.)*

6. Keep production records
7. Dress broilers
8. Prevent breast blisters
9. Stimulate egg production
10. Disinfect laying house
11. Grade eggs by candling
12. Size eggs
13. Debeak chicks
14. Vaccinate for fowl pox
15. Treat for external parasites
16. Worm poultry
17. Provide sanitary water
18. Feed balanced rations
19. Caponize cockerels
20. Provide litter

L. SHEEP

1. Select ram
2. Flush ewes
3. Treat for external parasites
4. Assist ewes at lambing
5. Creep feed lambs
6. Dock lambs
7. Castrate lambs
8. Artificially inseminate ewes
9. Ear tag
10. Shear sheep
11. Tie fleece
12. Cull farm flock
13. Determine estrus in ewes
14. Trim hooves
15. Worm for internal parasites
16. Keep production records

M. SWINE

1. Select herd boar
2. Select replacement gilts
3. Flush gilts
4. Vaccinate sows for leptospirosis
5. Provide farrowing stalls
6. Clean facilities before farrowing
7. Clean sow before farrowing
8. Assist sow at farrowing
9. Treat navels of baby pigs
10. Clip needle teeth
11. Creep feed pigs
12. Ring hogs
13. Treat for external parasites
14. Inject iron in baby pigs
15. Vaccinate for erysipelas
16. Vaccinate for brucellosis
17. Wean pigs at 4-6 weeks
18. Weigh pigs at 56 days
19. Formulate balanced ration
20. Castrate boar pigs

FIGURE 5-6

There are hundreds of supplementary skills which are useful for employment in agriscience (continued).

EXPLORING OPPORTUNITIES FOR SAEPs Students should use great imagination when considering an SAEP. Some complain that they don't have opportunities for meaningful SAEPs. Yet other students in similar circumstances find or create opportunities for effective programs. Seek the advice of your teacher for ideas. Also, observe what successful students have done in the agriscience program and in your community. Then develop an SAEP that provides the most opportunity to learn and earn.

Personal Interest Personal interest is an important factor in the success of an SAEP. Consider the kinds of activities you like to do and then build on those interests. A student interest survey or inventory should help you assess your natural interests and provide some guidance in developing an SAEP (Figure 5-7).

Resources Inventory A *resources inventory* is a summary of the resources that may be available for conducting SAEP activities. It includes information about your home, farm, work setting, and community that might be useful in considering your SAEP (Figure 5-8). Part of the inventory is a scale drawing of the property where you live or work. Making the scale drawing will help you realize what is available and help your teacher to suggest SAEP possibilities.

SELECTING AND IMPLEMENTING YOUR SAEP After completing the personal interest survey and the resources inventory, you should arrange a conference with your teacher. The conference should include discussions on your interests and a look at the possible production enterprises, improvement projects, and supplementary skills. After the conference you should record tentative plans for the SAEP (Figure 5-9). At this point you and your teacher should discuss the plan with your parents or guardians. If an employer is involved, he or she should become a partner in the planning process.

Securing a Job If placement on a farm or in an agribusiness is part of the SAEP plan, you will need a brief resumé (a one-page summary of information about a job applicant Figure 5-10). The prospective employer will be interested in your edu-

STUDENT INTEREST SURVEY

Place an 'X' in the blank by the tasks that you like to do or would like to learn to do.

Tasks Typical of Agribusiness

________ Delivering merchandise
________ Displaying merchandise
________ Driving trucks
________ Keeping records
________ Mowing lawns
________ Operating cash registers
________ Operating equipment
________ Pricing merchandise
________ Processing meat, milk, grains
________ Repairing equipment
________ Selling merchandise
________ Stocking shelves
________ Taking customer orders
________ Taking inventory
________ Taking telephone orders
________ Unloading trucks
________ Working outside
________ Working with people
________ Working with plants

Tasks Typical of Horticulture

________ Applying pesticides
________ Arranging flowers
________ Balling and burlapping trees
________ Building patios
________ Edging flower beds
________ Identifying plants
________ Lifting heavy materials
________ Making Christmas decorations
________ Making cuttings
________ Mowing lawns
________ Mulching beds
________ Operating power machinery
________ Planting bulbs
________ Planting grass
________ Planting seeds
________ Planting trees and shrubs
________ Protecting plants from weather
________ Pruning plants
________ Raking leaves
________ Selling plants
________ Watering plants
________ Weeding by hand
________ Working in various weather
________ Working with people

Tasks Typical of Production

________ Applying pesticides
________ Baling hay
________ Building fences and buildings
________ Castrating animals
________ Cleaning animals
________ Feeding animals
________ Getting up early
________ Handling manure
________ Harvesting crops
________ Helping parents
________ Keeping records
________ Lifting heavy materials
________ Milking cows
________ Operating machinery
________ Painting buildings
________ Planting crops
________ Plowing fields
________ Repairing buildings
________ Repairing machinery
________ Shearing sheep
________ Showing animals
________ Taking soil samples
________ Working in various weather
________ Working with animals

FIGURE 5-7
A student interest survey should be helpful to you in choosing SAEP activities.

RESOURCES INVENTORY

1. Name ______________________ Age ________ Class ________

2. Address ______________________ Phone ________

3. Parents' Name ______________________ Occupation ________

4. Number in my family ________ Boys________ Girls ________

5. I live: on farm ________ in town ________ on an acreage ________ .

6. Is land available for you to rent to produce crops? ________ yes ________ no

   a. If yes, how many acres? ________

   b. Which crop? ________

   c. Location of land? ________

7. Are facilities available for you to rent to produce livestock or livestock products? ________
   If so,

   a. What type of livestock? ________

   b. Number ________

   c. Location of facilities ________

8. Do you have available space for a garden? ________ yes ________ no

9. Do you have facilities for mechanical work? ________ yes ________ no

10. Do you have a greenhouse available for your use? ________ yes ________ no

11. Would you be interested in producing livestock or crops on the school farm?

    ________ yes ________ no If so, what type? ________

FIGURE 5-8A
The resources inventory is especially helpful in planning production enterprises and improvement projects.

Resources available to the student for the Supervised Agricultural Experience Program (SAEP).

A. MAP OF HOME FARM AND/OR BUSINESS

Let each square represent any convenient acreage or square footage, i.e., 20, 40, 80, 100, etc.

One square = ______________

LEGEND

Symbol	Meaning	Symbol	Meaning
———	Public Road	—>—	Stream
– – – –	Private Drive		Pond
	Farmstead		Trees
	Terrace	✕✕✕	Fence
	Terrace Outlet		Railroad
	Natural Drain	= =	Crossing
	Gullies		

FIGURE 5-8B

The resources inventory is especially helpful in planning production enterprises and improvement projects.

SELECTING A SUPERVISED AGRICULTURAL EXPERIENCE PROGRAM FOR

____
(Name of Student)

Instruction: Use this form to tentatively decide on a beginning vocational agriscience SAEP. This information will be used in vocational agriscience classes to develop detail plans for obtaining experiences.

My stated interest in agriscience/horticulture was ______________________________

____

Based upon my interest and the opportunities available to me to get practical experience in agriscience, I plan to include the following in my vocational agriculture SAEP.

1. Production or Productive enterprises (examples: beef, dairy, nursery production, Christmas trees)

   ____

   ____

2. Placement in an agribusiness (examples: supply store, florist shop, nursery, golf course, landscape contracting)

   ____

   ____

3. Improvement projects (examples: landscape your home, fertilize your lawn, plant trees)

   ____

   ____

4. Supplementary skills (example: change spark plugs, weld, change the oil in small engines)

   ____

   ____

5. Other activities (example: projects on school facility)

   ____

   ____

FIGURE 5-9

Enterprises, improvement projects, and supplementary skills should be selected and recorded early in the school year.

Personal Resumé

Name: John Smith

Address: 1234 Honeylocust Drive
Frederick, Maryland 21701

Telephone: (301)662-0000

Education: Junior at Frederick High School
Majoring in Landscaping

Age: 17 | Health: Excellent

Weight: 180 | Marital status: Single

Height: 6 ft.

Subjects Studied:
- Horticulture I
- Horticulture II
- Landscaping I
- Landscaping II
- Typing I

Student Activities:
- President: Future Farmers of America
- Editor of school yearbook
- Tennis team and softball team

Special Skills:
- Can ball and burlap trees, use cash register, water plants, transplant plants, and operate a tractor.

Employment Experience:
- Hardees Fast Food: Worked as a cashier and cook
- Landscaped neighbors yard

References:
- Mr. Ralph Rece, Principal Frederick County Vo-Tech, Frederick, MD.
- Mrs. Holly Deane, Instructor Frederick County Vo-Tech, Frederick, MD.
- Mr. Bernard Rose, Manager Hardees Fast Food, Frederick, MD.

Date Compiled ____________________

Signature ________________________

FIGURE 5-10
A personal resume will help when seeking a job.

cational and occupational background, which will help determine your qualifications and experiences. Be sure to ask a minimum of three adults who know your character and qualifications if you may list them as references. Your prospective employer will probably want to contact them.

When your resumé is complete, be sure your teacher has approved the final version. Then you are ready to approach employers for a job that will help achieve the objectives of your SAEP plan. Your approach is critical, since first impressions are lasting impressions. It is important to dress in a businesslike manner, be courteous and confident, and conduct yourself according to standard interview procedures (Figure 5-11).

Refining the Plan Plans should be worked out for each production enterprise. It is important to estimate the anticipated expenses and income, which will help you make financial arrangements to conduct the project. Also, some goals for the enterprise should be set, including the size of the project and some efficiency factor goals (Figure 5-12).

Employers are generally impressed with students who are eager to learn. In this regard, both

PREPARING FOR THE JOB INTERVIEW

I. What Employers Look for in an Employee

A. Attitude—The prospective employee should have a positive attitude about the job. He/she should show enthusiasm and a willingness to learn and work. Employers stress this as being one of the most important qualities they look for in prospective employees.

B. Experience—Previous experience of the prospective employee is important. However, employers are usually willing to train the person with a positive attitude.

C. Appearance—The prospective employee should be neat and clean, have hair combed and be well dressed. It is better to be overdressed than underdressed for an interview.

D. Posture—It is important to stand and sit up straight. The employer will be observing the way you carry yourself and will make judgments accordingly.

E. Mannerisms—Mannerisms are gestures that are made which may be annoying or could be welcomed. However, one should be aware of mannerisms. Do you:
1. Yawn a lot? If so, others will think you're bored or worse—lazy.
2. Fidget? Squirming may indicate lack of confidence or disinterest in the job.
3. Daydream? Give your full attention to the interviewer.

F. Handshake—Have a firm handshake; not bone crushing and not limp.

II. What Questions Should be Asked? The following questions may be asked if the information is not provided by the interviewer.

A. What type of jobs or tasks are to be done?

B. What are the policies and procedures for workers?

C. What are the working hours?

D. What is the rate of pay?

E. What arrangements are needed for time off?

F. If you are uncertain about something that has been discussed in the interview you should ask the employer to clarify or explain.

G. IN SUMMARY, BE POLITE AND ATTENTIVE DURING YOUR JOB INTERVIEW.

FIGURE 5-11

Preparation for the job interview will permit you to relax and give your total attention to the interviewer.

SETTING GOALS FOR PRODUCTION IN THE SAEP

I. What goals should be set for the Supervised Agricultural Experience Program?

A. Definition: A goal is the hoped for end result of hard work and should be challenging and realistic.

B. Goals should be challenging!

C. Goals should be reachable!

D. SAEP goals should focus on scope, learning opportunities and production efficiency factors.

E. Parents, employers, agriculture teachers and other qualified adults should help develop SAEP goals.

F. Goals should be recorded

G. Goals should be analyzed and evaluated periodically and new goals developed.

H. Goals provide direction and organization.

I. Setting realistic goals should help increase profits.

II. What are efficiency factors?

A. Definition: Efficiency factors are measures of production success and encourage enterprise improvement and profit. Examples of efficiency factors are as follows.

B. Size of Enterprise. For animal weight or livestock products produced.

C. Rate of Gain and Production.

1. Beef: $\text{Percent of calf crop} = \dfrac{\text{Calves born alive}}{\text{Cows bred}}$

2. Poultry: $\text{Percent of egg production} = \dfrac{\text{Average eggs per hen}}{\text{Number days in production}}$

3. Sheep: $\text{Percent of lamb crop} = \dfrac{\text{Lambs born alive}}{\text{Ewes bred}}$

4. Swine: $\text{Pigs farrowed per litter} = \dfrac{\text{Live pigs farrowed}}{\text{Sows bred}}$

   $\text{Weight produced per litter} = \dfrac{\text{Total production}}{\text{Number of litters}}$

FIGURE 5-12

Goals should state the number, size, time lines, and efficiency factors which you hope to achieve.

SETTING GOALS FOR PRODUCTION IN THE SAEP *(cont'd.)*

D. Returns and Feed Costs. Round total income and value of feed fed to the nearest whole dollar.

1. $\text{Returns per \$100 feed fed} = \frac{\text{Total Income}}{\text{Dollars worth of feed fed}} \text{ X } 100$

2. $\text{Returns per \$100 invested} = \frac{\text{Total Income}}{\text{Total Expenses}} \text{ X } 100$

3. $\text{Expense Per Cwt. of Production} = \frac{\text{Total Expenses}}{\text{Total Production}} \text{ X } 100$

4. $\text{Average weight sold} = \frac{\text{Total sales weight}}{\text{Animals sold}}$

5. $\text{Average price received} = \frac{\text{Total sales value}}{\text{Units sold}}$

E. Feeding Efficiency

Note: Convert all corn to shelled corn basis (56 lb. per bu.) before figuring efficiency factors.
Note: Poultry—1 unit is equal to 1 dozen eggs or 1.5 lb. of weight.
Note: Dairy—1 unit is equal to 1000 lb. milk or 100 lb. weight.

1. Feed cost per Cwt. or per unit

   a. $\text{For swine or beef cattle} = \frac{\text{Total feed cost}}{\text{Lb. weight produced}} \text{ X } 100$

   b. $\text{For sheep} = \frac{\text{Total feed cost}}{\text{Lb. wool + lb. weight}} \text{ X } 100$

   c. $\text{For dairy or poultry} = \frac{\text{Total feed cost}}{\text{Units of production}}$

2. Feed per Cwt. produced

   a. $\text{For hogs, beef cattle, or sheep} = \frac{\text{Lb. of feed fed}}{\text{Lb. weight produced}}$

   (For sheep, include wool with weight as in lb. above)

   b. $\text{For dairy or poultry} = \frac{\text{Lb. of each feed fed}}{\text{Units of production}}$

F. Death Loss

1. $\text{\% death loss} = \frac{\text{Weight of dead animals}}{\text{Total animal weight produced}}$

2. A low percent of death loss means a high enterprise rating for this item.

FIGURE 5-12
Goals should state the number, size, time lines, and efficiency factors which you hope to achieve (continued).

EXPERIENCE INVENTORY

Directions: Complete the following information sheet by listing any experiences you have had or would like to gain in the field of agriculture.

Tasks or Jobs	Can perform without help	Can perform with help	Can help perform	Can not or have not performed	Would like to learn to perform	How or where to obtain experience
Examples						
a. Drive tractor		X				Landscaping business
b. Take cuttings					X	Vo. Ag. Class
c. Keep records				X		My own & in class
1.						
2.						
3.						
4.						
5.						
6.						

FIGURE 5-13

The experience inventory is a device to plan and record experiences you hope to gain. It is also a mechanism for recording how well you have learned them.

the student and the employer can benefit from a carefully thought out statement of skills to be developed on the job. You should develop the list with the guidance of your teacher and prepare an experience inventory to record the completion of tasks or jobs (Figure 5-13). It should be frequently updated. The experience inventory helps the student, teacher, and employer track progress in achieving the set goals.

The special placement agreement is a document that helps finalize the plans for placement on a job (Figure 5-14). Such agreements need the signature of the student, parents or guardians, employer, and teacher. Once all parties are in agreement regarding the student's placement experiences, the chances for success are very high.

Another document that will help plan and conduct the SAEP is the improvement project plan and summary (Figure 5-15). This plan directs the student to describe the conditions found, plans for improvement, and estimated value of the improvement when completed. The summary is filled out as the improvement project progresses and serves as the record when finished.

Finally, a supplementary skills plan should be developed. These skills are chosen from lists that apply to the student's community and are recorded on a supplementary agriscience skills plan and record (Figure 5-16). The date completed should be added when the skill is acquired.

SUMMING UP A carefully planned supervised agricultural experience program is both challenging and rewarding. The student learns by doing and earns while learning. Learning is at the highest level, since the laboratory is the real world. The teacher benefits from having highly motivated students, since the learning is interesting and rewarding. The employer benefits by having better-trained employees. And the community benefits from the improvement projects and better-educated citizens.

PLACEMENT AGREEMENT

To provide a basis of understanding and to promote business-like relationships, this memorandum is established on ____________ , 19 ____ . This work will start on ____________ , 19 ____ , and will end on or about ____________ , 19 ____ , unless the arrangement becomes unsatisfactory to either party before the ending date. Person (employer) responsible for training ____________

The usual working hours will be as follows:

1. While attending school working hours shall be ____________
   When not attending school working hours shall be ____________
2. Provisions for overtime ____________
3. Provisions for time off ____________
4. Liability insurance coverage (type and amount) ____________

Wages will be at the following rate(s): ____________
Trial period ____________
Remainder of the agreement period ____________
And will be paid (when?) ____________

A. IT IS UNDERSTOOD THAT THE EMPLOYER WILL (check the items that apply):

______ Provide the student with opportunities to learn how to do well as many jobs as possible, with particular reference to those contained in the planned program;
______ Coach the student in methods found desirable in implementing project activities and handling management problems;
______ Help the student and teacher make an honest appraisal of the student's performance;
______ Avoid subjecting the student to unnecessary hazards;
______ Notify the parents and the school immediately in case of accident or sickness and if any other serious problem arises;
______ Assign the student new responsibilities when he/she can handle them;
______ Cooperate with the teacher in arranging conferences with the student on supervisory visits; and/or
Other:
____
____

B. THE STUDENT AGREES TO (check the items that apply):

______ Do productive work, recognizing that the employer must profit from the student's labor in order to justify employment;
______ Keep the employer's interest in mind and be punctual, dependable, and loyal;
______ Follow instructions, avoid unsafe acts, and be alert to unsafe conditions;
______ Be courteous and considerate of the employer, the family, and others;
______ Keep such records of work experience and make such reports as the school may require.
______ Develop plans for management decisions with the employer and teacher; and/or
Other:
____
____

FIGURE 5-14

A placement agreement states what every party is expected to do. It promotes good planning and reduces misunderstandings and conflicts.

PLACEMENT AGREEMENT ***(cont'd.)***

C. THE TEACHER, IN BEHALF OF THE SCHOOL, AGREES TO (check the items that apply):

______ Visit the student on the job at frequent intervals for the purpose of instruction and assurance that the student gets the most education out of the experience;

______ Show discretion in the time and circumstances of these visits, especially when the work is pressing; and/or

______ Provide appropriate job-related instruction at school; and/or

Other:

____

____

D. THE PARENTS AGREE TO (check the items that apply):

______ Assist in promoting the value of the student's experience by cooperating with the employer and the teacher of agriscience;

______ Satisfy themselves in regard to the living and working conditions made available to the student; and/or

Other:

____

____

E. ALL PARTIES AGREE TO:

______ An initial trial period of ______ working days to allow the student to adjust and prove himself/herself;

______ Discuss any issues concerning the job with the teacher before ending employment and/or

Other:

____

____

STUDENT's Signature ______________
Address ______________
____
Telephone Number ______________
Social Security No. ______________

PARENT's Signature ______________
Address ______________
____
Telephone Number ______________

EMPLOYER's Signature ______________
Address ______________
____
Telephone Number ______________

TEACHER's Signature ______________
School Address ______________
____
Telephone Number ______________
School Telephone Number ______________

FIGURE 5-14

A placement agreement states what every party is expected to do. It promotes good planning and reduces misunderstandings and conflicts (continued).

IMPROVEMENT PROJECT PLAN AND SUMMARY

Improvement Project No. ____________

A. Conditions found

____
____
____
____
____

B. Plans for improvement (including costs):

____
____
____
____
____

C. Value of improvement when completed:

____
____
____
____
____

AGRICULTURAL IMPROVEMENT PROJECT SUMMARY

Started: ____________________, 19 ___ Completed: ____________________, 19 ___

Date	Jobs Done	Hours of Labor	Cost of Materials & Equipment

FIGURE 5-15
Improvement projects should be planned and records kept on the jobs, hours, and cost involved.

SUPPLEMENTARY AGRISCIENCE SKILLS PLAN AND RECORD

Directions: Using the list of Supplementary Practices supplied by the instructor, complete the chart below by choosing skills you would like to include as part of your SAEP.

Skills, Practices, Job, or Experience	Place to obtain skill	Date planned to obtain skill	Date completed
Ex. 1. Operate Cash Register	On-Job	Sept. 16	Sept. 16
Ex. 2. Bud-Graft	School Farm	February	Feb. 20
1.			
2.			
3.			
4.			
5.			
6.			
7.			

FIGURE 5-16

Supplementary agriscience skills should be selected at the beginning of the year with plans for times and places to complete each item.

AGRI-PROFILE

CAREER AREA: SUPERVISED AGRISCIENCE EXPERIENCE

Learning by doing is generally accepted as the best way to become proficient in complex procedures and psychomotor skills. ***(Courtesy of the National FFA Organization)***

The phrase *supervised agriscience experience* has three important elements. Experience suggests hands-on or real-life activities in the work place. Attitudes, knowledge, and skills gained here will generally be salable in the future. *Agriscience* means the setting and skills will be in the area of plant or animal sciences, or management or mechanics related thereto. And *supervised* means experienced persons such as your teacher, parents, and/or employer recognizes your interest in learning and will help direct the experience.

Agriscience experience is obtained in many ways. Generally, the agriscience teacher helps the student develop an understanding of the process and find a suitable location for a productive experience with cooperation from the parents. However, sometimes sympathetic parents and supportive school programs are not available. In such cases, students must develop an extra measure of determination and seek experiences on their own.

Experiences may be for pay, or, simply for experience. Many schools now provide supervised experience programs in school laboratories. Whether for pay or not, the experiences obtained are worth the effort and give the participant an edge in the job market.

Student Activities

1. Define the "Terms To Know"
2. Describe the relationships between 1) classroom instruction and supervised agricultural experience and 2) the FFA program and supervised occupational experience.
3. Construct a bulletin board showing the relationships presented in Figure 5-1.
4. Study Figure 5-4 and write 5 ideas for production projects or enterprises to discuss with your teacher.
5. Discuss Figure 5-5 with your parents/guardians and select 2 or 3 improvement projects that you would like to conduct.
6. Review the examples presented in Figure 5-6 and choose 20 supplementary agriscience skills from enterprises other than your production and improvement projects.
7. Examine the tasks in the Student Inventory (Figure 5-7). Determine whether your interests are more in agribusiness, horticulture, or production agriscience.
8. Using Figure 5-8 as a guide, draw a map to scale of your home, farm, or business where you can conduct an SAEP. Make an inventory of the resources that may be available for you to use in conducting an SAEP.
9. Talk with your teacher and parents/guardians and write the names of the projects and supplementary skills that you definitely plan to do during the current year (Figure 5-9).
10. Prepare a personal resumé.
11. Apply and interview for a job.
12. Develop an experience inventory (Figure 5-13).
13. Work out a placement agreement with an employer (Figure 5-14).
14. Set goals for production projects (Figure 5-12).
15. Make detailed plans for improvement projects (Figure 5-15).
16. Make definite selection of supplementary agriscience skills and write down where and when you plan to accomplish the skills (Figure 5-16).

Self-Evaluation

A. MULTIPLE CHOICE

1. Conducting an activity in the daily routine of our society is said to be

   a. laboratory experience.
   b. real-world experience.
   c. simulation.
   d. supervised occupational experience.

2. Which is not a purpose or benefit of SAEPs?

   a. Become established in an agriscience occupation.
   b. Permit early graduation.
   c. Permit individualized instruction.
   d. Provide educational and practical experiences.

3. Which is not a major component of a comprehensive agriscience program?

a. Classroom/laboratory instruction.
b. FFA.
c. Memorization and recitation.
d. Supervised occupational experience.

4. SAEPs should be planned

a. at home with parents/guardians.
b. in the classroom.
c. on the job with employers.
d. all of the above.

5. A student-drawn map of the home property is an important part of a

a. job interview.
b. placement agreement.
c. resources inventory.
d. resumé.

6. A production or productive project

a. is the same as an improvement project.
b. may also be recorded as a supplementary skill.
c. may involve either ownership or placement for experience.
d. must be done without pay or profit.

7. Supplementary skills should be selected

a. as part of a placement agreement.
b. early in the school year.
c. from occupations outside of agriscience.
d. from the same enterprise as the production project.

8. Improvement projects

a. are always connected with employment.
b. focus primarily on leadership development.
c. must improve some part of the instructional program.
d. should be without pay.

B. MATCHING

_______ **1.** Enterprises
_______ **2.** Experience
_______ **3.** Interest survey
_______ **4.** FFA
_______ **5.** Improvement project
_______ **6.** Agricultural
_______ **7.** Production/Productive
_______ **8.** Program
_______ **9.** Project
_______ **10.** Resumé
_______ **11.** Supervised
_______ **12.** Supplementary skills

a. Series of related activities
b. Types of plants or animals
c. Total SAEP activities
d. Should result in wages or profit
e. Observed, done, or lived through
f. Assess interests
g. Leadership development
h. Beauty, safety, convenience, or value
i. Broad and flexible part of SAEP
j. Looked after or directed
k. Agriculture, agribusiness, and renewable natural resources
l. Information about a job applicant

UNIT 6 Leadership Development in Agriscience

OBJECTIVE To develop basic leadership skills.

Competencies to Be Developed

After studying this unit, you will be able to

- ☐ define leader and leadership.
- ☐ explain why effective leadership is needed in agriscience.
- ☐ state some major traits of good leaders.
- ☐ describe the opportunities for leadership development in FFA.
- ☐ demonstrate positive leadership skills.

TERMS TO KNOW

Leadership
Lead
Plan
Manage
Citizenship
Integrity
Knowledge
Courage
Tact
Enthusiasm
Unselfishness
Loyalty
Cooperative Extension Service
4-H
Girl Scout
Boy Scout
Extemporaneous
Parliamentary procedure
Business meeting
Presiding officer
Secretary
Minutes
Order of business
Executive meeting
Gavel
Motion
Main motion
Amend
Refer
Lay on the Table
Point of Order
Adjourn

MATERIALS LIST

paper
pencil or pen
bulletin board materials

What is leadership in agriscience? Agriscience has been described as a broad and diverse field. It is not just horticulture or supplies and services; not just professions or products, processing, and distribution; not just mechanics or forestry; and not just renewable natural resources or production. Agriscience is all of these. Then what is leadership in agriscience?

LEADERSHIP DEFINED

Leadership is defined as the capacity or ability to lead. To *lead* is to show the way by going in advance or to guide the action or opinion of. To do this in agriscience, you must have knowledge of technical information and people. Similarly, you must know how to organize and manage activities. Most jobs are too big for one person. We can only do part of what needs to be done. Therefore, we need to use the help of others.

A leader uses the knowledge and skills of others to achieve a common goal. For instance, a quarterback on a football team will use leadership skills to coordinate the team players to achieve a touchdown. Similarly, the wise batter in baseball will hit the ball in a place that not only gets the batter on first base, but permits other runners to advance around the bases. That properly placed hit supports the common goal of achieving runs. A single base hit, where everyone advances and no one gets out, may be the best for the team. On the other hand, a line drive, which doesn't quite make the fence, may result in a third out with no chance for other players to score.

WHY LEADERSHIP IN AGRISCIENCE? Agriscience is a highly organized industry. It involves people and complex processes. Leadership skills are necessary whenever people are assembled. Those who teach in agriscience are part of a team of teachers, principals, supervisors, community advisory groups, and others. Those in agribusiness are typically parts of teams consisting of the manager, office staff, sales representatives, field personnel, and board of directors (Figure 6-1). Those on farms may be the owner, manager, spouse, children, hired help, or neighbors who assist at times. The manager of a farm or business must *plan* (to think through, determine procedure, assemble materials, and train staff to do a job). Once a job is planned it may be accomplished through management. To *manage* is to use people, resources, and processes to reach a goal. A manager uses leadership skills continuously in working with others on a day-to-day basis (Figure 6-2).

A *landscaper* is one who plans, plants, builds, or maintains outdoor ornamental plants and landscape structures. At times he works alone; however, he also works with others and is a leader in many respects. When developing a plan for a customer, the landscaper exerts leadership. The landscaper knows the names, function, and performance of various plants. This information is used to develop an acceptable plan. Since the customer has personal ideas about landscaping, a professional has to consider these in the plan. It may test the landscaper's leadership skills to lead the customer to an acceptable plan with plants that survive the weather and conform to acceptable landscape practices.

The landscaper and other agriscience personnel may be called upon as officers or members of professional organizations to give testimony before legislators or other public officials on the need for laws, regulations, or other actions that affect their work. Or, it may be quite possible to end the day presiding over a meeting or taking minutes at a professional or civic meeting.

For the agriscience student, the need for leadership skills is also apparent. Having confidence to participate fully in class is important (Figure 6-3). To meet prospective employers and conduct a supervised agricultural experience program requires leadership skills. Functioning in communities, participating in group meetings, and making friends readily, all require acceptable leadership skills.

FIGURE 6-1
Agribusinesses rely on team efforts to achieve goals. ***(Courtesy Anderson Valley High School Agriculture Education Department)***

FIGURE 6-2
Managers are leaders.

FIGURE 6-3
Class participation is a form of leadership.

To be a good citizen, you must earn your way in life without infringing on the rights of others. Useful citizenship utilizes leadership to promote the common good in society. *Citizenship* means functioning in society in a positive way.

TRAITS OF GOOD LEADERS

Good leaders must have *integrity* (honesty). Without it, others cannot entrust an individual with the power to manage or control, even in minor things. A leader must have *knowledge,* which means familiarity, awareness, and understanding. Good leaders are dependable and have the *courage* (willingness to proceed under difficult conditions) and initiative to carry out personal and group decisions. To lead one must communicate. This requires good speaking and listening skills.

In working with others, *tact,* or the skill of encouraging others in positive ways, is useful. Similarly, a sense of justice to ensure the rights of others is important. *Enthusiasm,* or energy to do a job and inspiration to encourage others, is useful. *Unselfishness* means placing the desires and welfare of others above yourself. It, too, is an important quality for good leadership. These things permit *loyalty,* which means reliable support for an individual group or cause. These and other traits are achieved through effective leadership development.

LEADERSHIP DEVELOPMENT OPPORTUNITIES

Modern schools provide extensive opportunities for agriscience students to develop leadership skills. Students develop leadership in school organizations, athletics, and in classroom and shop situations. Some become leaders at home, on the job, in their church, or in community organizations.

4-H Clubs

The *Cooperative Extension Service* is an educational agency of the USDA and an arm of your state university. It provides educational programs for both youth and adults. Its programs include personal, home and family, community, and agriscience resources development. The Cooperative Extension Service also sponsors 4-H clubs. The *4-H* network of clubs is directed by Cooperative Extension Service personnel to enhance personal development and provide skill development in many areas, including agriscience (Figure 6-4). The four Hs in 4-H stand for head, heart, hands, and health. These provide the basis for the 4-H pledge which is, "I pledge my head to clearer thinking, my heart to greater loyalty, my hands to larger service, and my health to better living for my Club, my community, my country, and the world."

Most communities in the United States have 4-H clubs. They encourage personal projects and group activities. Such activities help develop leadership and agriscience skills. Since 4-H makes extensive use of volunteers, older members frequently become junior leaders and many become adult 4-H leaders.

Scout Organizations

Girl Scout and Boy Scout organizations provide opportunities for leadership development and skill development in agriscience and other areas. Scouts focus heavily on the out-of-doors and provide excellent leadership development and natural resources skills. They provide recognition through a system of merit badges, which are earned by learning skills and obtaining experiences in agriscience, as well as other areas.

FFA

As defined in Unit 5, the FFA is a youth organization which was developed specifically to expand the opportunities in leadership and agriscience skill development for students in public secondary schools. Only students enrolled in agriscience courses are eligible for membership in FFA.

Aim and Purposes

The FFA is part of the agriscience curriculum in most schools where agriscience is offered. It is an important teaching tool. It serves as a laboratory for developing leadership and citizenship skills. These, in turn, are helpful in learning agriscience skills. The primary aim of the

FIGURE 6-4
Leadership skills are developed through 4-H.

FFA is the development of agriscience leadership, cooperation, and citizenship. The specific purposes of the FFA may be paraphrased as follows.

1. To develop competent, aggressive, rural and agriscience leadership.
2. To create and nurture a love of country life.
3. To strengthen vocational agriscience students' confidence in themselves and their work.
4. To create more interest in the intelligent choice of agriscience occupations.
5. To encourage members in the development of individual farming programs and establishment of agriscience careers.
6. To encourage members to improve the home and its surroundings.
7. To participate in worthy undertakings for the improvement of agriscience.
8. To develop character, train for useful citizenship, and foster patriotism.
9. To participate in cooperative effort.
10. To encourage and practice thrift.
11. To encourage improvement in scholarship.
12. To provide and encourage the development of organized rural recreational activities.[1]

AGRI-PROFILE

CAREER AREA: LEADERSHIP DEVELOPMENT
The art of effective speaking is believed by many to be one of the most powerful tools of a leader. *(Courtesy of the National FFA Organization)*

Leadership development may be a career area or specialty for teachers, consultants, personnel managers, coaches, and others. However, many agriscience positions require good leadership capabilities as a tool for everyday use. Auctioneers, salespersons, managers, entrepreneurs, corporate executives, politicians, and anyone who directs others or routinely meets the public must have good leadership skills.

Leadership involves good planning, goal setting, and the ability to inspire others to work toward a common goal. Such skills as committee interaction, parliamentary procedure, and self-expression are important leadership techniques which are developed through study and practice. Such skills are used in church, civic, and community organizations as well as in work places. Group projects and club activities in agriscience provide excellent opportunities for leadership development.

The Emblem The FFA emblem contains five major symbols which help demonstrate the structure of the organization (Figure 6-5). They are as follows.

1. Eagle—The emblem is topped by the eagle and other items of our national seal. The eagle was placed in the emblem to represent the national scope of the organization. It could also represent the natural resources in agriscience.
2. Corn—Corn is grown in every state in the United States. It reminds us of common interest in agriscience, regardless of where we live.
3. Owl—The owl represents knowledge and wisdom. Use of this symbol in the emblem recognizes the fact that people in agriscience need a good education and that education must be tempered with experience to be of greatest usefulness.
4. Plow—The plow has been used to represent work-labor-effort. These qualities are needed to cause things to happen and to get results in agriscience.
5. Rising Sun—The rising sun is a symbol of the progressive nature of agriscience. It is symbolic of the need for workers in agriscience to cooperate and work toward common goals.

The FFA emblem may be constructed one symbol at a time. When assembled and dissembled in this manner, it is a good device to help others understand the FFA.

The Colors The official FFA colors are blue and gold. The shade of blue is national blue. The shade of gold is the yellow color of the corn. Therefore, the colors are called national blue and corn gold.

Motto The FFA motto contains phrases that describe the philosophy of learning and development in agriscience. The motto is:

□ Learning to Do

□ Doing to Learn

□ Earning to Live

□ Living to Serve

"Learning to Do" emphasizes the practical reasons for study and experience in agriscience. It also suggests ambition and willingness to productively use the hands as well as the mind. "Doing to Learn" describes procedures used in agriscience instruction at the doing level. Doing or experiencing results in the most permanent of learning. "Earning to Live" suggests that FFA members intend to develop their skills and support themselves in life. And "Living to Serve" indicates an intention to help others through personal and community service.

Salute The Pledge of Allegiance to the American flag is the official FFA salute. The words of

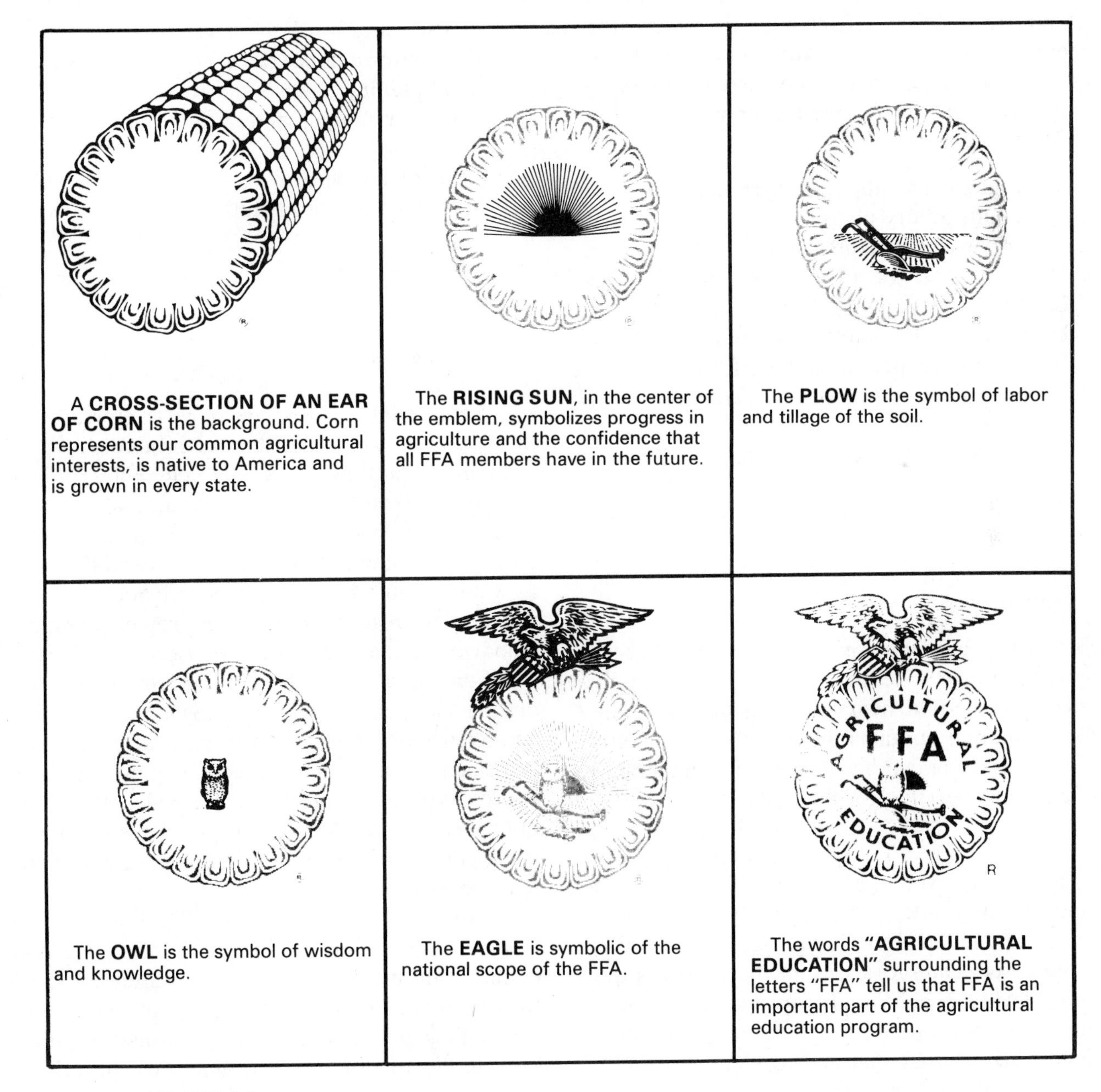

FIGURE 6-5
The FFA emblem contains five important symbols. *(Courtesy of the National FFA Organization)*

the pledge are: I pledge allegiance to the flag of the United States of America and to the Republic for which it stands, one nation under God, indivisible, with liberty and justice for all."

Degree Requirements

The FFA has four degrees, which indicate the progress a member is making. These are Greenhand, Chapter, State, and National degrees. The Greenhand and Chapter degrees are awarded by the FFA chapter in the agriscience department of the school. The State degree is awarded by the State Association and the National degree is awarded by the National FFA.

The Greenhand Degree is so named to indicate the member has learned beginning skills in FFA and is studying agriscience. To receive the Greenhand degree the member must meet the requirements spelled out in the current FFA *Official Manual*[2]. In general, the requirements for the Greenhand degree are:

1. Be enrolled in an approved agriscience course.
2. Have satisfactory plans for a supervised agricultural experience program.
3. Have learned the FFA Creed, Motto, and Salute.
4. Describe the FFA emblem, colors, and symbols.
5. Explain the proper use of the FFA jacket.
6. Have satisfactory knowledge of the history of the organization.
7. Know the duties and responsibilities of members.
8. Own or have access to a copy of the *Official Manual.*
9. Submit a written application for the Greenhand degree.

The requirements for the other three degrees help the member to learn and grow professionally from one level to the next in the organization. The *Official Manual* contains the exact requirements for all degrees and other details of membership and chapter operation for the FFA.

Contests

The FFA sponsors numerous contests at various levels. The first level is in the local chapter in the school. The second level is the district or regional level within the state. The third is the State Association level, and the fourth is the national level. Local FFA advisors determine which contests are appropriate for students in their programs. The contests that are conducted at each level generally reflect the content of the instructional programs.

The purpose of FFA contests is to encourage agriscience students to develop technical and leadership skills, and practice these skills in friendly competition with other FFA members. Some examples of contests that are typically for FFA groups are:

Agriscience business management
Agriscience mechanics
Agriscience production bowl
Dairy cattle judging and evaluation
Dairy foods
Floriculture
Forestry
Horses
Horticulture bowl
Horticulture happenings
Land judging
Livestock judging and evaluation
Meats
Nursery/Landscape
Parliamentary procedure
Poultry
Public speaking
Small engine service
Tractor operation
Vegetables

Some FFA contests require contestants to know how to grade agriscience products such as eggs, meat, poultry, fruits, and vegetables. Others require students to evaluate live animals such as beef and dairy cattle, horses, poultry, rabbits, sheep, and swine. Some require mechanical abilities such as welding, plumbing, electrics, irrigation, surveying, engine trouble-shooting and repair, painting, woodworking, and general tool use.

Some contests such as Floriculture and Horticulture Happenings require knowledge of the art and science of floral arrangement. Horticulture and Agriscience Production Bowls have questions about approved practices in agriscience which are tossed out to teams of FFA members. They vie for points using game buzzers, and official judges determine team scores.

Contests such as Forestry, Vegetables, and Nursery/Landscape require students to identify plants, plant materials, insects, and diseases. Land Judging involves evaluation of soil and land. Parliamentary procedure teams demonstrate their knowledge of correct parliamentary practices. These procedures are used to open meetings, conduct business, close meetings, and write minutes according to acceptable practice in the real world. Various types of speaking contests help individuals to sharpen their speaking skills.

Contests are organized so the participants are competing as individuals, teams of three or four participants, or chapters (Figure 6-6). Some examples of chapter contests are Safety, Building Our American Communities, and Chapter Award. These contests involve most or all of the students in the agriscience program, and provide valuable improvements to the program, school, and community.

FIGURE 6-6
Contests provide motivation, skill development, and recognition. ***(Courtesy Denmark High School Agriculture Education Department)***

Proficiency Awards The FFA has an extensive system of awards for individual members. These provide members with incentives and rewards for excellence in leadership and agriscience achievement. These awards change as the FFA expands to meet the changing nature of agriscience. Some proficiency awards available at the chapter, state, and national levels are:

Agriscience electrification
Agriscience mechanics
Agriscience processing
Agriscience sales and/or service
Beef production
Cereal grain production
Dairy production
Diversified crop production
Diversified livestock production
Feed grain production
Fiber crop production
Floriculture
Forage production
Forest management
Fruit and/or vegetable production
Home and/or farmstead improvement
Horse proficiency
Nursery operations
Oil crop production
Outdoor recreation
Placement in Agriscience production
Poultry production
Sheep production
Soil and water management
Specialty animal production
Specialty crop production
Swine production
Turf and landscape management
Wildlife management

FFA members may apply for the proficiency awards that fit their own situations. Medals and certificates are awarded to winners at the chapter level, while plaques and financial awards are given to those at the state and national levels. More information on proficiency and other award programs are described in the FFA *Student Handbook*[3].

PUBLIC SPEAKING Oral communication skills are important for good leadership. Effective leaders must speak with individuals, committees, small groups, or in large forums. The ability to relax, speak clearly, and state what is pertinent to the subject at hand is very useful. These skills are developed by applying basic principles of speech preparation and organization.

Speeches may be prepared or extemporaneous. An *extemporaneous* speech is a speech delivered with little or no preparation. Most speaking is extemporaneous. The ability to do effective extemporaneous speaking is enhanced by delivering planned speeches. Therefore, the planned speech is used to teach public speaking skills in FFA.

The Plan A speech should have at least three sections: the introduction, body, and conclusion. The plan should clearly identify the sections.

Introduction The introduction indicates the need for and importance of the speech. It should be carefully planned and spoken with confidence. The introduction may be in the form of statements or questions. If the introduction is not to the point, does not fit the occasion, or is not delivered in a spirited manner, the audience may not listen to the rest of the speech. The introduction might be only a few lines long, but it must capture the audience's attention.

Body The body of the speech contains the major information. It should consist of several major points which support one central theme or objective. Each major point is supported by additional information to explain, illustrate, or clarify the point. It is best to write the body of the speech in outline form (Figure 6-7).

INTRODUCTION

1. Honorable judges, instructors, and fellow students
2. Rabbits, cows, plants and plows—WE STILL NEED THEM!
3. When the family farm goes under, a piece of America goes under with it

BODY

1. The 1986 census of Agriculture revealed . . .
2. The midsize farm is likely to be the true family farm
   - Owned and operated by the family
   - Family receives benefit from their work
3. Some believe the family farm is a relic of the past!
   a. Press fascination with bankruptcy sales
   b. Farms not in view from interstate highways
   c. Small population on farms
   d. Animal rights groups and unionizing efforts
4. The family farm endures in the United States
   a. Better managed farms buy the weaker
   b. Disease epidemics threaten specialized operations
   c. Farm retailing is on the rise
   d. Small farms are becoming legitimate
5. Farm bankruptcy must be minimized
   a. Farm failure affects general businesses
   b. We will miss fresh farm produce
   c. We will see more pollution
   d. We will depend more on imports
6. Are there remedies? Yes!
   a. Remove politics from marketing
   b. Recognize farmers as astute business people
   c. See the total industry of agriscience
   d. Tax farm land for farm use
   e. Increase agriculture land preservation programs

CONCLUSION

1. Indeed!
   a. General George Washington nearly lost the continental army at Valley Forge from lack of food, shelter, and clothing.
   b. It can happen to us if we don't maintain a healthy farm situation in the United States today
   c. Don't give away our most valuable resource—the ability to feed, clothe, and shelter ourselves!

FIGURE 6-7
An outline of a winning speech used to convey the importance of family farms.

An outline will help direct the thought and delivery of the speech. After outlining the speech, a carefully worded narrative may be written to help the speaker fully develop the content of the speech. However, it must be emphasized that a speech should be given from an outline to avoid the temptation to memorize the speech. Memorization of a speech has two serious pitfalls. One, there is the danger it will sound like someone else's words and lack authenticity. Two, if the line of thought is lost during delivery, the speaker may not be able to find the location in the narrative. This can greatly damage the quality of the speech.

Conclusion The conclusion should remind the audience of the major theme or central point of the speech and briefly restate the major points. Further, the conclusion should leave the audience feeling like they want to take action to implement or adopt what you have said. Some speeches call for action, while others call for changes in attitude or perception. The more powerful speeches move people to action. The words needed to do such a big job must be carefully planned.

Giving the Speech Giving the speech can be fun and provide a lot of satisfaction (Figure 6-8). However, this fun and satisfaction does not come free. The speaker must have prepared the plan well and practice the speech extensively. Practicing the speech until the content becomes very familiar helps speaking become nearly automatic. The speech should be given orally to yourself several times. Then practice in front of a mirror to observe

FIGURE 6-8
Speaking can be enjoyable when the speaker has been properly trained. ***(Courtesy of the National FFA Organization; Bill Stagg, Photographer)***

National Public Speaking Contest ***Judges' Score Sheet***

PART I. FOR SCORING CONTENT AND COMPOSITION

Items To Be Scored	Points Allowed	Points Awarded Contestant												
		1	2	3	4	5	6	7	8	9	10	11	12	13
Content of Manuscript	200													
Composition of Manuscript	100													
Score on Written Production	300													

PART II. FOR SCORING DELIVERY OF THE PRODUCTION

Items To Be Scored	Points Allowed	Points Awarded Contestant												
		1	2	3	4	5	6	7	8	9	10	11	12	13
Voice	100													
Stage Presence	100													
Power of Expression	200													
Response to Questions	200													
General Effect	100													
Score on Delivery	700													

PART III. FOR COMPUTING THE RESULTS OF THE CONTEST

Items To Be Scored	Points Allowed	Points Awarded Contestant												
		1	2	3	4	5	6	7	8	9	10	11	12	13
Score on Written Production	300													
Score on Delivery	700													
TOTALS	1000													
*Less Overtime Deductions, for each minute or major fraction thereof	20													
*Less Undertime Deductions, for each minute or major fraction thereof	20													
GRAND TOTALS														
Numerical or Final Placing of Contestants														

*From the Timekeeper's record.

FIGURE 6-9A
A score sheet for evaluating speeches *(Courtesy of the National FFA Organization)*

Explanation of Score Sheet Points

Part I-For Scoring Content and Composition

1. *Content of the manuscript* includes:
   - Importance and appropriateness of the subject
   - Suitability of the material used
   - Accuracy of the statements included
   - Evidence of purpose

   - Completeness and accuracy of bibliography
2. *Composition of the manuscript* includes:
   - Organization of the content
   - Unity of thought
   - Logical development
   - Language used
   - Sentence structure
   - Accomplishment of purpose-conclusions

Part II-For Scoring Delivery of Production

1. *Voice* includes:
   - Quality
   - Pitch
   - Articulation
   - Pronunciation
   - Force

2. *Stage Presence* includes:
   - Personal appearance
   - Poise and body posture
   - Attitude
   - Confidence
   - Personality
   - Ease before audience

3. *Power of expression* includes:
   - Fluency
   - Emphasis
   - Directness
   - Sincerity
   - Communicative ability
   - Conveyance of thought and meaning

4. *Response to questions* includes:
   *Ability to answer satisfactorily the questions on the speech which are asked by the judges indicating originality, familiarity with subject and ability to think quickly.

5. *General effect* includes:
   Extent to which the speech was interesting, understandable, convincing, pleasing and held attention.

*NOTE: Judges should meet prior to the contest to prepare and clarify the questions to be asked.

FIGURE 6-9B
A score sheet for evaluating speeches *(Courtesy of the National FFA Organization)*

facial expressions, posture, and gestures. Finally, give the speech in front of others and invite them to make suggestions to improve the delivery (Figure 6-9).

Books have been written on techniques to enhance speech delivery. However, for the beginner, a few basic and time-tested procedures should be helpful for effective speaking. Some suggested procedures for giving speeches are:

- □ Have your teacher read and make suggestions on the content of your speech.
- □ Learn the content thoroughly through repeated thought and practice.
- □ Record the speech on a tape recorder and observe the sound, speed, power, and effectiveness of your voice. Make corrections to improve the delivery.
- □ Practice the speech in front of a mirror to observe posture, hand gestures, and facial expressions. Your posture should be erect and natural, with hands at your side or resting lightly on the edges of the podium. They should be used occasionally for gestures that emphasize a point, show direction, or indicate count.
- □ Ask your teacher for a score sheet for judging speeches. Deliver the speech in front of a trusted person who can check your delivery against the score sheet and provide suggestions for improvements. This may be a friend, relative, or teacher.

- □ Deliver your speech in front of your class for experience and suggestions.
- □ Ask your teacher to critique your speech for final approval.
- □ Deliver your speech in front of civic groups and/or in FFA public-speaking contests.

PARLIAMENTARY PROCEDURE

What is parliamentary procedure? Why are so many people familiar with it? Why is it important? Why should agriscience students be interested in learning parliamentary procedure? *Parliamentary procedure* is a system of guidelines or rules for conducting meetings. Most Americans who are influential in their communities are familiar with parliamentary procedure.

Parliamentary procedure is used to guide the meetings conducted by school groups, church groups, and civic organizations such as Lions, Rotary, and Ruitan clubs. Agriscience students should be interested in learning this procedure so they can have their opinions heard and influence decisions which affect their lives. Parliamentary procedure is important because it permits a group to:

- □ Discuss one thing at a time.
- □ Hear everyone's opinion in a relaxed, courteous atmosphere.
- □ Protect the rights of minorities.
- □ Make decisions according to the wishes of the majority of the group.

Requirements for a Good Business Meeting

A good *business meeting* is a gathering of people working together to make wise decisions. Wrong decisions cause unhappiness, loss of income, inefficiency in business and social activities, injury, and other problems. The most common outcome of poorly run business meetings is the waste of time and lack of results. Meetings run by groups of individuals who know and use parliamentary procedure are smooth, efficient, orderly, brief, and get a lot done. Some requirements for a good business meeting follow.

A Good Presiding Officer

A *presiding officer* is a president, vice president, or chairperson who is designated to lead a business meeting. He/she should be committed to the goals of the organization and want to lead the group into good group decisions. A good presiding officer must know and use proper parliamentary procedure.

A Good Secretary

A *secretary* is a person elected or appointed to take notes and prepare minutes of the meeting. *Minutes* is the name of the official written record of a business meeting. Minutes should include the date, time, place, presiding officer, attendance, and motions discussed at the meeting. They should be written clearly and kept in a permanent secretary's book.

Informed Members

Informed members are members who are active in the organization and want to be part of the group. They give previous thought to issues to be discussed and gather useful information about the issues. They share these thoughts with others in the meeting. This permits everyone to have the benefit of the best thinking in the group. It permits the best decisions to be made. Effective members know and use parliamentary procedure.

A Comfortable Meeting Room

The meeting place must be comfortable and free of distractions. A moderate temperature and good lighting are essential. Members should be seated so they can hear and see each other. For small groups, seating at a table or in a circle works well. Large groups must rely on a good sound system for the presiding officer to be heard and members must speak clearly with good volume in order to be heard. Good public speaking skills and thorough knowledge of parliamentary procedure helps the members considerably.

Conducting Meetings

The Order of Business

The *order of business* refers to the items and sequence of activities conducted at a meeting. The order of business is usually made up by the secretary. This generally grows out of an executive meeting. An *executive meeting* is a meeting of the officers to conduct the business of the organization between regular meetings. They also consider what needs to be discussed by the total membership at the regular meeting. The essential items in an order of business are:

- □ Call to order
- □ Reading and approval of minutes of the previous meeting
- □ Treasurer's report
- □ Reports of other officers and committees
- □ Old business

- □ New business
- □ Adjournment

Other items that are frequently used in orders of business are a program, speaker, or entertainment.

Parliamentary Practices

Use of the Gavel

The *gavel* is a small wooden hammerlike object used by the presiding officer to direct a meeting (Figure 6-10). It is used to call the meeting to order, announce the result of votes, and adjourn the meeting. It is also used to signal the members to stand, sit down, or reduce the noise level of the group. The gavel is a symbol of the authority of the office of president or chairperson, and should be respected by all attending the meeting.

In some organizations like the FFA, a system of taps is used to signal the audience to do certain things. In FFA meetings, the gavel is used as follows:

- □ One tap—the outcome of or decision about the item under consideration has been decided as announced by the presiding officer.
- □ Two taps—the meeting will come to order, members should sit down if standing, or members should be quiet except when recognized.
- □ Three taps—members should stand up.

Obtaining Recognition and Permission to Speak

For a meeting to be orderly, members must speak one at a time and in some logical and fair sequence. The presiding officer is regarded as the "traffic controller," and calls on members as they request to be recognized or according to certain rules. To be recognized, the member should raise the hand to get the presiding officer's attention. The presiding officer should call the member by name; then the person should stand and address the presiding officer as Madame or Mr. Chairperson, or Madame or Mr. President. The individual should then proceed to speak.

FIGURE 6-10
A gavel is the symbol of authority of the presiding officer. ***(Courtesy Denmark High School Agriculture Education Department)***

Presenting a Motion

A *motion* is a proposal, presented in a meeting, to be acted upon by the group. To present a motion, the member raises a hand and is recognized by the presiding officer. Then the member states, "Madame/Mr. President, I move that . . ." (and continues with the rest of the motion). The words "I move" are important to say when beginning the motion. Otherwise, you will be regarded as incorrect and not using good parliamentary procedure. In order for a motion to be discussed by the group, at least one other member must be willing to have the motion discussed. That second individual expresses this willingness by saying, "Madame/Mr. President, I second the motion."

Some Useful Motions

There are dozens of motions, but a few are basic ones that are generally known and widely used. Some of these are:

- □ *Main motion*—a basic motion used to present a proposal for the first time. The way to state it is to get recognized and then say, "I move . . ."
- □ *Amend*—a type of motion used to add to, subtract from, or strike out words in a main motion. The way to present an amendment is to say, "I amend the motion by . . ."
- □ *Refer*—a motion used to refer some other motion to a committee or person for finding more information and/or taking action on the motion. The way to state a referral is to say, "I move to refer this motion to . . ."
- □ *Lay on the Table*—a motion used to stop discussion on a motion until the next meeting. The way to table a motion is to say, "I move to table the motion."
- □ *Point of order*—a procedure used to object to some item in or about the meeting that is not being done properly. The procedure to use is to stand up and say, "Madame/Mr. President, I rise to a point of order!" The presiding officer should then recognize the member by saying,

"State your point." The member then explains what has been done incorrectly.

□ *Adjourn*—a motion used to close a meeting. The procedure is to say, "I move to adjourn."

The FFA *Student Handbook*[4] provides a listing of additional motions and how to use parliamentary procedure for more effective meetings (Figures 6-11 & 6-12).

SUMMARY OF MOTIONS

Classification	Kind	Second Required	Debatable	Amendable	Vote Required	Can Be Reconsidered
Privileged	1. Adjourn	Yes	No	No	Majority	No
	2. Question of Privilege	No	No	No	No	None
Incidental	3. Point of Order	No	No	No	None	No
	4. Appeal	Yes	Yes	No	Majority	Yes
	5. Suspend the Rules	Yes	No	No	2/3	No
	6. Division of the House	No	No	No	None	No
	7. Parliamentary Inquiry	No	No	No	None	No
	8. To Withdraw a Motion	No	No	No	Majority	Yes
Subsidiary	9. Lay on the Table	Yes	No	No	Majority	No
	10. Previous Question	Yes	No	No	2/3	Yes
	11. Postpone Definitely	Yes	Yes	No	Majority	Yes
	12. Limit Debate	Yes	No	No	2/3	Yes
	13. Refer to Committee	Yes	Yes	Yes	Majority	Yes
	14. Amend	Yes	Yes	Yes	Majority	Yes
	15. Postpone Indefinitely	Yes	Yes	No	Majority	Yes
Main	16. Main Motion	Yes	Yes	Yes	Majority	Yes
Other	17. Take from the Table	Yes	No	No	Majority	No
	18. Reconsider	Yes	Yes	No	Majority	No
	19. Rescind	Yes	Yes	Yes	2/3	No

FIGURE 6-11
Parliamentary skills useful in FFA and other organizations. *(Courtesy of the National FFA Organization)*

Agriscience students are encouraged to develop effective leadership skills. The ability to work effectively as a member of a group is essential for all. The ability to function as a chairperson or officer creates more opportunities to serve and influence the environment in which we live. The development of self confidence is essential. Self-confidence is a product of knowledge and skill. Therefore, all should strive to learn to speak well, learn to function in groups through parliamentary procedure, and utilize the opportunities in FFA for personal development.

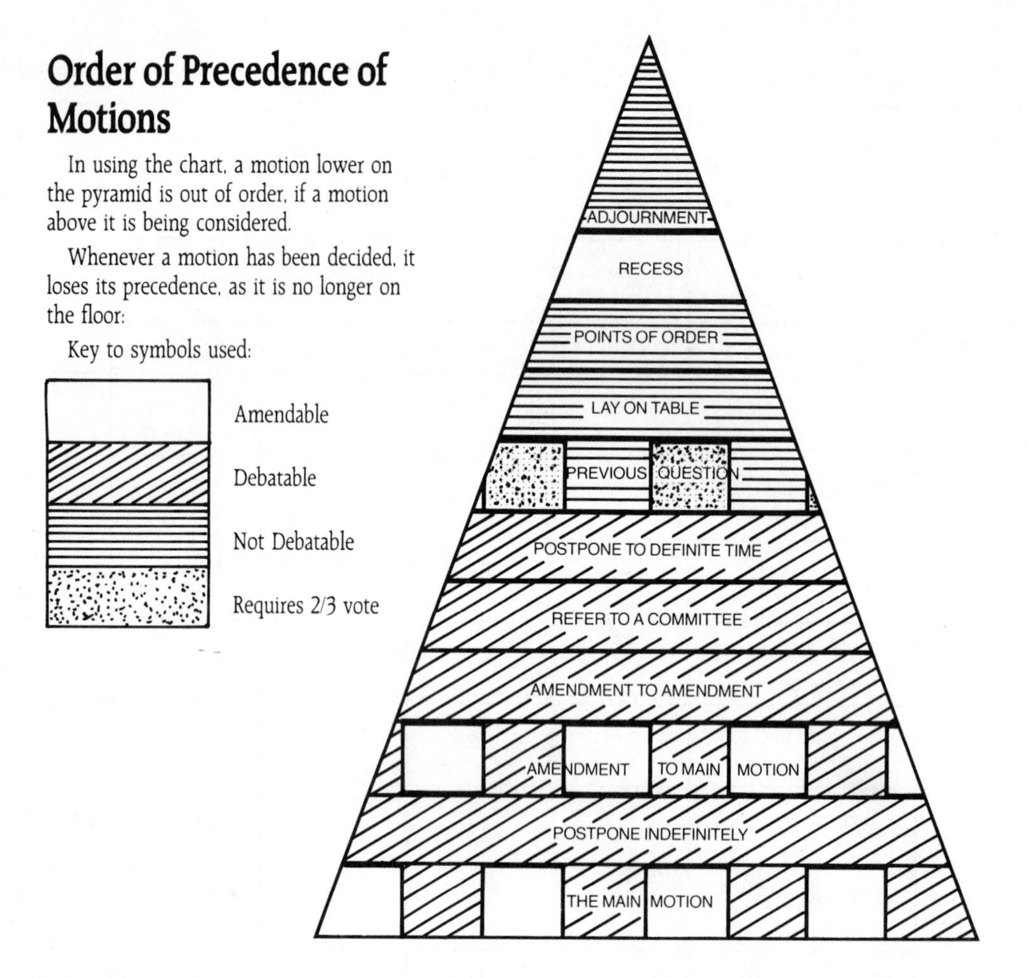

FIGURE 6-12

Correct order of precedence of motions. ***(Courtesy of the National FFA Organization)***

Student Activities

1. Define all of the "Terms To Know."
2. Make a list of the many ways that you exercise leadership in your family, school, and community.
3. Develop a bulletin board showing the symbols of the FFA emblem.
4. Make a bulletin board illustrating the purposes of the FFA. Include the FFA colors.
5. Write down 5 contests and 5 proficiency awards for which you would like to try out. Discuss these with your classmates and teacher.

6. Prepare and present a three-minute speech on an agriscience topic to your class.
7. Ask your teacher to let you use a gavel and direct a mock class or FFA meeting to learn parliamentary skills.
8. Form a parliamentary procedure team of 6 or 8 classmates and demonstrate various parliamentary skills to the class.

Self-Evaluation

A. MULTIPLE CHOICE

1. To lead is to
   a. manage.
   b. organize.
   c. show the way.
   d. all of the above.
2. Which is not a trait of a good leader?
   a. Courage
   b. Integrity
   c. Selfishness
   d. Tact
3. A leadership organization of the Cooperative Extension Service is
   a. Boy Scouts.
   b. FFA.
   c. Girl Scouts.
   d. 4-H.
4. Membership in FFA is limited to youth who
   a. are in the country.
   b. enroll in an agriscience program in school.
   c. plan careers in agriscience.
   d. seek leadership training.
5. Which is not a purpose of FFA?
   a. Develop leadership
   b. Intelligent choice of agriscience occupations
   c. Promote scholarship
   d. Promote self above others
6. The symbol which signifies that the FFA is a national organization is the
   a. corn.
   b. eagle.
   c. owl.
   d. rising sun.
7. The first line of the FFA motto is
   a. Doing to learn.
   b. Earning to live.
   c. Learning to do.
   d. Living to serve.
8. One requirement for the Greenhand Degree is
   a. a plan for supervised agricultural experience.
   b. $70 earned from agriscience experience.
   c. school grades of C or above.
   d. unselfish attitude in FFA activities.
9. An FFA activity not generally organized as a contest is
   a. dairy foods.
   b. forestry.
   c. land judging.
   d. wildlife management.

10. One of the last items in an order of business is

a. new business.
b. officer reports.
c. reading of the minutes.
d. treasurer's report.

11. The largest part of a speech is the

a. body.
b. conclusion.
c. introduction.
d. summary.

12. The only acceptable way to start a motion is to say

a. "I believe . . ."
b. "I make a motion that . . ."
c. "I move . . ."
d. "I think . . ."

B. MATCHING

______	**1.** Adjourn	**a.** Present a new proposal
______	**2.** Amend	**b.** Leave it to a committee
______	**3.** Lay on the table	**c.** Correct some procedure
______	**4.** Main motion	**d.** Close the meeting
______	**5.** Point of order	**e.** Consider it next meeting
______	**6.** Refer	**f.** Change a motion
______	**7.** Three taps of gavel	**g.** Prepare minutes
______	**8.** Secretary	**h.** Members stand

Natural Resources Management

UNIT 7 Maintaining Air Quality

OBJECTIVE To determine major sources of air pollution and identify procedures for maintaining and improving air quality.

Competencies to Be Developed

After studying this unit, you will be able to

- ☐ define air and identify its major components.
- ☐ analyze the importance of air to humans and other living organisms.
- ☐ determine the characteristics of clean air.
- ☐ describe common threats to air quality.
- ☐ describe important relationships between plant life and air quality.
- ☐ list practices that lead to improved air quality.

TERMS TO KNOW

Air
Water
Soil
Habitat
Sulfur
Hydrocarbons
Nitrous Oxides
Tetraethyl lead
Carbon monoxide
Radon
Radioactive material
Chlorofluorocarbons
Ozone
Pest
Pesticide
Toxic
Asbestos
Photosynthesis

MATERIALS LIST

pencil and paper
several specimen aerosol cans

Life as we know it on our planet requires a certain balance of air, water and soil. *Air* is a colorless, odorless, tasteless mixture of gases. It occurs in the atmosphere around the earth and is comprised of approximately 78% nitrogen, 21% oxygen and 1% argon, carbon dioxide, neon, helium, and other gases. *Water* is a clear, colorless, tasteless, and near odorless liquid. It's chemical makeup is two parts hydrogen to one part oxygen. *Soil* is the top layer of the earth's surface, suitable for the growth of plant life.

AIR QUALITY Without a reasonable balance of air, water and soil, most organisms would perish. Slight changes in the composition of air or water may favor some organisms, while causing others to diminish in number or health. Unfavorable soil conditions usually mean inadequate food, water, shelter, and other factors of *habitat* (the area or type of environment (Figure 7-1) in which an organism or biological population normally lives).

Threats to Air Quality

The mixture of gases we call air is absolutely essential for life. The air we breathe should be healthful and life supporting. Air must contain at least approximately 21% oxygen for human survival. If a human stops breathing and no life-supporting equipment or procedures are used, the brain will die in approximately 4-6 minutes. Further, air may contain poisonous materials or organ-

FIGURE 7-1
Clear air and water as well as productive soil are necessary for good habitat for plants, animals and humans.

isms which can decrease the body's efficiency, cause disease, or cause death through poisoning.

The circumference or distance around the Earth at its largest part is 24,000 miles. Yet abuse of the atmosphere in one area frequently damages the environment in distant parts of the world. Air currents flow in somewhat normal patterns and air pollution will move according to those patterns. However, when warm and cold air meet, the exact air movement will be determined by the differences in temperature, the terrain, and other factors. Therefore, humans have an obligation to keep the air clean for their own benefit as well as that of society at large.

There are major worldwide threats to air quality. These include sulfur compounds, hydrocarbons, nitrous oxides, lead, carbon monoxide, radon gas, radioactive dust, chlorofluorocarbons, pesticide spray materials, and others (Figure 7-2). Most of these are poisonous to breathe and have other damaging effects.

FIGURE 7-2
Automobiles, trucks, homes, and factories burn gasoline, oil, coal, and wood. These three products of combustion reduce air quality.

Sulfur It is present in coal and crude oil. It also combines with oxygen to form harmful gases, such as sulfur dioxide, when these and other fuels are burned. Therefore, most smoke or exhaust from homes, factories, or motor vehicles contains some harmful *sulfur* (a pale-yellow element occurring widely in nature) compounds unless special equipment is used to remove them. Once these invisible gases are in the air, they combine with moisture to form sulfuric acid, which falls as acid rain. Acid rain has been found to damage and kill trees and other plants. It also has a corrosive effect on metals.

Hydrocarbons They have become serious problems as factories and motor vehicles have become so prevalent during the 20th century. In the United States, *hydrocarbon* emissions (by-products of combustion or burning) are held in check by special emission-control equipment on automobiles and special equipment called stack scrubbers in large industrial plants. Hydrocarbon output is controlled on automobiles by crankcase ventilation, exhaust gas recirculation, air injection, and other procedures. Without this equipment, it would be impossible to breathe in our major cities and in heavy stop-and-go traffic on major highways.

Nitrous Oxides and Lead *Nitrous oxides* are compounds containing nitrogen and oxygen. They make up about 5% of the pollutants in automobile exhaust. Yet they are damaging to the atmosphere and must be removed from exhaust gases. This group of chemicals was the most difficult and perhaps the most costly group to remove from automobile emissions. Scientists and engineers solved the problem partly by installing catalytic converters.

Hot exhaust gases from the engine flow through a honeycomb of platinum in the catalytic converter. The reaction converts the nitrous oxides into harmless gases. Until recently, all gasoline contained *tetraethyl lead,* a colorless, poisonous, oily liquid. This lead improved the burning qualities of gasoline and improved the control of engine knocking. However, tetraethyl lead ruined catalytic converters. So catalytic converters could not be used until a substitute for tetraethyl lead was found in the 1970s.

Tetraethyl lead is a poisonous product. Unfortunately, it is still used in regular gasoline in the trucks and tractors where catalytic converters are not used. Therefore, lead and nitrous oxides are still major pollutants of the atmosphere.

Carbon Monoxide It is one of the automotive exhaust gases that cannot be removed with our present technology. It is the poisonous gas that kills people in automobiles with leaking exhaust systems. A similar hazard exists when engines are run in closed areas without adequate ventilation. Victims fall asleep and die. *Carbon monoxide* emissions (a colorless, odorless, and highly poisonous gas) may be reduced by keeping engines in good repair and properly tuned.

Radon It has become a hazard in homes in many parts of the United States in the last decade or two. *Radon* (a colorless, radioactive gas formed by disintegration of radium) moves up through the soil and flows into the atmosphere at low and usually harmless rates.

However, if a house or other building is constructed over an area where radon gas is being emitted, a hazardous condition can result. Radon gas can accumulate in buildings with cracks in the basement floors or walls. The problem can be prevented by tightly sealing all cracks and/or providing continuous ventilation either below the basement floor or throughout the building (Figure 7-3).

Radioactive Dust and Materials

Radioactive material is material that is emitting radiation. Dust resulting from atomic explosion or other nuclear reaction and materials contaminated by atomic accidents or wastes are of growing concern. The damage from radioactivity ranges from skin burns through sickness and hereditary damage to death. The possibility of worldwide contamination and other hazards from serious nuclear accidents has caused a reduction in plans for construction of atomic electrical-generation plants (Figure 7-4).

FIGURE 7-3
Keep quality control systems in good working order.

Chlorofluorocarbons

Chlorofluorocarbons are a group of compounds consisting of chlorine, fluorine, carbon, and hydrogen. They are used as aerosol propellants and refrigeration gas. These materials are very stable. Once released from an aerosol can or cooling system, they bounce around and eventually float upward into the high atmosphere. It is believed that chlorofluorocarbons will survive in the upper atmosphere for about 100 years. Meanwhile, their chlorine atoms destroy *ozone* without themselves being destroyed.

Ozone is a compound that exists in limited quantities about 15 miles above the earth's surface. It filters out harmful ultraviolet rays from the Sun. There is evidence that the ozone layer is being damaged and that increased problems with skin cancer and damage to the body's immune system is likely to occur. Most living organisms will be exposed to the damaging effect of ultraviolet rays.

In 1987, there was an international conference where at least 37 nations eventually agreed to schedule cutbacks in the production of chlorofluorocarbons. However, given the possible damage by this pollutant, it may make sense to stop all production immediately.

Pesticide Spray Materials

A *pest* is a living organism that acts as a nuisance. *Pesticide* is a material used to control pests. Many pesticides are chemicals mixed with water so they can be sprayed on plants, animals, soil, or water to kill or otherwise control diseases, insects, weeds, rodents, and other pests. Spray materials are pollutants if they carry *toxic* (poisonous) materials or are harmful to more than one organism. Spray materials used to

FIGURE 7-4
Avoid use of aerosols which contain chlorofluorocarbons.

control pests such as diseases, insects, weeds, rodents, and others are generally harmful to the air if they are not used exactly as specified by the government and the manufacturer. Poisons may be thinned out or diluted by air movement, but excessive toxic materials can overburden the atmosphere's ability to cleanse itself. The continued use and abuse of chemicals to control pests is an area of growing concern in maintaining air quality.

AGRI-PROFILE

CAREER AREA: AIR QUALITY CONTROL
Weather balloons are among the arsenal of tools used to monitor air quality. *(Courtesy USDA Agricultural Research Service)*

Careers in air quality are available with the weather services of local, state, and national agencies. Local and network radio and television stations all utilize weather reporters and meteorology forecasters. Of equal importance are those who monitor and help to improve air quality. Technicians collect and chemists analyze samples of air taken from various places in the atmosphere, buildings, and homes. Employees of environmental protection agencies and environmental advocacy groups are important links in our efforts to maintain a healthful environment.

Air quality specialists advise and assist industry in reducing harmful emissions from motor vehicles and industrial smoke stacks. Not to be overlooked are the entomologists who monitor the winds for signs of invading insects, and plant pathologists who watch for airborne disease organisms. Career opportunities in air quality maintenance and improvement will undoubtedly increase in the future.

Asbestos It was used extensively in the past for brake and clutch linings of vehicles, shingles for house siding, steam and hot-water pipe insulation, ceiling panels, and others.

Unfortunately, *asbestos* (a heat- and friction-resistant material) fibers are very damaging to the lungs and cause disease and death. Therefore, there are now state and national laws and codes in place to remove asbestos from public buildings, industrial settings, and general use.

Air and Living Organisms Oxygen in the air is consumed by plants and animals when they breathe. Animals, including humans, use oxygen to convert food into energy and nutrients for the body. Further, animals breathe out or exhale carbon dioxide gas. It seems we would eventually run out of oxygen. Fortunately, plants give off oxygen during the day. They create oxygen through the process of *photosynthesis* (a process where chlorophyll in green plants enables those plants to utilize light, carbon dioxide, and water to make food and release oxygen).

Maintaining and Improving Air Quality Air quality can be improved by reducing or avoiding the release of pollutants into the air and removing the pollutants that are there. Some specific practices to reduce air pollution are:

1. Stop using aerosol products containing chlorofluorocarbons.
2. Provide adequate ventilation in tightly constructed and heavily insulated buildings.
3. Have buildings checked for the presence of radon gas.
4. Use exhaust fans to remove cooking oils, odors, solvents and sprays from interior areas.
5. Clean and service furnaces, air conditioners, and ventilation systems regularly.
6. Maintain all sawdust, wood chip, paint spray, welding fume, dust removal, and general ventilation systems to function at their most efficient levels.
7. Keep gasoline and diesel engines properly tuned and serviced.
8. Keep all emissions systems in place and properly serviced on motor vehicles.

9. Observe all codes and laws regarding outdoor burning.
10. Report any suspicious toxic materials or conditions to the police or other proper authorities.
11. Avoid the use of pesticide sprays as much as possible.
12. Use pesticide spray materials strictly according to label directions.

Student Activities

1. Define the "Terms To Know."
2. Make a pie chart illustrating the components of air.
3. Stand in a safe location about 5 ft. behind and to the side of a parked truck or bus with its gasoline engine running. Notice the smell of the exhaust. Stand in the same relative position from an automobile that is less than 3 years old or has less than 40,000 miles on it. What differences do you observe in the truck and automobile exhausts? Why? Which do you think causes more pollution of the air?
4. Examine the written material on all of the gas pumps at a filling station or store. Which types of gas contain lead? Which types of vehicles can use gas containing lead? Which should not use leaded gasoline?
5. Examine 3 different aerosol cans. Which products use chlorofluorocarbons for the propellant? Why is it unwise to use such products?
6. Talk with an automobile tune-up specialist about the effect of engine adjustments on the content of exhaust gases.
7. Ask your teacher to invite in an air-pollution specialist to discuss the problems of air pollution in your town, county, or state.

Self-Evaluation

A. MULTIPLE CHOICE

1. Air is

   a. 78% argon.
   b. 21% nitrogen.
   c. 21% oxygen.
   d. 10% carbon dioxide.

2. Water is

   a. a mixture of gases.
   b. colorless.
   c. one part hydrogen to two parts oxygen.
   d. tasteless.

3. Without proper air to breathe, a human can only survive about

   a. 6 minutes.
   b. 12 minutes.
   c. 2 hours.
   d. 12 hours.

4. Radon gas is a widespread threat to air quality

   a. on the highway.
   b. in factories.
   c. in homes.
   d. in wooded areas.

5. Radioactive dust is likely to be caused by

a. improperly adjusted furnaces.
b. cracks in basement floors.
c. damaged ozone layer.
d. nuclear reactions.

6. Chemicals used to kill insects are always called

a. pests.
b. pesticides.
c. pollutants.
d. toxic materials.

7. One ingredient *not* associated with photosynthesis is

a. carbon dioxide.
b. oxygen.
c. radon.
d. water.

8. Chlorofluorocarbons have been found to damage

a. aerosol sprays.
b. ozone layer.
c. refrigeration units.
d. water pumps and equipment.

9. Poisonous gas we cannot remove from auto exhaust is

a. carbon monoxide.
b. hydrocarbons.
c. nitrous oxides.
d. radon.

10. The most reliable source of information on the use of a pesticide is

a. experienced applicators.
b. extension service.
c. personal experience.
d. product label.

B. MATCHING

______	**1.** Carbon monoxide	**a.** Tetraethyl
______	**2.** Chlorofluorocarbons	**b.** Pale yellow
______	**3.** Hydrocarbons	**c.** 5% of auto exhaust
______	**4.** Lead	**d.** Damages ozone layer
______	**5.** Nitrous oxides	**e.** Filters ultraviolet rays
______	**6.** Ozone	**f.** Diseases, insects, weeds
______	**7.** Radon	**g.** Chlorophyll, light, carbon dioxide
______	**8.** Pests	**h.** Causes death from auto exhaust
______	**9.** Photosynthesis	**i.** Leaks into houses
______	**10.** Sulfur	**j.** Pollutant from autos and factories

UNIT 8 Water and Soil Conservation

OBJECTIVE To determine the relationships between water and soil in our environment and the recommended practices for conserving these resources.

Competencies to Be Developed

After studying this unit, you will be able to

- ☐ define water, soil, and related terms.
- ☐ cite important relationships between land characteristics and water quality.
- ☐ discuss some major threats to water quality.
- ☐ describe types of soil water and their relationships to plant growth.
- ☐ cite examples of enormous erosion problems worldwide.
- ☐ describe key factors affecting soil erosion by wind and water.
- ☐ list important soil and water conservation practices.

TERMS TO KNOW

Potable
Fresh water
Tidewater
Domestic
Food chain
Universal solvent
Water cycle
Deserts
Irrigation
Precipitation
Evaporation
Water table
Fertility
Saturated
Free water
Gravitational water
Capillary water
Hygroscopic water
Purify
No till
Contour
Cover crop
Erosion
Port
Delta
Aquifer
Gully erosion
Sheet erosion
Mulch
Conservation tillage
Plant residue
Contour practice
Strip cropping
Crop rotation
Organic matter
Aggregates
Lime
Fertilizer
Grass waterway
Terrace
Overgrazing
Conservation plan

MATERIALS LIST

three growing plants in 4-6″ pots for student activity
kitchen scale or laboratory balance scale
plant watering containers
plant growing area

As you fly across the vast continents of the Americas, Europe, Asia, and Africa, there is a feeling that land is an endless resource (Figure 8-1). The great oceans of the world combine to provide an even larger area covered by water than the land masses of the world. Yet both resources have become limited and there is real concern that we are rapidly depleting them. In developed countries a safe and adequate water supply is generally threatened only by temporary shortages. Such shortages usually create only mild inconveniences such as restrictions on the use of water for nonessential tasks such as washing cars and watering lawns. But in Third World countries, a safe water supply is a luxury and sufficient volume for household use is unusual. While there seems to be a sufficient volume of water in most areas of the world, supplies of usable water are generally insufficient due to misuse, poor management, waste, and pollution.

Similarly, productive land is becoming a scarce commodity and ownership is very expensive. Good land is sought by individuals for homes, farms, and recreation; by businesses for banks, stores, warehouses, car lots, and others. Industry needs land for factories and storage areas. And government needs land for roads, bridges, buildings, parks, recreation, and military facilities. Most parties look for the best land; land that is level, with deep and productive soil. Such land is in great demand because it provides a firm foundation for roads and buildings, grows crops, supports trees and shrubs, and is easy to modify according to the desired use (Figure 8-2).

THE NATURE OF WATER AND SOIL

Water We live on the water planet! Most of the Earth's surface is covered with water. The oceans are vast and most Americans live near an ocean, river, lake, or stream. Those who do not have access to water pumped from deep wells, rivers, or lakes. Our bodies and the bodies of plants and animals are about 90% water. Therefore, we can survive only a few short days if our supply of *potable* (drinkable; that is, free from harmful chemicals and organisms) water is cut off.

Water is an essential nutrient for all plant and animal life. Water transports nutrients to living cells and carries away waste products. Water cools the body. It serves so many useful functions that a suf-

FIGURE 8-1
The vastness of land resources on the earth.

FIGURE 8-2
Good land is in high demand.

ficient supply of potable water is one of the first considerations for a healthy community.

Fresh Versus Salt Water   Most of the water on the earth is not fresh water and is not suitable for humans to drink or use except for transportation. *Fresh water* refers to water that flows from the land to oceans and contains little or no salt. The water in our oceans contains heavy concentrations of salt. Similarly, our bays and tidewater rivers contain too much salt for *domestic* (of or pertaining to the household) use. *Tidewater* refers to the water that flows up the mouth of a river as the ocean tide rises or comes in. Salt water is not fit for animal consumption, nor for plant irrigation.

Food Chains   The bays and oceans provide excellent support for bacteria, algae, and certain water plants. These, in turn, are consumed by insects and small fin- and shellfish. Insects and small fish are then consumed by the larger fish. The large fish, finfish and shellfish, in turn, are eaten by people, animals, and birds. Similar relationships exist among plants and animals on land and in streams and rivers. The interdependence of plants and animals for food constitutes what is known as a *food chain* (Figure 8-3).

The Universal Solvent   Water has been described as the *universal solvent* (a material that dissolves or otherwise changes most other materials). Nearly every substance will rust, corrode, decompose, dissolve, or otherwise yield to the presence of water. Therefore, water is seldom seen in its pure form. It generally has something in it.

Some minerals in water are healthful and give the water a desirable flavor. However, water is frequently carrying toxic or undesirable chemicals or minerals. Further, water may contain decayed plant or animal remains, disease-causing organisms, or poisons. Ocean water may be described as a thin soup. It is like our blood. It gathers and transports nutrients; is the habitat for microorganisms; carries life-supporting oxygen; cleanses and purifies; and neutralizes and removes wastes.

But scientists tell us that our rivers, bays, and seas have reached their breaking point. People are now polluting faster than nature can cleanse and

AGRI-PROFILE

CAREER AREAS: SOIL CONSERVATION/ HYDROLOGY

Salt-laden waste water flows from irrigated land in California's Imperial Valley into the Salton Sea. ***(Courtesy USDA Agricultural Research Service)***

Some major challenges in soil and water conservation are to keep rain water on the land where it falls, keep soil in place and stabilized with plants, minimize pollution of fresh water, and wisely manage land and water to maintain ecological balance. For instance, soil and water scientists at the University of California are developing ways to reuse salt-laden irrigation water. It is estimated that California alone may be generating upwards of 1.6 trillion gallons of sub-surface drainage water annually.

Soil and water conservation career options provide extensive opportunities for inside and outside work. Work may occur in Soil Conservation Service (SCS) or Agricultural Stabilization and Conservation Service (ASCS) offices. Or, one may work in the field served by such offices in every county of the United States. Typical job titles are conservation technician, farm planner, soil scientist, and soil mapper.

The United States Department of Interior, state departments of natural resources, city and county governments, industry, and private agencies hire people with soil and water conservation expertise. Such workers manage water resources for recreation, conservation, and consumption. They may work as consultants, law enforcement officers, technicians, administrators, heavy equipment operators, and the like.

purify (Figure 8-4). The sea contains all of the dissolved elements carried down by the Earth's rivers through all time to the low places of the crust. The water itself may evaporate and be carried back to the land as moisture in the clouds. Eventually it may fall to the land again as rain or snow (Figure 8-5). However, the minerals remain in the ocean for eternity. Many of those substances may be toxic or otherwise threatening to the living organisms of the sea. The water that falls as rain or snow would be pure if it was not contaminated by pollutants as it falls through the air.

The Water Cycle Moisture evaporates from the earth, freshwater sources, and the seas to form clouds. Clouds remain in the air until warm air masses meet cold air masses. This causes the water vapor to change to a liquid and fall as rain, sleet, or snow. The cycling of water between the water sources, atmosphere, and surface areas is called the *water cycle.*

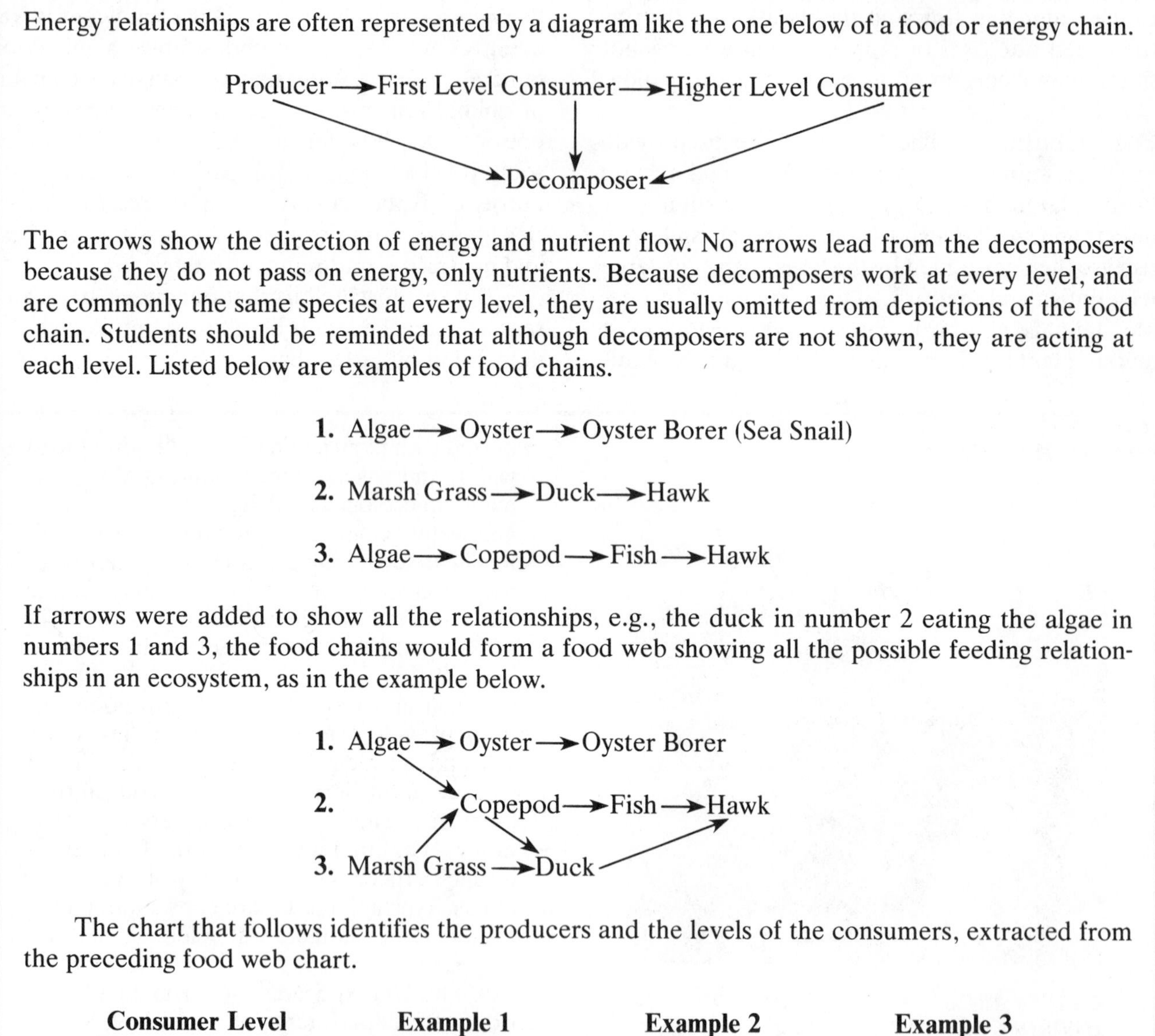

Energy relationships are often represented by a diagram like the one below of a food or energy chain.

The arrows show the direction of energy and nutrient flow. No arrows lead from the decomposers because they do not pass on energy, only nutrients. Because decomposers work at every level, and are commonly the same species at every level, they are usually omitted from depictions of the food chain. Students should be reminded that although decomposers are not shown, they are acting at each level. Listed below are examples of food chains.

1. Algae → Oyster → Oyster Borer (Sea Snail)
2. Marsh Grass → Duck → Hawk
3. Algae → Copepod → Fish → Hawk

If arrows were added to show all the relationships, e.g., the duck in number 2 eating the algae in numbers 1 and 3, the food chains would form a food web showing all the possible feeding relationships in an ecosystem, as in the example below.

The chart that follows identifies the producers and the levels of the consumers, extracted from the preceding food web chart.

Consumer Level	Example 1	Example 2	Example 3
Third		Hawk	
Second	Oyster Borer	Fish	Hawk
First	Oyster	Copepod	Duck
Producers	Algae	Algae	Marsh Grass

FIGURE 8-3
Food chains demonstrate the interdependance of plants and animals on each other.

BAY FACTS

SIZE: 195 miles long, 3.5 to 30 miles wide. Average depth: 24 feet.

DRAINAGE BASIN: 64,000 square miles, 50 major tributaries feed it.

WATER: Fresh water from tributaries mixes with salt water from Atlantic. Bay acts like sink, trapping pollutants. Only 1 percent are flushed out to sea.

PLANTS, ANIMALS: More than 2,000 species in bay, shoreline.

1 INDUSTRIAL WASTES: Thousands of commercial, industrial facilities discharge water containing toxic chemicals, metals, nitrogen, phosphorus into bay. Also: cooling needs can lead to corrosion, chlorine contamination.

2 MUNICIPAL SEWAGE: Water discharged from treatment plant contains nitrogen, phosphorus, toxic chemicals.

3 RAIN RUNOFF: Sediment from farms, forests, urban areas carries fertilizers, pesticides, herbicides. Over past 30 years, farmers have doubled the amount of fertilizers they use. Since 1960, herbicide use has tripled.

4 NUTRIENTS: High nutrient levels are most severe on northern, middle bay areas and tributaries, leading to excessive algae growth and less light filtering down to allow grasses to grow below water surface. This gives waterfowl less food. Algae dies quickly, using up oxygen in the water and making it unsuitable for aquatic life. This leads to fewer fish, shellfish, oysters.

5 TOXIC CHEMICALS: Rain washes sediment into bay. This often includes chemicals, herbicides, pesticides, toxic wastes from industry.

6 CHLORINE: It's used in bay for disinfecting drinking water, sewage, industrial processes. It's suspected of hindering spawning runs of migrating fish.

7 HEAVY METALS: Cadmium, mercury, copper from industrial wastewater, sewage treatment plants can be toxic. Lead from auto exhaust, iron and zinc from industrial discharge, shore erosion.

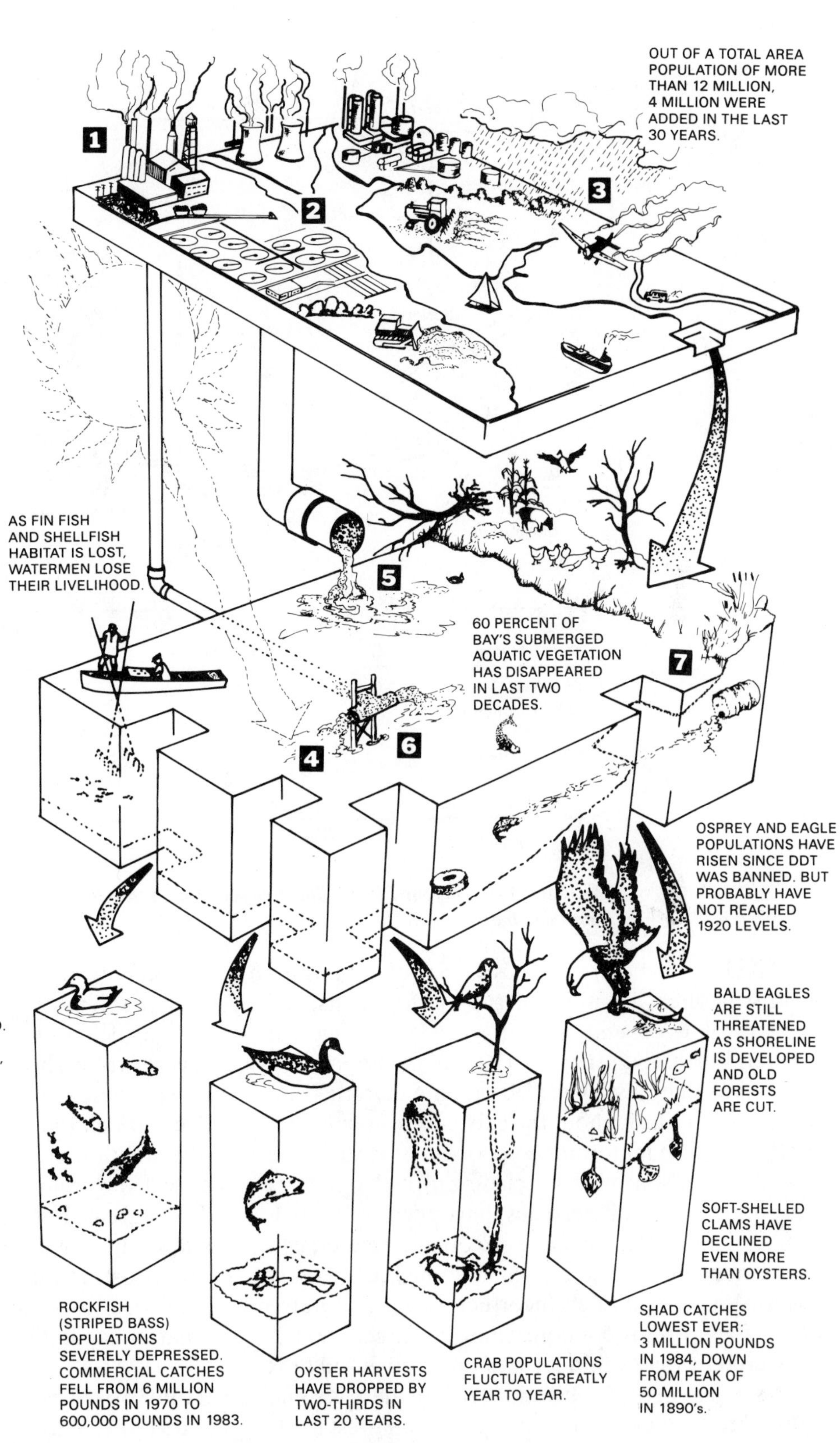

FIGURE 8-4

Water resources of the world are losing their productivity. *(Adapted from, and used with permission from The Washington Post "MD Magazine," Autumn 1988)*

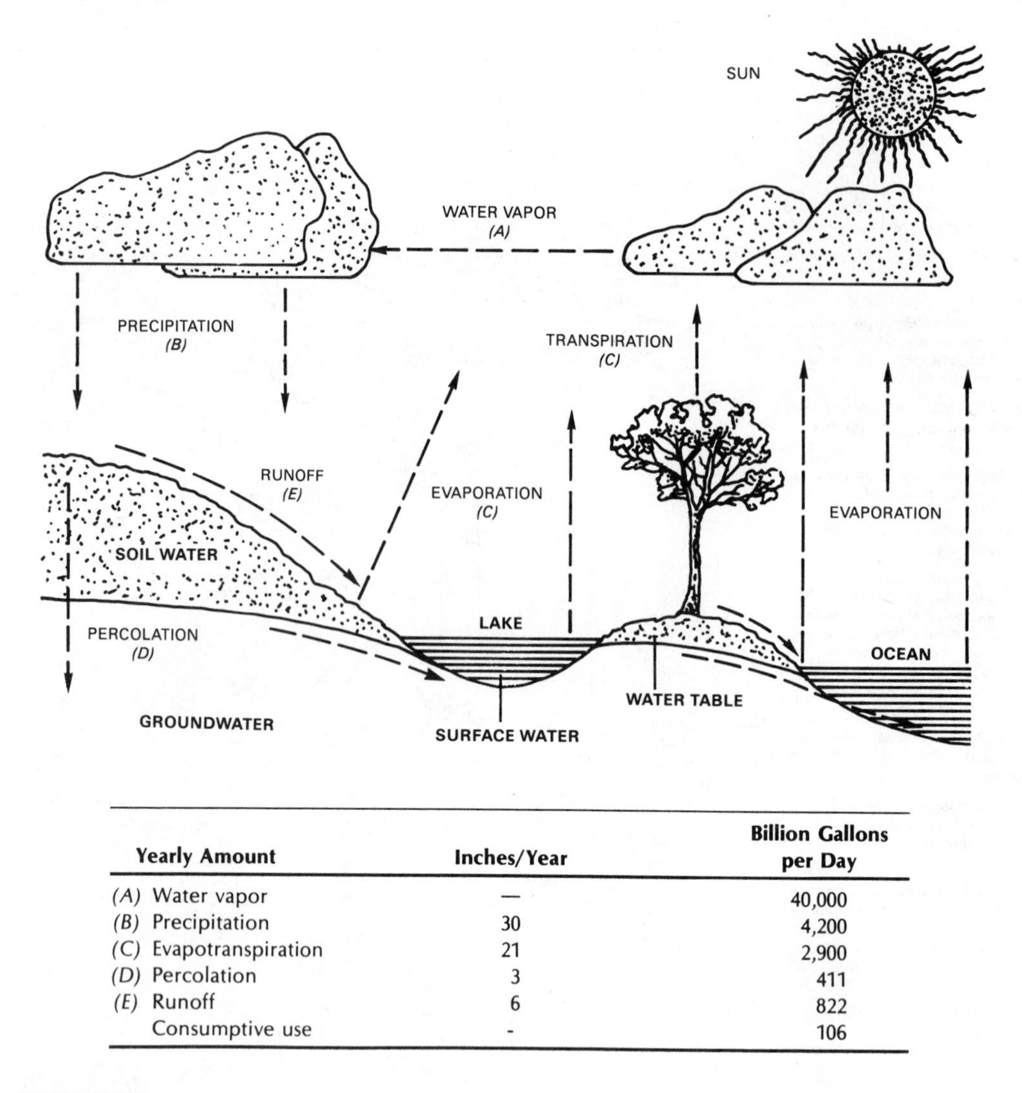

Yearly Amount	Inches/Year	Billion Gallons per Day
(A) Water vapor	—	40,000
(B) Precipitation	30	4,200
(C) Evapotranspiration	21	2,900
(D) Percolation	3	411
(E) Runoff	6	822
Consumptive use	-	106

FIGURE 8-5
The water cycle ***(From Plaster/Soil Science and Management, copyright 1987 by Delmar Publishers Inc.)***

LAND Land provides us solid foundations for our structures, nutrition and support for plants, and space for work and play. It provides storage for water. Further, it serves as a heat and compression chamber that has converted organic material into coal and oil throughout the existence of the Earth.

Soil is an important component of land. To be productive, soil must be made up of the correct proportions of soil particles, have the correct balance of nutrients, contain some organic matter, and have adequate moisture. Much of the Earth's crust is too rocky or has an incorrect balance of nutrients for crop production. Similarly, much land is covered by only a thin layer of productive soil or is too steep to permit cultivation. Where there is some useful soil, however, trees may survive. Forest lands provide lumber, poles, paper, and other products. Here too, humankind can benefit from the pleasures of wildlife and recreation.

Large areas of the Earth have a usable balance of soil particles and minerals for plant growth, but insufficient water. Areas with continuous, severe water shortages are called *deserts*. Fortunately, some desert areas have been made productive through the use of modern *irrigation* practices (the addition of water to plants to supplement that provided by rain or snow). Many great nations have desert areas, but do not have the money, technical knowledge, or water to make their deserts productive. Most nations of the world have such limited land resources that all land must be used to its greatest capacity.

Relationships of Land and Water

Precipitation Land and water are related to each other in many ways. Land in cold regions or

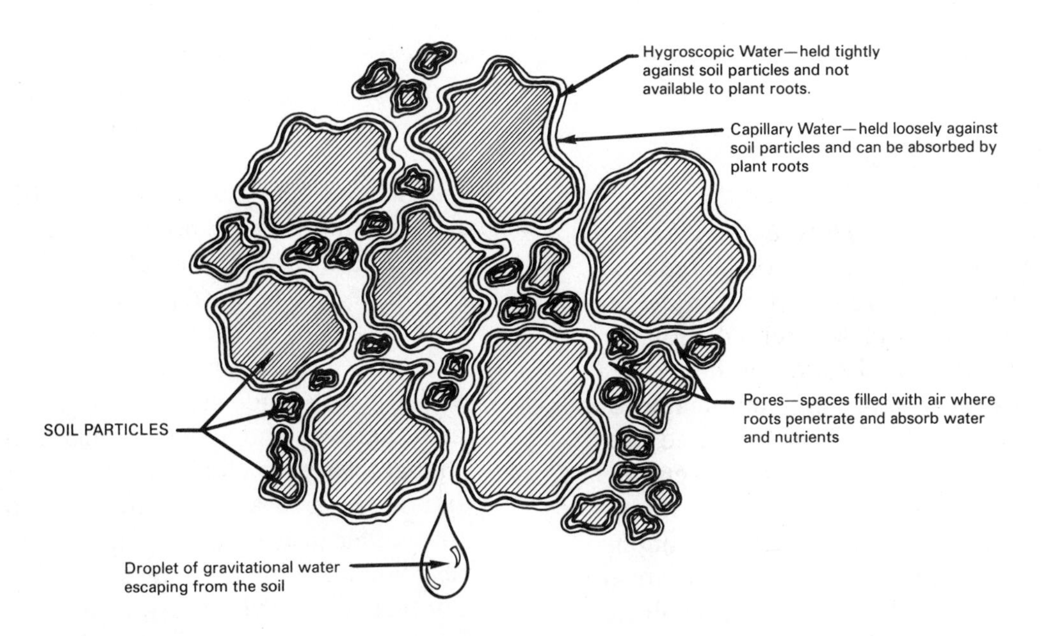

FIGURE 8-6
Plants use only capillary water. However, hygroscopic water contributes to soil structure and gravitational water is held in reserve for future use of plants or animals.

high altitudes holds moisture on its surface in the form of snow. This moisture is then released gradually to feed the streams and rivers after the *precipitation* (the formation of rain and snow) has stopped. Precipitation is caused by the change of water in the air from a gaseous state to a liquid state. It then falls to the land or bodies of water. Moisture-laden, warm air clouds contact cold air masses in the atmosphere and precipitation results. The clouds are formed by water changing from a liquid to a gas when it is evaporated by air movement over land and water. *Evaporation* means changing from a liquid to a vapor or gas.

Land as a Reservoir

Land serves as a container or reservoir. Water soaks down into the soil and forms a *water table* (the level below which soil is saturated or filled with water). Water held below the water table may run out onto the surface in the form of springs. These create or feed streams which, in turn, form rivers and flow into bays and oceans. Since ancient times, people have known to dig wells below the water table so water could be extracted for human needs.

Water moves upward in the soil from the water table to provide water for plant roots. From the roots, water travels throughout the plant and much of it evaporates from the leaves to contribute to the moisture supply in the atmosphere. Water helps soil by improving its physical structure. It is essential for microorganisms that live in the soil and contributes to the soil's fertility (the amount and type of nutrients in the soil) and structure.

Types of Ground Water

Soil is *saturated* when water is added until all the spaces or pores are filled. Plants will die from lack of air around their roots if soil remains saturated for very long. The water that drains out of soil after it has been wetted is called *free* or *gravitational water.* Gravitational water is what feeds wells and springs. After gravitational water leaves the soil, there is moisture left. Plant roots can absorb or take up this moisture called *capillary water.* Water that is held too tightly for plant roots to absorb is called *hygroscopic water* (Figure 8-6).

Benefits of Living Organisms

Both soil and water benefit from living organisms. Plants break the fall of raindrops and reduce damage to the soil from the impact of water. When plants drop their leaves and plant materials accumulate, they provide a rain-absorbing layer on the surface. When plant materials decay, they improve the structure of the soil. Plants, therefore, contribute to water absorption and reduce soil erosion. Further, decayed plant and animal matter contribute substantially to the nutrient content of the soil. Plants assist in moving water around by taking it up through the roots and releasing it in the atmosphere. Worms, insects, bacteria, and other small and microscopic plants and animals contribute by decomposing plant and animal matter. They also contribute in other important ways.

CONSERVING WATER AND IMPROVING WATER QUALITY

How can we reduce water pollution? How can soil erosion be reduced? What is the most productive use of water and soil without polluting or losing these essential resources? These are important questions deserving correct answers now! Farmers have long appreciated the value of these resources and generally used them wisely. However, economic conditions, government policies, production costs, farm income, extent of personal knowledge, and other factors influence the extent of conservation practices. Every citizen, business, agency, and industry impacts on air and water quality. Similarly, all have some influence on the use of our land and water resources.

The improvement of water quality can be achieved by proper land management, careful water storage and handling, and appropriate use of water. Some practices that help reduce water pollution are as follows.

1. Save clean water. Whenever we permit a faucet to drip, leave water running while we brush our teeth, or flush toilets excessively we waste water. Further, taking long showers, leaving water running while we wash automobiles or livestock, and using excessive water for pesticide washup wastes clean water. Once clean water is mixed with contaminants, it is not safe for use for other purposes until it is cleaned or repurified. *Purify* means removal of all foreign material.
2. Dispose of household products carefully. Many products under the kitchen sink, in the basement, or in the garage are threats to clean water. Never pour paints, wood preservatives, brush cleaners, or solvents down the drain. They will eventually enter the water supply, river, or ocean. Use all products sparingly and completely. Put solvents into closed containers and permit the dissolved materials to settle out. Then reuse the solvent. Stuff empty containers with newspaper to absorb all liquids before discarding. Then send the containers to approved disposal sites.
3. Care for lawns, gardens, and farmland carefully. Improve soil by adding organic matter. Mulch lawn and garden plants. Use proper lime and fertilizer. Only till soil that will not erode excessively. Cover exposed soil with a new crop immediately. Water soil only when it is dry and then water until the soil is soaked to a depth of 4–6″.
4. Practice sensible pest control. Most insecticides kill all insects—both harmful and beneficial ones. Insecticides also pollute water. Therefore, use cultural practices such as crop rotations and resistant varieties instead of insecticides whenever possible. Encourage beneficial insects and insect-eating birds around lawn, garden, and fields. Eliminate pools of stagnant water to eliminate mosquitoes. Keep untreated wood away from soil to avoid termite, other insect, and rot damage. Follow all pesticide label instructions exactly.
5. Control water run-off from lawns, gardens, feedlots, and fields. Keep soil covered with plants. Construct livestock facilities so manure can be collected and spread on fields. Use no-till cropping. *No-till* means planting crops without plowing or disking the soil. Alternate strips of close-growing crops (like small grains or hay) with row crops (like corn or soybeans). Farm on the *contour* (following the level of the land around a hill). Plant cover crops where regular crops do not protect the soil. A *cover crop* is a close-growing crop planted to temporarily protect the soil.
6. Control soil-erosion. Reduce the volume of rain water carried away by minimizing the amount of blacktop or concrete surfaces constructed. Use grass waterways in low areas of fields. Add manure and other organic matter to soil to increase water-holding capacity. Construct terraces on long or steep slopes. Leave steep areas in trees and sloping areas in close-growing crops.
7. Avoid spillage or dumping of gasoline, fuel, or oil on the ground. Turn over used petroleum products to truck and automobile service centers for disposal.
8. Keep chemical spills from running or seeping away. Do not flush chemicals away. The chemicals will damage lawns, trees, gardens, fields, and ground water. Sprinkle spills with an absorbent material such as soil, kitty litter, or sawdust. Then place the chemical-laden absorbent material in a strong plastic bag and discard it according to local recommendations.
9. Properly maintain your septic system (if your home has one). Avoid using excessive water. Do not flush improper materials down the toi-

let. Do not plant trees where their roots might interfere with sewer lines, septic tanks, or field drains. Don't run tractors and other heavy equipment over field drains.

The above recommendations are some practices that should help increase water supplies and decrease water usage and pollution.

LAND EROSION AND SOIL CONSERVATION

The Problem

Land Erosion—A Worldwide Problem

Land erosion (wearing away) is a serious problem world-wide (Figure 8-7). Both wind and water are capable of wearing away soil. The food and fiber production capabilities of large nations are being compromised due to extensive damage from soil erosion. Numerous cases can be cited from history where soil erosion has caused enormous problems. Yet, in the 1990s, the world is making the same mistakes in many areas that caused extensive losses in the past.

Consider, for example, an important colonial *port* (a town having a harbor for ships to take on cargo) along the eastern coast of what is now the United States. The colony was established in the year 1634. By 1706, the port had been replaced by another town as the central port for the colony. The harbor area of that original port was rapidly filling with soil, which was washed in from the surrounding area. The water was no longer deep enough to handle the ships.

FIGURE 8-7
Erosion of soil means loss of productive soil, damage to machinery, additional costs of production and pollution of streams, rivers, bays and oceans. *(Courtesy of USDA, Soil Conservation Service)*

Consider the massive size of the Mississippi River Delta (land created by the soil that water deposits at the mouth of a river). The Mississippi River Delta is over 15,000 square miles in area. In order for the Mississippi River Delta to be formed, the river had to deposit soil on the bottom of the Gulf of Mexico and continue until it reached the surface of the water. The soil gets into the river from smaller rivers and streams receiving run-off water from fields in the heartland of America. Therefore, one must conclude that the Mississippi River Delta was built from the bottom to the surface of the Gulf with America's best topsoil. Yet, in these modern times, it is estimated that the amount of soil being dumped by the river into the Gulf every day would fill a freight train 150 miles long. Further, it has been estimated that the amount deposited at the mouth of the Mississippi in one year is sufficient to cover Connecticut with soil 1″ deep, or over 8 ft. in 100 years! Similar deltas exist at the mouth of the Nile and other great rivers of the world.

During the late sixties and early seventies in the Peoples Republic of China, much of the forested land was cleared. It soon became apparent that much of that land depended on the trees to prevent soil erosion. Seeing the extent of damage from soil erosion, the country has responded with a national policy of rapid reforestation. There are over 1 billion people in China and all are expected to contribute to the tree planting effort. However, it will take many years for newly planted trees to grow sufficiently to protect the soil again. Unfortunately, the soil washed away during the absence of trees can never be returned to the land. Soil scientists report that it takes 300 to 500 years for nature to develop one inch of topsoil from bedrock.

Another serious threat to large areas of the world is the farming system known as "slash and burn." This is used in the tropical rain forests of South America, Africa, and Indonesia. In these areas, impoverished farmers cut or slash the jungle growth and burn the plant residues. The extensive burning causes serious air pollution and destroys useful fuel for cooking. Then they raise crops for several years until the soil is drained of its meager fertility and soil erosion takes its toll. They then move on to uncleared land and repeat the cycle. Sadly, once the land is cleared, it becomes difficult for plant growth to become re-established and the land is left permanently ruined. Reportedly, there are 50 acres of tropical rain forest lost every minute through this process. It is said that, in some regions, 80% of the rainfall is attributed to transpiration from trees. When the trees are removed, such areas are subject to drought.

National Problems

Each year, about 1.6 billion tons of soil are worn away from 417 million acres of U.S. farmland into lakes, rivers, and reservoirs. One ton equals 2,000 lbs. While some soils are deep and can tolerate a certain amount of erosion, many fragile soils cannot. The USDA's National Resources Inventory indicates 41 million acres or about 10% of our nation's cropland is highly erodible at rates of 50 or more tons per acre per year.

Of growing concern is the contamination of ground water under large areas of the United States. Ground-water pollution emerged as a public issue in the late 1970s. The first reports documented sources of contamination associated with the disposal of manufacturing wastes. By the early 1980s, several incidences of ground-water contamination by pesticides used on field crops were confirmed. Ground-water contamination can threaten the health of large populations. For instance, in New York State, the vast *aquifer* (a water-bearing rock formation) that underlies Long Island represents the only supply of drinking water for more than 3 million people. In the United States, major aquifers underlie areas that contain thousands of square miles of land, which often includes several states. Contamination of the aquifer in one place means contamination of the water for large areas.

Soil Conservation

Preventing and Reducing Soil Erosion

Soil erosion can be reduced or stopped by good land management. Fortunately, management practices that reduce soil erosion increase water absorption and retention. This, in turn, increases the supply of fresh water available for crops, livestock, wildlife, and people. Further, it reduces dryness and dustiness which are threats to good air quality. Most soil conservation methods are based on either (1) reducing raindrop impact, (2) reducing or slowing the speed of wind or water moving across the land, (3) securing the soil with plant roots, (4) increasing absorption of water, or (5) carrying run-

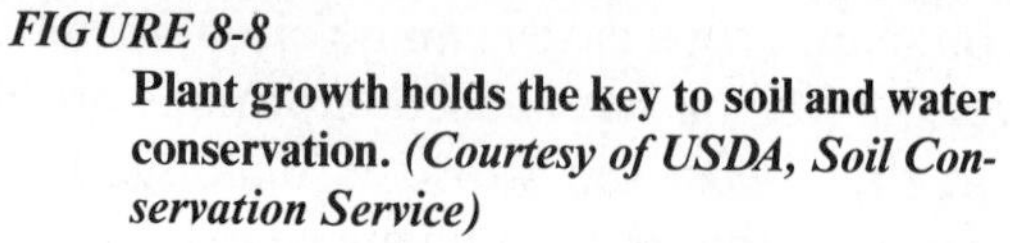

FIGURE 8-8
Plant growth holds the key to soil and water conservation. *(Courtesy of USDA, Soil Conservation Service)*

off water safely away. If water is free to run down a hill, it increases in volume and speed, picks up and pushes soil particles, and carries them off the land. This results in sheet and *gully erosion* (removal of soil to form relatively narrow and deep trenches known as gullies) and deposits soil, nutrients, and chemicals into streams. *Sheet erosion* is removal of even layers of soil from the land.

The following are some recommended practices to reduce or prevent wind and water erosion.

1. Keep soil covered with growing plants. Plants reduce the impact of raindrops, reduce the speed of wind and water across the land, and hold soil in place when threatened by moving wind or water (Figure 8-8).
2. Cover the soil with a mulch. *Mulch* is a material placed on soil to break the fall of raindrops, prevent weeds from growing, and/or improve the appearance of the area.
3. Utilize conservation tillage methods. *Conservation tillage* means using techniques of soil preparation, planting, and cultivation that disturb the soil the least and leave the maximum amount of plant residue on the surface. *Plant residue* is the plant material left when a plant dies or is harvested.
4. Use contour practices in farming, nursery production, and gardening. *Contour practice*

4. means conducting all operations such as plowing, disking, planting, cultivating, and harvesting across the slope and on the level. Any grooves or ridges created by machinery go around the slope or hill. When water tries to run down the hill, it encounters the grooves and ridges. Since these are level, the water tends to soak into the soil. This holds the water for future use.
5. Use strip cropping on hilly land. *Strip cropping* means alternating strips of row crops with strips of close-growing crops. Examples of close-growing crops are hay, pasture, and small grains such as wheat, barley, oats, and rye. Examples of row crops are corn, soybeans, and most vegetables. The strips of close-growing crops capture run-off water from the row crops and prevent it from entering streams.
6. Rotate crops. *Crop rotation* is the planting of different crops in a given field every year or every several years. Crop rotation permits close-growing crops to retain water and soil, and tends to rebuild the soil after losses incurred while row crops occupied the land.
7. Increase organic matter in the soil. *Organic matter* is dead plant and animal tissue. Nonliving plant leaves, stalks, branches, bark, and roots decay and become organic matter. Similarly, animal manure and dead insects, worms, and animal carcasses decompose to make organic matter. The decomposed organic matter forms a gel-like substance which holds soil particles into absorbent granules called *aggregates.* An aggregated soil is a water-absorbing and nutrient-holding soil. Organic matter also releases nutrients which improve plant growth.
8. Provide the correct balance of lime and fertilizer. *Lime* is a material that reduces the acid content of soil. It also supplies nutrients such as calcium and magnesium to improve plant growth. *Fertilizer* is any material that supplies nutrients for plants.
9. Establish permanent grass waterways. A *grass waterway* is a strip of grass growing in the low area of a field where water can gather and cause erosion.
10. Construct terraces. A *terrace* is a soil or wall structure built across the slope to capture water and move it safely to areas where it will not cause erosion.
11. Avoid overgrazing. *Overgrazing* refers to damage to plants or soil due to animals eating too much of the plants at one time. This reduces the plant's ability to hold soil or recover after grazing.
12. Use land according to a conservation plan. A *conservation plan* is a plan developed by soil and water conservation specialists to use land for its maximum production and water conservation without unacceptable damage to the land.

There are many careers in the field of water and soil conservation. The material in this unit hints at the problems created by humans as they use the natural resources around them to provide food, water, shelter, recreation, and other resources for living. The health, wealth, peace, and general welfare of humanity depends upon our skill in conserving these most basic resources.

Student Activities

1. Define the "Terms To Know."
2. Collect a sample of drinking water from each of 5 sources as follows: a safe spring or well, bottled pure water, and faucets attached to (1) galvanized pipe, (2) plastic pipe, and (3) copper pipe. (Do not run off the water before obtaining the samples from the faucets.) Taste a small amount of each sample and describe the taste. Do the samples have different tastes? If so, what causes the differences?
3. Study the eating habits of one species of birds in your community. Describe the food chain that accounts for the survival of that species. What effect did the use of DDT as an insecticide have on that species of birds before DDT was taken off the market?

4. Obtain a rain gauge and record the precipitation on a daily basis for several months. What variations did you observe from week to week? How do you explain the variations? What effect did these variations have on the agricultural activities of the community?

5. Conduct an experiment to determine the effect of soil water on plant growth. Obtain 3 inexpensive pots of healthy flowers or other plants. Each pot must be the same size and type, have the same amount and type of soil, and contain the same size and number of plants. Use the following procedure:
   a. Mark the pots "1," "2," and "3."
   b. Plug the hole in the bottoms of pots 2 and 3 so water cannot drain from the pots. Leave pot 1 unplugged so it has good drainage.
   c. Add water to pot 1 as needed to keep the plant healthy for several weeks. Use it as a comparison specimen (called the "control").
   d. Add water slowly to pot 2 until water has filled the soil and the water is just level with the surface of the soil. All pores of the soil are now filled and the soil is "saturated." Weigh the pot and record the weight as "A."
   e. Remove the plug from the bottom of pot 2 and permit the water to drain out. After water has stopped flowing from the drain hole, immediately weigh the pot again. Record the weight as "B." The water that flowed from pot 2 when the plug was removed is the free or gravitational water. Weight of the gravitational water, "D," should be calculated and recorded using the formula A − B = D. Do not add any more water to pot 2 for two weeks.
   f. With the hole plugged in pot 3, add water slowly until the water is just level with the top of the soil. Do not remove the plug and keep pot 3 filled with water to the saturation point for two weeks.
   g. Keep all three pots in a good growing environment for two weeks and record all observations.
   h. When the plant in pot 2 wilts badly due to lack of water, weigh the pot and record the weight as "C."

      Make the following calculations:
      Weight of the gravitational water (D) = A − B
      Weight of the capillary water (E) = B − C
      The weight of the hygroscopic water can only be determined by driving the remaining water from the soil by heating the soil in an oven.

6. Ask your teacher to help you design and conduct a project that demonstrates some or all of the following:
   a. Effect of the force of raindrops on soil.
   b. Effect of soil aggregation on absorption.
   c. Effect of slope on erosion.
   d. Effect of living grass on erosion control.
   e. Effect of plant residue on erosion control.

Self-Evaluation

A. MULTIPLE CHOICE

1. Most of the Earth's surface is covered with

   a. crops.
   b. farms.
   c. trees.
   d. water.

2. The bodies of plants, animals and humans consist of about what percent water?

   a. 10%
   b. 40%
   c. 70%
   d. 90%

3. The universal solvent is

   a. gasoline.
   b. paint thinner.
   c. varsol.
   d. water.

4. The content of ocean water may be likened to

   a. fresh water.
   b. pure water.
   c. thin soup.
   d. varsol.

5. The interdependence of plants and animals on each other for food is known as

   a. domestic.
   b. food chain.
   c. symbiosis.
   d. universal relationship.

6. The land serving as a heat and compression chamber gives us

   a. building foundations.
   b. coal and oil.
   c. crops.
   d. wildlife habitat.

7. Ground water which is available for plant root absorption is called

   a. capillary.
   b. free.
   c. gravitational.
   d. hygroscopic.

8. Improvement of water quality can be achieved by

   a. appropriate use of water.
   b. careful water storage and handling.
   c. proper land management.
   d. all of the above.

9. The problem of land erosion is

   a. characteristic of developed countries only.
   b. found world-wide.
   c. mostly found in poor countries.
   d. not a substantial problem in view of food surpluses.

10. A conservation plan

    a. conserves soil and water.
    b. is prepared by professionals.
    c. maximizes productivity of land.
    d. all of the above.

B. MATCHING

_______ **1.** Contour
_______ **2.** Delta
_______ **3.** Evaporation
_______ **4.** Fertility
_______ **5.** Irrigation
_______ **6.** No-till
_______ **7.** Port
_______ **8.** Precipitation
_______ **9.** Purify
_______ **10.** Water table

a. Amount and type of nutrients
b. Without plowing or disking
c. Town with a harbor
d. Removal of foreign material
e. Soil deposited by water
f. Top of saturated soil
g. Rain and snow
h. Supplemental water
i. On the level
j. Change from liquid and gas

C. COMPLETION

1. An ______ is a water-bearing rock formation.
2. The process of creating narrow and deep trenches eroded in soil is known as ______ ______ .
3. Material placed on soil to break the fall of raindrops is called ______ .
4. Alternating row crops with close-growing crops is known as ______ ______ .
5. ______ is used to reduce acid in soil.
6. One inch of topsoil may be formed from bedrock in about ______ years.
7. ______ ______ is dead plant and animal material in soil.
8. A ______ ______ may be planted to temporarily protect soil from erosion.
9. In the United States about ______ tons of soil are worn away from farmland each year.
10. About ______ acres or 10% of the U.S. cropland is highly erodible.

UNIT 9 Soils and Hydroponics Management

OBJECTIVE To determine the origin and classification of soils, and identify effective procedures for soils and hydroponics management.

Competencies to Be Developed

After studying this unit, you will be able to

- ☐ define terms in soils, hydroponics, and other plant-growing media management.
- ☐ identify types of plant-growing media.
- ☐ describe the origin and composition of soils.
- ☐ discuss the principles of soil classification.
- ☐ determine appropriate amendments for soil and hydroponics media.
- ☐ discuss fundamentals of fertilizing and liming materials.
- ☐ identify requirements for hydroponics plant production.
- ☐ describe types of hydroponics systems.

TERMS TO KNOW

Media
Medium
Hydroponics
Decomposed
Leaf mold
Compost
Sphagnum
Peat moss
Bogs
Water-logged
Perlite
Vermiculite
Leached
Microbes
Parent material
Horizon
Profile
Residual soils
Alluvial deposits
Lacustrine deposits
Loess deposits
Colluvial deposits
Glacial deposits
Percolation
Permeable
Capability classes
Capability subclasses
Capability units
Organic matter
Mineral matter
Horizon A
Tillable
Topsoil
B Horizon
C Horizon
Bedrock
Coarse-textured (sandy) soil
Medium-textured (loamy) soil
Fine-textured (clay) soil
Structure
Aggregates
Crumbs
Decompose
Amendment
pH
Acidity
Alkalinity
Neutral
Primary nutrients
Complete fertilizer
Fertilizer grade
Active ingredient
Broadcasting
Incorporated
Band application
Side dressing
Top-dressing
Starter solutions
Foliar sprays
Knife application
Legumes
Nitrate
Aggregate culture
Water culture
Aquaculture

Nutriculture
Aeroponics
Continuous-flow systems
Inert

The role of plants in our environment and their importance in our lives has been discussed in previous units. Plants are necessary to nourish the animals of the world and maintain the balance of oxygen in our atmosphere. However, they depend upon soil, water, and air as their media for support. *Media* is plural for medium. *Medium* is a surrounding environment in which something functions and thrives.

PLANT-GROWING MEDIA

For discussion in this unit, the word media will be used to mean the material that provides plants with nourishment and support through their root systems.

Types of Media

Media comes in many forms. The oceans, rivers, land, and man-made mixtures of various materials are the principal types of media for plant growth. Seaweed, kelp, plankton, and many other plants depend upon water for their nutrients and support. It is only recently that we have come to understand the tremendous amount of plant life in the sea. The plant life in oceans and rivers are important for feeding the animal life of the sea and bodies of fresh water. Humankind has used fish as a staple food since the beginning of time.

It has long been known that water could be used to promote new root formation on the stems of certain green plants and completely support plant growth on a limited basis. Recently, however, it has been found that food crops can be grown efficiently in structures where plant roots are submerged in or sprayed with solutions of water and nutrients. These solutions feed the plants, while mechanical apparatus provide physical support. The practice of growing plants without soil is called *hydroponics.* Hydroponics has become an important commercial method of growing green plants. This will be discussed later in this unit.

Soil

Soil is defined as the top layer of the Earth's surface, suitable for the growth of plant life. It has long been the predominant medium for cultivated plants (Figure 9-1). In early years, humans accepted the soil as it existed. They planted seeds using primitive tools and did not know how to modify or enhance the soil to improve its plant-supporting performance.

Ancient civilizations discovered that plant-growing conditions were improved on some land where deposits were left after river waters flowed over the land during flood season. Similarly, other land was ruined by flood waters. Therefore, early efforts to improve plant-growing media was a matter of moving to better soil. Obviously, good soil was a valuable asset and, therefore, was the cause of intense personal disputes and wars among nations.

Other Media

In addition to water and soil, certain other materials will hold water and support plant growth. Ironically, some of the best nonsoil and nonwater media are partially decomposed (decayed) plant materials. One common material available around most homes is leaf mold and compost. *Leaf mold* is partially decomposed plant leaves. *Compost* is a mixture of partially decayed organic matter such as leaves, manure, and household plant wastes. It is mixed with lime and fertilizer in correct proportions to support plant growth (Figure 9-2).

There is a group of pale or ashy mosses called *sphagnum.* These are used extensively in horticulture as a medium for encouraging root growth and growing plants under certain conditions. *Peat moss* consists of partially decomposed mosses which have accumulated in water-logged areas called *bogs.* *Water-logged* means soaked or saturated with water.

FIGURE 9-1
Soil has long been the predominant medium for plant growth.

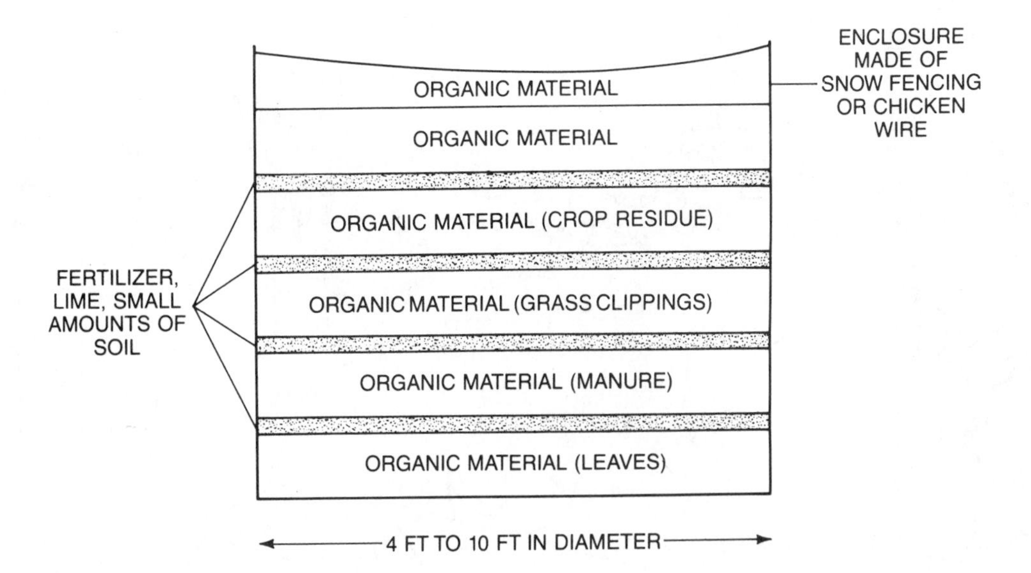

FIGURE 9-2
Compost is excellent organic matter *(From Plaster/Soil Science and Management, copyright 1987 by Delmar Publishers Inc.)*

Both substances have excellent air- and water-holding qualities.

Many other sources of plant and animal residues may become plant-growing media. For instance, a fence post may rot on the top and hold moisture from rainfall. In time, a plant seed may be deposited in the rotted wood medium, germinate, root in the wood, and start to grow. Even horse manure mixed with straw is used extensively as a medium for growing mushrooms. In this instance, both animal residue (manure) and plant residue (straw) combine to make an effective medium.

Some mineral matter can also become nonsoil and nonwater media. For instance, volcanic lava and ash remain black, gray, and barren only briefly. Soon the layer cools and cracks and seeds settle on it. Then moisture causes the seeds to germinate and roots penetrate the volcanic residue. Soon the area is covered with plant life.

Horticulturists use certain mineral materials in plant-growing areas, too. *Perlite* is a natural volcanic glass material having water-holding capabilities. Perlite is used extensively for starting new plants. *Vermiculite* is mineral matter from a group of mica-type materials that is also used for starting plant seeds and cuttings.

ORIGIN AND COMPOSITION OF SOILS

Factors Affecting Soil Formation Productive soils develop on the Earth's surface as the atmosphere, sunlight, water, and living things meet and interact with the mineral world (Figure 9-3). If soil is suitable for plant growth to a depth of 36″ or more, that soil is regarded as "deep." Many soils of the earth are much more shallow than this. Plants attach themselves to the soil by their roots, grow, manufacture food, and give off oxygen. Plants and animals of various sizes live on and in the soil, utilizing carbon dioxide, oxygen, water, mineral matter, other plants and animals, and products of decomposition.

Soils vary in temperature and organic matter, and in the amount of air and water they contain. The kinds of soils formed at a specific site are determined by the forces of climate, living organisms, parent soil material, topography, and time. Each factor varies by local conditions.

Climate Climatic factors, such as temperature and rainfall, greatly affect the rate of weathering. When temperature rises, the rate of chemical reactions increases and growth of fungi, organisms (such as bacteria), and plants increase. The rate and amount of rainfall in a locality greatly affect the soil. In areas of high rainfall, the soils usually are leached and acid. *Leached* means having certain contents removed by water. If the land is covered by trees, the action of high temperatures and moisture on leaf residues generally creates an acid soil. Rainfall during cold weather has less effect on the soil than during warm or hot weather.

Slope and location of a field also affect erosion and drainage and thus influence soil formation. Moreover, free water in the soil carries fine

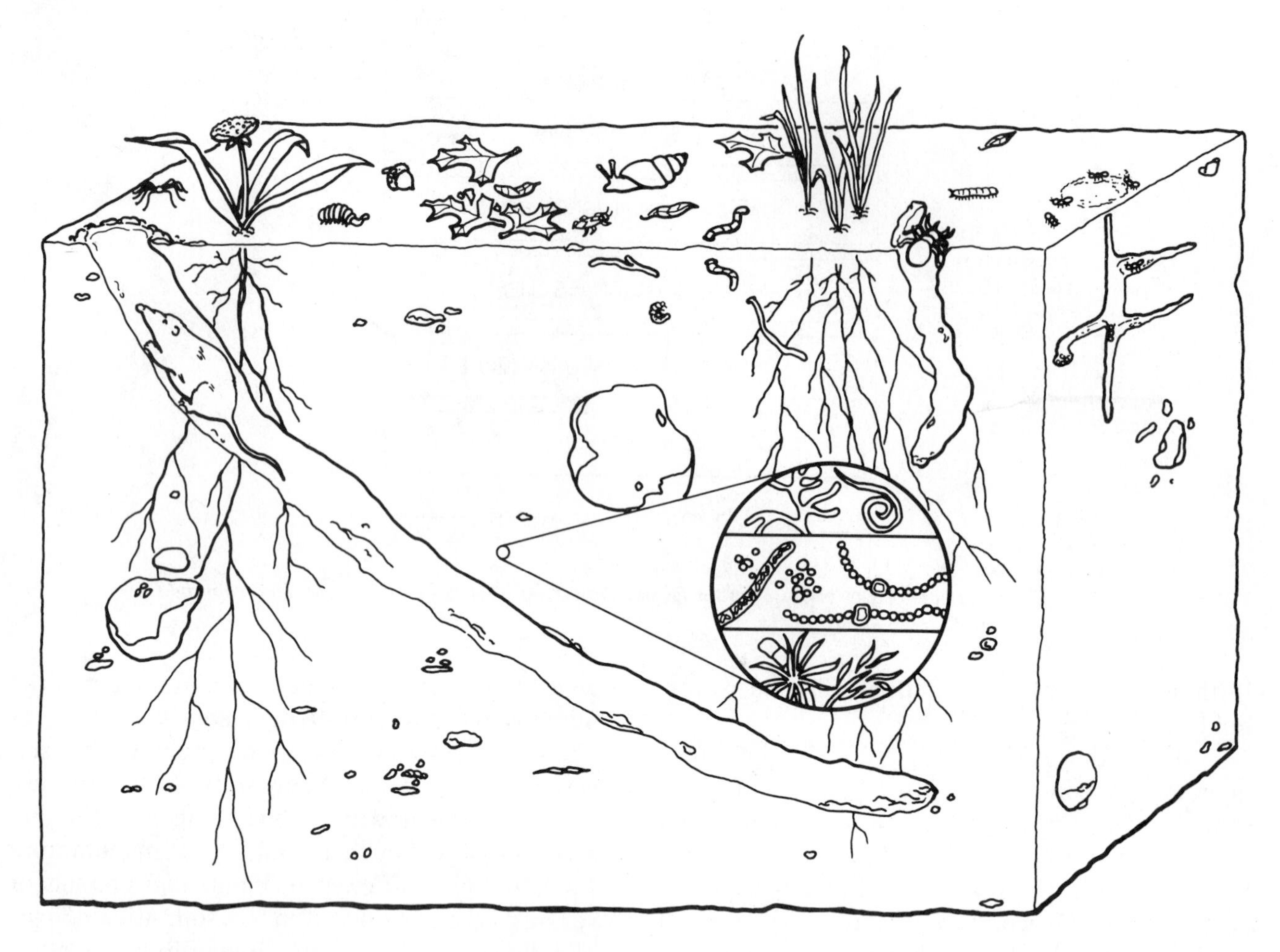

FIGURE 9-3
A soil ecosystem.

particles to the deeper layers and tends to produce "layering." Too much water prevents or retards microbial growth and may exclude air by waterlogging. Water and temperature also have the effect of swelling and contracting soil particles.

Living Organisms Living organisms such as microbes, plants, insects, animals, and humans exert considerable influence on the formation of a soil. Certain types of soil bacteria and fungi aid in soil formation by causing decay or breakdown of the plant and animal residues in the soil. Carbon dioxide and other compounds essential to soil formation are released by microbic activity. *Microbes* are microscopic plants and animals. Without soil microbes, organic materials would not decay.

Numerous insects, worms, and animals contribute to the formation of soil by mixing the various soil materials. Earthworms consume and digest certain soil substances and discharge body wastes. This aids decomposition and soil mixing. All such dead organisms add to soil organic matter.

Human activity also influences soil formation as cultivation, bulldozing, and construction projects disturb the surface layer, which influences soil-forming processes. The clearing of land removes native plant life and greatly modifies soil-forming activities (Figure 9-4).

Parent Material *Parent material* is the horizon of unconsolidated material from which a soil develops. *Horizon* means layer. Parent materials comprise the C horizon of a typical soil profile. *Profile* means a cross-sectional view of soil (Figure 9-5).

Parent materials formed in place are called *residual soils.* Other soils are transported and deposited by water, wind, gravity, or ice. *Alluvial deposits* are transported by streams, and *lacustrine deposits* are left by lakes. *Loess deposits* are left by wind, *colluvial deposits* by gravity, and *glacial deposits* by ice.

The kind of parent material from which a soil is formed influences the many characteristics of

FIGURE 9-4
Human activities greatly influence soil development of land being cleared or moved.

that soil. Natural fertility and texture are influenced greatly by the parent material of a profile.

Topography Slope and drainage affect soil formation both directly and indirectly. On a steep slope, loose material is moved downward by runoff water, gravity, and movement of humans and other animals. This movement not only breaks up soil materials and adds them to the lower levels, but it exposes subsoil materials along the upper slopes. The movement of soil materials has a pulverizing effect on the material being moved as well as on the material left behind.

Slope affects the distribution of water that falls on the Earth's surface. On level areas, the water soaks in and moves through the soil in a process called *percolation.* On nonlevel land, the water tends to run off and moves some surface soil with it. Soils that develop on level land tend to be poorly drained, while soils on level to gently-rolling slopes are better drained and more productive (Figure 9-6).

Drainage (or lack of it) affects the water table in a particular field or area. The water table has a direct bearing on soil formation, especially if it is near the surface. When a soil is saturated with water, little or no air can penetrate it. The lack of air reduces the action of fungi, bacteria, and other soil-forming activities in the soil.

A wet soil, therefore, is a slow-forming soil and usually is low in productivity. Because of the lack of air, undecomposed organic matter will accumulate in a wet soil. This organic matter generally causes the soil to be a blackish color. Poor drainage, accompanied by free water in the soil, reduces or retards plant growth and affects soil formation.

FIGURE 9-5
A soil profile.

Time Soils form by the chemical and physical weathering of parent material, over time, as affected by climate, living organisms, and topography. Therefore, time itself is regarded as a factor in soil formation. Chemical weathering is the result of the chemical reactions of water, oxygen, carbon dioxide, and other substances which act upon the rocks, minerals, organic matter, and life that compose the soil. The leaching action of water hastens the weathering process by removing soluble materials, and chemicals react with each other to form new chemicals in the soil.

Weathering Weathering refers to mechanical forces caused by temperature change such as heating and cooling, freezing and thawing. As these processes occur, rocks, minerals, organic matter, and other soil-forming materials are broken into smaller and smaller particles until soil is formed. Soils at different stages of weathering will differ widely. Weathering causes soils to develop, mature, and age much as people do.

Soils develop rapidly, mature, and then develop certain characteristics of age. Plant nutrients are released quickly from the minerals, plant growth increases, and organic matter accumulates. Soils age more slowly as they continue to weather.

Eventually, nutrients in the soil are depleted. Water moving through the soil leaches away many soluble portions. At this stage many soils are acid because the limestone originally in them is gone. As the supply of nutrients in the soil decreases, the amount of plant growth is reduced to the point where the organic matter decomposes faster than it is produced. When soils become acid and have lost their native fertility, they require expensive amendments to keep them productive.

In *permeable* soils (permitting movement), the fine clay particles tend to move downward from the surface soil into the subsoil during the weathering process. This movement, together with further

FIGURE 9-6
Topography greatly influences soil formation.

AGRI-PROFILE

CAREER AREAS: AGRONOMY/HYDROPONICS
Christopher Colancino, research assistant at a major university, uses a lysimeter to check soil moisture and salinity for corn plants grown in a special research container. ***(Courtesy USDA Agricultural Research Service)***

Electronic devices have become tools of the trade for soil and water research and management. Career options in soil and hydroponics management overlap those in soil and water conservation to a certain extent. However, additional opportunities occur in management as producers on farms and in hydroponics greenhouses. The practice of hydroponics is not new, but hydroponics for commercial production has captured the imagination of the world. The recent popularity of hydroponics operations provides new career opportunities, especially in urban areas.

Soil and water management specialists are in demand on the global scene. Progressive third world countries need help in policy development, education, and project management. They hope to leap from their primitive agriculture of the past to the agriscience of the present in a few short years.

breakdown of the rock material, accounts for the fact that many soil types have a higher percentage of clay in the subsoil than in the surface soil.

SOIL CLASSIFICATION

Soil scientists have developed a system for mapping soils according to the physical, chemical, and topographical aspects of the land. Such maps have lines showing the outline of soil types and provide numerical codes keyed to large amounts of information about the land. The experienced soil technician can obtain a wealth of information about the land by consulting a soils map and accompanying material.

Land Capability Maps

Soil mapping and land classification have been a priority of the U.S. Soil Conservation Service (SCS) throughout most of this century. Now the SCS can provide maps and classification information for almost any area in the United States, down to small areas on individual farms. Soil Conservation personnel work with local farmers to develop farm plans. These provide recommendations for land use, cropping systems, crop production practices, and livestock systems. These plans are designed to maximize production while controlling soil erosion and enhancing productivity.

Land capability maps indicate (1) capability class, (2) capability subclass, and (3) capability unit.

Capability Class

Capability classes are the broadest groups on a soils map and are designated by Roman numerals I through VIII (Figure 9-7). The numerals indicate progressively greater limitations and narrower choices for practical use of the land as follows.

Class I—soils have few limitations that restrict their use.

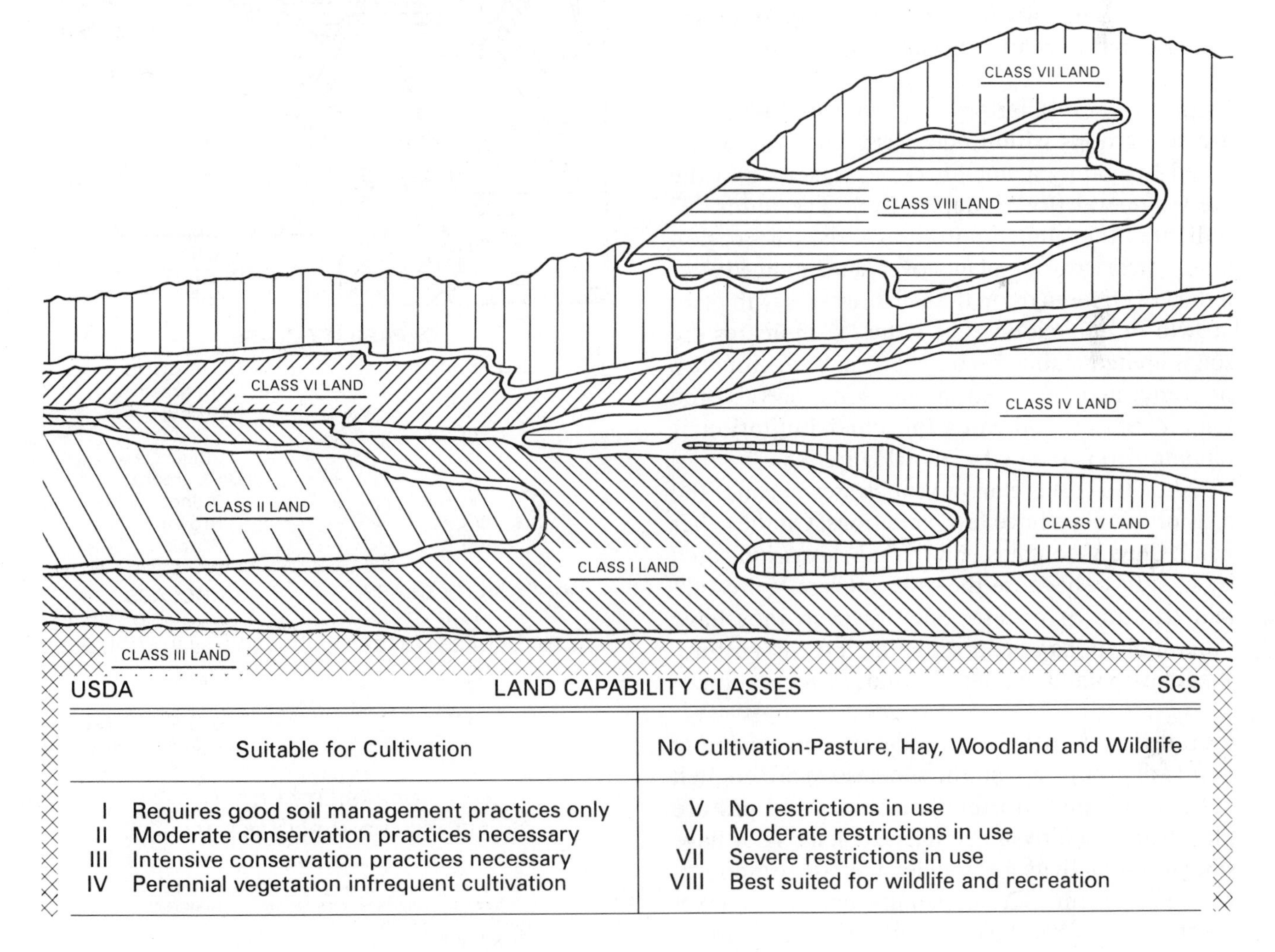

Suitable for Cultivation		No Cultivation-Pasture, Hay, Woodland and Wildlife	
I	Requires good soil management practices only	V	No restrictions in use
II	Moderate conservation practices necessary	VI	Moderate restrictions in use
III	Intensive conservation practices necessary	VII	Severe restrictions in use
IV	Perennial vegetation infrequent cultivation	VIII	Best suited for wildlife and recreation

FIGURE 9-7
The eight classes of land in the United States.

- Class II—soils have moderate limitations that reduce the choice of plants or require moderate conservation practices.
- Class III—soils have severe limitations that reduce the choice of plants, require special conservation practices, or both.
- Class IV—soils have very severe limitations that reduce the choice of plants, require very careful management, or both.
- Class V—soils are not likely to erode but have other limitations, impractical to remove, that limit their use.
- Class VI—soils have severe limitations that make them generally unsuitable for cultivation.
- Class VII—soils have very severe limitations that make them unsuitable for cultivation.
- Class VIII—soils and landforms have limitations that nearly prevent their use for commercial plants.

Land capability maps are usually color-coded for ease in differentiating capability classes.

Capability Subclasses *Capability subclasses* are soil groups within one class. They are designated by adding a small letter e, w, s, or c to the class numeral (for example, IIe.). The letter "e" indicates the main limitation is risk of erosion unless close-growing plant cover is maintained; "w" indicates water in or on the soil interferes with plant growth or cultivation. The letter "s" indicates the soil is limited mainly because it is shallow, droughty, or stony; and "c," used in only some parts of the United States, indicates the chief limitation is climate—too cold or too dry (Figure 9-8).

In Class I there are no subclasses because the soils of this class have few limitations. On the other hand, Class V contains only the subclasses indicated by w, s, or c, because the soils in Class V are subject to little or no erosion. However, they have other limitations which restrict their use to pasture, range, woodland, wildlife habitat, or recreation.

Capability Units The soils in one *capability unit* (soil groups within the subclasses) are enough alike to be suited to the same crops and pasture plants, and require similar management. They have similar productivity capabilities and other responses to management. The capability unit is a convenient grouping for making many statements about the management of a soil.

Capability units are generally designated by adding (numbers 0 to 9) to the subclass symbol (for example, IIIe4 or IIw2). Thus, in one symbol, the Roman numeral designates the capability class or degree of limitation; the small letter indicates the subclass or kind of limitation; and the Arabic numeral specifically identifies the capability unit within each subclass. A map legend for each soil grouping is included with the soil and land capa-

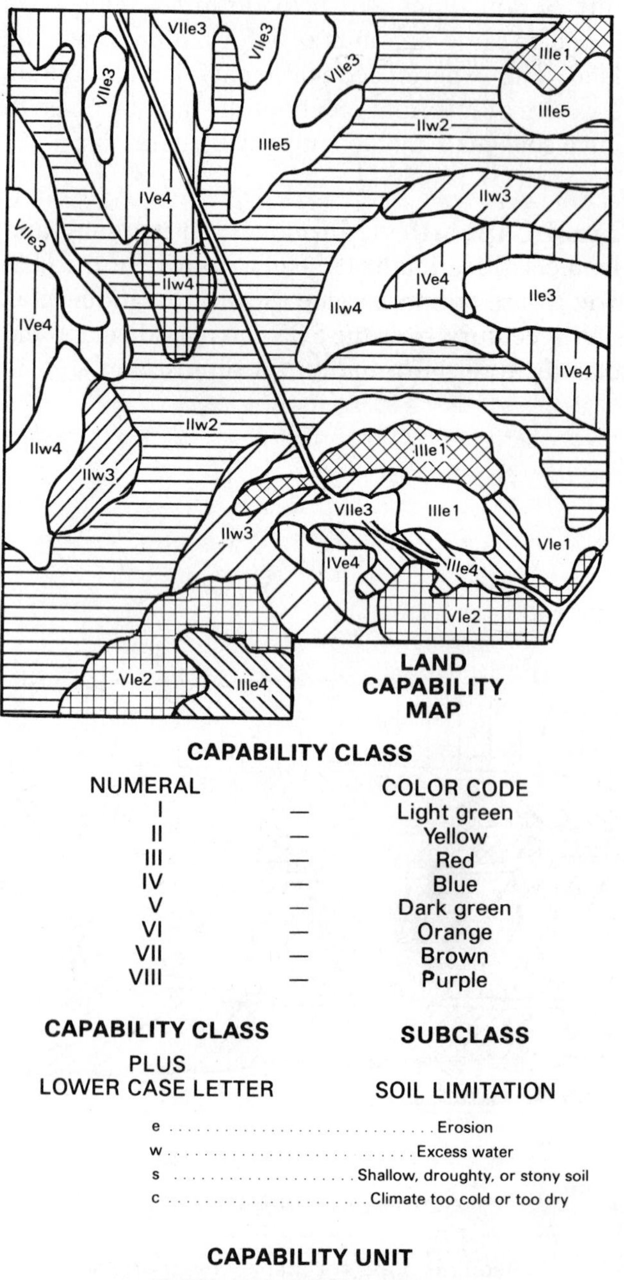

FIGURE 9-8
Land cauability map.

bility map. This type of map is also an example of how symbols can be effectively used. They make it possible to place a lot of information on a small space of the map by use of a code.

When the landowner and the soil conservationist start planning for the most intensive use of a farm, they need a soil and land capability map. This will help them prepare a conservation plan map or proposed land-use map. The soil conservationist and the landowner, through the use of the soil and land capability map, discuss the kinds of soil on the farm. The present and original land uses are also discussed and noted on the map. In developing a proposed land-use map, the soil conservationist must know the personal goals or objectives of the landowner and plans for developing the land.

Many things can be involved in reaching land use decisions, field by field. Field boundaries may need to be changed so all the soil in each field is suited for the same purpose and management practices. The desired balance between cropland, pasture, woodland, and other land uses will need to be considered. Appropriate livestock enterprises should also be considered to match the land's characteristics and potential. If there is a good potential for income-producing recreation enterprises in the community, an area may be used for hunting, campsites, or fishing.

The landowner and soil conservationists must consider how to treat each field to get the desired results. The SCS conservationist can give many good suggestions, but the landowner must decide what to do and when and how to do it. As planning decisions are made, the conservationist will record them in narrative form and make them part of the plan map. This becomes the farm conservation plan. It is the guide for the farming operation in the years ahead.

Workers from the Soil Conservation Service are also available to give on-site technical assistance in applying and maintaining the farm conservation plan. They provide management advice and services to nonfarm landowners, developers, strip-mining operations, and other activities where soil is involved (Figure 9-9).

FIGURE 9-9
Soil Conservation Service personnel must advise on the many uses of land.

PHYSICAL, CHEMICAL, AND BIOLOGICAL CHARACTERISTICS

Soil Profile Undisturbed soil will have four or more horizons in its profile. These are designated by the capital letter O, A, B, and C (Figure 9-10). The O horizon is on the surface and is comprised of organic matter and a little mineral matter. *Organic matter* originates from living sources such as plants, animals, insects, and microbes. *Mineral matter* is derived from nonliving sources such as rock materials.

Horizon A is located near the surface and is mineral matter and organic matter. It contains desirable proportions of organic matter, very fine mineral particles called clay, medium-sized mineral particles called silt, and larger mineral particles called sand. The appropriate proportion of these creates soil that is *tillable*, or workable with tools and equipment. With the presence of desirable plant nutrients, chemicals, and living organisms, the A horizon generally supports good plant growth. The A horizon is frequently called *topsoil*.

The *B horizon* is below the A horizon and is generally referred to as subsoil. The mineral content is similar to horizon A, but the particle sizes and properties will differ. Since organic matter comes from decayed plant and animal materials, the amount naturally decreases as distance from the surface increases.

The *C horizon* is below the B horizon and is comprised mostly of parent material. Horizon C is important for storing and releasing water to the upper layers of soil but does not contribute much to plant nutrition. It is likely to contain larger soil particles and may have substantial amounts of gravel and large rocks. The area below horizon C is called *bedrock*.

Texture Texture refers to the proportion and size of soil particles (Figures 9-11 & 9-12). Texture

Horizon	Name	Colors	Structure	Processes Occurring
O	Organic	Black, dark brown	Loose, crumbly, well broken up	Decomposition
A	Topsoil	Dark brown to yellow	Generally loose, crumbly, well broken up	Zone of leaching
B	Subsoil	Brown, red, yellow, or gray	Generally larger chunks, may be dense or crumbly, can be cement-like	Zone of accumulation
C	Parent material (slightly weathered material)	Variable—depending on parent material	Loose to dense	Weathering, disintegration of parent material or rock

FIGURE 9-10
Characteristics of soil horizons.

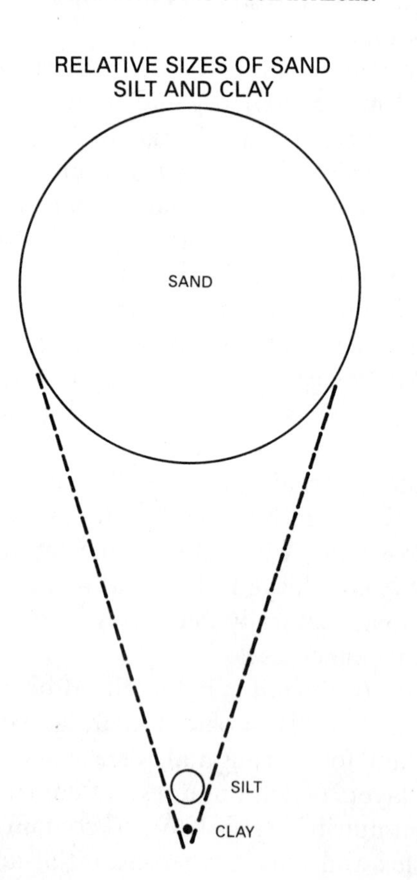

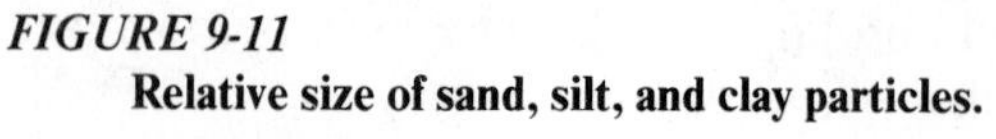

FIGURE 9-11
Relative size of sand, silt, and clay particles.

can be determined very accurately in the laboratory. However, it can also be judged or estimated by the feel of the soil as follows (Figure 9-13).

SIZE OF SOIL PARTICLES

Name	Size, Diameter in Millimeters
Fine gravel	2–1
Coarse sand	1.00–0.50
Medium sand	0.50–0.25
Fine sand	0.25–0.10
Very fine sand	0.10–0.05
Silt	0.05–0.002
Clay	less than 0.002

FIGURE 9-12
Ranges of sizes of soil particles.

1. Make a stiff mud ball.
2. Rub the mud ball between the thumb and forefinger.
3. Note the degree of coarseness and grittiness due to the sand particles.
4. Squeeze the mud between the fingers and then pull your thumb and fingers apart.
5. Note the degree of stickiness due to the clay particles.
6. Make the soil slightly more moist and note that the clay leaves a "slick" surface on the thumb and fingers.

The outstanding physical characteristics of the important textural grades, as determined by the

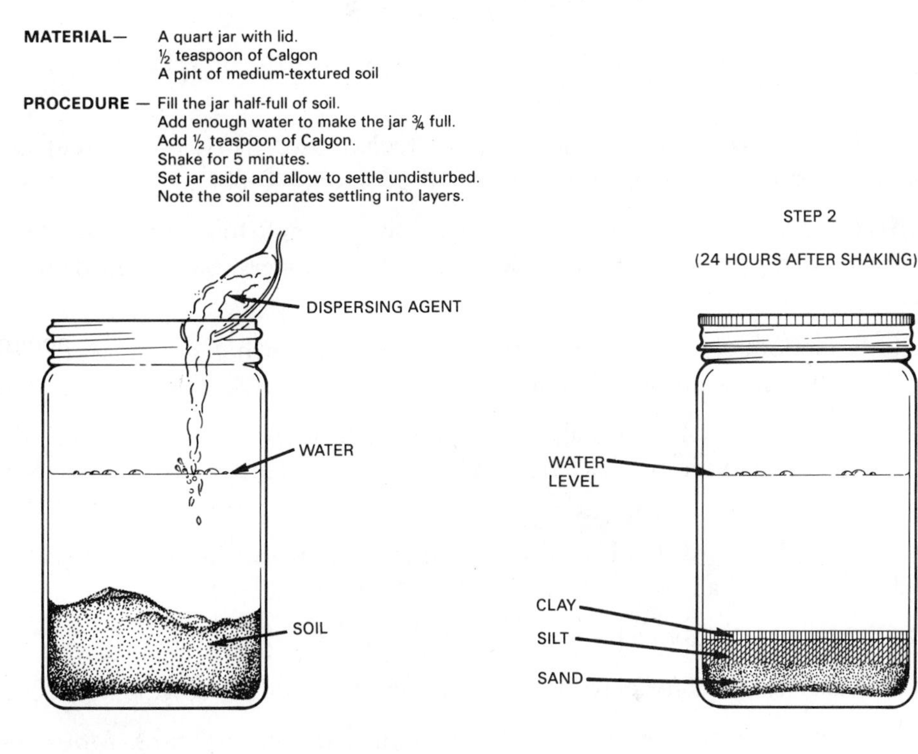

FIGURE 9-13
Mechanical analysis of soil to determine texture.

"feel" of the soil, are as follows (Figure 9-14).

Coarse-textured (sandy) soil is loose and single grained. The individual grains can be seen readily or felt. Squeezed, when dry in the hand, it will fall apart when the pressure is released. Squeezed when moist, it will form a cast, but will crumble when touched.

Medium-textured (loamy) soil has a relatively even mixture of sand, silt, and clay. However, the clay content is less than 20%. (The characteristic properties of clay are more pronounced than those of sand.) A loam is mellow with a somewhat gritty feel, yet fairly smooth and highly plastic. Squeezed when moist, it will form a cast which can be handled quite freely without breaking.

Fine-textured (clay) soil usually forms very hard lumps or clods when dry and is quite plastic. It is usually very sticky when wet. When the moist soil is pinched between the thumb and fingers it will form a long, flexible "ribbon." A clay soil leaves a "slick" surface on the thumb and fingers when rubbed together with a long stroke and a firm pressure. The clay tends to hold the thumb and fingers together due to its stickiness.

Structure Soil *structure* refers to the tendency of soil particles to cluster together and function as soil units called aggregates (Figure 9-15). *Aggregates* or *crumbs* contain mostly clay, silt, and sand particles held together by a gel-type substance formed by organic matter.

Aggregates absorb and hold water better than individual particles. They also hold plant nutrients and influence chemical reactions in the soil.

Another major benefit of a well-aggregated soil or a soil with good structure is its resistance to damage by falling raindrops. When hit by falling rain, the aggregate stays together as a water-absorbing unit, rather than separating into individual particles. When aggregates on the surface of soil dry out, they remain in a crumbly form and permit good air movement. However, dispersed soil particles run together when dry and form a crust on the surface. The crust prevents air exchange between the soil and atmosphere and decreases plant growth. The process and benefits of aggregation is applicable mostly to fine- and medium-textured soils.

Organic Matter As seen from the previous discussion, organic matter plays an important role in soil structure. Soil is a living medium with a great variety of living organisms. Some groups of organisms of the plant kingdom are:

SOIL TEXTURAL CLASSES

SAND–*Dry*: loose and single grained; feels gritty. *Moist*: will form very easily-crumbled ball. Sand—85-100%, Silt—0-15%, Clay—0-10%.

LOAMY SAND–*Dry*: silt and clay may mask sand; feels loose, gritty. *Moist*: feels gritty; forms easily-crumbled ball; stains fingers slightly. Sand—70-90%, Silt—0-30%, Clay—0-15%.

SANDY LOAM–*Dry*: clods easily broken; sand can be seen and felt. *Moist*: moderately gritty; forms ball that can stand careful handling; definitely stains fingers. Sand—43-85%, Silt—0-50%, Clay—0-20%.

LOAM–*Dry*: clods moderately difficult to break; mellow, somewhat gritty. *Moist*: neither very gritty nor very smooth; forms a firm ball; stains fingers. Sand—23-52%, Silt—28-50%, Clay—7-27%.

SILT LOAM–*Dry*: clods difficult to break, when pulverized feels smooth, soft and floury, shows fingerprints. *Moist*: has smooth or slick "buttery" or "velvety" feel; stains fingers. Sand—0-50%, Silt—50-88%, Clay—0-27%.

CLAY LOAM–*Dry*: clods very difficult to break with fingers. *Moist*: has slightly gritty feel; stains fingers; ribbons fairly well. Sand—20-45%, Silt—15-53%, Clay—27-40%.

SILTY CLAY LOAM–same as above but very smooth. Sand—0-20%, Silt—40-73%, Clay—27-40%.

SANDY CLAY LOAM–same as for clay loam. Sand—45-80%, Silt—0-28%, Clay—20-35%.

CLAY–*Dry*: clods cannot be broken with fingers without extreme pressure. *Moist:* quite plastic and usually sticky when wet; stains fingers. (A silty clay feels smooth, a sandy clay feels gritty.) Sand—0-45%, Silt—0-40%, Clay—40-100%.

FIGURE 9-14
Major soil textural classes.

A. Roots of higher plants
B. Algae: green, blue-green, diatoms.
C. Fungi: mushroom fungi, yeasts, molds.
D. Actinomycetes of many kinds: aerobic, anaerobic, autotrophic, heterotrophic.

Some examples of groups of organisms from the animal kingdom that are prevalent in soils are:

A. Those that subsist largely on plant material: small mammals, insects, millipedes, sow bugs (wood lice), mites, slugs, snails, earthworms.
B. Those that are largely predators: snakes, moles, insects, mites, centipedes, spiders.
C. Microanimals that are predatory, parasitic, and live on plant tissues: nematodes, protozoans, rotifers.

FIGURE 9-15
Good soil structure is very important in medium and heavy soils.

Organic Living organisms excrete cell or body wastes which become part of the organic content of soil. Further, the microbes of the soil and the remains of larger plants and animals decompose or decay into soil-building materials and nutrients.

People who grow indoor plants at home, garden, farm, produce greenhouse crops, or grow nursery stock generally find it useful to add organic materials to the soil. Popular sources of organic matter for soil amendments are peat moss, leaf mold, compost, livestock manure, sawdust, and others.

Some important benefits or functions of organic matter in soil are:

1. it makes the soil porous.
2. it supplies nitrogen and other nutrients to the growing plant.
3. it helps hold water in the soil so that growing plants are protected against droughts.
4. it aids in managing the soil moisture content.
5. it furnishes food for soil organisms.
6. it serves as a store house for mineral nutrients.
7. it minimizes leaching.
8. it serves as a source of nitrogen and growth promoting substance.
9. it stabilizes the soil structure.

Other Properties of Soils

There are numerous properties of soils which soil scientists, managers, technicians, and operators utilize in their work with soils. Water was discussed in the previous unit. Additionally, consideration of external factors such as land position, slope, and stoniness may be important.

The soil profile has many factors which cannot be considered in the scope of this text. Some of these are soil color, depth, drainage, permeability, and erosion.

MAKING AMENDMENTS TO PLANT-GROWING MEDIA

The term *amendment* is used here to mean addition to or change in. Most soil amendments are made to add organic matter, add specific nutrients, or modify soil pH. The *pH* is a measure of the degree of acidity or alkalinity. *Acidity* is sometimes referred to as sourness and *alkalinity* is referred to as sweetness. The pH scale ranges from 0 (maximum acidity) to 14 (maximum alkalinity). The midpoint of the scale is 7 which is *neutral*, meaning neither acid nor alkaline (Figure 9-16).

Crops grow best in media with a narrow pH range unique to that plant species (Figure 9-17A). Most plants require a pH somewhere between 5.0 and 7.5. Some crops, such as potatoes and blueberries, prefer a soil pH around 5.5. Alfalfa requires a pH of 6.5 to 7.0 for maximum growth.

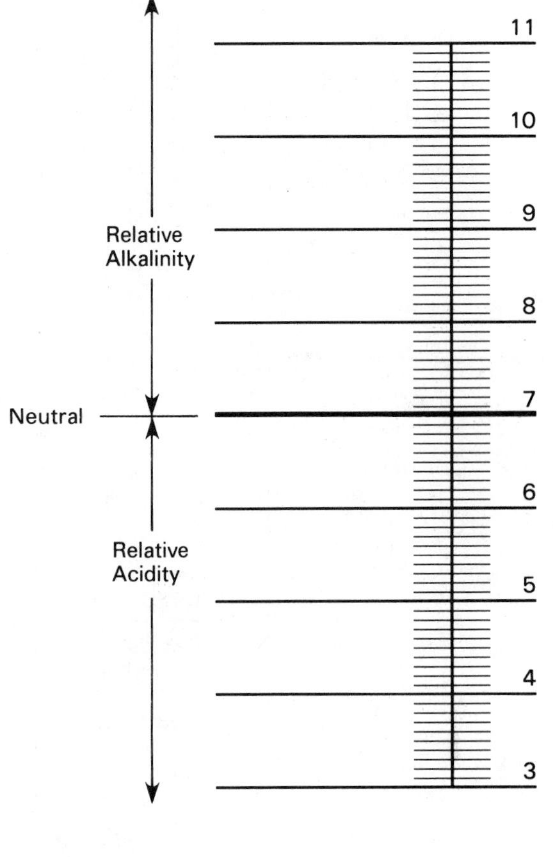

FIGURE 9-16
pH scale.

Liming

Soils that were historically covered by trees, such as the northeastern, western, and northwestern parts of the United States, develop acidic soils. When cleared of trees, such soils need lime added to raise the pH for the efficient production of most farm crops.

A pH test can be made with a test kit on-site by the plant technician or soil samples can be sent to a university or commercial laboratory for analysis. The local cooperative extension agent can make suggestions in this regard. Laboratory services generally include an analysis of nitrogen, phosphorus, and potash, as well as pH. Liming and fertilizing recommendations are also provided by many testing laboratories (Figure 9-17B).

HOW TO TAKE A SOIL SAMPLE IN YOUR LAWN OR GARDEN

1. Select an appropriate sampling tool (spade, auger or probe).
2. Make a sketch dividing your yard into sampling areas—for example, front lawn, garden, flower bed, slope, back lawn. Appropriately label each area. See Figure 9-17B, view #1.
3. When taking your samples, avoid wet or bare spots. Soils that are substantially different in plant growth or past treatment should be sampled separately, provided their size and

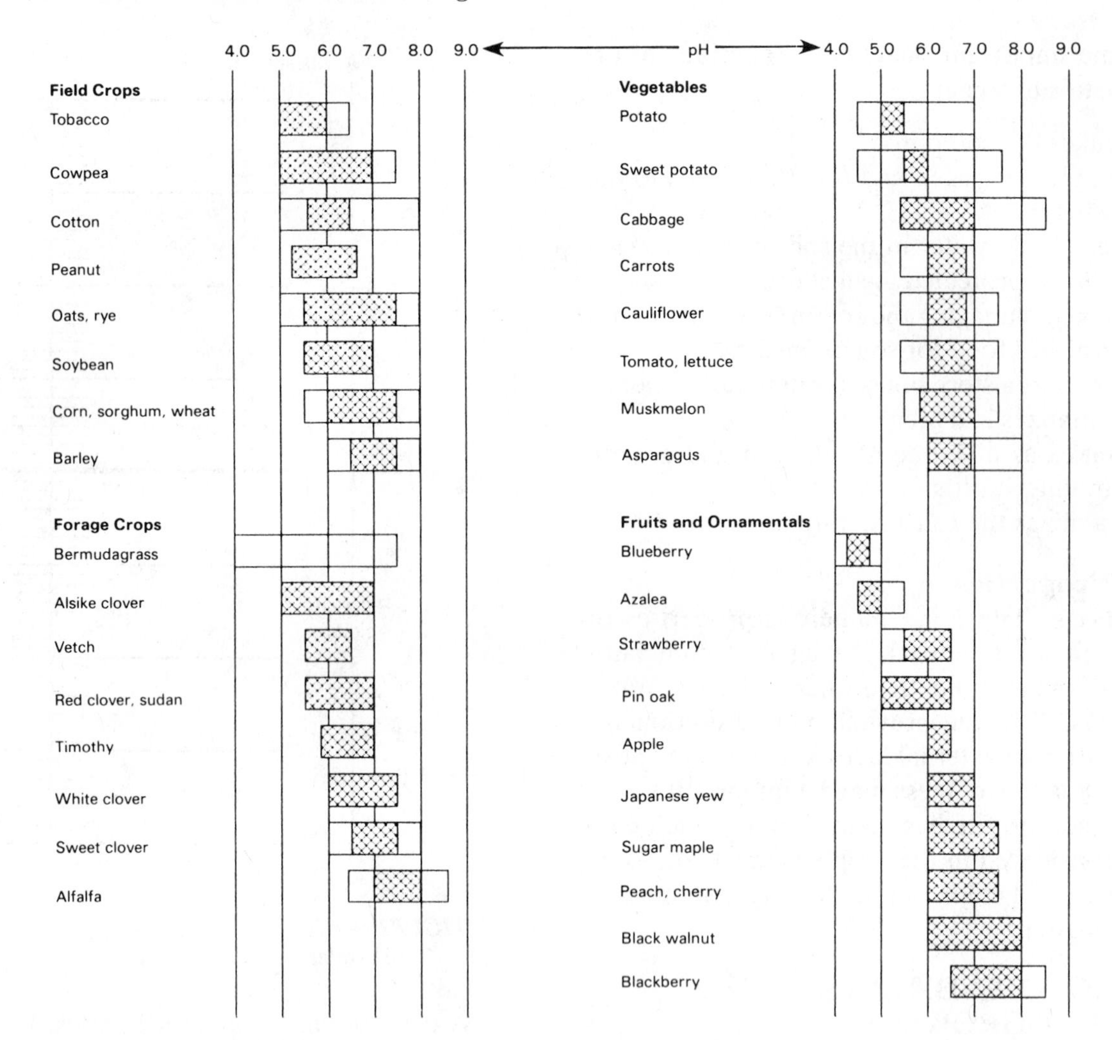

FIGURE 9-17A
Plants grow in pH ranges from approximately 4 (acid) to 8.5 (alkaline).

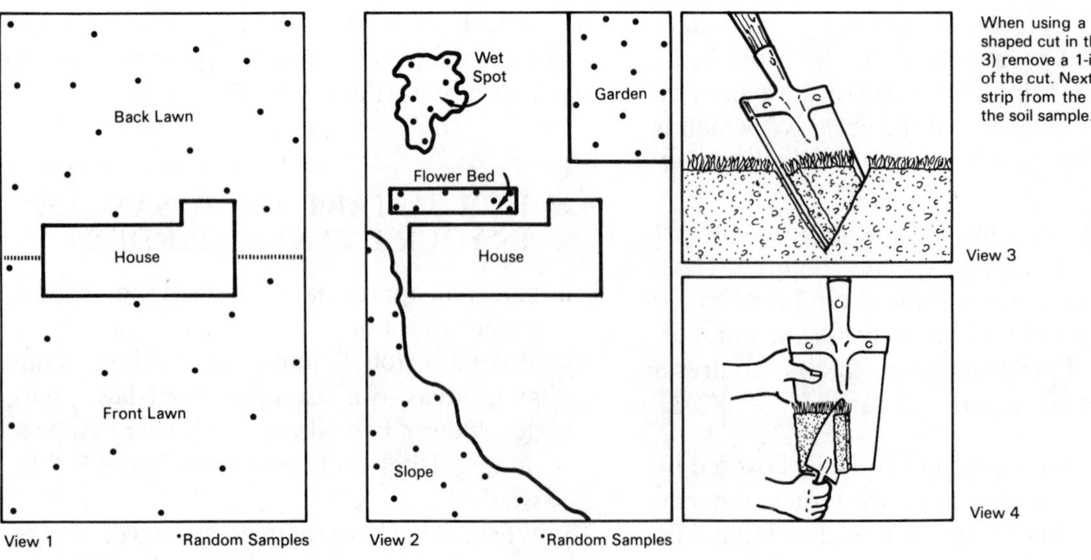

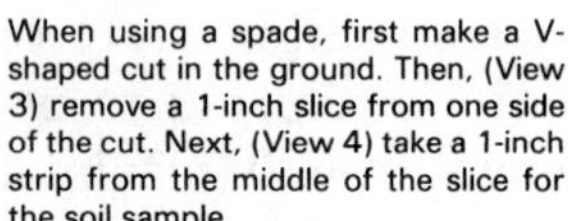
When using a spade, first make a V-shaped cut in the ground. Then, (View 3) remove a 1-inch slice from one side of the cut. Next, (View 4) take a 1-inch strip from the middle of the slice for the soil sample.

FIGURE 9-17B
An accurate soil test is dependent on a representative sample of soil taken from a uniform area.

nature makes it feasible to fertilize or lime each area separately.

4. After removing surface litter, take your sample from the correct depth; this is 2″ in established lawns, and about 6″ in gardens, flower beds, new lawns to be established or other areas to be tilled.
5. Submit a separate composite sample for each significantly different area of your lawn—front lawn, back lawn, flower bed, etc. Your composite sample for each area should include a small amount of soil taken from 15 to 20 randomly selected locations in the area represented, for each sample.
6. When using a spade, first make a "V" shaped cut. Then remove a 1-inch slice from one side of the cut. See Figure 9-17B, view #3. Now take a 1-inch strip from the middle of this slice. (See Figure 9-17B, view 4.) Use this little slice as a part of your composite sample for each area.
7. **Air-dry the soil; do not use heat.** Mix the soil from a single area in a clean bucket. Place about 1 pint of this mixture into the sample box. Use a separate box for each area. Fill in blanks on box or information sheet for each box.
8. Send soil sample(s) and information sheet(s) to the soil test laboratory.

Note: The Cooperative Extension Service in most counties and/or cities can arrange for soil testing.

The pH Test The amount of agricultural lime required to raise soil to the desired pH level is indicated by a pH test. The same pH value may require different amounts of agricultural lime. This is due to the fact that soils contain varying amounts of organic matter, clay, silt, and sand. The greater the organic matter and clay content, the greater the amount of lime required to correct the acidity.

Even though all the lime required by your soil is applied, do not expect the pH to move up quickly. It will generally require two to three years before the desired pH is reached.

Mix Well An acre's worth of soil, to a 6″ depth, weighs about 2 million pounds. Therefore, it takes a lot of mixing to distribute a relatively small amount of lime with a lot of soil. Your soil tests are based on a plow depth of about 7″. If you plow deeper you should increase the lime recommendation accordingly.

Standard Ground Limestone The lime recommendation is based on standard ground limestone which should contain a minimum of 50% lime oxides (calcium oxide plus magnesium oxide). About 98% should pass through a 20-mesh sieve, with a minimum of 40% passing through a 100-mesh sieve. The higher the percentage of limestone passing through a 100-mesh screen, the faster this limestone will correct soil acidity. The best limestone will have the greatest calcium and/or magnesium content and will be ground to a small particle size. It is more important to finely grind a high magnesium stone than a high calcium stone. A high magnesium or dolomitic limestone should always be used when a magnesium deficiency is indicated by a soil test.

(One ton [2000 lbs.] of standard ground limestone is approximately equivalent to 1500 lbs. of hydrated lime or 1100 lbs. of ground burned limestone.)

Correcting Excessive Alkalinity A reduction of soil alkalinity is desirable where soils have a high pH and the alkaline condition may be associated with unsatisfactory crop growth. In most cases, this condition will exist where heavy applications of lime are made at one time or where lime has been applied to a soil with a high pH.

To lower the pH value of soil, use sulfur or aluminum sulfate. Sulfur, at 1½ lbs. or aluminum sulfate at 5 lbs. per 100 ft.^{2}, will lower the alkalinity, under most conditions, by ½ pH. Aluminum sulfate acts rapidly, producing acidification in 10 to 14 days. Sulfur requires 3 to 6 months. Broadcast the material over the surface and thoroughly work it into the soil. For full benefit, sulfur should be applied in the fall, after garden crops are harvested. Aluminum sulfate may be applied in early spring.

Use chemicals cautiously to lower alkalinity. Before these measures are taken you should consider other factors that may be responsible for poor growth (such as drought, insect and disease injury, and fertilizer burning).

Fertilizers and Fertilizing Essential plant nutrients are discussed in Unit 16, "Plant Physiology." However, it should be noted that nitrogen, phosphorus, and potash are known as the three *primary nutrients.* These three ingredients must be present for a fertilizer to be called a *complete fertilizer.*

The proportions of nitrogen, phosphorus, and potassium are known as the *fertilizer grade*, expressed on a fertilizer container as percentages of the contents of the container by weight. Therefore, a 100-lb. container of fertilizer with a grade of 10–10–10 contains 10% nitrogen, 10% phosphorus, and 10% potassium.

If the total amount of fertilizer (100 lbs.) is multiplied by the percentage of each ingredient, the pounds of each ingredient may be calculated. Therefore: .10 (% of nitrogen) × 100 (total lbs. of fertilizer) = 10 lbs. of nitrogen. The amount of phosphorus and potassium is also 10 lbs. each.

Fertilizer is frequently shipped in 80-lb. bags. Therefore, one bag of 10–10–10 fertilizer would have 8 lbs. nitrogen, 8 lbs. phosphorus, and 8 '' s. potassium, Figure 9-18, for a total of 24 lbs. of *active ingredients* (a component which achieves one or more purposes of the mixture).

Some popular grades of fertilizer are 5–10–5, 5–10–10, 10–10–10, 6–10–4, 0–15–30, 0–20–20, 8–16–8, 8–24–8. These grades are formulated to meet the needs of various crops on various soils. The amount and grade of fertilizer to apply are determined by (1) the specific crop to be grown, (2) the desired yield or performance of the crop, (3) fertility of the soil, (4) physical properties of the soil, (5) previous crop, and (6) type and amount of manure applied. Therefore, decisions on rate of application must be made on a local basis.

FIGURE 9-18
Fertilizer grade.

Organic Fertilizers Included in this category are all kinds of animal manures, and compost made with manures and other plant or animal products. Organic commercial fertilizers include the dried and pulverized manures, bone meal, slaughterhouse tankage, blood meal, dried and ground sewage sludge, cottonseed meal, and soybean meal.

Organic fertilizers have certain definite characteristics: (1) Nitrogen is usually the predominating nutrient, with lesser quantities of phosphorus and potassium. One exception to this is bone meal, in which phosphorus predominates, and nitrogen is a minor ingredient. (2) The nutrients are only made available to plants as the material decays in the soil, so they are slow acting and long lasting. (3) Organic materials alone are not balanced sources of plant nutrients and their analysis in terms of the three major nutrients is generally low.

Inorganic Fertilizers Various mineral salts which contain the plant nutrients in combination with other elements are in this type of fertilizer. They all have certain characteristics quite strongly in contrast with organic fertilizers. (1) The nutrients are in soluble form, and quickly available to plants. (2) The soluble nutrients make them caustic to growing plants and can cause injury. Care must be used in applying them to growing plants. They should not come in contact with the roots or remain on plant foliage for any length of time. (3) Analysis of chemical fertilizers is relatively high in terms of the nutrients they contain.

Fertilizers are needed to replenish mineral nutrients depleted from a soil by crop removal or by such natural means as leaching. Some soils with high fertility may need only nitrogen or manure. Use of fertilizers that also contain small amounts of copper, zinc, manganese, boron, and other minor elements is not considered necessary for most soils, but may be needed in certain soils and for certain crops.

There are also unmixed fertilizers which carry only one element (Figure 9-19). Most important of these unmixed materials usually would be the nitrogen and phosphate carriers. Nitrogen carriers vary from 16 to 45% nitrogen. Be careful in using nitrogen materials. Too much may cause excessive and soft growth.

PLANT NUTRIENTS IN FERTILIZER MATERIALS

Materials	Nitrogen N%	Phosphorus P_2O_5%	Potassium K_2O%	Calcium %	Magnesium %	Sulphur %
Ammonium Nitrate	33.5	—	—	—	—	—
Ammonium Sulphate	21	—	—	—	0.3	23.7
Urea	45	—	—	1.5	0.7	.02
Sodium Nitrate	16	—	0.2	0.1	.05	.07
Calcium Nitrate	15	—	—	19.4	1.5	.02
Calcium Cyanamide	21	—	—	38.5	.06	0.3
Anhydrous Ammonia	82	—	—	—	—	—
Superphosphate	—	20	0.2	20.4	0.2	11.9
Liquid Phosphoric Acid	—	58	—	—	—	—
Ammonium Phosphate	11	48	0.2	1.1	0.3	2.2
Potassium Chloride	—	—	62	—	0.1	0.1
Potassium Sulphate	—	—	53	0.5	0.7	17.6

FIGURE 9-19
Plant nutrients in fertilizer materials.

Superphosphate fertilizers carry only phosphorus. Phosphorus promotes flower, fruit, and seed development. It also stiffens up stem growth and stimulates root growth. Superphosphate usually is applied to manure to give a better balance of nutrients for the plant (100 lbs. to each ton of horse or cow manure and 100 lbs. to each half-ton of sheep manure).

Fertilizer Applications

There are many ways to apply fertilizers. For the home owner's lawn, the most likely method is *broadcasting* (spreading evenly over the entire surface). In the case of gardens or fields, broadcasted fertilizer may be *incorporated* or mixed into the soil by spading, plowing, or disking.

Band application places fertilizers about 2″ to one side of and slightly below the seed. This method is used extensively for raw crops in gardens and fields.

Side dressing is done by placing fertilizer in bands about 8″ from the row sides of growing plants. This method is popular for raw crops such as corn and soybeans.

Top-dressing is a procedure where fertilizer is broadcast lightly over close-growing plants. Top-dressing is used for adding nitrogen to small grain crops after they are established.

Starter solutions are diluted mixtures of single or complete fertilizers used when plants are transplanted. Their purpose is to provide small amounts of nutrients that will not burn the tender roots of young plants.

Other methods of applying fertilizers include the application of *foliar sprays* directly onto the leaves of plants, and *knife application* of anhydrous ammonia gas in the soil. Further, the method of adding liquid fertilizer to irrigation water is used extensively in the United States.

Using Manure

Animal manure is a valuable product when handled properly (Figure 9-20). Its content of plant nutrients makes it a valuable fertilizer material. The organic matter aids in developing and maintaining structure in soils.

To obtain the most from manure these practices should be followed.

1. Use adequate bedding to absorb all of the liquids.
2. Reinforce with superphosphate at the rate of 2 lbs. per cow per day. Apply in gutter or

PLANT NUTRIENTS AVAILABLE FROM ONE TON OF WELL-PRESERVED MANURE DURING THE FIRST YEAR AFTER APPLICATION

Kind	Nitrogen	Phosphorus	Potassium
Cattle (Phosphated)	5 lb	5 lb	5.0 lb
Cattle (Non-Phosphated)	5 lb	5 lb	2.5 lb
Poultry	4 lb	10 lb	7.0 lb

Substituting manure for commercial fertilizer: The fertilizer recommendation is 80 lb. nitrogen plus 80 lb. phosphorus plus 80 lb. potassium. Where 10 tons of well-preserved phosphated cattle manure is applied per acre, it has an acre fertilizer value of 50 lb. nitrogen + 50 lb. phosphorus + 50 lb. potassium (10 tons x 5 + 5 + 5). The difference is 30 + 30 + 30 which can be obtained with fertilizer.

FIGURE 9-20
Animal manure is valuable for plant feeding as well as soil building.

before bedding in feeder sheds or loafing pens. (Super-phosphated manure is no more corrosive to machinery than nonphosphated manure.)

3. Spread evenly. An 8–12 ton application per acre is recommended. (Fifty bushels weigh about one ton.)
4. When possible, incorporate into the soil immediately after spreading.
5. Do not spread on steep slopes when the ground is frozen.
6. When storing manure, keep compact and under cover.
7. Apply to crops which will give best response such as corn, sorghums, potatoes, and tobacco.

Legume Crops *Legumes* are plants in which certain bacteria utilize nitrogen gas from the air and convert it to *nitrates* (the form of nitrogen used by plants). Some examples of legumes are beans, peas, clovers, and alfalfa. The process of converting nitrogen gas to nitrate by bacteria in the roots of legumes is called nitrogen fixation.

Nitrogen fixation reduces or eliminates the need for adding expensive nitrogen fertilizer to legume crops. Further, when the roots of legume plants decay, large amounts of nitrates are left behind for the next crop. Hence corn, which requires large amounts of nitrogen, should follow alfalfa or clover in a field, since the legumes leave unused nitrogen behind.

Rotation Fertilization Many crop rotations start with a small grain crop. The preparation of the seedbed for small grains provides an excellent opportunity to put lime and fertilizer into the feeding zone for roots of the sod crop that normally follow. Lime and phosphorus move very slowly in the soil, and unless the roots contact these elements, they can not provide the seedling's high requirements for phosphorus in the critical first year of growth.

A properly limed and fertilized sod crop is the backbone of a successful crop rotation. The growth of nutrient-enriched grass and legume roots throughout the soil provides a most favorable medium for natural soil-building processes. The addition of organic matter, the movement of fertilizer elements by the roots into the soil, and the production of root channels by sod produce soils that absorb water better, erode less, are easier to work, and grow better crops.

Phosphorus and potassium added to sod will benefit the sod and, in addition, will be placed in the best location and be in the best form to supply these elements to the long-season row crops which follow. Nitrogen produced by the legume or added to grass will be present in organic form and will be released in the best possible way for the row and small grain crops.

The composition of soils is complex and depends on many factors. Soil is dynamic and changing all the time. Nature has provided many cycles to help provide for soil renewal. However, the scientific management of soils makes them more productive and helps to assure productive soils for efficient production for future generations.

HYDROPONICS

The term hydroponics refers to a number of types of systems used for growing plants without soil. Some major ones are:

1. *Aggregate culture*, where a material such as sand, gravel, or marbles supports the plant roots.
2. *Water culture, aquaculture*, or *nutriculture*, in which the plant roots are immersed in water containing dissolved nutrients.
3. *Aeroponics*, in which the plant roots hang in the air and are misted regularly with a nutrient solution.

4. *Continuous-flow systems*, in which the nutrient solution flows constantly over the plant roots. This system is the one most commonly used for commercial production.

Hydroponics is growing in importance as a means of producing vegetables and other high-income plants. In areas where soil is lacking or unsuitable for growth, hydroponics offers an alternative production system. Equally good crops can generally be produced in a greenhouse in conventional soil or bench systems.

When plants are grown hydroponically, their roots are either immersed in or coated by a carefully controlled nutrient solution. The nutrients and water are supplied by the solution alone and not by aggregates or other inert material that support the roots. *Inert* means inactive.

Plant Growth Requirements Hydroponically grown plants have the same general requirements for good growth as soil-grown plants. The major difference is the method by which the plants are supported and the nutrients are supplied for growth and development.

Water Providing plants with an adequate amount of water is not difficult in a water culture system. During the hot summer months a large tomato plant may use one-half gallon of water per day. However, quality can be a problem. Water with excessive alkalinity or salt content can result in a nutrient imbalance and poor plant growth. Softened water may contain harmful amounts of sodium. Water that tests above 320 parts per million of salts are likely to cause an imbalance of nutrients.

Oxygen Plants require oxygen for respiration to carry out their functions. Under field and normal greenhouse conditions, oxygen is usually adequate as provided by the soil. When plant roots grow in water, however, the supply of dissolved oxygen is soon depleted, and damage or death soon follow unless supplemental air is provided. Where supplemental oxygen is needed, a common method of supplying oxygen is to bubble air through the water. It is not usually necessary to provide supplementary oxygen in aeroponic or continuous-flow systems.

Mineral Nutrients Green plants must absorb certain minerals through their roots to survive. These minerals are supplied by soil. The elements needed in large quantities are nitrogen, phosphorus, potassium, calcium, magnesium, and sulfur. The nutrients needed in tiny amounts are iron, manganese, boron, zinc, copper, molybdenum, and chlorine. An over-supply of these nutrients is toxic to plants. In hydroponic systems, all nutrients normally supplied by soil must be included in the water to form the solution or media (Figure 9-21).

Light All vegetable plants and many flowers require large amounts of sunlight. Hydroponically grown vegetables, like those grown in a garden, need at least 8 to 10 hours of direct sunlight each day to produce well. Artificial lighting is a poor substitute for sunshine since most indoor lights do not provide enough intensity to produce a crop. Incandescent lamps supplemented with sunshine or special plant-growth lamps can be used to grow transplants but are not adequate to grow the crop to maturity. High-intensity lamps such as high-pressure sodium lamps can provide more than 1,000 foot-candles of light. They may be used successfully in small areas where sunlight is inadequate. However, these lights are too expensive for commercial operations.

Spacing Adequate spacing between plants will permit each plant to receive sufficient light in the greenhouse. Tomato plants, pruned to a single stem, should be allowed 4 ft^2 per plant. European seedless cucumbers should be allowed 7–9 ft^2, and seeded cucumbers need about 7 ft^2 Leaf lettuce plants should be spaced 7–9″ apart within the row and 9″ between rows. Most other vegeta-

FINAL PPM	NUTRIENT
210 ppm	Nitrogen (N)
57 ppm	Phosphorus (P)
252 ppm	Potassium (K)
201 ppm	Calcium (Ca)
50 ppm	Magnesium (Mg)
5 ppm	Iron (Fe)
0.12 ppm	Molybdenum (Mo)
0.6 ppm	Manganese (Mn)
0.5 ppm	Zinc (Zn)
0.18 ppm	Copper (Cu)
0.6 ppm	Boron (B)

FIGURE 9-21
A nutrient solution for hydroponics growing of specialty lettuce and herbs.

bles and flowers should be grown at the same spacing as recommended for a garden.

Greenhouse vegetables will not do as well during the winter as in the summer. Shorter days and cloudy weather reduce the light intensity and thus limit production.

Temperature Plants grow well only within a limited temperature range. Temperatures that are too high or too low will result in abnormal development and reduced production. Warm-season vegetables and most flowers grow best between 60° and 75–80°F. Cool-season vegetables, such as lettuce and spinach, should be grown between 50–70°F.

Support In the garden or field, plants are supported by roots anchored in soil. A hydroponically grown plant must be artificially supported with string, stakes, or other means.

Future of Hydroponics Hydroponics is increasing in use commercially and will undoubtedly become increasingly important in the future. Research is expanding and new techniques are being developed. The use of nutrient solutions as media for growing plants will be an important part of agriscience in the future.

Student Activities

1. Define the "Terms To Know."
2. Observe soils and other media used in flower pots, trays of vegetable seedlings, greenhouses, gardens, road banks, construction sites, and other places.
3. Examine the living organisms in a shovel full of soil taken from an outdoor area that is damp, moist, and high in organic matter.
4. Invite a Soil Conservation Service professional in to discuss soil mapping and land-use planning.
5. Obtain a land-use map from SCS for your home, farm, or other area. Study the material and report your findings to the class.
6. Observe a profile at a cut in a road bank, stream bank, or hole. Identify the O, A, B, and C horizons.
7. Do a mechanical analysis of a sample of soil using a fruit jar, water, and a dispersing agent such as calgon (Figure 9-13).
8. Observe the feel of some of the soil used in number 7.
9. Obtain a pH test kit and test samples of soil from home.
10. Do research on models and procedures for a home or school hydroponics unit. Plan, build, and use the unit to experiment with hydroponics production of various plants.

Self-Evaluation

A. MULTIPLE CHOICE

1. An example of plant-growing media is

   a. soil. c. perlite.
   b. water. d. all of these.

2. Which is not organic matter?

   a. leaf mold c. sphagnum moss
   b. peat moss d. vermiculite

3. Decay of organic matter is caused by

a. large animals.
b. microbes.
c. rodents.
d. water.

4. Which is not a factor affecting soil formation?

a. hydroponics
b. gravity
c. ice
d. water

5. Humans affect soil formation by

a. acid.
b. alkaline.
c. bulldozing.
d. weathering.

6. The land class with the fewest limitations is

a. Class I.
b. Class III.
c. Class VI.
d. Class VIII.

7. The land classes suitable for field crop production are

a. I, II, IV, VI.
b. I, II, III, IV.
c. I, IV, V, VIII.

8. The horizon that is most supportive of plant growth is

a. horizon O.
b. horizon A.
c. horizon B.
d. horizon C.

9. The smallest soil particle is

a. clay.
b. gravel.
c. sand.
d. silt.

10. Good soil structure means

a. aggregated.
b. crumbly.
c. raindrop endurance.
d. all of the above.

11. Soil pH is generally raised by adding

a. sulfur.
b. nitrogen.
c. lime.
d. complete fertilizer.

12. Hydroponics refers to

a. aggregate culture.
b. aquaculture.
c. nutriculture.
d. all of the above.

B. MATCHING (Group I)

______	**1.** Colluvial	**a.**	Deposited by ice
______	**2.** Alluvial	**b.**	Deposited by lakes
______	**3.** Glacial	**c.**	Formed in place
______	**4.** Lacustrine	**d.**	Deposited by gravity
______	**5.** Loess	**e.**	Class II land with erosion problem
______	**6.** Leaching	**f.**	Cross section of soil
______	**7.** Organic matter	**g.**	Makes soil dark in color
______	**8.** Residual	**h.**	Removal of soluble materials
______	**9.** II.e	**i.**	Deposited by streams
______	**10.** Profile	**j.**	Deposited by wind

MATCHING (Group II)

______	1. Horizon A	a.	Mostly organic matter
______	2. Horizon B	b.	Parent material
______	3. Horizon C	c.	Topsoil
______	4. Horizon O	d.	Subsoil
______	5. Coarse	e.	5-10-5
______	6. Medium texture	f.	Equal parts of nitrogen, phosphorus, and potassium
______	7. Fine texture	g.	Sulfur
______	8. 10-10-10	h.	Loamy soils
______	9. Complete fertilizer	i.	Clay soils
______	10. Lowers pH	j.	Sandy soils

C. COMPLETION

1. Soil is defined as the top layer of the Earth's surface suitable for the growth of ______ .
2. Three groups of plants found in soil are ______ , ______ , and ______ .
3. Three groups or types of animals that live in the soil are ______ , ______ , and ______ .
4. Five important benefits of organic matter in soil are ______ , ______ , ______ , ______ , and ______ .

UNIT 10 Forestry Management

OBJECTIVE To determine the relationship of forests to our environment and the recommended practices for utilizing forest resources.

Competencies to Be Developed

After studying this unit, you will be able to

- ☐ define forest terms.
- ☐ describe the forest regions of the United States.
- ☐ discuss important relationships among forests, wildlife, and water resources.
- ☐ identify important types and species of trees.
- ☐ discuss important properties of wood.
- ☐ apply principles of good wood-lot management.
- ☐ describe procedures for seasoning lumber.

TERMS TO KNOW

Forest
Trees
Shrubs
Board foot
Lumber
Forestry
Evergreen
Conifers
Softwood
Deciduous
Hardwood
Pulpwood
Clearcut
Plywood
Veneer
Hardness
Shrinkage
Warp
Ease of working
Grain
Wood lot
Silviculture
Seedlings
Forester
Virgin

FIGURE 10-1
A forest contains trees, shrubs, and other plants as well as animal life. *(Courtesy of Jim Jones, Agriculture Instructor)*

MATERIALS LIST

bulletin board materials
samples of forest and wood products
tree leaf collection
forestry reference materials

There are nearly 500 million acres of productive forest land in the United States. A *forest* is a complex association of trees, shrubs, and plants, which all contribute to the life of the community (Figure 10-1). *Trees* are woody perennial plants, with a single stem that develops many branches. Trees usually grow to more than 10 ft. high. *Shrubs* are bushy woody plants with multiple stems. A productive forest is one that is growing trees for lumber or other wood products on a continuous basis. *Lumber* is boards that are sawed from trees. It is bought and sold by the board foot. A *board foot* is a unit of measurement for lumber that is equal to 1″ × 12″ × 12″ (Figure 10-2).

There are 235 million acres of forest land in the United States that are classified as unproductive. The 735 million acres of forests represent

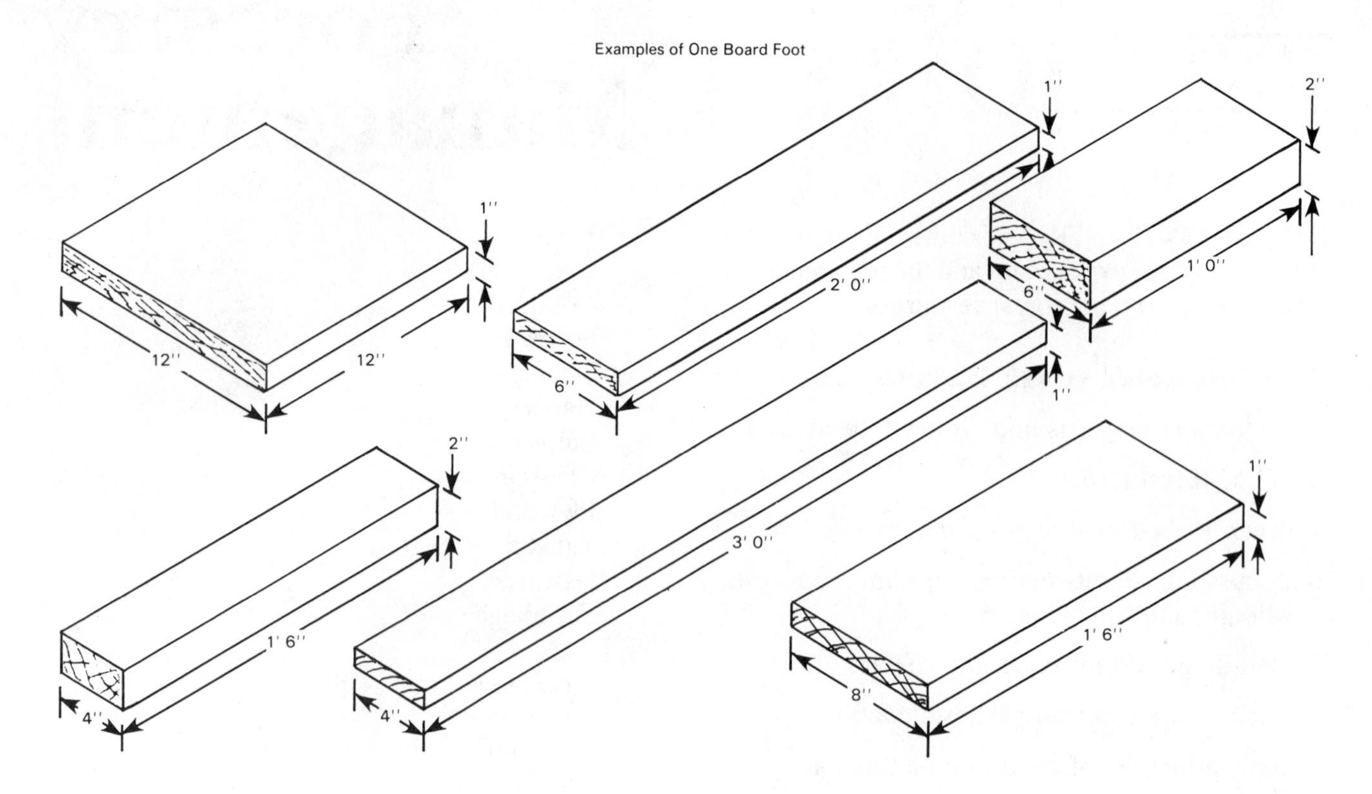

The formula for calculating board feet is —
bd. ft. = number of pieces × thickness in inches × width in inches × length in feet ÷ 12.
Calculate the board feet in the following:

Problem 1. 5 pieces 2″ × 4″ × 8′.

Solution:
$$\frac{5 \times 2'' \times 4'' \times 8'}{12} = 26\tfrac{2}{3} \text{ bd. ft.}$$

Problem 2. 6 pieces 1″ × 8″ × 10′ *or,* the same pieces dressed would be 6 pieces ¾″ × 7½″ × 10′. (Fractions from ½″ to 1″ are considered as 1″.)

Solution:
$$\frac{6 \times 1'' \times 8'' \times 10'}{12} = 40 \text{ bd. ft.}$$

Problem 3. 8 pieces 2″ × 6″ × 38″.
(If the length is in inches, divided the product by 144 instead of 12. Why?)

Solution:
$$\frac{8 \times 2'' \times 6'' \times 38''}{144} = 25\tfrac{1}{3} \text{ bd. ft.}$$

FIGURE 10-2
Lumber is bought and sold by the board foot. One board foot has a volume of 144 cubic inches. Any combination of dimensions which equals 144 cubic inches is one board foot.

nearly one-third of the total land in the United States. Unproductive forest land includes forests that will not produce an average of more than 100 board feet of lumber per acre per year and areas that are not practical to harvest. It also includes parks, wilderness land, and game refuges where harvesting of trees is not permitted.

When you consider that there are 860 species of trees in the United States, it is evident that forestry is an important part of the economy. *Forestry* is the management of forests.

Trees in the forests of the United States are divided into two general classifications: evergreen and deciduous. *Evergreen* trees do not shed their leaves on a yearly basis. Evergreen trees of commercial importance are mostly conifers. *Conifers* are evergreen trees that have needle-like leaves and produce lumber called *softwood. Deciduous* trees shed their needles every year and produce lumber called *hardwood.*

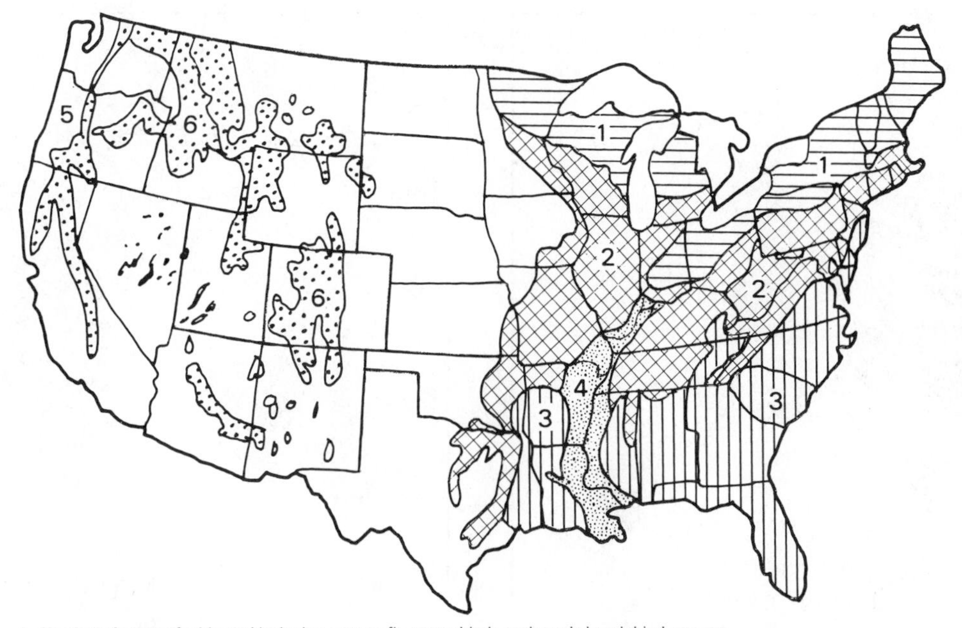

1. Northern forests of white-red-jack pine, spruce-fir, aspen-birch, and maple-beech-birch groups.
2. Central forests of oak-hickory.
3. Southern forests of: Oak-pine, loblolly-shortleaf pine, and longleaf-slash pine groups.
4. Bottom land forests of oak-gum-cypress.
5. West coast forests of: Douglas-fir, hemlock-sitka spruce, redwood, and some western hardwood groups.
6. Western interior forests of: Ponderosa pine, lodgepole pine, Douglas-fir, white pine, western larch, fir-spruce, and some western hardwood groups.

FIGURE 10-3
Primary forest regions in the United States *(Courtesy of USDA)*

FOREST REGIONS OF THE UNITED STATES

There are six major forest regions in the continental United States. They are the northern forest, central forest, southern forest, bottomland forest, West Coast or Pacific Coast forest, and western interior forests (Figure 10-3). There is also a small area of tropical or subtropical forest in southern Florida and southeastern Texas.

There are also two forest regions in Alaska and two in Hawaii. The Alaskan forest regions include coastal forests and interior forests. Hawaiian forests are classified as wet forest and dry forest (Figure 10-4).

Northern Forest

This region is located in the northeastern United States. It extends from Maine to southern Ohio and Indiana, and west to Wisconsin and Minnesota. The forests are mostly mixtures of hardwoods and softwoods although there are some pure stands of softwoods.

The forests of the northern or eastern region cover about 115 million acres. They account for about 15% of the total yearly lumber production in the United States. About 26% of the United States *pulpwood* (wood used for making fiber for paper and other products) production also comes from this region.

The conifers that are important in this forest region include white pine, spruce, jack pine, balsam fir, white cedar, and hemlock. Hardwoods of commercial value are beech, oak, aspen, birch, walnut, maple, gum, cherry, ash, and basswood.

Central Forest

This forest region is located mostly east of the Mississippi River and south of the northern forest region. It contains more varieties and species of trees than any other forest region. It is composed mostly of hardwood trees. Hardwoods of commercial importance in the central forest region include oak, hickory, beech, maple, poplar, gum, walnut, cherry, ash, cottonwood, and sycamore. The conifers that are of economic value in this region include Virginia pine, pitch pine, shortleaf pine, red cedar, and some hemlock.

Southern Forest

This region is located in the southeastern part of the United States. It extends south from Delaware to Florida, and west to Texas and Oklahoma. It is the forest region with the most potential for meeting the future lumber and pulpwood needs of the United States.

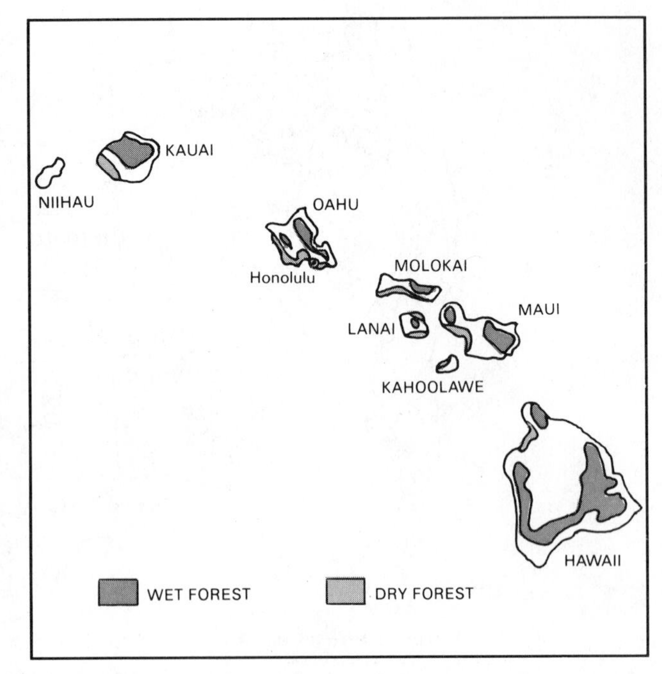

FIGURE 10-4
Forest regions of Alaska and Hawaii

The most important trees in the southern forest are conifers. They include Virginia, longleaf, loblolly, shortleaf, and slash pines. Oak, poplar, maple, and walnut are hardwood trees of economic importance.

Bottomland Forest These forests occur mostly along the Mississippi River. They contain mostly hardwood trees and are often among the most productive of the U.S. forest regions. This is due to the high fertility of the soils in this area. Oak, gum, tupelo, and cypress are the major hardwood species found there.

West Coast or Pacific Coast Forest This region is located in northern California, Oregon, and Washington. It is the most productive of the forest regions in the United States and has some of the largest trees in the world. The approximately 48 million acres of West Coast forest provides more than 25% of the annual lumber production in the United States. About 19% of the pulpwood and 75% of the plywood produced in the United States comes from trees grown in the West Coast forest region.

Trees in the West Coast forests include 300-ft.-tall redwoods and giant Sequoias that may be as much as 30 ft. in diameter. Douglas fir, ponderosa pine, hemlock, western red cedar, Sitka spruce, sugar pine, lodgepole pine, noble fir, and white fir are conifers that are important in this region. Hardwood species include oak, cottonwood, maple, and alder.

Western Interior Forest The interior forests are much less productive than the West Coast forest region. This region is divided up into many small areas and extends from Canada to Mexico. About 27% of the lumber produced in the United States comes from the 73 million acres in this forest region.

Most of the trees of commercial value in the western interior forest are the western pines. They are western white pine, ponderosa pine, and lodgepole pine. Spruce, fir, larch, western red cedar, and hemlock also grow there in smaller quantities. Aspen is the only hardwood of commercial importance in the western interior forest region.

Tropical Forest The tropical or subtropical forests of the continental United States are located in southern Florida and in southeastern Texas. It is the smallest forest region in the United States. The major trees in this region are mahogany, mangrove, and bay, which are unimportant commercially. However, they are very important ecologically.

Alaskan and Hawaiian Forests The Alaskan coastal forest region extends from southeastern Alaska to

Kodiak Island. Trees of importance are Sitka spruce and western hemlock. There are about 6 million acres included in the Alaskan coastal forest region.

The Alaskan interior forest consists of about 105 million acres of forest land. Most of the wood products harvested in this region are used locally because of the difficulty in transporting them to other areas. Trees of commercial importance are spruce, aspen, and birch.

The wet forest region of Hawaii produces ohia, boa, tree fern, kukui, tropical ash, mamani, and eucalyptus. Most of these woods are used in the production of furniture and novelties.

The dry forest region of Hawaii produces koa, haole, algaroba, monkey pod, and wiliwili. None of these are of commercial value.

RELATIONSHIPS BETWEEN FORESTS AND OTHER NATURAL RESOURCES

The relationships between forests and other natural resources, such as water and wildlife, are important to the overall well-being of the ecological system.

Forests filter the rain as it falls and help reduce erosion of the soil. They trap soil sediment and help maintain water quality. Trees and shrubs of the forest are also instrumental in removing much of the pollutant materials from the air and from water run-off. The roots of trees filter excess nutrients from cropland's water run-off. They also help reduce the harmful effects of excess fertilizer nutrients which enter streams and underground water systems.

The relationships between forests and many types of wildlife are numerous. Forests provide food, shelter, protection, and nesting sites for many species of birds and animals. They also help maintain the quality of streams so that fish and other types of aquatic life can live and thrive. The shade provided by forests helps to maintain proper water temperatures for the growth and reproduction of aquatic life.

The wildlife found in the different types of forests varies considerably (Figure 10-5). Some species must have the open areas that occur naturally or that have been clearcut by man. A forest that has been *clearcut* has had all of the marketable trees removed from it. Other wild birds and animals require mature forests to provide their needs in life. As forests change naturally, or as a result

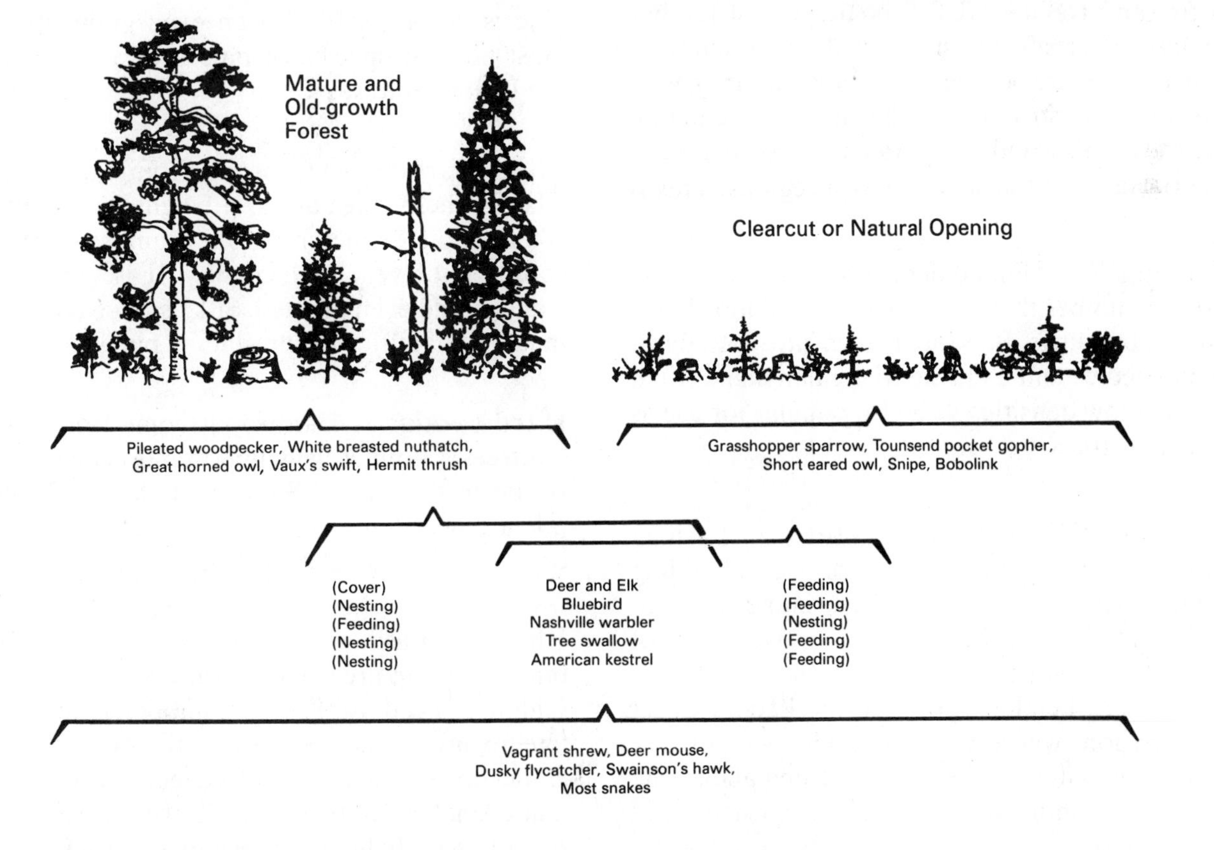

FIGURE 10-5

Preference in forest habits for some wildlife species

of harvesting, the types of wildlife that inhabit it also change.

IMPORTANT TYPES AND SPECIES OF TREES IN THE UNITED STATES

Trees may be described in terms of the lumber they produce. Characteristics such as hardness, weight, tendency to shrink and warp, nail- and paint-holding ability, decay resistance, strength, and surface qualities are utilized to evaluate the usefulness of lumber.

Softwoods

The softwoods, or needle type evergreens, that are important commercially in the United States include Douglas fir, balsam fir, hemlock, white pine, cedar, southern pine, ponderosa pine, and Sitka spruce.

Douglas Fir

This fir is probably the most important species of tree in the United States today. It grows to a height of more than 300 ft. and a diameter of more than 10 ft. About 20% of the timber harvested in the United States each year is Douglas fir. One-hundred-year-old stands of Douglas fir can produce 170,000 board feet of lumber per acre. This is five to six times the production of most other softwood species. Douglas fir is very popular as construction lumber and for the manufacture of plywood. *Plywood* is a construction material made of thin layers of wood glued together.

Balsam Fir

Found in the forests of the Northeast, the lumber that it produces is used mostly for framing buildings. Balsam fir trees have soft, dark-green needles and a classic triangular shape when grown at low densities. It is also popular for use as Christmas trees.

Hemlock: Eastern and Western

Eastern hemlock is strong and popular for use as building material. At times, it can be brittle and difficult to work with. Eastern hemlock grows over most of the northern forest range.

Western hemlock grows in the Pacific Coast forest region, where yearly rainfall averages 70″. Western hemlock lumber is very strong and is one of the most important sources of construction-grade lumber. It is also important to the pulpwood industry.

White Pine

The lumber is soft, light, and straight-grained. It has less strength than spruce or hemlock, and is more popular as a wood for cabinet-making. White pine grows from Maine to Georgia.

Cedar: Eastern Red, Eastern White, and Western Red

Eastern red cedar is used for fence posts because it is very resistant to decay. It is also used for the lining of chests and closets because its odor repels many insects. White cedar is a swamp tree with decay-resistant wood that is often used for shingles and log homes. Western red cedar resembles redwood in appearance. It is used where decay resistance, rather than strength, is important.

Southern Pine

Included in the category of southern pine is longleaf pine, shortleaf pine, loblolly pine, and slash pine. The southern pines grow in the southern and south Atlantic states. Lumber from southern pine is used for construction. It is also used for pulpwood and plywood.

Ponderosa Pine

These pines grow from Oregon to Mexico. It may live for 600 or more years and is very drought tolerant. Wood from ponderosa pine is usually soft with an even grain. It is the most popular species of pine used for lumber in the United States.

Sitka Spruce

Growing from California to Alaska, these trees attain a height of as much as 300 ft. and a diameter of 18 ft. Lumber from Sitka spruce is of very high quality. It is very strong, straight, and even-grained. Sitka spruce is also used in large quantities for producing pulpwood.

Hardwoods

Hardwoods come from deciduous trees. Commercially important species of hardwoods in the United States include birch, maple, poplar, sweetgum, oak, ash, beech, cherry, hickory, sycamore, walnut, and willow.

Birch

Easily recognized by their white bark, birch grows in areas where summer temperatures seldom exceed 70°F. Birch lumber is dense and fine-textured. It is used for furniture, plywood, paneling, boxes, baskets, and veneer, as well as for many small novelty items. *Veneer* is a very thin sheet of wood glued to a cheaper species of wood and used in paneling and furniture making.

Maple The lumber is classified as either hard or soft. Hard maple lumber is heavy, strong, and very hard. It is used for butcher blocks, workbench tops, flooring, veneer, and furniture. Soft maple is only about 60% as strong as hard maple, and is used in the same applications. Some species produce sweet sap that is made into maple sirup.

Poplar This tree grows over most of the eastern United States. Poplar is classified as a hardwood due to its deciduous structure. However, lumber from poplar trees is soft, light, and usually knot-free. Poplar lumber may be white, yellow, green, or purplish in color and can be stained to resemble most of the fine hardwoods. Furniture, baskets, boxes, pallets, and building timbers are frequent uses of poplar lumber.

Sweetgum The tree is easily recognized by its star-shaped leaves and distinctive ball-shaped fruit. Sweetgum trees grow to as much as 120-ft. tall and 3–5 ft. in diameter. Lumber from sweetgum trees has interlocking grain and is used for trim in houses, furniture, pallets, railroad ties, boxes, and crates. The gum that comes from wounds in trees can be used as natural chewing gum or as a flavoring or perfume.

Oak: White and Red There are two general types of oaks in the United States: white and red. White oak lumber is very hard, heavy, and strong. Its pores are plugged with membranes that make it nearly waterproof. It is used for structural timbers, flooring, furniture, fencing, pallets, and other uses where strength of wood is a necessity.

Red oak is similar to white oak except that it is very porous. It is not very resistant to decay and must be treated with wood preservatives when used outside. Chief uses of red oak include furniture, veneer, and flooring.

Aspen This hardwood tree grows in the Northeast, Great Lake states, and the Rocky Mountains. It is very rapid-growing, making lumber that tends to be weaker than most construction-grade timber. It is used for pulpwood and for lumber (in applications where strength is not of prime importance).

Ash The lumber is heavy, hard, stiff, and has a high resistance to shock. It also has excellent bending qualities. It is popular for use in handles, baseball bats, boat oars, and furniture, where it resembles oak in appearance.

AGRI-PROFILE

CAREER AREAS: DENDROLOGY/ SILVICULTURE/FORESTRY

Estimating the amount of lumber in a tree and on a tract of wooded land is an important forest management and marketing skill. ***(Courtesy of the National FFA Organization)***

Forestry is known as a career area for rugged individuals preferring the outdoors and to work in relative isolation. However, many jobs in forestry are in urban areas and involve a lot of indoor work. The United States Forestry Service hires large numbers of forestry technicians and managers. Many forestry jobs do involve an extensive amount of outdoor work, but most jobs provide a desirable mix of outdoor and indoor work.

Forestry includes the work of the dendrologist engaged in the study of trees; the silviculturist specializing in the care of trees; the forestry consultant who advises private forest land owners; lumber industry workers; government foresters; loggers; national and state forest rangers; and fire fighters. A relatively new position is that of the urban forester, responsible for the health and well-being of the millions of trees found in parks, along streets, and in other areas of our cities.

World concern exists for the slash and burn destruction of rain forests and the denuding of whole nations of trees as a matter of national policy or lack thereof. The foresters of the future will work with other specialists in soil, water, and wildlife conservation to manage the renewable natural resources of the world.

Beech Grown in the eastern United States, it is heavy and hard and is noted for its shock resistance. It is hard to work with and very prone to decay. Beech is used in veneer for plywood, flooring, handles, and containers.

Cherry This tree can be found from southern Canada through the eastern United States. Cherry wood is very dense and stable after drying. It is desirable and very popular in the production of fine-quality furniture. It is expensive and in limited supply, so it is used mostly for veneer and paneling; it may, however, be used for other woodworking purposes.

Hickory The eastern United States is also the area where hickory grows best. Hickory lumber is hard, heavy, tough, and strong. It is somewhat stronger than Douglas fir when used as construction lumber. Other uses for hickory include handles, dowel rods, and poles. Hickory is also popular for use as firewood and for smoking meat.

Sycamore The wood is used for flooring in barns, trucks, and wagons because of its strength and shock resistance. It grows from Maine to Florida, and west to Texas and Nebraska. Boxes, pallets, baskets, and paneling are other uses for sycamore.

Black Walnut A premier wood for the manufacture of fine furniture, it grows from Vermont to Texas. The wood has straight grain and is easily machined with woodworking tools. Because walnut is slow-growing and very desirable, it is often made into veneer to get more use from its chocolate-brown heartwood. It is also the source of black walnuts.

Black Willow Most of the black willow of commercial value is grown in the Mississippi River Valley. It is soft, light, and has a uniform texture. Willow is used mostly in construction for subflooring, sheathing, and studs. Some willow is also used for pallets and for interior components of furniture. Black willow is sometimes a low-cost substitute for walnut, since it has a similar brown appearance when finished.

There are many other domestic softwoods and hardwoods that grow in the United States. Many of these are important in local areas. The types discussed are but a sampling, rather than a definitive list, of the commercially important trees in the United States.

PROPERTIES OF WOOD

The various properties of wood determine the uses for which it is best adapted. It should be noted that there are wide variations within a specific species of tree and that the properties discussed here are general in nature.

Hardness The property of hardness refers to a wood's resistance to compression. Hardness determines how well it wears. It is also a factor in determining the ease of working the wood with tools. Splitting and difficulty in nailing are problems when wood is too hard.

Weight The weight of wood is a good indication of its strength. In general, heavy wood is stronger than light wood. The moisture content affects the weight of wood, so comparisons should always be made between woods with similar moisture content.

Shrinkage The change in dimensions of a piece of wood as it reacts to changes in humidity or temperature is referred to as *shrinkage*. The amount of expansion and contraction of wood greatly affects construction techniques used in building items from wood (Figure 10-6).

Warp The term refers to the tendency of wood to bend permanently due to moisture change. The tendency to warp is a problem in some types of wood. Warping is caused by uneven drying of wood across its three dimensions.

Ease of Working In wood, *ease of working* refers to the level of difficulty in cutting, shaping, nailing, and finishing the wood. It is influenced by the hardness of the wood and by the characteristics of the grain. The *grain* of wood is hard and

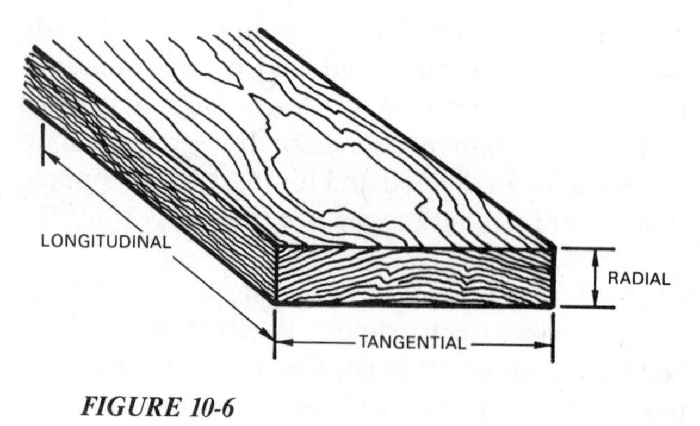

FIGURE 10-6
Dimensions of shrinkage of wood

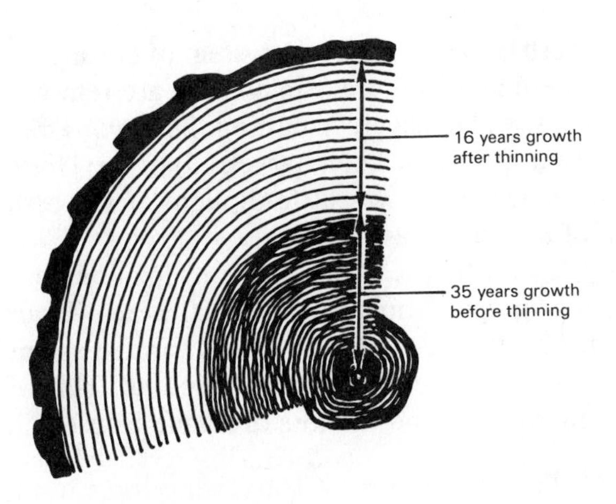

FIGURE 10-7
The annual rings seen in the cross section of a log or stump reflect the conditions of growth experienced by the tree on a year to year basis and determines the appearance of the grain. ***(From Cooper/Agricultural Mechanics: Fundamentals and Applications, copyright 1987 by Delmar Publishers Inc.)***

soft patterns caused by growth of the tree as it adds successive annual-growth rings (Figure 10-7). In general, soft wood is easier to work than hard, dense wood.

Paint Holding The ability of a type of wood to hold paint may be determined by the type of paint, surface conditions, and methods of application as well as characteristics of the wood itself. Moisture content, amount of pitch in the wood, and the presence of knots will also affect how well wood holds paint.

Nail Holding This ability refers to the resistance of wood to the removal of nails. In general, the harder and denser the wood, the better it holds nails.

Decay Resistance The resistance of wood to the microorganisms that cause decay is a chief factor in determining where many types of wood are used. Natural resistance to insects which live in, tunnel through, or devour wood also affects the choice of species for a given application.

Bending Strength The ability of wood to carry a load without breaking is a measure of bending strength. Bending strength is important when determining types and sizes of lumber to use for rafters, beams, joists, and other building applications.

Stiffness The resistance of wood to bending under a load is referred to as stiffness. When wood used in construction is not sufficiently stiff, ceilings and walls may flex and buckle and the wallboards crack under a load.

Toughness The ability of a wood to withstand blows is called its toughness. Hardwoods are much more shock-resistant than softwoods. This characteristic is particularly important in woods used for tool handles.

Surface Characteristics The appearance of the wood's surface is sometimes the determining factor in its use. The pattern of the grain greatly influences the staining characteristics and beauty of the finished product. The number and type of knots and pitch pockets are surface characteristics that need to be considered when selecting lumber for a project.

WOOD LOT MANAGEMENT

The proper management of a wood lot involves more than just the harvest of trees and the removal of unwanted species. A *wood lot* is a small, privately owned forest. The production of trees for harvest is a long-term investment and mistakes in management take a long time to correct.

Some of the factors that need to be considered in the management of wood lots include soil, water, light, type of trees, condition of trees, markets available, methods of harvesting, and replanting. Using scientific methods in the management of forests is called *silviculture*.

Restocking the Wood Lot The least expensive method of replacing trees harvested from the wood lot is natural seeding. Sources of seed for the desired species must be available in the forest and conditions must be right for seed germination in order for natural seeding to take place. If seed from natural sources is not available, seeds from other sources may be planted on the forest site.

A surer method of restocking the wood lot is to plant trees of the desired species rather than rely on seeds to do the reforestation. In most cases, *seedlings* (young trees started from seeds) are planted during late winter and early spring, before the new season's growth begins. Wood lots can be planted with one species of tree or a mixture of several compatible species.

Management of the Growing Wood Lot

Management of a wood lot is much more involved than just sitting back and watching it grow. Proper care and management is important if the forestry enterprise is to be successful.

Trees that are of no commercial value should be removed as soon as possible to eliminate competition for light, moisture, and nutrients. Because these "weed" trees are removed when they are small, there is seldom a market for them and they are left on the wood-lot floor to decay. If weed trees are of sufficient size, however, they may be used for firewood.

When all the trees of a wood lot are nearly the same age, they often need to be thinned at 15–30 years of age. Trees should be thinned any time that the crowns or branches occupy less than one-third of the height of the trees. Usually, about one-fourth to one-third of the trees in the wood lot are removed during thinning.

Trees that are being grown for lumber are often pruned of side branches to produce a better quality log. Prompt removal of side branches helps keep logs free of knots. Only rapidly growing trees should be pruned. Branches should be pruned flush with the trunk of the tree. Pruning is usually done during the fall and winter, when the trees are dormant.

Planning a Harvest Cutting

A wood lot should never be harvested without a plan. A harvesting plan can maximize the income from the wood lot over many years. The use of a forester's services is usually wise in developing a harvesting plan. A *forester* is a person who studies and manages forests.

There are several systems of harvesting a wood lot. They include clearcutting, seed-tree cutting, shelterwood cutting, diameter limit cutting, and selection cutting.

Clearcutting

This is a system of cutting timber where all of the trees in area are removed. It was used extensively in cutting the *virgin* forests (one that has never been harvested) of the United States when there was little thought of reforestation of the cut areas.

Clearcutting is usually done in small patches, ranging in size from ½ acre to 50 acres (Figure 10-8). Prompt reforestation with desirable species of trees on the clearcut areas is essential. This will prevent erosion and the loss of fragile forest soil.

Seed-Tree Cutting

Harvesting trees by the seed-tree cutting method is very similar to clearcutting. The primary difference is that enough seed-bearing trees are left uncut to provide seeds for reforestation (Figure 10-9). The trees left to produce seeds should be very representative of the desired species and free from insect, disease, or mechanical damage.

The removal of the seed trees usually takes place after the cut area is repopulated with desirable seedlings.

Shelterwood Cutting

In shelterwood cutting, enough trees are left standing after harvesting to provide for reseeding the wood lot. They will also protect the area until the young trees are well established. After the young trees become established, the residual trees are harvested.

Diameter Limit Cutting

When harvesting trees by the diameter limit method, all trees above a certain diameter are cut. Slow growing or diseased trees are often left standing when harvesting is done using this method. Rapid growing trees of desirable species may also be cut before they reach their production potential. Often this leaves a wood lot with only slow growing trees that may be undesirable. Diameter limit cutting is used advantageously to remove trees left from previous harvesting, which reduces competition with younger trees.

CLEAR CUT
(NO TREES)

FIGURE 10-8
Clearcutting is generally done in small patches. ***(From Camp/Managing Our Natural Resources, copyright 1988 by Delmar Publishers Inc.)***

FIGURE 10-9
Seed-tree cutting leaves some trees to produce seed for reforestation after harvest. ***(From Camp/Managing Our Natural Resources, copyright 1988 by Delmar Publishers Inc.)***

Selection Cutting This is usually the method of harvest used when the trees in the wood lot are of different ages. Selecting the trees for harvest can be a difficult task for the average person and obtaining the services of an experienced forester is usually wise. This system of harvesting allows for fairly frequent income and maintains the wood lot's esthetic value.

Protecting the Wood Lot The wood lot must be protected from fire, pests, and domestic animals if it is to yield up to its potential. It is estimated that more than 13 billion board feet of timber are destroyed each year by pests. This represents nearly 25% of the estimated net growth of forests and wood lots each year.

Fire Millions of dollars worth of timber each year are destroyed by fire. In addition to actually killing trees, fire slows the growth of others, and damages some so that insects and diseases may attack them. Fire also burns the organic matter on the forest floor, which takes nutrients away from the trees and exposes the soil to erosion.

To help prevent fires, debris should be removed from around trees. Weeds, brush, and other trash around the edges of the wood lot should be removed. Not allowing human use, such as hunting and fishing, during dry periods may also reduce the potential for fire. The construction of permanent fire breaks are useful for fire control.

A plan of how to deal with a fire is important in minimizing damage should a fire occur. State and county foresters can assist in developing fire prevention and control plans. These include planning for water storage, procedures for notifying proper authorities, and obtaining appropriate equipment and help.

Pests Insects and diseases cause more damage to existing forests than fires. Pests cause trees to be weak and deformed. They consume leaves, damage bark, and retard the growth of trees. They kill billions of trees each year. However, control of diseases and insects in forest lands is difficult and expensive.

Removal of dead, damaged, or weak trees may help reduce disease and insect problems. Not only are weak trees attacked, but healthy ones are sometimes favorite targets of pests. Prompt action in dealing with outbreaks of insects and diseases will do much to insure a profit at harvest time.

Domestic Animals The grazing of wood lots by cattle and sheep usually results in the destruction of all small seedlings in the forest. It also eliminates most of the wood-lot floor coverage. Livestock may also strip the bark from trees, causing them to die. Wood lots do not provide much food for grazing animals, so it is still wise to exclude livestock from them.

Harvesting the Wood Lot Before timber is harvested, a market for it must be found. Various methods of marketing the timber should be explored to determine the most profitable alternatives.

The actual harvest of the wood lot can be done by the owner or by a contractor skilled in timber harvest. Another alternative is to sell the standing timber to a company that later harvests and markets it.

Regardless of how the timber is harvested, it is extremely important that contracts covering all pertinent information be drawn up and signed by all parties involved.

SEASONING LUMBER

The proper seasoning of lumber is essential in order to protect it from damage. As soon as a tree is cut, it starts to lose moisture. If the wood dries too slowly, it may be subject to rot, stain, or insect damage. If it is allowed to dry too quickly, lumber may twist and warp or split. This may make it unsuitable for most uses.

Wood that is sawed for lumber should be stacked immediately after sawing to allow for even drying (Figure 10-10). The stacking should take place on a level, well-drained location. It should be off the ground and stacked so that air can circulate freely around each board. The stacked lumber should also be protected from weather. The amount of time required for lumber to dry depends on the thickness of the lumber and the species of tree from which it was cut (Figure 10-11).

The forestry industry in the United States produces about 16 billion cubic feet of wood products each year. The demand for wood products is almost 19 billion cubic feet per year. Only by carefully managing our forestry resources can the United States fulfill its need for wood and wood products in the future.

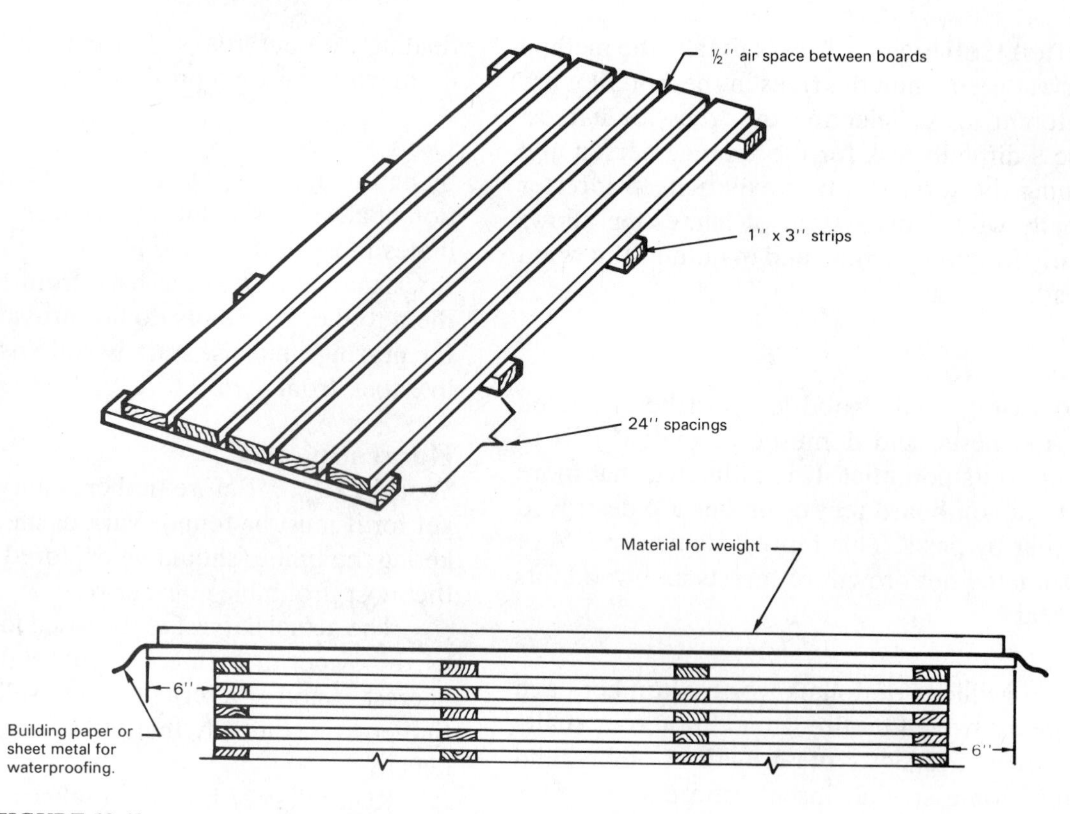

FIGURE 10-10
Freshly sawed lumber must be properly stacked to prevent warping and end splitting as the lumber dries.

Species	Number of Days
American Beech	70 to 200
Ash, White	60 to 200
Aspen	50 to 150
Basswood	40 to 150
Birch, Yellow, Sweet	70 to 200
Butternut	60 to 200
Cedar, Eastern Red	60 to 90
Cedar, Northern White	80 to 130
Cherry, Black	70 to 200
Cottonwood, Eastern	50 to 150
Fir, Balsam	150 to 200
Douglas-fir, Interior North	20 to 180
Hemlock, Eastern	90 to 200
Hickory	60 to 200
Locust, Black	120 to 180
Maple, Hard (Sugar, Black)	50 to 200
Maple, Soft (Red, Silver)	30 to 120
Oak, Red	70 to 200
Oak, White	80 to 250
Pine, Eastern White	60 to 200
Pine, Red (Norway)	40 to 200
Pine, Southern Yellow	30 to 150
Yellow-Poplar	40 to 150
Spruce, Red and White	30 to 120
Sycamore	30 to 150
Walnut	70 to 200

FIGURE 10-11
Time required for lumber of various trees to reach an air-dry condition

Student Activities

1. Define the "Terms To Know."
2. Make a collection of forest products.
3. Make a collection of tree leaves, bark, or twigs that are of economic importance in your area.
4. Make a bulletin board showing the many uses of forests and forest products.
5. Write a report on a species of trees or a type of forest product of interest to you.
6. Visit a local forest and identify the types of trees growing there.
7. Have a forester visit the class and speak about forest management.
8. Visit a wood-processing plant operation to learn how trees are processed into other products.
9. Write a report on a career in the forestry or wood products industry.
10. Study Figure 10-2 and learn how to calculate board feet (BF). Determine the number of board feet in the following:

    a. A board that is 1″ × 12″ × 12′. BF = ______
    b. A plank that is 2″ × 8″ × 16′. BF = ______
    c. A 4″ × 4″ × 8′ board. BF = ______
    d. Six 2″ × 4″ × 8′ boards. BF = ______
    e. Twenty 2″ × 10″ × 16′ boards. BF = ______

11. What is the total board feet in number 10? BF = ______
12. If the items in number 10 were rough sawed lumber and selling at $.20 per board foot, what would be the cost of the total order? $______

Self-Evaluation

A. MULTIPLE CHOICE

1. Trees that are used for making paper are called

    a. timber.
    b. lumber.
    c. pulpwood.
    d. veneer.

2. The scientific management of forests is

    a. silviculture.
    b. aboriculture.
    c. pomology.
    d. olericulture.

3. There are about ______ acres of productive forests in the continental United States.

    a. 105 million
    b. 235 million
    c. 500 million
    d. 735 million

4. The most important commercial species of trees in the United States is

    a. oak.
    b. Douglas fir.
    c. redwood.
    d. walnut.

5. About 75% of the wood for plywood is harvested from the ______ forest region.

   a. Northern
   b. Central
   c. Alaskan interior
   d. Pacific Coast

6. A forest that has never been harvested is called

   a. virgin.
   b. hardwood.
   c. clearcut.
   d. seedling.

7. The seed-tree method of harvesting

   a. cuts all trees over a certain diameter.
   b. cuts all trees under a certain diameter.
   c. cuts about one-third of the trees in the wood lot.
   d. cuts all but a few trees left for seed.

8. Which of the following was not stated as a property of wood?

   a. nail-holding ability
   b. bending strength
   c. color
   d. surface characteristics

9. The yearly demand for wood products in the United States is about ______ cubic feet.

   a. 19 billion
   b. 23 billion
   c. 190 billion
   d. 253 million

10. In a wood lot with even-aged trees, it usually needs to be thinned when the trees are ______ years old.

   a. 1–5
   b. 5–10
   c. 10–15
   d. 15–30

B. MATCHING

______ 1. Tree
______ 2. Shrub
______ 3. Wood lot
______ 4. Veneer
______ 5. Plywood
______ 6. Lumber
______ 7. Forestry
______ 8. Softwood
______ 9. Hardwood
______ 10. Conifer

a. Thin layers of wood, glued together
b. Wood from conifers
c. Small, multistemmed plant
d. Management of wooded land
e. Wood from deciduous trees
f. Thin slices of wood
g. Small forest
h. Tree with needle-like leaves
i. Woody, single-stem plant
j. Boards sawed from trees

C. COMPLETION

1. The air drying of lumber is referred to as ______ .
2. Nearly ______ % of the estimated net growth of forests each year is lost to fire and pests.
3. ______ is a type of harvesting where every tree over a certain size is cut.
4. Oak is a type of ______ or ______ tree.
5. Douglas fir is an example of an ______ or ______ tree.
6. Cutting all of the trees in an area is called ______ .

7. The ______ forest region is located along the Mississippi River.

8. The forest region with the most potential for meeting future needs for forest products is the ______ forest region.

UNIT 11 Wildlife Management

OBJECTIVE To determine the relationships between wildlife and the environment and approved practices in managing wildlife enterprises.

Competencies to Be Developed

After studying this unit, you will be able to

- ☐ define wildlife terms.
- ☐ identify characteristics of wildlife.
- ☐ describe relationships between types of wildlife.
- ☐ understand the relationships between wildlife and humans.
- ☐ describe classifications of wildlife management.
- ☐ identify approved practices in wildlife management.
- ☐ discuss the future of wildlife in the United States.

TERMS TO KNOW

Wildlife
Habitat
Vertebrates
Predators
Prey
Parasitism
Warm-blooded animals
Mutualism
Predation
Commensalism
Competition
Wetlands

MATERIALS LIST

bulletin board materials
reference materials on wildlife, wildlife management, and pollution

Wildlife has been part of the life of humans since the beginning of time. *Wildlife* are animals that are adapted to live in a natural environment without the help of humans. Early humans followed herds of wild animals and killed what they needed in order to live and survive. They observed what the animals did and what they ate to determine what was safe for human consumption. Early humans also used wildlife as models for their artwork and in many of their ceremonial rites.

As settlers came to the new world, wildlife often provided the bulk of the food available until food production systems could be developed. Supplies of wildlife seemed to be inexhaustible as the skies were blackened with the flight of millions of passenger pigeons, and herds of bison created dust storms as they migrated on the vast prairie.

Unfortunately, supplies of wildlife were and are not unlimited. Humans have destroyed wildlife *habitat* (the area where a plant or animal normally lives and grows). Humans have polluted the air and water supplies, killed wildlife in tremendous numbers, and generally disregarded the needs of wildlife. As a result, many species of wildlife in the United States can benefit greatly from proper management.

CHARACTERISTICS OF WILDLIFE

All *vertebrate* animals (animals with backbones) are included in the classification of wildlife. They have many of the same characteristics as humans. Growth processes, laws of heredity, and general cell structure are all characteristics that are common to both humans and animals. When populations become too dense, disease outbreaks occur, populations suffer from starvation, and disposal of waste becomes a problem.

The wildness of the animal itself is a characteristic that allows the animal to survive without interference or help from humans. The animal's wildness often contributes to the interest that humans have in wildlife. Characteristics identified as wildness are what attract hunters to hunting and fishermen to fishing. Bird watching and wildlife photography would be far less fascinating if wildlife were less wild and wary of humans.

With few exceptions, wildlife lives in an environment over which it has no control. Wildlife must be able to adapt to whatever it is presented with in terms of food and environment, or it will perish. It must also possess natural senses that allow it to avoid predators and other dangers. *Predators* are animals that feed on other animals. The animal being eaten by the predator is the *prey.*

The ability to avoid over-population is a characteristic of many groups of wildlife. Establishing and defending territories is one way that wildlife may naturally avoid overpopulation. The stress of overpopulation causes some animals to slow down or stop reproducing altogether.

WILDLIFE RELATIONSHIPS

Every type of wildlife is part of a community of plants and animals where all individuals are dependent on each other. Any attempt to manage wildlife must take into account the relationships that exist naturally. Because relationships within the wildlife community are constantly changing, it is very difficult to set standard procedures for their management.

The balance of nature is actually a myth because wildlife communities are seldom in a state of equilibrium. The numbers of various species of wildlife are constantly increasing and decreasing in response to each other and to many external factors such as natural disasters. These include fires, droughts, and disease outbreaks. The interference of humans also often upsets the natural balance of nature.

Some of the natural relationships that exist in the wildlife community include parasitism, mutualism, predation, commensalism, and competition.

Parasitism

The relationship between two organisms, either plants or animals, in which one feeds on the other without killing it is called *parasitism.* Parasites may be either internal or external. An example of a parasitic relationship is the wood tick, which may live on almost any warm-blooded animal. *Warm-blooded animals* have the ability to regulate their body temperature.

Mutualism

Mutualism refers to two types of animals that live together for mutual benefit. There are many examples of mutualism in the wildlife community. Tick pickers are birds that remove and eat ticks from many of the wild animals in Africa, to the mutual benefit of both. The wild animals have parasites removed from them and the birds receive nourishment from the ticks. A moth that lives only on a certain plant is also the only pollinator of that plant in several relationships. Some plant seeds will germinate only after having passed through the digestive tract of some bird or animal.

FIGURE 11-1
Bobcats are predators that feed on small rodents and birds over a wide range of the United States. ***(Courtesy of U.S. Fish and Wildlife Service)***

Predation

When one animal eats another animal, the relationship is called *predation* (Figure 11-1). Predators are often very important in controlling populations of wildlife. Foxes are necessary to keep populations of rodents and other small animals under control. Populations of predators and prey tend to fluctuate widely. When predators are in abundance, prey becomes scarce because of overfeeding. When prey becomes scarce, predators may starve or move to other areas, which allows for great increases in the populations of prey.

Commensalism

Commensalism refers to a plant or animal that lives in, on, or with another, sharing its food, but not helping or harming it (Figure 11-2). One species is helped but the other is neither helped nor harmed. Vultures waiting to feed on the leftovers from a cougar's kill is an example of commensalism.

Competition

When different species of wildlife compete for the same food supply, cover, nesting sites, or breeding sites, *competition* exists. When competition exists, one species may increase in numbers while the other declines. Often the numbers of both species decrease as a result of

FIGURE 11-2
Commensalism is a wildlife relationship which benefits one species and neither helps nor harms the other. *(Courtesy of U.S. Bureau of Sport Fisheries & Wildlife)*

FIGURE 11-3
Competition for the same food supply helps keep predators under control. *(Courtesy of U.S. Fish and Wildlife Service)*

competition. Owls and foxes both competing for the available supply of rodents and other small animals is an example of competition (Figure 11-3).

The various relationships that exist between species of wildlife make it necessary to consider more than just one species any time that management is contemplated. Understanding the relationships that exist in the entire wildlife community is essential if wildlife management programs are to be successful.

RELATIONSHIPS BETWEEN HUMANS AND WILDLIFE These relationships may be biological, ecological, or economic. Biological relationships exist because humans are animals that are very similar to wildlife. Relationships may be ecological because humans are but one species among nearly 1 million species of animals that inhabit the planet Earth.

The economic relationships that exist between humans and wildlife are important. Originally humans were dependent on wildlife for food, clothing, and shelter. Today, there are six positive values of wildlife's relationship with humans—commercial, recreational, biological, aesthetic, scientific, and social.

The harvesting and sale of wildlife and/or wildlife products is an example of commercial relationships between humans and wildlife. Raising wild animals for use in hunting, fishing, or other purposes also falls into this category.

Hunting and fishing, as well as watching and photographing wildlife, are included in the category of recreational relationships (Figure 11-4). Although it is estimated that more than $2 billion is spent each year on hunting and fishing and at least another $2 billion on other recreational uses of wildlife, many of the recreational values of wildlife are intangible.

The biological values of human and wildlife relationships are often difficult to measure. Some examples of biological relationships include pollination of crops, soil improvement, water conservation, and control of harmful diseases and parasites (Figure 11-5).

Aesthetic values refer to beauty. Watching a butterfly sipping nectar from a flower, a fawn grazing beside its mother, or a trout rising to a hatch of mayflies are all examples of the aesthetic values of wildlife. Wildlife also provides the inspiration for much artwork. Even though the aesthetic values of wildlife are not measurable in economic terms, they often contribute much to the mental well-being of the human race.

Using wildlife for scientific studies often benefits humans. Scientific relationships between humans and wildlife have existed from the beginning of time, when early humans watched wild animals to determine which plants and berries were safe to eat.

Social values of the relationship between humans and wildlife are also difficult to measure. Wildlife has the ability to enhance the value of

FIGURE 11-4
Hunting is a very popular recreational use of wildlife. ***(Courtesy of U.S. Bureau of Sport Fisheries & Wildlife)***

their surroundings simply by their presence. They provide the opportunity for variety in outdoor recreation, hobbies, and adventure. They also make leisure time much more enjoyable.

CLASSIFICATIONS OF WILDLIFE MANAGEMENT

Wildlife management can be divided into several classifications for ease in developing management plans. Techniques for management vary tremendously according to classification. Some wildlife management classifications are farm, forest, wetlands, stream, and lakes and ponds.

The management of farm wildlife is probably the most visible wildlife management classification. The development of fence rows, minimum tillage practices, improvement of woodlots, and controlled hunting are all techniques that have long been used to manage farm wildlife. Rabbits, quail, pheasants, doves, and deer are the types of wildlife most normally managed in this category.

Forest wildlife is more difficult to manage. Plans must be developed so that timber and wildlife can exist at desired populations and possibly be harvested. Management of forest wildlife may include population controls to prevent destruction of habitat. Deer, grouse, squirrels, and rabbits are usually wildlife species included in forestry wildlife management programs.

The most productive wildlife management areas are wetlands. *Wetlands* include all areas

FIGURE 11-5
Biological values of wildlife and human relationships include the pollination of crops by honeybees. ***(Courtesy of Honey Bee Research, Weslaco, TX)***

between dry upland and open water. Marshes, swamps, and bogs are all wetland areas. Because these areas are very sensitive to changes in environmental conditions, careful management of them for wildlife is essential. The wetlands are home for ducks, geese, muskrats, raccoons, deer, pheasants, grouse, woodcock, fish, frogs, and thousands of other species of wildlife.

The management of running water or streams is often a difficult task. Water pollution and the need for clean water for a growing human population continue to increase at a rapid pace. Potential damage to the wildlife in streams from chemical pollution, the building of dams, roads, home construction, and the drainage of swampland are critical considerations for the stream wildlife manager.

Management of wildlife in lakes and ponds is normally somewhat easier than it is in streams because water is standing rather than running. Population levels of pond wildlife, oxygen levels, pollutants, and the availability of food resources are all concerns of the pond and lakes wildlife manager.

APPROVED PRACTICES IN WILDLIFE MANAGEMENT

Farm Wildlife Management of wildlife on most farms is usually a byproduct of farming or ranching. It is often given little attention by the farmer or rancher except when wildlife causes crop damage and financial losses.

Much of the management of farm wildlife involves providing suitable habitat for them to live, grow, and reproduce. This may involve leaving some unharvested areas in the corners of fields, planting fence rows with shrubs and grasses that provide winter feed and cover, or leaving brush piles when harvesting woodlots.

AGRI-PROFILE

CAREER AREAS: WILDLIFE BIOLOGIST/ MANAGER/TECHNICIAN

Habitat analysis and game counts are techniques used by wildlife specialists to advise government agencies on setting hunting and fishing limits. ***(Courtesy USDA Agricultural Research Service)***

Wildlife biologists work with fish and game species of land, fresh water streams and lakes, tidal marshes, bays, seas, and oceans. Wildlife biologists generally have master's or doctor's degrees in biology. They tend to utilize the basic sciences in their work.

Wildlife managers typically have associate or bachelor's degrees. They work in government agencies advising land owners and managing game populations on public lands. Their work frequently requires the use of helicopters, small planes, snowmobiles, all terrain vehicles, horses, and land rovers, as well as time in the wild on foot.

Wildlife officers interact continuously with the hunting and fishing community. They advise governments in establishing fish and game laws and programs for habitat improvement. Wildlife officers have the backing of strict laws and stiff penalties for offenders. However, much of their time is spent on educating the public and obtaining private assistance in improving habitat and maintaining game populations.

The timing of various farming operations is also important in a farm wildlife management program. When possible, mowing should be delayed until after pheasant eggs have hatched in the spring. Crop residues should be left standing over the winter to provide food and cover. Planting crops attractive to wildlife on areas that are less desirable as cropland is an excellent farm wildlife management practice. Providing water supplies for wildlife during dry periods is often necessary to maximize the numbers of farm wildlife on the area being managed.

Harvesting farm and ranch wildlife by hunting has been shown, by extensive research, to have little impact on spring breeding populations (Figure 11-6). Excess populations of farm and ranch wildlife that are not harvested by humans usually die during the winter. Even heavy hunting pressures seldom result in severe damage to wildlife populations. The sale of hunting rights to hunters is a way to increase the income on many farms and ranches. In addition, it often means the difference between profit and loss in the farming enterprise.

Management of wildlife on game preserves or farms set up specifically for hunting often differs drastically from other wildlife management programs. Often species of animals and birds that are not native to the area are raised and released on the preserve. Native wildlife species may also be raised in pens and released to the farm or preserve expressly for harvest by hunters.

Forest Wildlife The types and numbers of forest wildlife in any specific woodland are dependent on many factors. They include type and age of the trees in the forest, density of the trees, natural

FIGURE 11-6

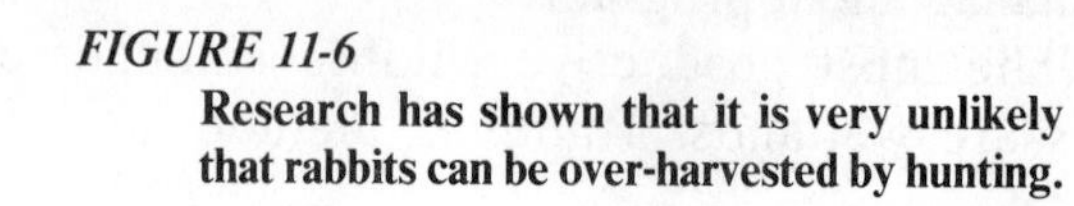

Research has shown that it is very unlikely that rabbits can be over-harvested by hunting.

forest openings, types of vegetation on the forest floor, and the presence of natural predators.

Management of forest wildlife is usually geared toward increasing numbers of desired species of wildlife. If desired populations of wildlife are present, the management goal is usually to maintain those populations. Sometimes, numbers of certain species of forest wildlife increase to the point where destruction of habitat occurs. When this happens, control measures may have to be instituted to restore proper balances.

The steps in developing a forest wildlife management plan should include taking an inventory of the types and numbers of wildlife living in the forest area to be managed. Goals for the use of the forest and the wildlife living in it need to be developed. The third step in the development of a forest wildlife management plan is determining the types and populations of wildlife that the forest area can support and how best to manage the forest so that required habitat is provided.

The requirements for forest wildlife include food, water, and cover. These necessities must be readily available to the desired species of forest wildlife at all times. Management practices that meet these requirements include making clearings in the forest so that new growth will make twigs available to deer to feed on. They also include selective harvesting so that trees of various ages exist in the forest to make a more suitable habitat for squirrels and many other species of forest wildlife. Leaving piles of brush for food and cover is also a management practice that leads to increased production of forest wildlife. Care in managing harvests of forest products so that existing supplies of water are not contaminated is also important in good wildlife management.

Deer, grouse, squirrels, and rabbits are the forest wildlife species that are usually targeted for management because they are valuable for recreational purposes, especially hunting (Figure 11-7). They may also be managed to prevent the destruction of valuable forest trees and other products.

It should be noted that during times of overpopulation of forest animals, especially deer, it is seldom a good idea to provide supplemental food. Natural losses should nearly always be allowed to occur, including starvation of excess animals or allowing heavier-than-normal hunting pressures. Artificial feeding of wildlife populations usually results in further population increases and an extension of the problem.

Wetlands Wildlife No area of American land is more important to wildlife than the wetlands. Wetlands include any land that is poorly drained—swamps, bogs, marshes, and even shallow areas of standing water (Figure 11-8). The wetlands are constantly changing as wet areas fill in with mud and decaying vegetation. They eventually become dry land that contains forests.

Wetlands provide food, nesting sites, and cover for many species of wildlife. Ducks and geese are probably the most economically important type of wildlife that needs the wetlands for survival (Figure 11-9). Other types of wildlife found in the

FIGURE 11-7
Management of deer is an important part of the forest wildlife manager's job. *(Courtesy USDA)*

FIGURE 11-8
American wetlands are home to hundreds of species of wildlife. *(Courtesy Denmark High School Agriculture Education Department)*

FIGURE 11-9
Wetlands are important nesting areas for millions of ducks and geese in the United States. *(Courtesy USDA)*

wetlands area include woodcock, pheasants, deer, bears, mink, muskrats, raccoons, and thousands of other lesser known species.

Management of wetlands for wildlife may include impounding water. Open water areas should occupy about one-third of the wetlands for optimum use by wildlife. The depth of the standing water should not be more than about 18″.

The management of the plant life in the wetlands is also important. This may include cutting trees to open up the wetland area. Many species of wildlife require large open areas in order to thrive. Care must be taken not to remove hollow trees that are used as nesting sites for some species of wildlife. Wetland areas can also be opened up by killing excess trees rather than cutting them. This provides resting areas for many types of wetlands wildlife.

Establishing open, grassy areas around wetlands also helps to attract many types of wildlife to the area. Planting millet, wild rice, and other aquatic types of plants in the wetlands also helps to attract wildlife.

A serious hazard to wetlands wildlife is pollution. Pollution of water flowing into the wetlands area may come from agriculture, industry, or the disposal of domestic wastes. Because pollutants are trapped in the mud and silt of the wetlands, the effects of pollutants are often long-lasting.

In areas lacking natural nesting sites, populations of some wildlife species can be greatly increased by providing artificial nesting sites. Wood duck boxes, old tires, and islands surrounded by open water provide safe nesting sites for many species of wetland wildlife.

FIGURE 11-10
Trout are the principle species of cold water fish of economic importance in streams. *(Courtesy U.S. Department of the Interior, Fish and Wildlife Service)*

Raising certain species of ducks in captivity and later releasing them in wetlands areas has been important in maintaining viable duck populations. This has been important as more and more natural duck nesting areas have been destroyed to meet the needs of the human race.

Stream Wildlife This type of wildlife can be divided into two general categories—warm-water and cold-water. These categories are based on the water temperatures at which the wildlife, primarily fish, can best grow and thrive. There is little or no difference in the management practices between the two categories of wildlife. In general, fish are the type of stream wildlife that is managed, although many other types of wildlife depend on streams for their existence (Figure 11-10).

As land is developed, forests are harvested, and civilization is expanded, streams and their wildlife populations come under increasing pressure. Because we cannot build new streams, it is essential that existing ones be managed properly.

Management practices for streams include preventing stream banks from being overgrazed by livestock. Fencing the stream to limit access by livestock is also wise to reduce pollution and the destruction of the stream banks.

Good erosion control practices on lands surrounding streams is important to help maintain clear, clean water. It is also important to prevent chemical pollutants from entering streams.

Maintaining stream-side forestation is important in regulating stream temperatures during the warm summer months. Some species of fish stop feeding and may even die when stream temperatures become too high. The amount of dissolved oxygen in warm water is also much lower than it is in cold water. Without that oxygen, aquatic wildlife dies.

Anything that impedes the flow of the stream also serves to change it, to the detriment of many species of stream wildlife and the benefit of some others. Trout must have swiftly moving, cool water in which to thrive, while catfish are perfectly happy in sluggish streams (Figure 11-11).

Care must be taken to maintain desirable species of wildlife in the stream. Introducing new species of wildlife in the stream may result in reducing native wildlife already in the stream.

FIGURE 11-11
Catfish are perfectly happy in a slow-moving stream and may destroy the habitat of other game fish. ***(Courtesy U.S. Department of the Interior, Fish and Wildlife Service)***

The maintenance of population levels of stream wildlife that are in balance with the available food supply is important. Too many fish for the available food supply normally results in stunted fish of no value to fishermen. This situation does provide an increased food supply for some types of birds and animals that use streams for their food supply. Overfishing of predatory species of fish such as bass or northern pike may allow perch or sunfish to overpopulate the stream and become stunted. Often the only way to restore streams to a desired mix of fish species is to remove the unwanted species. This is accomplished by netting, poisoning, or electric shocking. These techniques are generally legal only for authorized and specially trained personnel.

The artificial rearing and stocking of desired species of stream wildlife is a management practice that is important in many streams. Typically, game species of fish are stocked in streams for fishermen to catch and remove. Often few or no fish survive to reproduce, and stocking must take place each year.

The regulation of sport fishing is often necessary to maintain desirable populations of game fish. This may include closed seasons, minimum size limits, creel limits, and restricted methods of catching fish.

Lakes and Ponds

Wildlife The management practices for wildlife in lakes and ponds are usually very similar to those for managing streams wildlife. Pollution must be controlled. Wildlife populations must be managed to maintain desired mixes of species. Harvest and use must also be controlled to ensure wildlife for the future.

There are some differences between management of wildlife in streams and management of wildlife in ponds and lakes. Because the water in lakes and ponds is normally standing, the amount of oxygen available for aquatic life sometimes becomes critical in the hot summer months. In small ponds, artificial means of incorporating oxygen into the water may be used to prevent fish kills.

Water temperatures in lakes and ponds are more variable than they are in streams. This means that different species of fish are usually dominant in ponds and lakes. In many ponds and lakes, fish populations are predominantly largemouth bass and sunfish (Figures 11-12 and 11-13).

FIGURE 11-12
Largemouth bass are popular ponds and lakes game fish over much of the United States. ***(Courtesy American Fisheries Society, photo by John Scarola)***

FIGURE 11-13
Sunfish provide food for bass and much sport fishing in ponds and lakes. ***(Courtesy U.S. Fish and Wildlife Service)***

When it is necessary or desirable to rid a pond or lake of unwanted species of fish such as carp, it is often much easier to do so because the water is contained (Figure 11-14). Sometimes it is possible to drain the body of water to remove the unwanted species of wildlife. More often, the body of water is simply poisoned so that all fish and other species of pond wildlife are killed. The pond is then restocked with desirable wildlife species.

The management of wildlife is a very imprecise business. Often specific species of wildlife are managed to the detriment of others. It is reasonably clear that any human interference in the wildlife community results in changes that are not always to the benefit of much of that community.

FIGURE 11-14
Carp are often an unwanted species of wildlife in lakes and ponds because they stir up the bottom while searching for food. Many desirable species of fish do not survive well in muddy water. ***(Courtesy U.S. Fish and Wildlife Service)***

THE FUTURE OF WILDLIFE IN THE UNITED STATES

A bright future for all of the wildlife species is not assured in the United States because the needs of the human population continue to compete with those of the wildlife community. The outlook for wildlife in the future is not all bleak, however. Humans have recognized that it is possible to satisfy the needs of wildlife and the demands of humans if careful management practices are instituted.

Careful studies of how humans and wildlife can peacefully co-exist are being made. There are sincere attempts being made to reduce the pollution of the environment. Polluted areas are being cleaned up. More extensive testing of new chemicals and other pesticides and their environmental effects is being conducted. The effects of new construction on wildlife habitat is also studied before that construction begins.

Establishing large acreages in national parks and wildlife refuges is also important for the future well-being of the wildlife in the United States. More emphasis on management of wildlife resources, rather than simply exploitation, also bodes well for the future of the country's wildlife.

Realistically, some species of wildlife will decline and cease to exist in the future, while other species will proliferate. This has been the case in the past and will continue to be so in the future, with or without the interference or help of the human population.

Student Activities

1. Define the "Terms To Know."
2. Visit a wildlife area and identify the species of wildlife that you see there.
3. Construct a bulletin board showing the species of wildlife that are important to hunters and fishermen in your area.
4. Have a wildlife manager visit your class and explain management practices that he or she uses in the area that he or she manages.
5. Write a report on the effects of pollution on wildlife in some area of the United States.
6. Participate in a New Year's Day bird count.
7. Plant a feed patch area for wildlife.
8. Participate in a stream or other wildlife area cleanup program.
9. Visit a zoo and list the species of wildlife that are endangered.
10. Develop a list of endangered species of wildlife in your area and what is being done to prevent their extinction.
11. Write a report on a species of wildlife that interests you.
12. Have a local farmer or rancher speak to the class on what measures he or she is taking to prevent polluting the environment of wildlife.
13. Invite a birdwatcher or wildlife photographer to discuss their hobby or profession with the class.

Self-Evaluation

A. MULTIPLE CHOICE

1. ______ is a species of fish adapted to cold, running water in streams.

   a. Carp
   b. Catfish
   c. Trout
   d. Sunfish

2. Forest wildlife survive best in forests that are

   a. of mixed-age trees.
   b. deciduous.
   c. evergreen.
   d. of even-age trees.

3. Trees growing along streams help to

   a. regulate water flow.
   b. provide food for aquatic wildlife.
   c. provide oxygen for fish.
   d. regulate stream temperatures.

4. The most important wildlife management area is the

   a. farm.
   b. stream.
   c. wetlands.
   d. forest.

5. When two species of wildlife live together for the benefit of both, the relationship is called

   a. mutualism.
   b. predation.
   c. commensalism.
   d. competition.

6. The raising and sale of wildlife for hunting is an example of what type of relationship between humans and wildlife?

   a. biological
   b. recreational
   c. social
   d. commercial

7. One species of fish that is often undesirable in lakes and ponds is the

   a. bass.
   b. carp.
   c. sunfish.
   d. trout.

8. Rabbits, quail, pheasants, doves, and deer are wildlife often targeted for management in

   a. farm areas.
   b. forests.
   c. wetlands.
   d. none of the above.

9. Management of deer is a concern of ______ wildlife managers.

   a. forest
   b. farm
   c. wetlands
   d. all of the above

10. Wetlands should be made up of about ______ shallow, standing water for optimum wildlife use.

    a. one-fourth
    b. one-third
    c. one-half
    d. two-thirds

B. MATCHING

______ **1.** Predation
______ **2.** Parasitism
______ **3.** Mutualism
______ **4.** Commensalism
______ **5.** Competition

a. One type of wildlife living and feeding on another type
b. Two types of wildlife eating the same food
c. One type of wildlife eating another type of wildlife
d. One type of wildlife living in, on, or with another type but without helping or harming it
e. Two types of wildlife living together for the benefit of both

C. COMPLETION

1. Inspiration for artwork is a/an ______ value of wildlife.
2. Fishing is a/an ______ value of wildlife.
3. Pollination of crops is a/an ______ value of wildlife.
4. The ability of wildlife to increase the value of their surroundings simply by their presence is a/an ______ value of wildlife.
5. The observation of wildlife by early humans to determine what was safe to eat is a/an ______ value of wildlife.

UNIT 12 Aquaculture

OBJECTIVE To recognize the biological requirements necessary for the production of aquatic plants and animals.

Competencies to Be Developed

After studying this unit, you will be able to

- ☐ explain the food chain in a freshwater pond.
- ☐ discuss water quality and list eight measurable factors.
- ☐ describe three major aquaculture production systems.

TERMS TO KNOW

Aquaculture
Aquaculturist
Natural fisheries
Salinity
Gradient
Wetlands
Terrestrial
Amphibians
Bays
Estuaries
Saltwater marshes
Brackish water
Fry
Spawn
Photosynthesis
Gills
Shellfish
Crustaceans
Molt
Water quality
ppt
Dissolved oxygen
ppm
pH
Water hardness
Buffer
Turbidity
Ammonia/nitrite/nitrate
TAN
Toxins
Salmonids
Spats
Seining
Rolls over
Larvae

MATERIALS LIST

bulletin board materials
small freshwater aquarium (optional)
30 g of NaCl with 1,000 ml flask

Water covers three-quarters of the Earth's surface. This resource produces both plants and animals that are used to feed the world. *Aquaculture* is the management of this aquatic environment to increase the harvest of usable plant and animal products. The *aquaculturist* is a professional trained in aquaculture who must understand where the organism lives, and how it eats, grows, and reproduces. The manipulation of these factors determines the success of the agricultural production system, since natural fisheries have worked to their capacity to produce fish (Figure 12-1).

These production systems are part of an integrated industry that requires specialized products and services. Aquaculturists include nutritionists, feed mill operators, pathologists, fish hatchery managers, processing managers, researchers, and growers. We use these services to produce fresh and processed seafoods, shellfish, and ornamental fish and plants.

Natural fisheries occur in nature without humans. They include the rivers, oceans, continental shelves, reefs, bays, lakes, and rivers. These fisheries are currently farmed by sophisticated fishing fleets that are so efficient that yearly catches (yields) have leveled off or are decreasing. The population of the world continues to increase; therefore, aquaculture must be used to produce more aquatic plants and animals for food. Understanding the aquatic environment, the biology of the organisms, and how to control the production of aquatic plants and animals is essential.

THE AQUATIC ENVIRONMENT The oceans represent the largest expanse of water resources in the world. They represent a reservoir of water containing sol-

FIGURE 12-1
Natural fisheries have reached their capacity to produce more fish. *(Courtesy USDA)*

FIGURE 12-2
The natural water cycle

uble nutrients and materials washed from the land. Over time, the evaporation of water into the atmosphere has increased the concentration of these nutrients until the *salinity* of the water is high. We call this seawater. The concentrations of these nutrients or salts are so high that land plants and animals are unable to survive using this water for irrigation or drinking (Figure 12-2).

Rain contains only small amounts of salts. Therefore, accumulation of water on land and its flow into the seas generates a *gradient,* or measurable change over time or distance in salinity. This affects the types of organisms that can flourish.

The Salinity Gradient

Freshwater *wetlands,* such as marshes, ponds, and streams, are generated by rainwater. They accumulate the natural rainfall and provide an aquatic environment that represents the transition between aquatic and *terrestrial* for land plants and animals. *Amphibians,* such as frogs, turtles, and reptiles, are animals that live part of their lives in these fresh waters and the remaining period on land. Several plant species, such as cattails (*Typha spp.*), watercress (*Nasturtium officinale*), water spinach (*Ipomoea reptans*), and rice also require a transitional period of flooding and drainage to flourish (Figure 12-3).

As the water flows into large streams and lakes, this runoff accumulates more soluble nutrients and the salinity increases. The profile of plants and animals begins to change as other organisms are more adapted to this changed environment. Flows accumulate into rivers that empty into *bays, estuaries,* and *saltwater marshes.* Bays are open waters along coastlines where fresh water and salt water mix. Estuaries are ecological systems influenced by brackish or salty water. The saltwater marshes are lowlands influenced by tidal waters. The fresh water mixes with the seawater to create unique growing conditions for various fish, shellfish, and aquatic plants. This *brackish water* of fresh water and salt water fluctuates with the tide, flow of the rivers, and weather. Several types of seaweeds are commercially produced in the Orient. These include types of red, green, and brown algae. Similar plant growth of algae must be maintained in intensive fish systems to feed the small fry. *Fry* are small, newly hatched fish.

The migration of saltwater salmon upstream to *spawn* (lay eggs) in freshwater streams illustrates the gradient effect on the life cycle of aquatic organisms. Although adult salmon cannot survive in fresh water, the young eggs must hatch in low-salinity water and then migrate back to the ocean.

FIGURE 12-3
Freshwater wetlands serve as natural recyclers of organic wastes. ***(Courtesy USDA)***

FIGURE 12-4
The aquatic food chain in a freshwater pond

The Aquatic Food Chain The aquatic environment constantly changes to maintain a balance of organisms that function in the food chain or system. A simple illustration is the makeup of a freshwater pond. The food chain is fueled by sunlight. Green plants and algae use this energy to grow and utilize nutrients they absorb from the water. These plants are eaten by animals, which are preyed upon by larger animals (fish, reptiles, etc.) and sometimes caught by humans. These large animals return nutrients to the water as waste or carrion that are reabsorbed by plants for growth. The maintenance of this food chain supports life (Figure 12-4).

The aquaculturist must understand the affect of any management on this cycle and compensate with technology or design in an aquaculture production system.

General Biology The biology of aquatic plants and animals is very similar to that of terrestrial plants and animals. Both eat and sleep, reproduce, and interact with the environment. Green plants harvest energy from the sun through *photosynthesis* and absorb nutrients from the water to manufacture carbohydrates, proteins, and cellulose. They serve as the waste recyclers in the aquatic environment by constantly absorbing waste products (nutrients) and contributing to the food chain. Like any land plant, they respond to fertilization, shading, competition, insects, disease, and weather. Aquatic plants are composed of many parts, as discussed in Unit 15, "Plant Structures and Taxonomy." Certain parts help them compete within the aquatic environment. Green plants absorb carbon dioxide from the water and release oxygen during photosynthesis. During the night, plants reabsorb a smaller amount of oxygen and release carbon dioxide.

Aquatic animals, particularly fish, complement the relationship with plants by generating carbon dioxide during respiration and releasing soluble nutrients through waste products and decay. Figure 12-5 illustrates the anatomy of fish and shows the specialized *gills* or lungs that absorb oxygen and release carbon dioxide into the water.

In this competitive environment of predator and prey, the plants provide shelter and food that maintain the animals' life cycle.

Shellfish are aquatic animals with a shell or shell-like extensions. Nonmotile adult shellfish, including clams, mussels, oysters, and others, occupy a unique niche in the aquatic environment. Located on the bottom, these organisms have developed an efficient pumping mechanism that filters great quantities of water, filtering out edible microscopic plants and animals.

Crustaceans are a group of aquatic organisms with exoskeletons. These organisms *molt* or replace their outer shell as they grow. Saltwater lobsters (*Homarous spp.*), crawfish (*Procambarus spp.*), and the various crabs, shrimps, and prawns are important crustaceans. These mobile organisms are characterized by a hard exoskeleton that must soften and split as the animal *molts* into a larger shell.

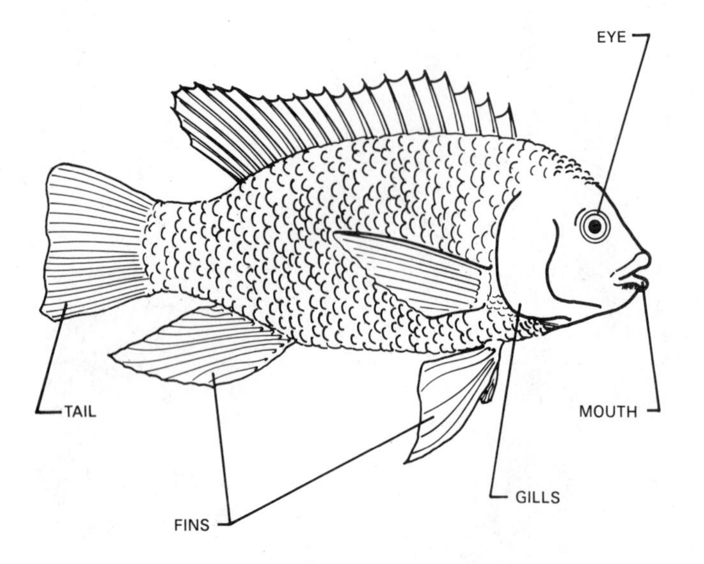

FIGURE 12-5
The anatomy of a fish

Aquaculture Production

The aquaculturist, in an attempt to increase the production of any aquatic organism, must monitor and maintain the optimum *water quality.* Water quality has several chemical and physical characteristics that interact within the water and must be measured and maintained within a narrow range to satisfy growth and development (Figure 12-6).

There are several different types of water-quality test kits available to test the following characteristics.

Salinity is the measurement of total mineral solids in water. It is measured by either electrical conductivity (ohms/cm) or is calculated against known standards and converted to *ppt* (parts per thousand). The higher the concentration of salts, the greater the conductivity reading. A seawater standard can be made with artificial sea salts or by dissolving 29.674 g of NaCl in one liter of H_2O. Measurements of salinity usually range from 0–32 ppt (fresh water to salt water).

Dissolved oxygen (oxygen in water) is a function of the temperature, pressure, and atmospheric oxygen. The cooler the water, the higher the pressure. The more contact with atmospheric oxygen, the higher and faster oxygen can be solubilized in water. Measured by oxygen probes or chemical tests, the results are reported as 0–10 *ppm* (parts per million). Water at 85° F (30° C) is saturated at about 8 ppm. Most fish can survive at levels of about 3 ppm but quickly become stressed and succumb to other problems. Rainbow trout must have

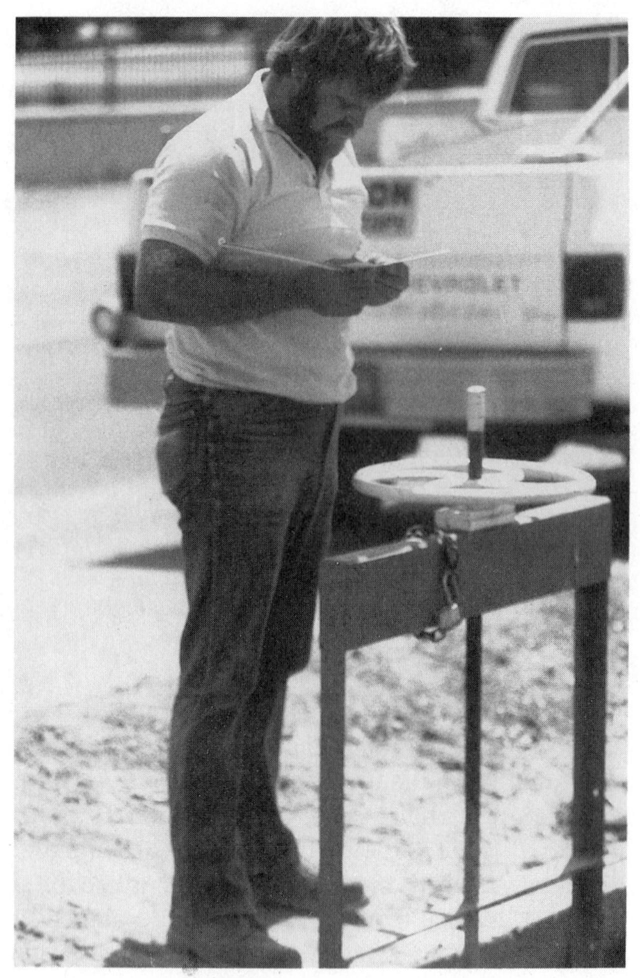

FIGURE 12-6
Water quality measurements must be taken by instruments or by chemical tests. ***(Courtesy C. W. Hardeman, Scottsboro High School Agriculture Education Department)***

excellent dissolved oxygen and can only be cultured in oxygen-saturated water.

The measurement of acidity or alkalinity in water is the *pH.* This factor affects the toxicity of soluble nutrients in the water. Using a pH meter or paper tape, this measurement is recorded as a number from 1 to 14. Numbers below 7 represent acidic solutions, 7 represents neutrality, and numbers above 7 represent alkalinity. Most aquatic plants and animals grow well in water with a pH between 7 and 8.

Water hardness is measured by chemical analysis and is expressed as ppm calcium. This element is essential in the development of the exoskeleton of shellfish and crustaceans. It also serves as a chemical *buffer* that stabilizes rapid shifts in pH.

Turbidity in water is caused by the presence of suspended matter. High turbidity limits photosynthesis and visibility. A simple method for estimating pond turbidity uses a white Secchi disc that is lowered into the water. When visibility is impaired, the depth is recorded. The greater the depth, the less turbid the water.

Temperature limits the adaptive range of almost all aquatic organisms. Sunlight warms the upper surface of open waters but does not penetrate it. Deep waters, cool-region currents, and melting winter covers can affect water temperature.

Ammonia/nitrite/nitrate is a group of nitrogen compounds generated by aquatic animals, first as urea and ammonia. These waste products are converted to nitrite by microscopic organisms in the water, and then to nitrate. They are ultimately converted to nitrogen gas or are absorbed by plants. The accumulation of both ammonia and nitrite is toxic to fish and often limits commercial production. Total ammonial nitrogen or *TAN* is recorded by chemical assay in ppm. This does not reflect the toxicity of the measured amount, since the toxicity of ammonia is dependent on the pH. Generally, levels of *TAN* are maintained below 1 ppm.

Toxins represent a host of materials that adversely affect the growth and development of aquatic plants and animals. This includes agricultural chemicals, pesticides, municipal wastes, and industrial sludges. Chemical analyses are difficult and often inconclusive.

Selection of Aquaculture Crops

The actual selection of aquatic crops that may be grown is dependent on the resources and experience of the aquaculturist. Like terrestrial crops, each species of aquatic crop has a particular set of water-quality needs to ensure survival and reproduction. A discussion of a few of the well-known aquatic crops should indicate the diversity of this commercial industry. Characteristics will also vary between species. Ideal conditions should be requested from your local county extension office and local aquaculturists.

Trout or *salmonids* are a high-quality fish product in high demand. The trout flourishes in high-quality water. Dissolved oxygen must be kept above 5 ppm. Salmonids also require low salinity, cool temperatures (15° C), and low turbidity. The TAN must be maintained at less than 0.1 ppm.

Catfish (*Ictalurus spp.*) represent one of the fastest growing aquaculture industries in the United States. Current figures project that over 130,000 acres of ponds are producing catfish annually. Catfish thrive at 24–26° C, a pH between 6.6 and 7.5, a water hardness of 10 ppm, and dissolved oxygen above 4 ppm.

Crawfish (*Procambarus spp.*) thrive in freshwater lakes and streams. Good growth occurs at 21–29° C, a pH close to neutrality, a water hardness of 50–200 ppm, salinity up to 6 ppt, and dissolved oxygen above 3 ppm.

AGRI-PROFILE

CAREER AREA: AQUACULTURE RESEARCHER

USDA geneticist Gary Carmichael and reproductive physiologist Cheryl Goudie search for ways to increase the ratio of male to female catfish, since males grow faster than females. ***(Courtesy USDA Agricultural Research Service)***

Catfish farming and other aquaculture enterprises are big business in many nations of the world. Similarly, catfish farming is one of the fastest growing food production enterprises in the United States. Cultured seafood production is rapidly approaching the volume of that taken from natural waters.

The rapid growth in aquaculture has spurred research and development activities. These in turn have stimulated career opportunities in animal science, nutrition, genetics, physiology, aquaculture construction, facility maintenance, pollution control, fish management, harvesting, marketing, and other.

As productive land becomes more scarce among the global resources, aquaculture will play an increasing role in providing high-quality food at affordable prices.

Clams and oysters are cultivated in bays and estuaries subject to tidal flows. Good growing conditions vary greatly with species, but include approximately 6–20 ppt salinity, 15–30° C, dissolved oxygen above 1 ppm, and adequate amounts of microscopic organisms (low turbidity). Production should improve on cultivated sites when improved *spats* are developed for stocking.

Shrimp and prawns are being cultured in brackish-water ponds and estuaries. Conditions for good growth are temperatures above 25° C, a salinity of 20%, high levels of microorganisms, and dissolved oxygen above 4 ppm. Hatchery techniques are complex and involve several distinct growing stages.

Production System The cultivation of any aquatic organism by the aquaculturist integrates the necessary cultural requirements with existing resources. This blend has developed three general production programs (Figure 12-7).

Open ponds, rivers, and bays are stocked with natural or cultured young and are maintained with densities that are balanced with the existing ecosystem. Competing species are controlled, and natural recycling techniques are encouraged. This form of aquaculture can use both natural and constructed ponds. Some care must be taken to prevent any peaks in TAN. Overfertilization stimulates rapid aglae blooms, with high dissolved oxygen during photosynthesis but very low levels during early morning hours. Stress leads to high levels of mortality. Even lower dissolved-oxygen levels and fish kills can occur. The aquaculturist must be careful to monitor incoming water for toxins and other suspended materials. Harvesting must involve draining or *seining* the entire production area. Seining is the removal of fish with nets.

FIGURE 12-7
Commercial fish ponds are subject to weather conditions. ***(Courtesy Denmark High School Agriculture Education Department)***

Caged Culture Caged culture represents a more capital-intensive program. Aquatic animals or plants are contained in a small area and waste products are removed by the flushing action of the natural waters (Figure 12-8). Confining growing fish in floating-pen cages or shellfish on suspended float tables are techniques for managing increasing densities of aquatic organisms. As is true on a cattle feedlot, the aquaculturist confines the crops in a limited area and provides the necessary feed and cultural management. The natural ebb and flow of the water removes waste products and replenishes dissolved oxygen.

Cage culture can be designed for both natural waters and newly constructed ponds. Aquaculturists have a better idea of growth rates and can adjust feeding ratios more economically. Some growers have reported problems when a pond *"rolls over"* (changes water quality suddenly during certain weather conditions and brings the less-oxygenated water to the surface). Fish in cages are unable to move and can be stressed or killed. In this intensive production system, the aquaculturist must ensure adequate nutrition, disease control, predatory control, and physical maintenance like any terrestrial animal producer. The young stock must be caught from natural waters or produced in controlled hatcheries. Successful operations include the production of Atlantic salmon off the coast of Norway, Nova Scotia, and Maine. Hybrids of striped bass have been cultured in cages in Maryland and California. Trout have been cultured in net pens suspended in mountain streams and ponds.

Shellfish growers in Japan and the United States have demonstrated the production of oysters, clams, and mussels on suspended float tables or ropes. Production is increased by higher water quality and ease of harvest.

Recirculating Tanks Many areas of the world lack sufficient water resources to maintain a viable aquaculture industry. Recirculating systems must circulate the waste water through a biological purifier and return it to the growing tank. This complicated process is similar to that in a city waste-treatment plant. The system must remove the solid fish wastes, soluble ammonia/nitrite/nitrates, and carbon dioxide, and replace dissolved

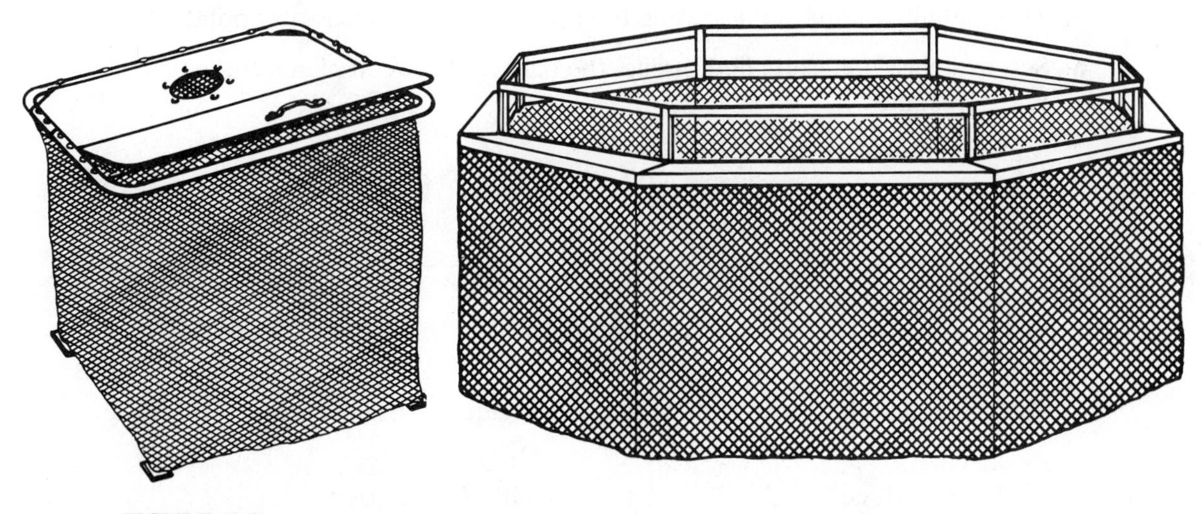

FIGURE 12-8
Fish can be raised in pens or cages.

oxygen. The pH must be maintained and integrated into the biological needs of the bacteria that populate the biological filters. As in any pond, if any single parameter is ignored, the organisms become stressed and production is decreased. The University of Wisconsin has been experimenting with yellow perch, the University of Maryland Eastern Shore with tilapia, and the Walt Disney World EPCOT Center with paddlefish.

Hatcheries The development of more intensive aquaculture systems will depend on a constant supply of high-quality young organisms. The natural fisheries are threatened by overfishing, pollution, and habitat destruction. Hatcheries are investigating the parameters that affect fish breeding habits and attempt to induce spawning, fry, or larvae production. *Larvae* are mobile aquatic organisms that grow into nonmobile adults.

These advances in fish and shellfish management allow the industry to develop improved breeds and hybrids to support improved production. Improved genetic lines of trout, catfish, and salmon are already commercially available. The increase of existing ornamental fish within a country also prevents the introduction of imported fish diseases and parasites.

Aquaculture and Resource Management The demand for aquaculture products will continue to increase. This demand will stimulate a tremendous growth in commercial aquaculture. During the expansion of the commercial aquaculture industry, the conflicts for scarce natural resources and clean water, the impact on recreational areas, and the potential pollution effects will have to be resolved by trained specialists.

Student Activities

1. Visit a local supermarket or seafood market and list the seafood products. Classify them as fish, shellfish, or crustaceans; freshwater or saltwater products; or imported products.
2. Describe why some of the above products might not be produced in your area.
3. Locate three local aquaculturists and discuss their production systems. How do they maintain water quality?
4. Visit a local pet store and then describe the various parts of a freshwater aquarium. How is the water quality maintained in this recirculating system?

5. Make a bulletin board illustrating the food chain of a freshwater pond.
6. Set up a class aquarium and discuss the balance between plants and animals in the system.
7. Make salt water.

Self-Evaluation

A. MULTIPLE CHOICE

1. The aquaculturist must understand how aquatic organisms

   a. eat. c. live.
   b. reproduce. d. all of the above.

2. The yearly catch of fish from natural waters is

   a. increasing. c. hard to determine.
   b. holding constant or decreasing. d. mostly catfish.

3. The highest salinity is measured in

   a. pond water. c. creeks.
   b. irrigation water. d. ocean water.

4. The accumulation of salts in water occurs most often when

   a. water runs across agricultural land.
   b. water settles in a pond.
   c. water collects in a drainage ditch.
   d. water is lost through evaporation.

5. Brackish water is

   a. colored black.
   b. located in tidal areas.
   c. collected from small creeks and branches.
   d. mostly high salinity (20–34 ppt).

6. Water quality is less affected by

   a. fish density. c. chemical runoff.
   b. weather. d. fish species.

7. The greater the density of fish in a system, the

   a. smaller the tank.
   b. larger the fish.
   c. the more difficult the management.
   d. the higher the temperature.

8. A fish kill can occur when a pond "rolls over"

   a. because of the temperature shock.
   b. because the cages sink to the bottom.
   c. because of low dissolved oxygen.
   d. because the fish turn upside down.

B. MATCHING

______ 1. Salinity
______ 2. Dissolved oxygen
______ 3. Turbidity
______ 4. Temperature
______ 5. Ammonia
______ 6. Hardness

a. Measured in ppm calcium
b. Recorded as degrees F or C
c. Measured as ppm or %
d. Depth of visual penetration
e. TAN
f. Conductivity or ppt

C. COMPLETION

1. Both plants and animals are part of the food ______ .
2. The ______ ______ are where fresh water and seawater mix.
3. The ______ of a fish absorb oxygen from the water.
4. Between 32 and 24 ppt is the salinity of ______ .
5. Trout need a ______ dissolved oxygen concentration than clams to grow and mature.
6. Crawfish must ______ or break out of their exoskeleton to grow.
7. Salmon must return to ______ water to spawn and complete their life cycle.
8. Pond culture relies mostly on ______ recycling of fish waste products.
9. The recirculating production systems must treat the fish wastes with ______ filters.

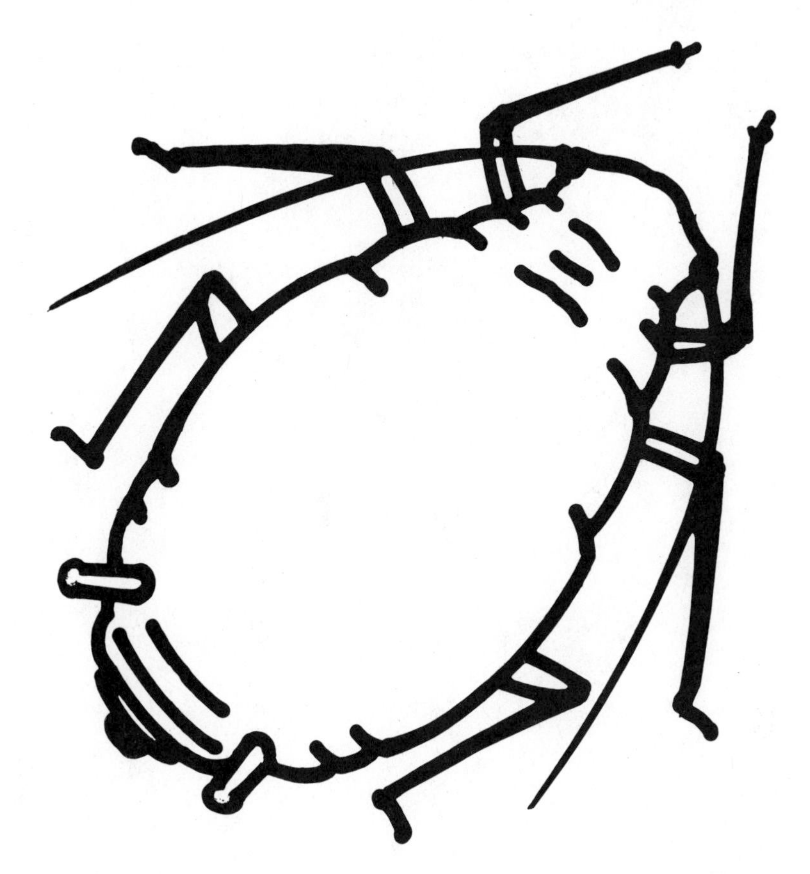

Integrated Pest Management

UNIT 13 Biological, Cultural, and Chemical Control of Pests

OBJECTIVE To develop an understanding of the major pest groups and the importance of effective pest management programs.

Competencies to Be Developed

After studying this unit, you will be able to

- ☐ define pest, disease, insect, weed, biological, cultural, chemical, and other terms associated with integrated pest management.
- ☐ know how the major pest groups adversely affect agriscience activities.
- ☐ describe weeds based on their life cycles.
- ☐ describe both the beneficial and detrimental roles that insects play.
- ☐ recognize the major components of disease and the causal agents.
- ☐ explain and understand the concept of integrated pest management.

TERMS TO KNOW

Diseases
Vector
Insect
Arachnid
Pest
Defoliate
Pesticide
Weeds
Pathogens
Annual weed
Biennial weed
Perennial plant
Rhizome
Node
Stolon
Meristematic tissue
Entomophagous
Exoskeleton
Metamorphosis
Instar
Plant disease
Causal agent
Disease triangle
Abiotic diseases
Biotic diseases
Fungi
Hyphae
Mycelium
Bacteria
Appressorium
Viruses
Symptom
Mosaic
Nematodes
Integrated pest management (IPM)
Key pest
Pest population equilibrium
Economic threshold level
Monitor
Scouts
Quarantine
Targeted pest
Eradication
Cultivar
Biological control
Predator
Parasites
Cultural control
Clean culture
Trap crop
Chemical control
Pesticide resistance
Pest resurgence

MATERIALS LIST

insect net
killing jar
insect mounting pins
insect specimen labels
pictures of pests

The ability to control pests by either chemical, cultural, or natural control methods has afforded the American people an unprecedented standard of living. We often take for granted an unlimited food supply, good health, a stable economy, and an aesthetically pleasing environment. Without effective pest control strategies, our standard of living would decrease.

Good pest management practices have resulted in dramatic yield increases for every major crop. A single farmer in 1850 could only support himself and four people, while in the 1980s a farmer could support more than 80 people. The ability to control plant and animal *diseases* or disorders vectored by insects or other arthropods has reduced the incidence of malaria, typhus, and Rocky Mountain spotted fever. A *vector* is a living organism that transmits or carries a disease organism. An *insect* is a six-legged animal with three body segments. An *arachnid* is an eight-legged animal, such as a spider and a mite.

The impact of pest management in maintaining a stable economy can be seen on a regional and national basis. The regional economy suffered shortly after the boll weevil's introduction into the United States in 1892. The weevil devastated much of the cotton crop in the early 1900s. Similarly, the potato blight disease in Ireland caused famine and mass migration of the Irish to other parts of the world in 1845.

TYPES OF PESTS

The word *pest* is a general name for any organism that may adversely affect human activities. We may think of an agricultural pest as one that competes with crops for nutrients and water, *defoliates* plants (to strip a plant of its leaves), or transmits plant or animal diseases. The major agricultural pests are weeds, insects, and plant diseases. However, other types of pests exist. Some examples, and the class of *pesticide* or chemical used for killing them, are listed below.

Type of Pest	Class of Pesticide
mites, ticks	acaricide
birds	avicide
fungi	fungicide
weeds	herbicide
insects	insecticide
nematodes	nematacide
rodents	rodenticide

Damage by pests to agricultural crops in the United States has been estimated to be one-third of the total crop production potential. Therefore, an understanding of the major pest groups and their biology is required to ensure success in reducing crop losses to pests.

Weeds

Weeds are plants that are undesirable and are often considered out of place (Figure 13-1). The definition of a weed is therefore a relative term. Corn growing in a soybean field or white clover growing in a turfgrass are examples of weeds.

Weeds can be considered undesirable for any of the following reasons:

1. They compete for water, nutrients, light, and space, resulting in reduced crop yields.
2. They decrease crop quality.
3. They reduce aesthetic value.
4. They interfere with maintenance along rights-of-way.
5. They harbor insects and disease *pathogens* (organisms that cause disease).

Weeds can be divided into three categories based on their life spans and their periods of vegetative and reproductive growth.

Annual Weeds

An *annual weed* is a plant that completes its life cycle within one year. Two types of annual weeds occur, depending upon the time of year in which they germinate. A winter annual germinates in the fall and will actively grow until late spring. It will then produce seed and die during periods of heat and drought stress. Examples of winter annuals are chickweed, henbit, and yellow rocket.

A summer annual germinates in the late spring, with vigorous growth during the summer months. Seed are produced by late summer and the plant will die during periods of low temperatures and frost. Examples of summer annuals are crabgrass, spotted spurge, and fall panicum.

Biennial Weeds

A *biennial weed* is a plant that will live for two years. In the first year the plant produces only vegetative growth, such as leaf, stem, and root tissue. By the end of the second year, the plant will produce flowers and seeds. This is referred to as reproductive growth. After the seed is produced, the plant will die. There are only a few plants that are considered biennials. Some examples are bull thistle, burdock, and wild carrot.

Perennial Weeds

A *perennial plant* can live for more than two years and may reproduce by seed and/or vegetative growth. The production of

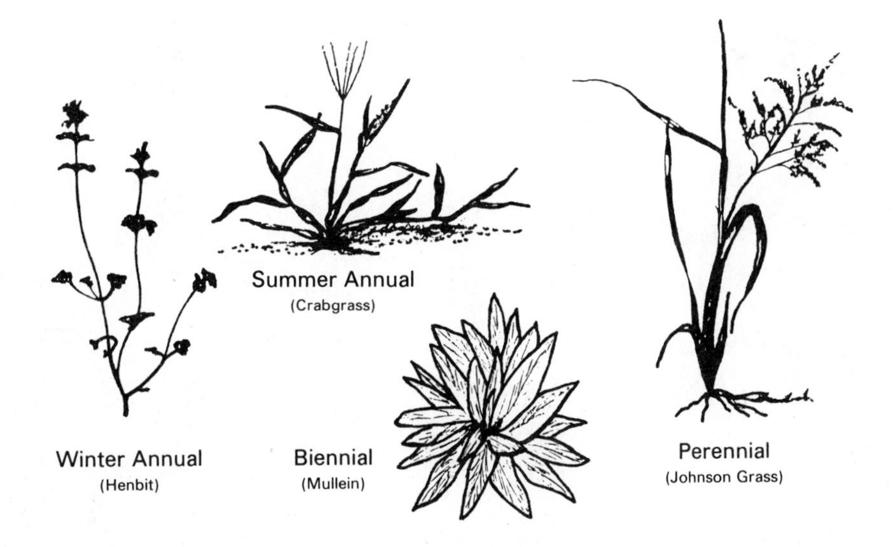

FIGURE 13-1
Different types of plants considered to be weeds and their life cycles. ***(Courtesy Maryland Cooperative Extension Service)***

rhizomes, stolons, and an extensive rootstock are ways in which perennial plants reproduce vegetatively. A *rhizome* is a stem that runs underground and gives rise to new plants at each joint or *node*. A *stolon* is a stem that runs on the surface and gives rise to new plants at each node. These plant parts have *meristematic tissue* (tissue capable of starting new plant growth). Examples of perennial weeds are dandelion, Bermuda grass, Canada thistle, and nutsedge.

Insects

Beneficial Insects Scientists estimate that there are more than 1 million species of insects that inhabit the world. A majority of them are beneficial, or helpful to humans. For example, insects are necessary for plant pollination. In the United States it is estimated that bees pollinate more than $1 billion worth of fruit, vegetable, and legume crops per year. Honey, beeswax, shellac, silk, and dyes are just a few of the commercial products produced by insects. Many insects are entomophagous and help in natural control of other insect species. *Entomophagous* insects feed on other insects. Insects that inhabit the soil, act as scavengers, or feed on undesirable plants all play important roles. These insects increase soil tilth, contribute to nutrient recycling, and act as biological weed control agents.

Finally, insects are at the lower levels of the food chain. Thus, they support higher life forms such as fish, birds, animals, and humans.

Insect Pests When compared to the total number of insect species, there are relatively few species that cause economic loss. However, it is estimated that crop losses due to such insects total more than $4 billion annually.

Insects can cause economic loss by feeding on cultivated crops and stored products. They can also vector plant and animal diseases, inflict painful stings or bites, or act as nuisance pests.

Insect Anatomy Insects are considered to be one of the most successful groups of animals present on Earth. Their success in numbers and species is attributed to several characteristics, which include their anatomy, reproductive potential, and developmental diversity.

Insects are in the class Insecta and are characterized by the following similarities (Figure 13-2):

1. They have an *exoskeleton*, which is the body wall of the insect. It provides protection and support for the insect.
2. The exoskeleton is divided into three regions: head, thorax, and abdomen.
3. There are segmented appendages on the head called antennae, which act as sense organs.
4. Three pairs of legs are attached to the thorax of the body.
5. Wings are present (one or two pairs) in a majority of species. This allows for dispersal and greater utilization of their habitat.

Feeding Damage Insects have either chewing or sucking mouthparts. Damage symptoms caused

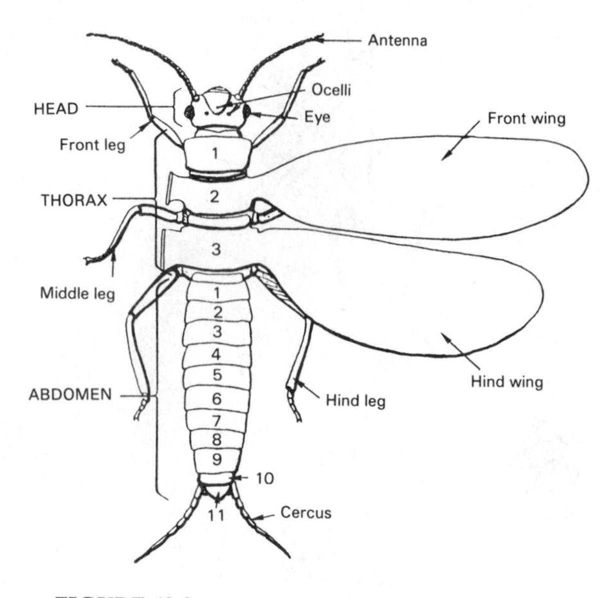

FIGURE 13-2
A diagram of an adult insect

by chewing insects are leaf defoliation, leaf mining, stem boring, and root feeding. Insects with sucking mouthparts produce distorted plant growth, leaf stippling, and leaf burn (Figure 13-3).

Development As insects grow from an egg to an adult, they pass through several growth stages. This growth process is known as *metamorphosis*. The two types of metamorphosis are gradual and complete.

Gradual metamorphosis consists of three life stages: egg, nymph, and adult (Figure 13-4). As a nymph, the insect will grow and pass through several *instars* (the stage of the insect between molts). Each time the insect sheds its exoskeleton, or molts, it passes into the next instar. For example, chinch bugs have five instars before they reach the adult form, but will vary in size, color, wing formation, and reproductive ability. Once the insect reaches the adult stage, no further growth will occur.

Complete metamorphosis consists of four life stages: egg, larva, pupa, and adult (Figure 13-5). The larval stage is the period when the insect grows. As larvae molt, they pass to the next larval instar. A Japanese beetle will have three larval instars before developing to the pupa stage. The pupa is a resting period. It is also a transitional stage of dramatic morphological change from the larva to adult.

Plant Diseases A *plant disease* is any abnormal plant growth. The occurrence and severity of plant disease is based on these three factors:

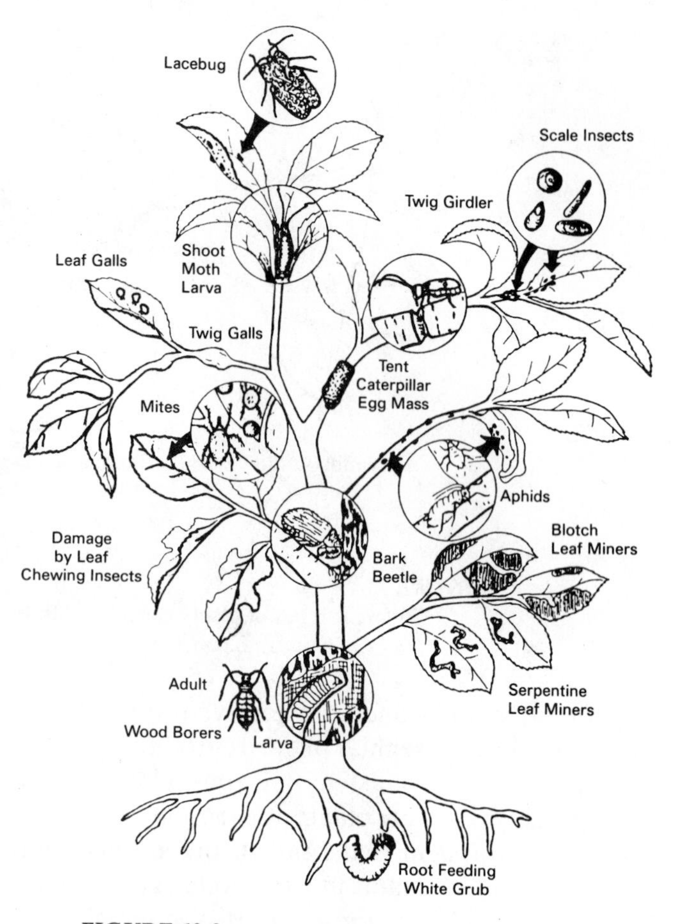

FIGURE 13-3
The different types of insect damage *(Courtesy Maryland Cooperative Extension Service)*

1. A susceptible plant or host must be present.
2. The disease organism or causal agent must be present. A *causal agent* is an organism that produces a disease.
3. Environmental conditions conducive for the causal agent must occur.

The relationship of these three factors is known as the *disease triangle* (Figure 13-6). Disease control programs are designed to affect each or all of these three factors. For example, if crop irrigation is decreased, a less favorable environment will exist for the disease organism. Breeding programs have introduced disease resistance into new plant lines for many different crops. Pesticides may also be used to suppress and control the disease organism.

Causal Agents for Plant Disease

Diseases may be incited by either abiotic factors or biotic agents. *Abiotic (nonliving) diseases* are caused by environmental or manmade stress. Examples of abiotic diseases

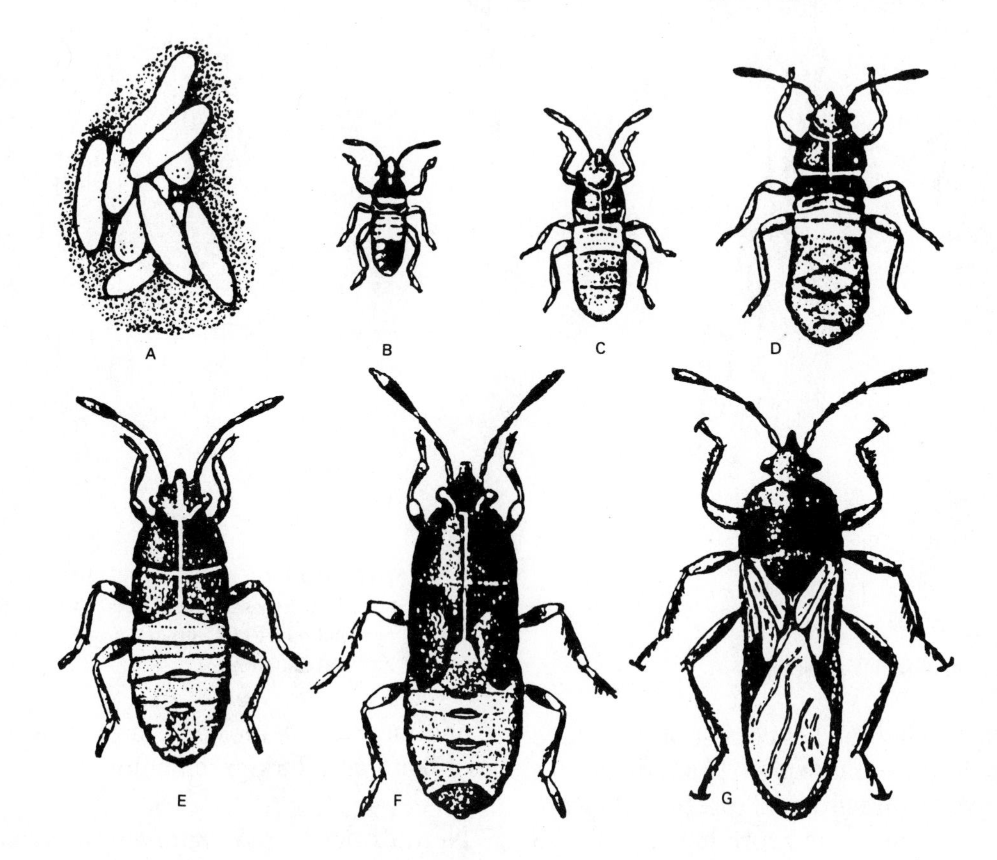

FIGURE 13-4
Gradual metamorphosis of the chinch bug

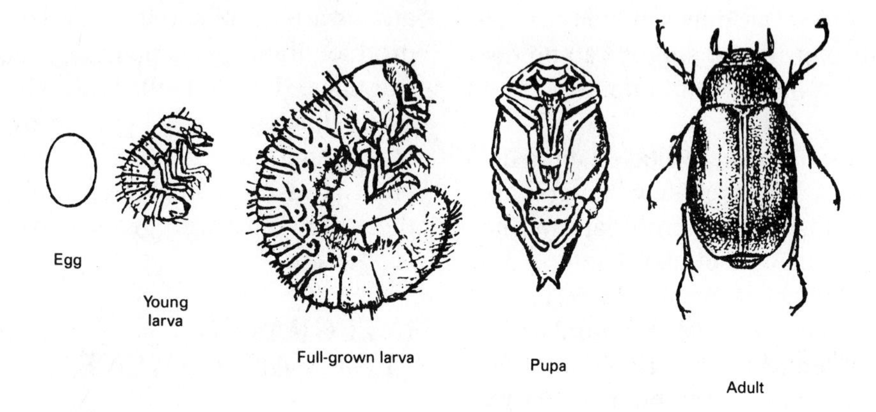

FIGURE 13-5
Complete metamorphosis of the June beetle

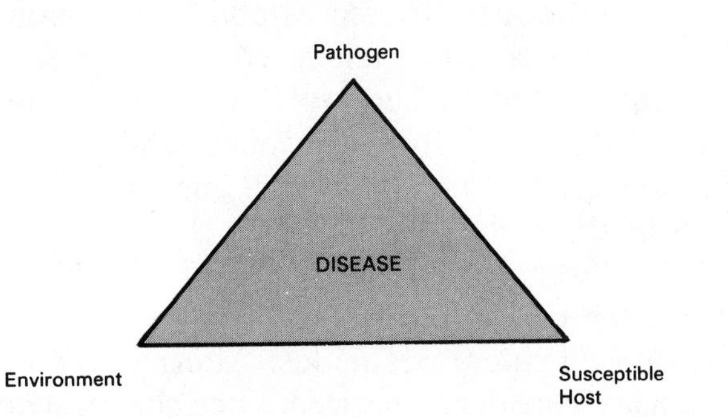

FIGURE 13-6
The components of the disease triangle ***(Courtesy Kevin Mathias, PhD)***

are nutrient deficiencies, salt damage, air pollution, and temperature and moisture extremes.

Biotic means living. *Biotic diseases* are caused by living organisms (Figure 13-7). Examples of causal agents or organisms are fungi, bacteria, viruses, nematodes, and parasitic seed plants. These organisms are parasites if they derive their nutrients from another living organism. Examples and a discussion of causal agents for plant diseases follow.

Fungi Fungi (plural for fungus) are the principal causes of plant disease. *Fungi* are plants that

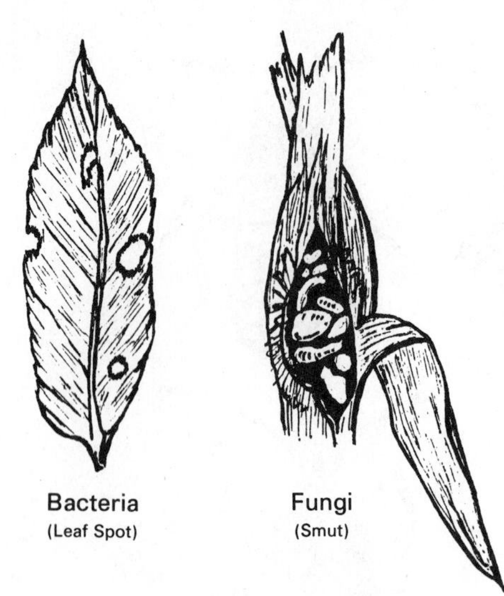

FIGURE 13-7
Examples of biotic diseases

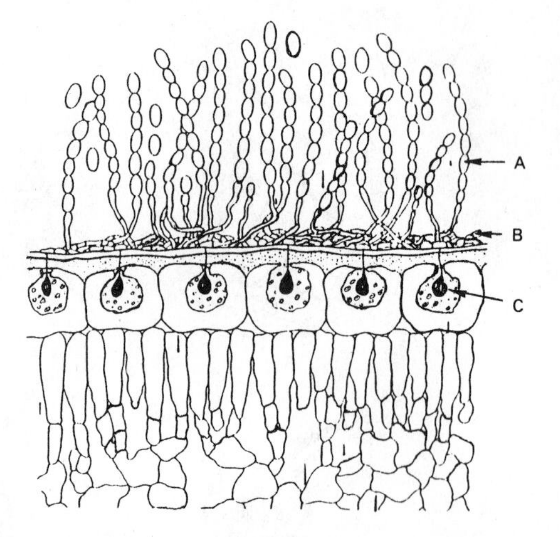

FIGURE 13-8
Powdery mildew of rose. (A) conidia (a sexual spore) (B) mycelium of fungus (C) fungal haustoria

lack chlorophyll. Their bodies consist of threadlike vegetative structures known as *hyphae* (Figure 13-8). When hyphae are grouped together, they are called *mycelium*. Fungi can reproduce and cause disease by producing spores or mycelia. Spores can be produced asexually or sexually by the fungus. For example, a mushroom produces millions of sexual spores under its cap. These spores can be dispersed by wind, water, insects, and humans.

Bacteria *Bacteria* are one-celled or unicellular microscopic plants. Relatively few bacteria are considered plant *pathogens* (a microorganism that will cause disease). Being unicellular, bacteria are among the smallest living organisms. Bacteria can only enter a plant through wounds or natural openings, such as a stem lenticel. Bacteria can be disseminated in ways similar to fungi, but they do not produce an appressorium. An *appressorium* is the swollen tip of a hypha by which the fungus attaches itself to the plant. Several important bacterial diseases are fire blight of apples and pears and bacterial soft rot of vegetables.

Viruses Plant viruses are pathogenic entities. *Viruses* are composed of nucleic acids surrounded by a protein sheath. They are capable of altering a plant's metabolism by affecting protein synthesis. Plant viruses are transmitted by seeds, insects, nematodes, fungi, grafting, and mechanical means, including sap contact. Viral diseases produce several well-known symptoms. A *symptom* is the visible change to the host caused by a disease. These symptoms are ring spots, stunting, malformations, and mosaics. A *mosaic* symptom is a leaf pattern of light- and dark-green color.

Nematodes *Nematodes* are roundworms that may live in the soil or water, within insects, or as parasites of plants or animals. Plant parasitic nematodes are quite small, often less than 4 mm, and produce damage to plants by feeding on stem or leaf tissue (Figure 13-9). The main symptom of nematode damage is poor plant growth, which results from nematodes feeding on the roots. The major plant parasitic nematodes are included in one of the following three groups: root-knot, stunt, or root-lesion.

INTEGRATED PEST MANAGEMENT

History *Integrated pest management (IPM)* is a pest control strategy that relies on multiple control practices. It also establishes economic threshold levels in determining control actions. The concept of integrated control is not new. Entomologists had developed an array of cultural and natural controls for the boll weevil and other insect pests by the early 1900s.

However, our approach to pest management during the period 1940–69 moved to a major reliance on chemical pesticides. Alternate control strategies were deemphasized, since chemical control gave excellent results at a low cost.

It was not until 1972 that a major change in policy occurred in the United States to encourage other pest control strategies. Natural, biological,

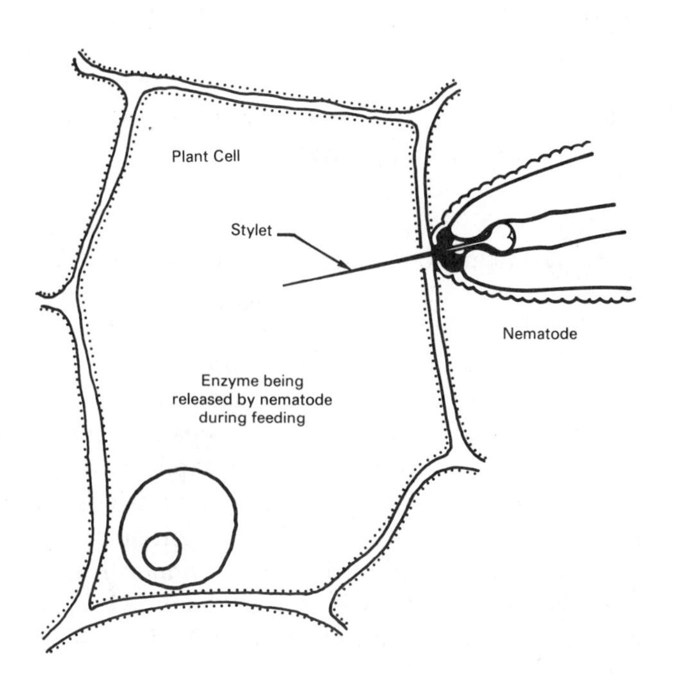

FIGURE 13-9
A nematode feeding on a plant cell. The stylet is a tiny speck-like feeding structure. *(Courtesy IMS, Texas A & M University)*

and cultural control programs began to be introduced as alternatives to chemical pest control.

The more recent trend toward reduced usage of chemicals for pest control was triggered by a book published in 1962 entitled, *Silent Spring*, by biologist Rachel Carson. After 1962, heavy reliance on chemical pest management began to be questioned. Carson's book created a public awareness of the environmental pollution that results from the overuse of pesticides. Other adverse side effects from pesticide misuse and/or overuse were beginning to occur as well. These included reports of pest resurgence, resistance to pesticides, and concern over human health.

Since the 1970s, great strides have been made in the development and implementation of IPM programs. The end result has been to lessen dependency on chemical use, while still achieving acceptable pest control.

Principles and Concepts of IPM The following concepts or principles are important in understanding how IPM programs should operate.

AGRI-PROFILE

CAREER AREA: ENTOMOLOGY/ PLANT PATHOLOGY
Entomologist Sammy Pair ponders this moth stage of a corn earworm retrieved from an insect trap. *(Courtesy USDA Agricultural Research Service)*

The work of entomologists and plant pathologists is never done. They do battle in the laboratory, in buildings, or in the field against insects and diseases that consume or ruin much of what we produce. The advice and service of these specialists is sought to control diseases and damaging insects, as well as to encourage beneficial ones.

Entomologists and plant pathologists attempt to control or reduce the buildup of damaging insect and disease populations. Such work may include assessing damage; attracting, trapping, counting and observing insects; and advising, directing, and assisting those who attempt to control insects and plant diseases. The mysteries of some insects are so great that scientists must specialize on just a few insects to be truly knowledgeable about them. Gone are the days when chemicals were our chief means of controlling insects and diseases. Rather, we now utilize a variety of techniques collectively known as integrated pest management.

Career opportunities exist for field and laboratory technicians as well as for degree-holding specialists. Neighborhood jobs may include termite control, scouting, spraying, crop dusting, inspecting, monitoring, selling, and managing field research projects. Honey bee specialists may manage hives to pollinate crops for improved seed and fruit production.

Key Pests A *key pest* occurs on a regular basis for a given crop (Figure 13-10). It is important to be able to identify key pests and to know their biological characteristics. Further, the weak link in each pest's biology must be found if management of the pest is to be successful.

Crop and Biology Ecosystem The integrated pest manager must learn the biology of the crop and its ecosystem. The ecosystem of the crop consists of the biotic and abiotic components that are present within the crop. The biotic components of the ecosystem are the living plants and animals. The abiotic components are nonliving factors, such as soil and water. Examples of human-managed ecosystems are a field of soybeans, a turfgrass area, or a poultry production operation.

Pest	% of total arthropod pests
Lacebugs	21
Mites	19
Scales	13
Borers	7
Leaf miners	7
Japanese beetle	4
Aphids	4
Bagworms	4
Galls	3
Weevils	+ 1
Total	**83**

FIGURE 13-10
Key pests in six Maryland communities in 1982. ***(Courtesy University of Maryland, Entomology Department)***

Ecosystem Manipulation With IPM, an attempt is made to understand the influence of ecosystem manipulation on lowering pest populations (Figure 13-11). To illustrate this concept, the manager must ask, What would happen to the pest population equilibrium if a disease-resistant plant were introduced? *Pest population equilibrium* occurs when the number of pests stabilizes or remains steady. The introduction of disease-resistant plants should lower the pest population below the economic threshold level. The *economic threshold level* is the point where pest damage is great enough to justify the cost of additional pest control measures.

Threshold Levels The level of a pest population is important. For instance, the mere presence of a pest may not warrant any control measure. But, at some point, the damage created by insects will be great enough to warrant additional control measures. Various threshold levels are developed to determine if and when a control measure should be implemented. This will avoid excessive economic loss to plants from pest damage (Figure 13-12).

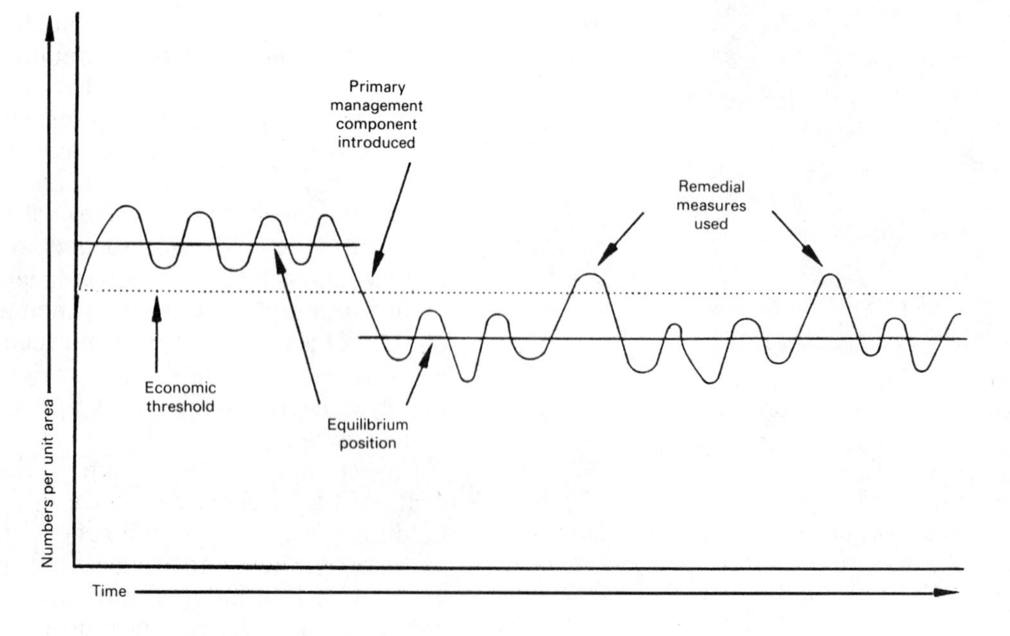

FIGURE 13-11
The effect of lowering the equilibrium position of a pest ***(Courtesy University of Maryland, Entomology Department)***

Economic Injury Threshold for Alfalfa Weevil; Number of Larvae From 30-Stem Sample

How to use table below:

1. Use plant height category that fits the field.
2. Estimate the value of crop in dollars per ton of hay equivalent and the cost to spray an acre.
3. From monitoring the field, find the number of alfalfa weevil larvae from a sample of 30 stems.
4. The number in each small box indicates the number of larvae per 30-stem sample that is required before a spray application would be profitable under these conditions.

Example: Plants in the field are 20 inches high (use Category II), hay is valued at $80 per ton, cost to spray is $8.00 per acre, and you collected 40 larvae from the sample of 30 stems. The number in the box common to $80 and $8 is 75. This means that under these conditions, 75 larvae are needed before a spray would be profitable. Since you collected only 40 larvae, a spray at this time will not be profitable.

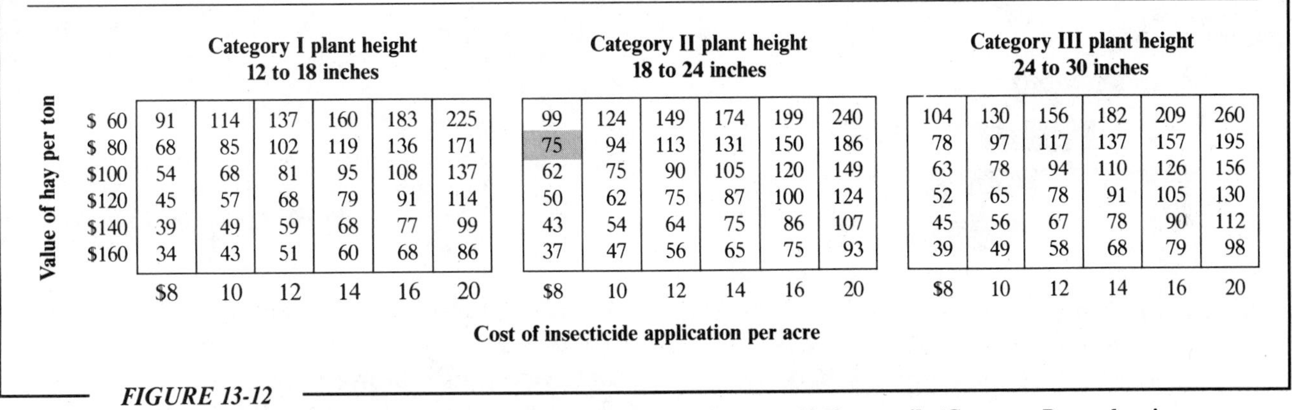

Category I plant height 12 to 18 inches

Value of hay per ton	$8	10	12	14	16	20
$ 60	91	114	137	160	183	225
$ 80	68	85	102	119	136	171
$100	54	68	81	95	108	137
$120	45	57	68	79	91	114
$140	39	49	59	68	77	99
$160	34	43	51	60	68	86

Category II plant height 18 to 24 inches

Value of hay per ton	$8	10	12	14	16	20
$ 60	99	124	149	174	199	240
$ 80	75	94	113	131	150	186
$100	62	75	90	105	120	149
$120	50	62	75	87	100	124
$140	43	54	64	75	86	107
$160	37	47	56	65	75	93

Category III plant height 24 to 30 inches

Value of hay per ton	$8	10	12	14	16	20
$ 60	104	130	156	182	209	260
$ 80	78	97	117	137	157	195
$100	63	78	94	110	126	156
$120	52	65	78	91	105	130
$140	45	56	67	78	90	112
$160	39	49	58	68	79	98

Cost of insecticide application per acre

FIGURE 13-12
Chart to determine economics threshold level for the alfalfa weevil *(Courtesy Pennsylvania State University)*

Economic threshold levels have been determined by first developing a pest-damage index (Figure 13-13). It is crucial in the decision-making process to know the level of pest infestation that will cause a given yield reduction. Pest populations are measured in several different ways. They can be counted in number of pests per plant or plant part, number of pests per crop row, or number of pests per sweep with a net above the crop.

Monitoring

For IPM to be successful, a *monitoring* (checking) or scouting procedure must be performed. Different sampling procedures have been developed for various crops and pest problems (Figure 13-14). The presence or absence of the pest, amount of damage, and stage of development of the pest are several visual estimates a scout must make. The method used must be speedy and accurate.

Scouts are people who monitor fields to determine pest activity. They must be well trained in entomology, pathology, agronomy, and horticulture.

PEST CONTROL STRATEGIES

Pest control programs can be grouped into several broad categories. These include regulatory, biological, cultural, physical and mechanical, and chemical.

Regulatory Control

Federal or state governments have created laws that prevent the entry or spread of known pests into uninfested areas. Regulatory agencies also attempt to contain or eradicate certain types of pest infestations. The Plant Quarantine Act of 1912 provides for inspection at ports of entry. Plant or animal quarantines are implemented if shipments are infested with targeted pests. A *quarantine* is the isolation of pest-infested material. A *targeted pest* is a pest which, if introduced, poses a major economic threat.

If a targeted pest becomes established, an eradication program will be started. *Eradication* means total removal or destruction of a pest. This type of pest control is extremely difficult and expensive to administer. In California, the Mediterranean

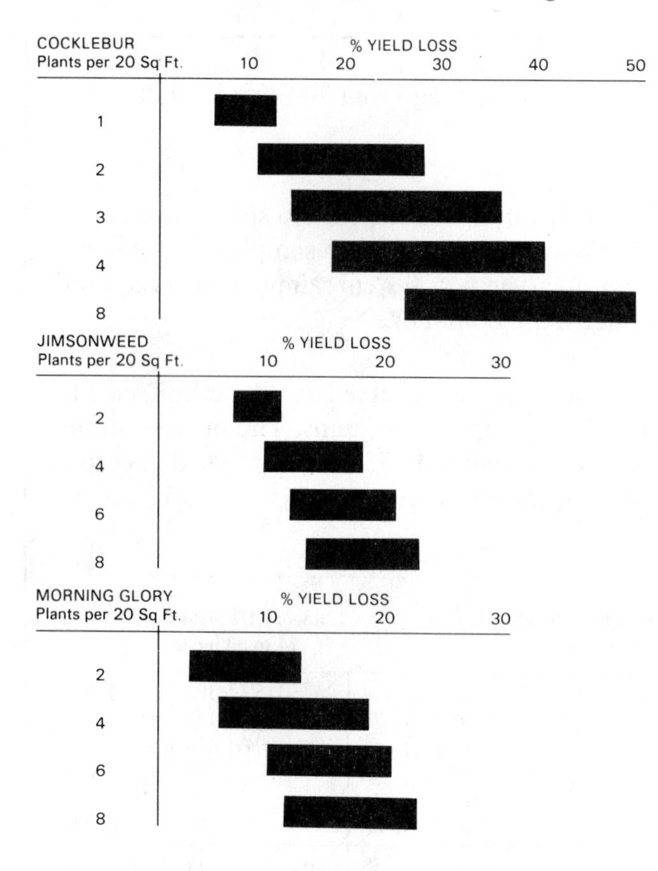

FIGURE 13-13
The pest-damage index for several weeds in soy beans *(Courtesy University of Maryland)*

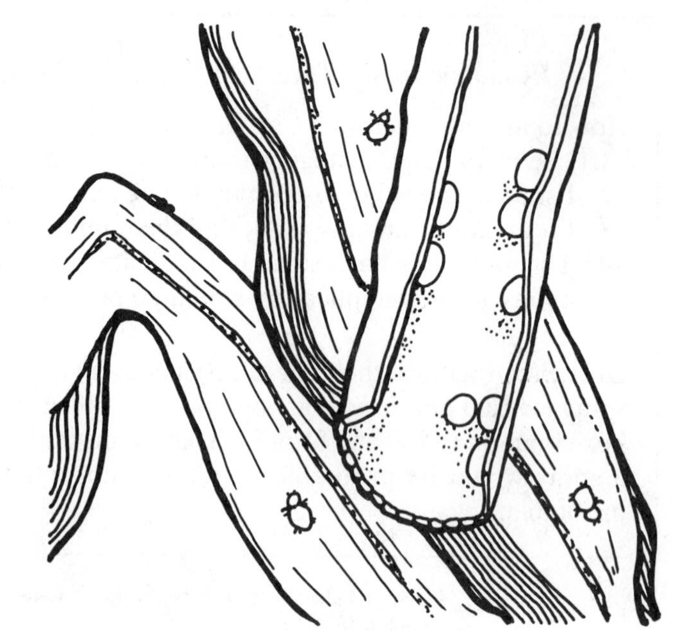

FIGURE 13-14
Random sampling of a plant stem and leaf to determine pest populations and damage *(Courtesy Maryland Agriculture Extension Service)*

fruit fly was eradicated at a cost of $100 million in 1982. This program relied on chemical spraying, sanitation, sterile male releases, and pheromone traps to ensure complete eradication.

Host Resistance The development of plants having pest resistance is an extremely effective control practice. The advantages of such a system are as follows:

1. Low cost
2. No adverse effect to the environment
3. A significant reduction in pest damage
4. Ability to fit into any IPM program

Breeding programs attempt to identify and select plants with pest resistance. Presently, new plant cultivars with improved resistance to pests are released annually. A *cultivar* is a plant developed by humans, as distinguished from a natural variety.

Biological Control *Biological control* means control by natural agents. Such agents may be predators, parasites, and pathogens. A *predator* is an animal that feeds on a smaller or weaker organism. An example of a predator is the lady beetle. Aphids are the lady beetle's principal prey. *Parasites* are organisms that live in or on another organism. The Braconid wasp is parasitic on the caterpillars of many moths and butterflies. *Pathogens* are organisms that will produce disease within their host. For example, the bacteria *Bacillus popilliae* is a pathogen, since it causes the milky spore disease in Japanese beetle grubs.

Successful biological control programs reduce pest populations below economic thresholds and keep the pests in check. Such programs require a thorough understanding of the biology and ecology of the beneficial organism, as well as of the pest.

Cultural Control *Cultural control* is the attempt to alter the crop environment to prevent or reduce pest damage. It may include such agricultural practices as soil tillage, crop rotation, adjustment of harvest or planting dates, and irrigation schemes. Other practices that are considered cultural control are clean culture and trap crops. *Clean culture* refers to any practice that would remove breeding or overwintering sites of a pest. This may include removal of crop leaves and stems, destruction of alternate hosts, or pruning of infested parts. A *trap crop* is a susceptible crop planted to attract a pest into a localized area. The trap crop is then either destroyed or treated with a pesticide.

Physical and Mechanical Control These control programs use direct measures to destroy pests. Examples of such practices are steam sterilization, hand removal, cold storage, and light traps. Implementation of these control practices are costly and provide varying pest control results.

Chemical Control *Chemical control* is the use of pesticides to reduce pest populations. Chemical control programs have been very cost effective. However, various problems occur if this practice is misused or overused. Problems that can develop are environmental pollution, pesticide resistance, and pest resurgence. *Pesticide resistance* is the ability of an organism to tolerate a lethal level of a pesticide. *Pest resurgence* refers to a pest's ability to repopulate after its natural control has been eliminated by a pesticide application.

Integrated pest management seems to be our best defense against pests. Biological and cultural controls are favored when they are effective. However, we cannot control certain pests without the use of chemical pesticides. Under such circumstances, it is very important to use chemical pesticides safely.

Student Activities

1. Define the "Terms To Know."
2. Use reference materials to find and list 10 examples of beneficial insects and 10 examples of insect pests.
3. Name five insects that are beneficial some of the time and pests some of the time.
4. Make an insect collection, with the insects properly identified and named.
5. Make a drawing of an insect and label the various body parts, appendages, and mouth parts.
6. Research a major crop in your community and discuss the key pests and measures recommended to control those key pests.
7. Develop a collage showing pests of plants in your community.
8. Make a weed collection and identify each. Your instructor may have a weed identification key for this purpose.

Self-Evaluation

A. MULTIPLE CHOICE

1. A biennial weed will live for

   a. one year. c. three years.
   b. two years. d. more than three years.

2. The major causal agent of plant disease is

   a. nematodes. c. viruses.
   b. bacteria. d. fungi.

3. The number of insect species in the world is estimated to be

   a. 100,000. c. 1 million.
   b. 500,000. d. none of the above.

4. The term instar refers to the development stage of

   a. plants. c. bacteria.
   b. fungi. d. insects.

5. Plant diseases are vectored by

   a. wind.
   b. rain.
   c. insects.
   d. none of the above.

6. A nematode is a type of

   a. fungus.
   b. roundworm.
   c. annual plant.
   d. insect.

7. A type of regulatory control is

   a. plant quarantine.
   b. sanitation.
   c. crop rotation.
   d. soil tillage.

8. The control practice that relies on the introduction of parasites and predators is

   a. cultural.
   b. chemical.
   c. biological.
   d. host resistance.

9. A threshold level is also known as the

   a. pesticide residues.
   b. control program.
   c. degree of pest control.
   d. pest concentration.

10. A pesticide used to control diseases is a/an

   a. fungicide.
   b. nematacide.
   c. insecticide.
   d. acaricide.

B. MATCHING

_____ 1. Biotic
_____ 2. Pest
_____ 3. Summer annual
_____ 4. Acaricide
_____ 5. Instar
_____ 6. Appressorium
_____ 7. Abiotic
_____ 8. Entomophagous
_____ 9. Symptom
_____ 10. IPM

a. The insect stage between molts
b. A chemical used to control mites and spiders
c. Integrated pest management
d. A visible change to the host caused by pests
e. Adversely affects human activities
f. Diseases caused by living organisms
g. A plant that germinates in the summer and lives for only one year
h. Structure by which the fungus attaches itself to a plant
i. Nonliving factors
j. Insects on which other insects feed

C. COMPLETION

1. Soil tillage is an example of _____ control.
2. An _____ will control weeds.
3. A _____ lives in or on another organism.
4. Insects have _____ pairs of legs.
5. Complete metamorphosis consists of the following stages: egg, _____ , _____ , and adult.

6. ______ resistance is when a plant has developed its own defensive response to pests.
7. ______ control will utilize quarantine practices.
8. Monitoring is essential for ______ programs to work successfully.
9. ______ pests occur on a regular basis for a given crop.
10. Causal agents of disease are ______ , ______ , ______ , and ______ .

UNIT 14 Safe Use of Pesticides

OBJECTIVE To determine the nature of chemicals used to control pests, know important terms regarding chemical safety, and practice the safe use of pesticides.

Competencies to Be Developed

After studying this unit, you will be able to

- ☐ describe the past and present trends of pesticide usage in the United States.
- ☐ recognize the popular classes of chemicals used for pest management and their role in pest control.
- ☐ read and interpret information on pesticide labels.
- ☐ state the components of proper dress for individuals handling pesticides.
- ☐ describe appropriate storage and handling of pesticides.
- ☐ describe the environmental and health concerns relating to pesticide use.

TERMS TO KNOW

Element
Compound
Inorganic compound
Organic compound
Selective herbicide
Nonselective herbicide
Contact herbicides
Systemic herbicide
Xylem tissue
Phloem tissue
Preemergence herbicide
Postemergence herbicide
Mode of action
Photodecomposition
Volatilization
Insecticides
Protectant fungicide
Eradicant fungicide
General-use pesticide
Restricted-use pesticides
Trade name
Formulation
Common name
Net contents
Signal words
Symbols
LD_{50}
LC_{50}
Misuse statement
Toxicity
Acute toxicity
Chronic toxicity
Carcinogen
Drift
Vapor drift

The development and use of pesticides has provided many benefits for both the producer of agricultural commodities and the consumer. However, there are also risks associated with pesticide use. These risks will be reduced when pesticides are properly applied, stored, and disposed of. Improper use of pesticides will increase the risk of environmental contamination and any adverse effects on human health.

Pesticide usage is a very controversial issue in America today. It is important to balance objectively the benefits and the risks associated with pesticide use. The Environmental Protection Agency (EPA) will soon conduct a benefit-risk assessment on each pesticide currently registered. When a pesticide is registered for use and is applied according to label directions, the benefits greatly outweigh the risks.

The benefits of pesticide use are summarized below.

1. Increase in yield of food and fiber
2. Reduction in loss of stored products
3. Increase in crop quality
4. Economic stability
5. Better health
6. Environmental protection and conservation

The increase in crop and animal yields, and improved quality of crops and livestock products resulting from the use of pesticides, have been adequately documented (Figure 14-1). It has been estimated that the average total income spent on food in the United States would increase from 17% to

Crop	Estimated Percent Increase	Major Pests Controlled
Corn	25	Weeds, rootworms, corn borers, seedling blights
Cotton	100	Pink bollworms, boll weevils, nematodes, boll rots
Alfalfa seed	160	Weeds, alfalfa weevils
Potatoes	35	Tuber rots, blackleg, soft rots, blights
Onions	140	Botrytis blights, neck rot, smut, onion maggots

FIGURE 14-1
Yield increases from pesticide applications ***(Courtesy "The New Pesticide User's Guide")***

30% without the protection offered by pesticides. Benefits to human health are best illustrated where insecticides are used to control insects that carry and spread malaria and typhus diseases. Minimum tillage or no-tillage practices have reduced soil erosion. These practices would not be possible without the use of pesticides. Many other substantial benefits to the quality of life resulting from the proper use of pesticides could be cited.

HISTORY OF PESTICIDE USE

The use of chemicals to control pests is not new. Elements such as sulfur and arsenic were among the first chemicals used for this purpose. An *element* is a uniform substance that cannot be further decomposed by ordinary chemical means. Homer, about 1000 B.C., wrote that sulfur had pest control ability. The Chinese, about A.D. 900, discovered the insecticidal properties of arsenic sulfide, a chemical *compound* (a chemical substance that is composed of more than one element).

Until the late 1930s, pest control chemicals or pesticides were mainly limited to *inorganic compounds* (any compound that does not contain carbon). Examples of other inorganic pesticides are mercury and Bordeaux mixture. Bordeaux mixture is a combination of copper sulfate and lime. It is used for plant disease control.

A majority of today's pesticides are synthetically produced *organic compounds* (a compound that contains carbon). The organic chemistry involved in pesticide production is often complex and extremely diverse. However, a classification system for pesticides that is based on the type of

PESTICIDE TYPE	TARGETED PESTS
acaricide	mites, ticks
algaecide	algae
attractant	insects, birds, other vertebrates
avicide	birds
bactericide	bacteria
defoliant	unwanted plant leaves
desiccant	unwanted plant tops
fungicide	fungi
growth regulator	insect and plant growth
herbicide	weeds
insecticide	insects
miticide	mites
molluscicide	snails, slugs
nematicide	nematodes
piscicide	fish
predacide	vertebrates
repellents	insects, birds, other vertebrates
rodenticide	rodents
silvicide	trees and woody vegetation
slimicide	slime molds
sterilants	insects, vertebrates

FIGURE 14-2
Classification of pesticides based on the target pests ***(Courtesy "The New Pesticide User's Guide")***

pest being controlled is useful (Figure 14-2). The major pesticide groups are herbicides, insecticides, and fungicides.

In the United States, pesticide use totaled more than 820 million pounds and cost $6 billion in 1988. Presently, EPA has registered more than 600 chemicals that are formulated into some 30,000 products for pest control. Of the three major pesticide categories, the largest volume was for herbicides, followed by insecticides and fungicides.

HERBICIDES

Herbicides can be grouped into several major categories based on application method, type of control, and chemical structure. The terminology for herbicide usage, type of control, and chemical family follows.

Selective Herbicides

A *selective herbicide* kills or affects a certain type or group of plants. The selectivity of a herbicide can be caused by many different factors. Some of these factors include:

1. differences in herbicide chemistry, formulation, and concentration.
2. differences in plant age, morphology, growth rate, and plant physiology.
3. environmental differences concerning temperature, rainfall, and soil types.

Nonselective Herbicides

A *nonselective herbicide* controls or kills all plants. These herbicides are used for many different purposes. Examples of their use are for railroad rights-of-way, industrial areas, fence rows, irrigation and drainage banks, and renovation programs.

Contact Herbicides

Contact herbicides will not move or translocate within the plant. They only affect that part of the plant with which they come in contact. They are often used for controlling annual weeds.

Systemic Herbicides

A *systemic herbicide* is absorbed by the plant and is then translocated in either xylem or phloem tissue to other parts of the plant. *Xylem tissue* is where water and minerals are transported within the plant. *Phloem tissue* is responsible for transporting carbohydrates within the plant.

Preemergence and Postemergence Herbicides

A *preemergence herbicide* is applied prior to weed or crop germination. A *postemergence herbicide* is applied after the weed or crop is present.

CHEMICAL FAMILIES OF HERBICIDES

Herbicides are chemicals that are used to control weeds. These compounds can affect plant growth in many different ways. Presently, more than 23 chemical families of herbicides have been developed. Each one has a unique chemical structure and a different site or mode of action. *Mode of action* is a term used to describe the way in which herbicides adversely affect plant growth. Several of the more important chemical herbicide families and their characteristics are described below.

Acetanilides

Acetanilides interfere with cell division and protein synthesis. They are applied as either preemergence or preplant applications for control of annual grasses and some annual broadleaf weeds. Two popular herbicides within this category are Lasso and Dual. They are used extensively for weed control in corn and soybeans.

Dinitroanilines

These act on root tissue, preventing root development in seedling plants. They are preemergence herbicides applied to prevent weed germination and should be incorporated into the soil. The dinitroanilines are deactivated very quickly by *photodecomposition* (chemical breakdown caused by exposure to light), *volatilization* (changing to gases), and chemical processes. Examples of these herbicides are Balan and Prowl. They are used for control of annual grasses and broadleaf weeds in many different crops.

Phenoxy

Phenoxy herbicides affect plants by causing overstimulation of growth. They perform best when applied as postemergence foliar sprays. They are selective herbicides that affect broadleaf weeds in grass crops. The herbicide 2, 4-D was first used in 1942, and is still widely used.

Triazines

These are photosynthetic inhibitors that interfere with the process of photosynthesis. They are preemergence herbicides, used to control both annual and broadleaf weeds in grass crops. Aatrex and Princep are two examples of these herbicides. They are primarily used in corn weed-control programs.

INSECTICIDES

Insecticides are chemicals used to control insects. They can affect the insect in many different ways. The classification system of insecticides may be based on chemical structure and/or their mode of action, as shown in the table on page 183.

The botanical, inorganic, and oil insecticides are some of the original chemicals used for insect control. Sulfur, for example, may be used to control mites. Its effectiveness as an insecticide was discovered thousands of years ago.

Rotenone is a chemical present in the roots of certain species of legume plants. It affects insects by inhibiting respiratory metabolism and nerve transmission. Rotenone was first used in 1848 and is a botanical insecticide still used today. It is a contact and stomach-poison insecticide. Rotenone is principally used in vegetables for controlling flea beetles, loopers, Japanese beetles, and many other insects.

Superior oils are highly refined and are applied to ornamentals and citrus crops to control scale insects, mites, and other soft-bodied insects. Oils act by excluding oxygen from the insect, thus causing suffocation. It is also believed that oils may destroy cell membrane function. Oils were first used in the early 1900s to control San Jose scale in fruit orchards.

CHEMICAL GROUP	MODE OF ACTION	COMMON/TRADE NAMES
A. Inorganics	Protoplasmic Poisons Physical Poisons	Sulfur Silica Aerogel
B. Oils	Physical Poisons	Superior Oils
C. Botanicals	Metabolic Inhibitors	Pyrethrum Rotenone
D. Synthetic Organics		
1. Chlorinated Hydrocarbons	Nerve Poison	Lindane Endosulfan-Thiodan
2. Organophosphates	Nerve Poison	Diazinon-Spectracide Parathion-Thiophos
3. Carbamates	Nerve Poison	Carbofuran-Furadan Carbaryl-Sevin
4. Pyrethroids	Nerve Poison	Permethrin-Ambush Fluvalinate-Mavrik
5. Insect Growth Regulators (IGRs)	Alter Insect Growth	Methoprene-Altosid Kinoprene-Enstar
E. Biorational/Microbial Insecticides	Wide Range of Activity	*Bacillus thuringiensis*

The synthetic organic group of insecticides was principally developed after 1940. This group includes the chlorinated hydrocarbons, organophosphates, carbamates, pyrethroids, insect growth regulators, and other minor classes. A majority of these insecticides adversely affect nerve transmission. The chlorinated hydrocarbons were the first to be synthesized (manmade). Released in 1940, DDT was the first and most popular chlorinated hydrocarbon. This chemical had excellent insecticidal properties and was used extensively. However, environmental and health problems were linked to DDT and other insecticides within this group. Because of these risks, many of the chlorinated hydrocarbons (such as DDT, aldrin, dieldrin, and chlordane) are banned for use in the United States.

The organophosphate and carbamate insecticides are the principal insecticide used to control insects today. Approximately 60% of all insecticides produced in the United States are from these two groups. They control insects by affecting the nervous system. The potential for pesticide poisoning of people and livestock is high for these insecticides. They will affect the nervous systems of humans and animals in a manner similar to the way they affect the insects they are designed to kill. The dose or the amount of insecticide applied is the discriminating factor with respect to insect control or human poisoning.

FUNGICIDES Fungicides are chemicals used to control plant diseases caused by fungi. Chemical control of plant diseases is more difficult than it is for weed and insect control. Fungi are plants without chlorophyll. They are parasites of other plants. The fungicide must be selective enough to control the fungus but not to adversely affect the host. Also, fungi have many generations each growing season. Therefore, reapplication of the fungicide will be required to provide effective control.

Protectant Fungicide A *protectant fungicide* is applied prior to disease infection. This will provide a chemical barrier between the host and the germinating spores. However, this barrier will be broken down by environmental weathering of the fungicide. Rainfall, sunlight, temperature, and plant growth are the major causes of this breakdown. The fungicide will have to be reapplied if adequate disease control is expected.

Eradicant Fungicide An *eradicant fungicide* can be applied after disease infection has occurred. These fungicides act systemically and are translocated by the plant to the site of infection. They offer a longer control period than do protectant fungicides, since they are not so prone to environmental weathering.

Chemical Structure Fungicides, like insecticides and herbicides, can also be classified into different chemical families. The following text on page 184 lists some of the major groups of fungicides.

CHEMICAL FAMILY	FUNGICIDE ACTIVITY	COMMON/ TRADE NAME
Benzimidazoles	Eradicant	Benomyl-Tersan 1991
Dicarboximides	Protectants	Captan-Captane
Dithiocarbamates	Protectants	Mancozeb-Dithane M45
Oxathiins	Eradicants	Carboxin-Vitavax

Inorganic Fungicides The elements sulfur, copper, mercury, and cadmium, or mixtures of them, are some of the oldest pesticides used by humans. They protect many ornamental and turfgrass from diseases.

Organic Fungicides This is a newer group of fungicides, which are used both as protectants and as eradicants. Examples of some of the chemical families within this group are listed at the top of this page.

PESTICIDE LABELS

The information on a pesticide-container label instructs the user on the correct procedures for application, storage, and disposal of the pesticide. The pesticide label is a summary of information gathered by the pesticide manufacturer and is required for product registration. The registration process is estimated to take 8–10 years and costs up to $20 million per pesticide. Several of the studies required to meet federal standards are summarized below.

A. Chemical and Physical Properties: water solubility, volatility, movement in soils, stability to heat and light and other factors affecting environmental stability.
B. Toxicology studies: determine acute oral, dermal, and inhalation toxicity to various animals; evaluate chronic toxicity, including any effect on reproduction or the ability of the pesticide to be a carcinogen.
C. Residue analysis: determine the amount of pesticide residues at the time of harvest, develop safe tolerance levels for any pesticide residue.
D. Metabolism studies: determine application exposure, determine consumer exposure to pesticide residues, establish safety practices that minimize exposure.

The label is a legal document indicating proper and safe use of the product. Pesticide use that differs from that specified on the label is a misuse. Such use is illegal, and the offender can be charged with civil or criminal penalties. When improperly applied, a pesticide can pose danger to the applicator, the environment, and other people. Therefore, it is important to read, understand, and follow the information on the pesticide label.

The pesticide label and other labeling information must meet federal standards. The label consists of a front panel and a back panel on the product (Figures 14-3 and 14-4). If there is insufficient room on the panels, additional labeling information will be attached, in booklet form, to the product. An outline of a label follows.

Use Classification Pesticides are classified as either general use or restricted use. A *general-use pesticide* poses minimal risk when applied according to label directions.

Restricted-use pesticides pose a greater risk to humans and the environment. Therefore, anyone applying a restricted-use pesticide must be properly trained and certified. Applicator certification is administered by each state. An applicator must meet a minimum set of standards and is often evaluated by tests in order to be certified.

Trade Name A *trade name* is the manufacturer's name for its product. It appears on the label as the most conspicuous item. The manufacturer will use the name in all of its promotional campaigns. The same chemical may have several different trade names, depending on the type of formulation and patent rights.

Formulation *Formulation* refers to the physical properties of the pesticide. The pesticide chemical or active ingredient will often have to be modified to allow for field use. These modifications may include adding inert ingredients such as solvents, wetting agents, powders, or granules to the pesticide. This will result in different formulations. Some examples of the different types of formulations and their abbreviated label names are as follows on page 187.

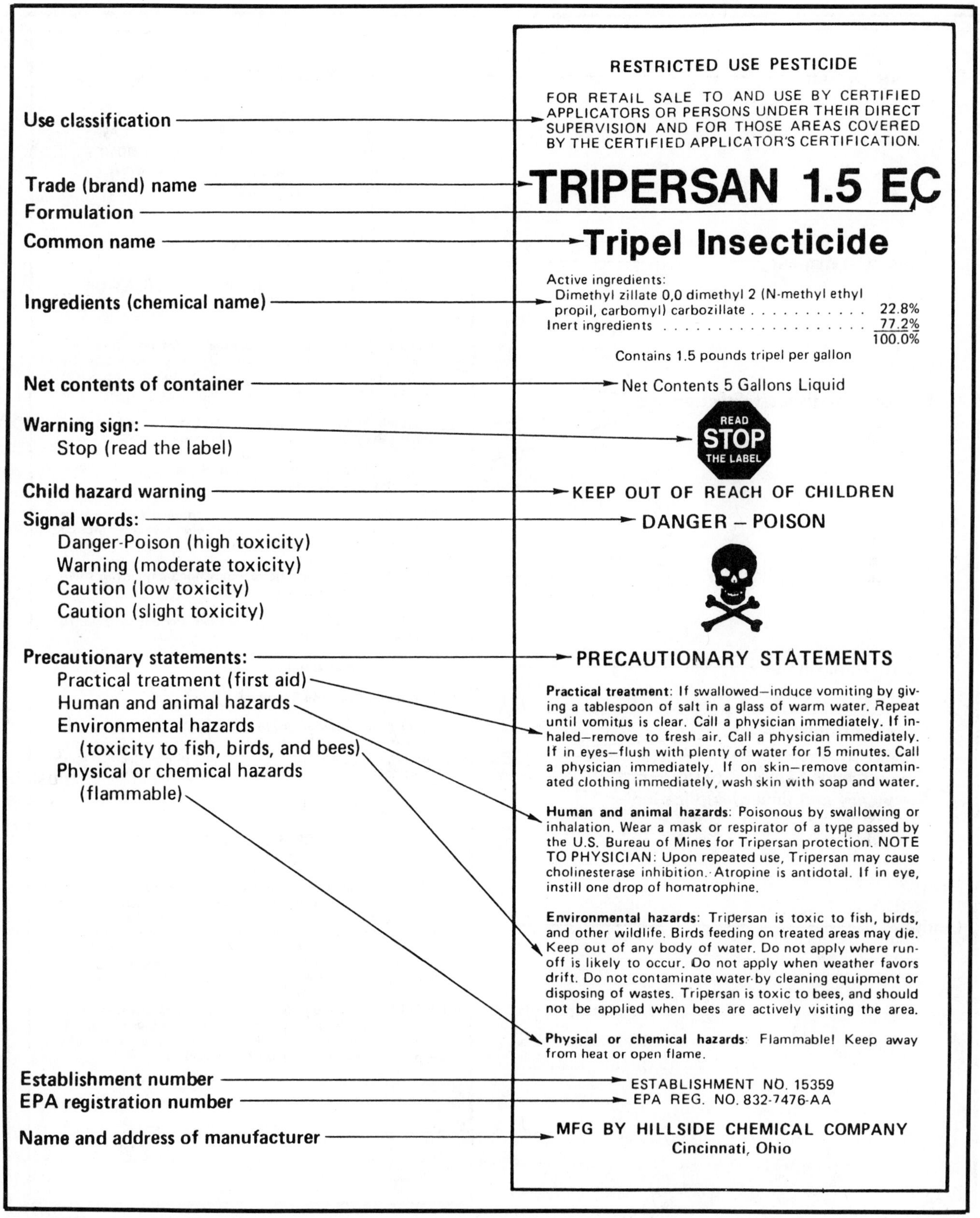

Use classification → RESTRICTED USE PESTICIDE

FOR RETAIL SALE TO AND USE BY CERTIFIED APPLICATORS OR PERSONS UNDER THEIR DIRECT SUPERVISION AND FOR THOSE AREAS COVERED BY THE CERTIFIED APPLICATOR'S CERTIFICATION.

Trade (brand) name → **TRIPERSAN 1.5 EC**

Formulation → EC

Common name → **Tripel Insecticide**

Ingredients (chemical name) →

Active ingredients:

Dimethyl zillate 0,0 dimethyl 2 (N-methyl ethyl propil, carbomyl) carbozillate	22.8%
Inert ingredients	77.2%
	100.0%

Contains 1.5 pounds tripel per gallon

Net contents of container → Net Contents 5 Gallons Liquid

Warning sign:
Stop (read the label) → READ STOP THE LABEL

Child hazard warning → KEEP OUT OF REACH OF CHILDREN

Signal words: → DANGER – POISON

- Danger-Poison (high toxicity)
- Warning (moderate toxicity)
- Caution (low toxicity)
- Caution (slight toxicity)

Precautionary statements: → PRECAUTIONARY STATEMENTS

- Practical treatment (first aid)
- Human and animal hazards
- Environmental hazards (toxicity to fish, birds, and bees)
- Physical or chemical hazards (flammable)

Practical treatment: If swallowed—induce vomiting by giving a tablespoon of salt in a glass of warm water. Repeat until vomitus is clear. Call a physician immediately. If inhaled—remove to fresh air. Call a physician immediately. If in eyes—flush with plenty of water for 15 minutes. Call a physician immediately. If on skin—remove contaminated clothing immediately, wash skin with soap and water.

Human and animal hazards: Poisonous by swallowing or inhalation. Wear a mask or respirator of a type passed by the U.S. Bureau of Mines for Tripersan protection. NOTE TO PHYSICIAN: Upon repeated use, Tripersan may cause cholinesterase inhibition. Atropine is antidotal. If in eye, instill one drop of homatrophine.

Environmental hazards: Tripersan is toxic to fish, birds, and other wildlife. Birds feeding on treated areas may die. Keep out of any body of water. Do not apply where runoff is likely to occur. Do not apply when weather favors drift. Do not contaminate water by cleaning equipment or disposing of wastes. Tripersan is toxic to bees, and should not be applied when bees are actively visiting the area.

Physical or chemical hazards: Flammable! Keep away from heat or open flame.

Establishment number → ESTABLISHMENT NO. 15359

EPA registration number → EPA REG. NO. 832-7476-AA

Name and address of manufacturer → MFG BY HILLSIDE CHEMICAL COMPANY
Cincinnati, Ohio

FIGURE 14-3
Front panel of a sample pesticide label ***(Courtesy of Vo-Ag Services, University of Illinois)***

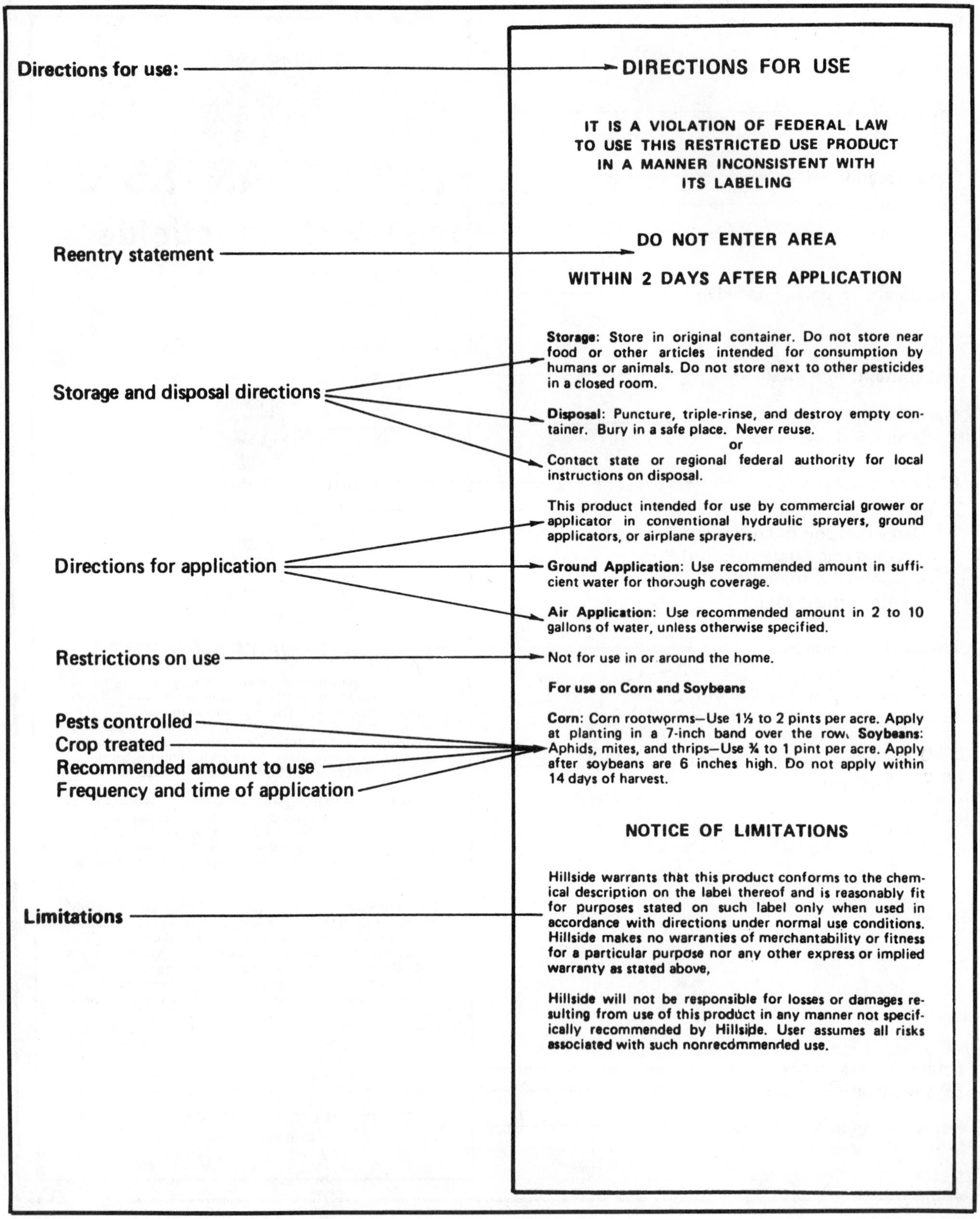

DIRECTIONS FOR USE

IT IS A VIOLATION OF FEDERAL LAW TO USE THIS RESTRICTED USE PRODUCT IN A MANNER INCONSISTENT WITH ITS LABELING

DO NOT ENTER AREA

WITHIN 2 DAYS AFTER APPLICATION

Storage: Store in original container. Do not store near food or other articles intended for consumption by humans or animals. Do not store next to other pesticides in a closed room.

Disposal: Puncture, triple-rinse, and destroy empty container. Bury in a safe place. Never reuse.

or

Contact state or regional federal authority for local instructions on disposal.

This product intended for use by commercial grower or applicator in conventional hydraulic sprayers, ground applicators, or airplane sprayers.

Ground Application: Use recommended amount in sufficient water for thorough coverage.

Air Application: Use recommended amount in 2 to 10 gallons of water, unless otherwise specified.

Not for use in or around the home.

For use on Corn and Soybeans

Corn: Corn rootworms—Use 1½ to 2 pints per acre. Apply at planting in a 7-inch band over the row. **Soybeans**: Aphids, mites, and thrips—Use ¾ to 1 pint per acre. Apply after soybeans are 6 inches high. Do not apply within 14 days of harvest.

NOTICE OF LIMITATIONS

Hillside warrants that this product conforms to the chemical description on the label thereof and is reasonably fit for purposes stated on such label only when used in accordance with directions under normal use conditions. Hillside makes no warranties of merchantability or fitness for a particular purpose nor any other express or implied warranty as stated above.

Hillside will not be responsible for losses or damages resulting from use of this product in any manner not specifically recommended by Hillside. User assumes all risks associated with such nonrecommended use.

FIGURE 14-4

Back panel of a sample pesticide label *(Courtesy of Vo-Ag Services, University of Illinois)*

Toxicity Rating	Label Signal Words	Oral LD_{50} (mg/kg)	Dermal LD_{50} (mg/kg)	Inhalation LC_{50} (μg/l or ppm)	Lethal Oral Dose, 150-pound Man
high	Danger-Poison	0–50	0–200	0–2,000	few drops to 1 teaspoon
moderate	Warning	50–500	200–2,000	2,000–20,000	1 teaspoon to 1 ounce (2 tablespoons)
low	Caution	500–5,000	2,000–20,000	200,000+	1 ounce to 1 pint+ or 2 pounds
very low	Caution	5,000+	20,000+	....	1 pint+ or 2 pounds+

FIGURE 14-5
The toxicity ratings and signal words for pesticides *(Courtesy College of Agriculture, University of Illinois, Urbana-Champaign)*

Granules—G
Solutions—S
Wettable Powders—WP or W
Soluble Powders—SP
Dry Flowables—DF
Emulsifiable Concentrates—EC or E
Dusts—D

Common Name The *common name* is given to a pesticide by a recognized authority on pesticide nomenclature. A pesticide is identified by a trade name, common name, and chemical name. It may have several trade names, but will only have one common and one chemical name. The common name or generic name identifies the active pesticide ingredient and can be used for comparison shopping.

Ingredients The percentages by weight of both the active and inert ingredients are stated on the label. The active ingredient is identified by its chemical name. This is the name of the chemical structure for that pesticide. The inert or inactive ingredients do not have to be listed by their chemical names. The label must state only the total percentage of inert material.

Net Contents The label will give the amount of product in the container. This is referred to as *net contents.* This quantity can be expressed in gallons, quarts, pints, and pounds.

Signal Words and Symbols The *signal words* and *symbols* describe acute toxicity of the pesticide. The different categories are based on LD_{50} and LC_{50} values and on skin and eye irritation. The signal words used on pesticide labels are (1) danger—poison, (2) warning, and (3) caution. These words are used to alert the person handling or using the pesticide of the poisoning effect of contact with the chemical.

Pesticides with high toxicity, where only a few drops to one teaspoon will kill a 150 lb. person, are labeled "DANGER—POISON." Pesticides with moderate toxicity, where one to two tablespoons will kill a 150 lb. person, are labeled "WARNING." Those restricted-use pesticides requiring more than two tablespoons of the chemical to kill a 150 lb. person are labeled "caution." Obviously, even the pesticides with "CAUTION" as the signal word must be handled with extreme care.

Acute toxicity is measured by determining LD (lethal dose)$_{50}$ values when the pesticide is absorbed through the skin or is ingested orally. These values are determined by inhalation studies. *LD_{50}* is the amount or dose of the pesticide that is required to kill 50% of test populations. It is expressed in milligrams (mg) of pesticide per kilogram (kg) of body weight. The lower the LD_{50} value, the more toxic a pesticide.

LC_{50} is the lethal concentration of the pesticide in the air that is required to kill 50% of test populations. It is expressed in micrograms per liter ($\mu g/\ell$) or in parts per million (ppm). The lower the LC_{50} value, the more toxic a pesticide is.

All labels must bear the statement, "KEEP OUT OF REACH OF CHILDREN," regardless of pesticide toxicity. Figure 14-5 shows the toxicity ratings for the various signal words.

Precautionary Statements These statements on the label will list any known hazards to humans, animals, and the environment. They will advise the user how to minimize any adverse effect that the pesticide may have. The categories that are normally listed are as follows:

Hazards to Humans and Domestic Animals
Statement of Practical Treatment
Environmental Hazards
Physical and Chemical Hazards

Establishment and EPA Registration Numbers The establishment number establishes the manufacturer. The EPA registration number indicates that the product has passed the review process imposed by the EPA.

Name and Address of Manufacturer All pesticide labels must contain the name and address of the company that manufactures and distributes the pesticide.

Directions for Use The correct amount, timing, and mixing of the pesticide is given under the directions for use section of the label. The label will also list the different pests that are controlled, the application technique, and any other specific directions for optimum control.

Misuse Statement The *misuse statement* appears on the label to remind the user to apply the pesticide according to label directions. Problems associated with pesticides, whether they involve environmental pollution or human poisoning, often occur due to pesticide misuse.

Reentry Information Specific directions on reentering a treated area will appear under this heading. Only a few pesticides require reentry times of more than a day after application. However, even if the pesticide label does not contain specific restrictions, no one should ever be allowed to enter a treated area until the pesticide has dried.

Storage and Disposal Directions This section will describe the proper storage and disposal of pesticides. It is recommended that you purchase only the amount of pesticides needed for the current season. Stockpiling them will only increase storage risks and, ultimately, the problem of pesticide disposal if they can no longer be used.

Notice of Limitations The manufacturer guarantees that the product will perform as the label states. The company also conveys inherent risks to the user if the pesticide is applied in a manner inconsistent with the label. The manufacturer will limit its liability in case of lawsuits stemming from misuse of the pesticide.

AGRI-PROFILE

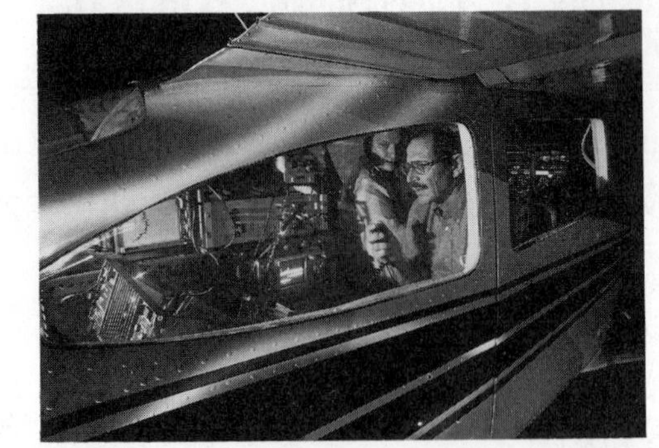

CAREER AREAS: PESTICIDE APPLICATOR/ PESTICIDE SPECIALIST/CHEMIST

Airplane and helicopter crews are important members of the teams that apply chemical and biological pesticides to large crop, range, and forest areas. *(Courtesy USDA Agricultural Research Service)*

Tens of thousands of chemical formulations have been developed in recent years to control insects, diseases, weeds, nematodes, rodents, and other pests. We must use chemicals to help control pests. However, these materials are likely to be hazardous to the operator, plants, animals, or the environment if not properly used.

Pesticide applicators are employed by farmers and ranchers, lawn service companies, farm and garden supply firms, termite control companies, highway departments, and railroads, as well as self-employed persons. Special training and licensing is required for handling most pesticides.

Pesticide applicators may work in homes, buildings, fields, forests, and even the holds of ships. They may dust, spray, bait, or fumigate depending on the setting and the pest. Pesticide applicators must always use protective clothing and other devices to protect against accidental poisoning. Their tools may be as simple as an aerosol can or as complex as an orchard spray rig or specially equipped helicopter.

RISK ASSESSMENT AND MANAGEMENT

Risk Measurement An experienced pesticide applicator who understands the hazards of pesticides will take steps to reduce risks. The relationship of risk or hazard in pesticide applications has been expressed by the following:

$$\text{Risk (Hazard)} = \text{Toxicity} \times \text{Exposure}$$

The hazard or risk is the relationship between the toxicity of the pesticide and the exposure or use of the pesticide. *Toxicity* is a measure of how poisonous a chemical is. These data may be expressed in several ways. *Acute toxicity* describes the immediate effects (within 24 hours) of a single exposure to a chemical. Acute toxicity data based on dermal (skin), oral (by mouth), and inhalation (breathing) exposure routes have been determined. Signal words on pesticide labels indicate acute toxicity values. *Chronic toxicity* measures the effect of a chemical over a long period of time. To determine this information, the chemical is administered at low levels, with repeated exposures to the test animals. The effect of the chemical on reproduction or as a potential carcinogen or any other adverse effects are evaluated.

Limiting Exposure A pesticide's toxicity cannot be changed, but risk can be managed by addressing the exposure component of the formula. Many things can be done to reduce exposure. Examples of practices that can reduce exposure are as follows:

1. Select a pesticide formulation with a lower exposure potential. For example, granule formulations have a much lower exposure potential than do emulsifiable concentrates.
2. Use protective clothing and other safety equipment during the time of pesticide mixing, application, and disposal.
3. Apply pesticides during weather conditions that will not cause pesticide drift and those that provide for the most effective control.
4. Check all application equipment for proper working condition before applying pesticides.
5. Store pesticides and application equipment properly.

The pesticide label will provide guidance concerning acceptable protective clothing or gear for the application of the pesticide. Recommended protective clothing and gear will vary according to the toxicity of the pesticide. Even if no special gear is required, it is best to minimize your exposure to all pesticides by selecting appropriate clothing.

Appropriate clothing includes long pants, a long-sleeved shirt, nonabsorbent shoes, and socks. Avoid the use of any leather clothing, particularly shoes, since leather will absorb pesticides. The use of heavy denim clothing will provide good repellency to any pesticide and can be washed to remove any pesticide residue.

Special Gear

Gloves and Boots Unlined rubber or neoprene gloves and boots will significantly reduce pesticide exposure (Figure 14-6). Any type of cloth-lined or leather boot or glove will only increase exposure if the lining becomes contaminated. Gloves should be tucked inside sleeves if you are working below the waist. They should be left outside your sleeves if you are working above the waist. Pants legs should be placed over the boots. By following these rules, you can prevent material from entering the inside of protective clothing or gear.

Hat and Coveralls Absorption of pesticides through the skin and into the body is highest in the scrotal area, ear canal, forehead, and scalp (Figure 14-7). The use of an appropriate hat and coveralls will minimize the exposure of pesticides in these areas. Lightweight, one-piece, repellent coveralls with hoods are available, and they provide excellent protection.

Apron During the mixing and loading operations, the applicator is exposed to pesticide concentrates. The use of a rubber or neoprene apron at this time will prevent pesticide concentrates from splashing onto the chest, waist, and legs of the applicator.

Goggles and Face Shield Eyes are extremely sensitive to many pesticides. The use of goggles and face shields is recommended when mixing pesticide concentrates or working in a spray, dust, or fog.

Respirators This type of gear reduces the inhalation of pesticide fumes and/or dust. A recommended respirator for pesticide applications is a cartridge respirator, which will absorb toxic fumes and vapors and filter any dust particles in the air. These respirators are often used during the mixing and application of a pesticide.

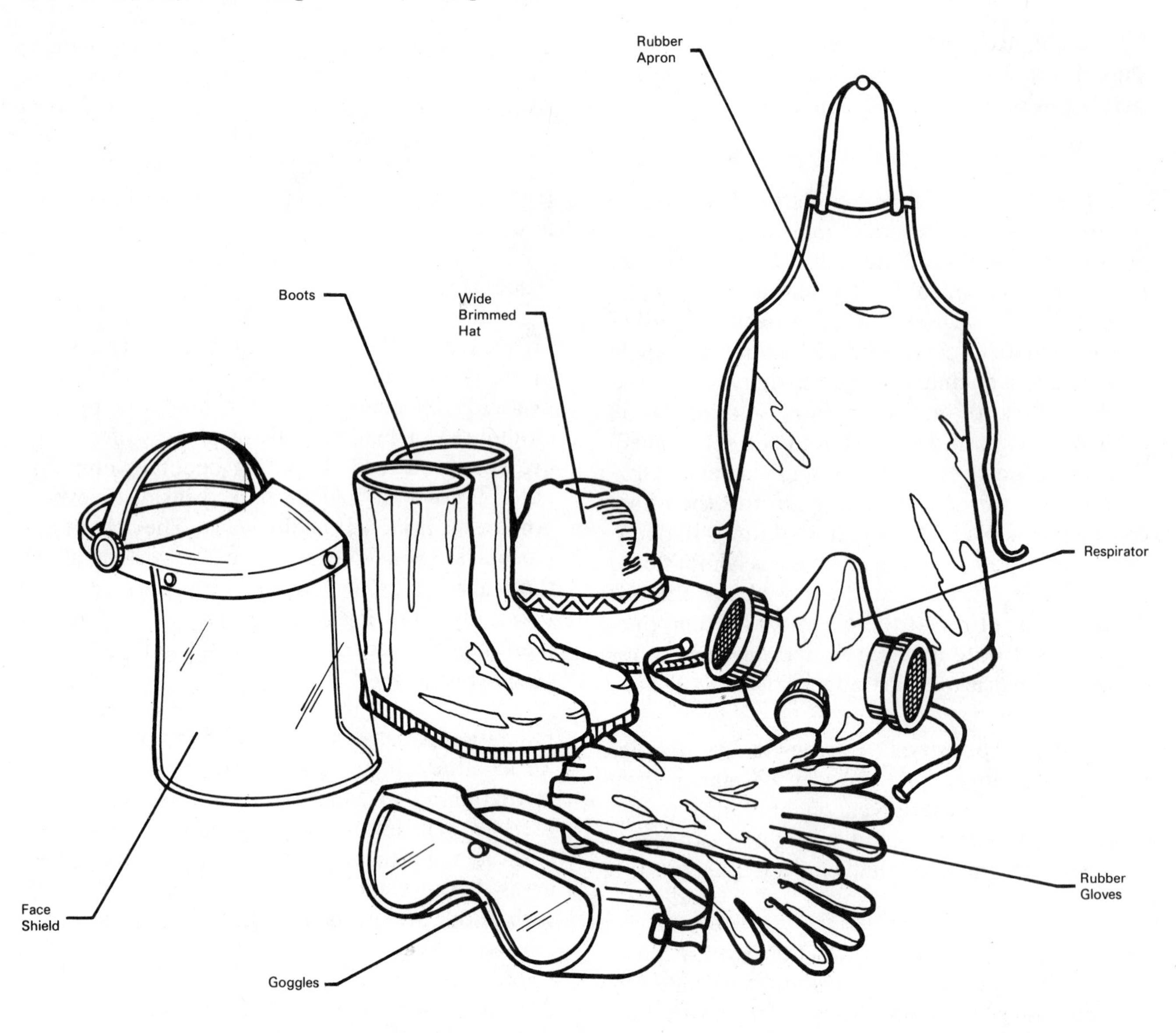

FIGURE 14-6
Safety equipment for pesticide mixing and application

Personal Hygiene Dermal exposure is the principal method of pesticide entry for the applicator. Personal hygiene can drastically reduce this type of exposure. Washing or showering at the end of the work day will remove any pesticide residue on the body. In case of a pesticide spill or splash at the work site, water can be used to immediately remove the material from the skin. After pesticide use, washing your hands prior to eating or using the restroom will further decrease pesticide exposure. Cleaning protective gear and clothing should also be done to prevent any residual exposure to pesticides on these objects.

PESTICIDE STORAGE

Improper storage of pesticides can pose as much danger to the applicator, other people, animals, and the environment as misapplication of pesticides. Some important considerations in selecting a storage facility are the site location and building specifications.

Ideally, the site should be separate from other equipment or material storage facilities. This will reduce risk by decreasing exposure to individuals not involved in pesticide applications. The building should not be located on a floodplain where flooding will introduce pesticides into surface water. It should be built to prevent any runoff or drainage from the site onto sensitive areas. Spill and drainage containment for large storage facilities is highly recommended. Containment systems would trap the pesticides and aid in emergency situations to minimize any environmental damage if the pesticides were to move from the site.

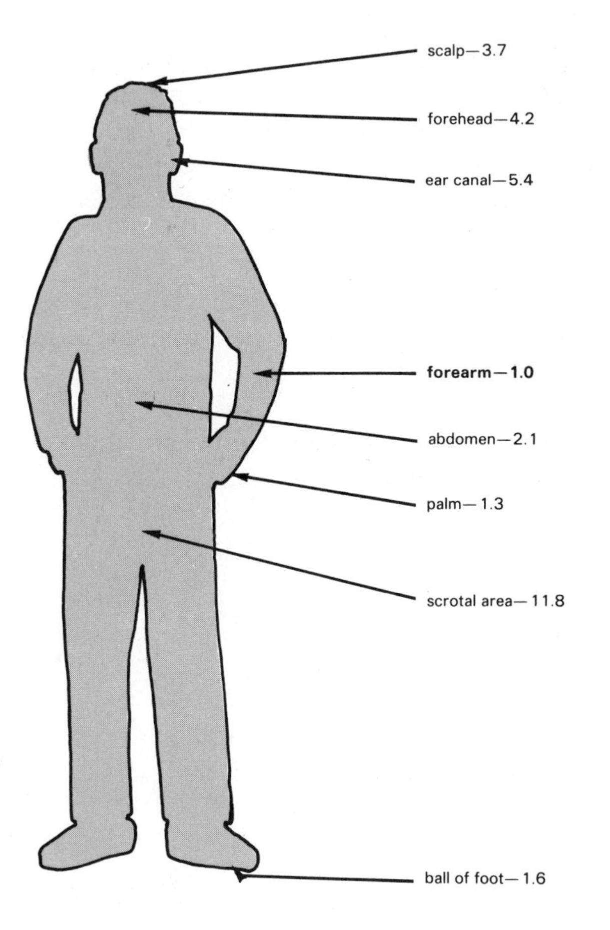

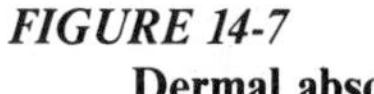
FIGURE 14-7
Dermal absorption sites and rates of pesticides into the body. Comparisons are based on forearm absorption rates.

A well-planned storage building should be well ventilated, have a source of heat and water, be fireproof, have a secure locking system, and have sufficient storage area. The storage area should be well marked with placards indicating the presence of pesticides. The arrangement of the pesticides within the storage area should allow for ease in handling and safety. Tips to provide good storage conditions are:

1. Separate each pesticide class for storage on its own shelf.
2. Keep products off the floor.
3. Store containers so the labels remain in good condition and the containers remain orderly.
4. Practice good housekeeping.

HEALTH AND ENVIRONMENTAL CONCERNS

The current use pattern of pesticides has caused a heightened awareness of their risks. Human health and environmental quality are

Time Period	Lethal Pesticide-Related Accidents per Year
1956	152
1961	111
1968-1970	55
1971-1973	38
1974-1976	32
1977-1979	31
1980-1982	28
1983-1984	27

FIGURE 14-8
The number of lethal pesticide-related accidents in the U.S. from 1956 ***(Courtesy of Science of Food and Agriculture, Cast)***

the major issues in assessing the hazards of pesticide use. The Environmental Protection Agency must conduct a benefit-risk assessment for each pesticide that is registered or reregistered for use. This is an extremely controversial issue. The EPA presently defines acceptable risk to the public at one death per million due to pesticide exposure.

Human Health The number of lethal pesticide-related poisonings in the United States has decreased over the years (Figure 14-8). In 1984, the total number of accidental deaths from all causes in the United States was 92,000. Deaths attributed to pesticide poisoning only numbered 27. However, more than 100,000 nonfatal pesticide poisoning cases per year have been estimated. A majority of the reported deaths were children involved in accidental ingestion of pesticides in the home. The cause has been traced to improper handling and storage practices by homeowners and other consumers.

The residues of a few pesticides used on food crops can pose potential health problems as carcinogens. A *carcinogen* is a material capable of producing a cancerous tumor. In 1987, the National Academy of Science estimated that certain types of pesticide residues in food crops "may" cause up to 20,000 cancers per year. This estimate was based on a "worst-case" scenario. It assumes that exposure to the pesticide would be continuous over 70 years, with the maximum allowable pesticide residue on the food when eaten. Obviously, careful handling and preparation of food eliminates most of this risk.

Environmental Concerns

After a pesticide is applied, not all of the pesticide reaches or remains in the target area (Figure 14-9). When this happens, the pesticide is often considered an environmental pollutant.

The movement of a pesticide from the designated area may occur in several ways. Drift, soil leaching, runoff, improper disposal and storage, and improper application are some of the major causes for a pesticide's becoming an environmental pollutant. Our natural resources that can be contaminated are ground and surface water, soil, air, fish, and wildlife.

Recent surveys have shown that more than 50% of the counties in the United States have potential groundwater contamination from agrichemicals. The three main factors affecting groundwater contamination by agrichemicals are:

Soil type and other geological characteristics
The pesticide's persistence and mobility within the soil
The production and application methods of pesticide users

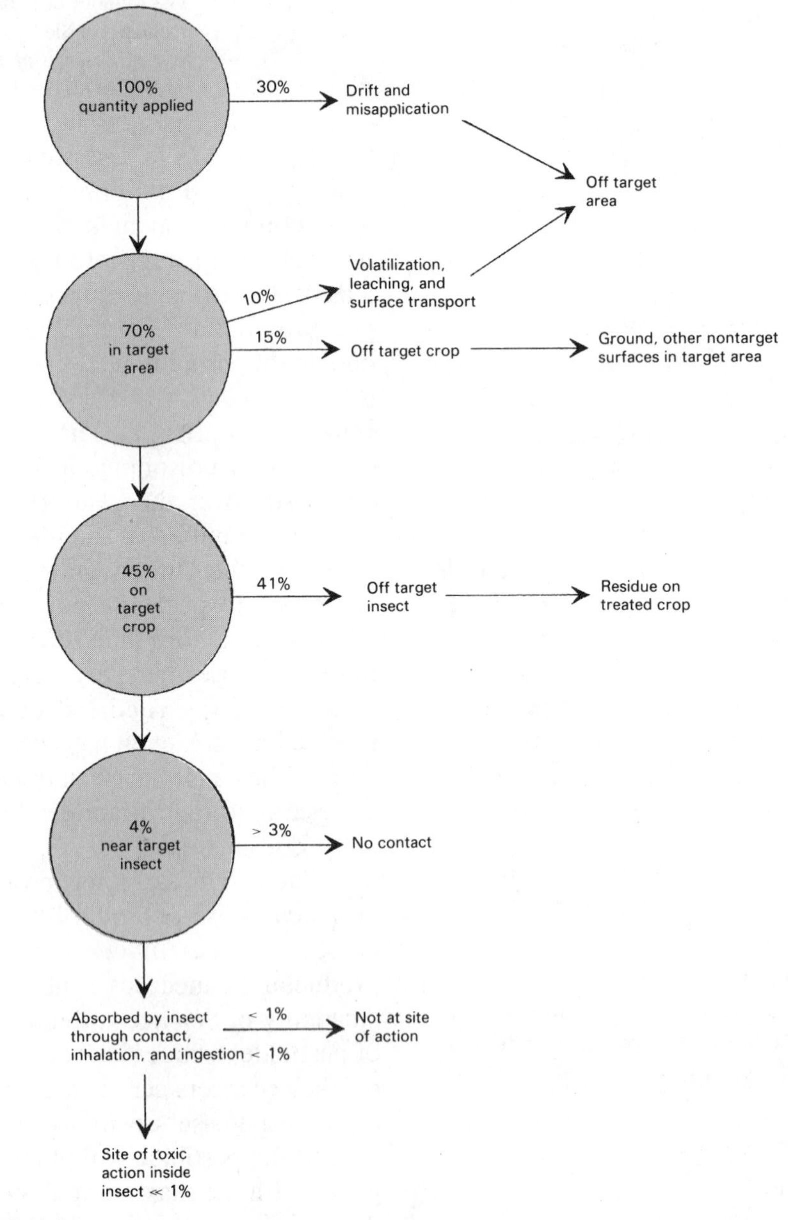

FIGURE 14-9

The movement of a pesticide after discharge from an aerial spray plane *(Used with permission of Flint, M. L. and R. Van den Bosch. From Introduction to Integrated Pest Management. 1981. Plenum Press, New York. 240 pp.)*

Pesticide drift is a major cause of soil and air contamination. *Drift* is the movement of a pesticide through the air to nontarget sites. It will occur at the time of pesticide application when small spray particles are moved by air currents to nontargeted areas. Also, vapor drift of a pesticide may occur after an application. *Vapor drift* is movement of pesticide vapors due to chemical volatilization of the product.

The adverse effects of pesticides on fish and wildlife may directly result in animal mortality. They may also indirectly influence animal feeding or reproduction. Pesticide labeling will indicate any potential harm to wildlife, and this information should be heeded to minimize risk. Fish, birds, bees, and other animals will be affected when pesticides reach them or their habitat.

Environmental contamination by agrichemicals can be decreased by several management practices. The use of integrated pest management (IPM) programs can reduce pesticide use. If pesticides are used, then proper mixing, application, and storage and disposal must be done. These practices will decrease any adverse effect on the environment. An attempt must also be made to minimize any effect that temperature, soil type, rainfall, and wind patterns may have on a pesticide's becoming an environmental pollutant.

Chemical pesticides are an important part of our food- and fiber-production capability. They are necessary to maintain our current standard of living in the United States and the world. However, they do create risks if used by improperly trained or careless individuals. Therefore, it is important that those using pesticides be properly informed and follow label instructions.

It is important that consumers wash and handle food products to minimize the intake of pesticide residues on food. While the government does extensive research to arrive at safe pesticide-residue levels, it is in the best interest of the consumer to wash fruits, vegetables, and all plant parts that may contain pesticide residues. This precaution will further reduce the hazard of pesticides.

Similarly, the wise person will become familiar with the use of pesticides. Such uses include pesticides to control roaches, termites, flies, insects, and diseases of lawn and garden plants; insect repellants and insecticides used for outdoor camping and recreation; and other everyday common practices. The benefits of pesticides far outweigh the risks if sensible use is observed.

Student Activities

1. Define the "Terms To Know."
2. Ask your instructor to show examples of an approved pesticide applicator's respirator, goggles, gloves, boots, clothing, and other protective materials.
3. Prepare and present a class demonstration on the proper use of one or more items of protective clothing and devices for safe handling of pesticides.
4. Do some research and prepare to debate the issue of pesticide residues on food in the United States. Arrange for at least three classmates to make similar preparation for a debate.
5. Work with the classmates arranged for in item 4 and present a debate in class. Have two people take the position that pesticide residues on food are dangerous and must be reduced. Have the other two persons defend the position that current levels of pesticide residues are safe and the benefits of using pesticides outweigh the risks to consumers.
6. Ask a professional pesticide applicator with a special interest in pesticide safety to demonstrate safe pesticide application principles to the class.
7. Collect newspaper and magazine articles on accidents involving pesticides. Study the articles and determine how each accident could have been avoided. Report your conclusions to the class. *Note:* Many pesticide accidents occur in or on the home, lawn, and garden—don't overlook such cases.

8. Conduct a survey of pesticide storage and use practices in your home, farm, and place of employment. Correct all safety violations.

Self-Evaluation

A. MULTIPLE CHOICE

1. An example of an inorganic pesticide is
   a. pyrethrum.
   b. rotenone.
   c. Bordeaux mixture.
   d. organophosphate.

2. The total amount of pesticides used in the United States is
   a. 820 million pounds.
   b. 420 million pounds.
   c. 620 million pounds.
   d. 1.2 billion pounds.

3. A preemergence herbicide is applied
   a. after the weed or crop is present.
   b. before the weed or crop is present.
   c. at the time of planting.
   d. none of the above.

4. Which herbicide family inhibits photosynthesis?
   a. acetanilines
   b. dinitroanilines
   c. phenoxys
   d. triazines

5. An example of botanical insecticide is
   a. 2, 4-D.
   b. rotenone.
   c. diazinon.
   d. sulfur.

6. Pesticide registration often takes
   a. 1-2 years.
   b. 3-4 years.
   c. 8-10 years.
   d. 15 years.

7. Pesticide risk can be decreased by
   a. proper pesticide exposure.
   b. reading the label.
   c. minimizing pesticide exposure.
   d. all of the above.

8. The signal word for a highly toxic pesticide is
   a. CAUTION.
   b. WARNING.
   c. DANGER-POISON.
   d. none of the above.

9. An example of protective clothing or gear that will minimize inhalation of a pesticide is
   a. respirator.
   b. boots.
   c. gloves.
   d. coveralls.

10. The number of lethal pesticide poisoning cases per year is
    a. less than 20.
    b. 20-30.
    c. 40-60.
    d. more than 100.

B. MATCHING

_______ 1. Carcinogen
_______ 2. Chronic toxicity
_______ 3. Water
_______ 4. DF
_______ 5. Signal word
_______ 6. Acute toxicity
_______ 7. SP
_______ 8. Drift
_______ 9. Protectant
_______ 10. Eradicant

a. The effect of a single exposure to a pesticide
b. Movement of a pesticide through the air to nontarget sites
c. Repeated exposures to a low dose of a pesticide
d. A soluble powder pesticide formulation
e. A material capable of producing a tumor
f. A fungicide used after disease infection
g. Used to remove pesticides from the body
h. A dry flowable pesticide formulation
i. Describes the acute toxicity of a pesticide
j. A fungicide used prior to disease infection

C. COMPLETION

1. An ______ will control insects.
2. A ______ herbicide will control all types of plants.
3. A systemic herbicide will be translocated in the ______ and ______ tissue.
4. The amount of money spent on pesticides in 1988 was ______ billion dollars.
5. A pesticide will only have one ______ and one ______ name.
6. To reduce exposure to pesticides, the use of ______ is recommended for clothing and boots.
7. Environmental and health hazards of a pesticide are listed under the ______ ______ of the label.
8. The ______ number identifies where the pesticide was manufactured.
9. "LD" refers to the ______ ______ of a pesticide.
10. The ______ name is the manufacturer's name for its product.

Plant Sciences

UNIT 15 Plant Structures and Taxonomy

OBJECTIVE To identify major parts of plants and state the important functions of each.

Competencies to Be Developed

After studying this unit, you will be able to

- ☐ draw and label the major parts of plants.
- ☐ describe the major functions of roots, stems, fruits, and leaves.
- ☐ draw and label the parts of a typical root, stem, flower, fruit, and leaf.
- ☐ explain some of the variations found in the structure of root systems, stems, flowers, fruits, and leaves.
- ☐ describe the relationship of plant parts to fruits, nuts, vegetables, and crops.

TERMS TO KNOW

Adventitious roots
Taproot
Ornamental
Fibrous roots
Root cap
Area of cell division
Area of cell elongation
Xylem
Phloem
Area of cell maturation
Root hairs
Stems
Woody
Herbaceous
Bulbs
Corms
Rhizomes
Tubers
Vascular bundle
Node
Internode
Axillary bud
Axil
Lenticels
Terminal bud
Vegetative bud
Flowering bud
Leaf
Margins
Simple leaf
Compound leaf
Leaf blade
Petiole
Photosynthesis
Cuticle
Epidermis
Palisade cells
Spongy layer
Chloroplasts
Mesophyll
Stoma
Guard cells
Flower
Bract
Stamen
Filament
Anther
Pollen
Pistil
Stigma
Style
Ovary
Ovules
Perfect flower
Imperfect flower
Pollination
Petals
Corolla
Sepals
Caylx
Fruit
Vegetable
Nut
Taxonomy
Genus
Species
Binomial
Variety

MATERIALS LIST

plant collection materials
bulletin board materials

Plants are a basic part of the food chain. Without plants the web of life cannot exist, and most animals and humans will die. A knowledge of plant growth is essential. To have a better understanding of plants, it is necessary to identify the parts that make up the plant. The casual observer sees stems, branches, leaves, and possibly flowers, and some nuts or fruits. The agriscience technician, however, will see a series of interconnected tissues and organs that depend on each other to function. The technician knows that all of the organs do not need to be present at one time, but is aware that each cell or organ (composed of cells) has an important role in the successful growth of the plant. The technician or plant scientist is concerned with the efficient growth of plants. A plant may become stressed if one or more parts are absent or not functioning properly.

The basic industry of agriculture is dependent on the proper function of plants. The animal grower needs many kinds of plants to feed livestock and poultry. The plant industry needs plants for seeds to sell to feed mills and to farmers for growing crops and other plants. The horticulture industry needs plants for seeds and cuttings, and plants for food such as fruits, vegetables, and nuts. Plants are also needed for landscaping the inside and outside of homes and office buildings.

To be successful with plants, you must have a knowledge of plant parts and how they function. Such a knowledge is essential, whether you are growing, selling, or using plants.

THE PLANT Plants are composed of many parts. Each part is important in the overall life and function of a plant. The root system is normally under the ground and is responsible for anchoring the plant and supplying water and nutrients. The stem or trunk is normally above the ground and functions as a support system for the rest of the aboveground parts. Leaves constitute the food-manufacturing parts of plants. Flowers come in many sizes, colors, and shapes, and function as the seed-producing part of the plant. Healthy plants produce seeds, nuts, fruits, and vegetables. These parts are popular foods for animals and humans. They may also be used for reproduction of the plant (Figure 15-1).

FIGURE 15-1
Seeds, nuts, and fruits are plant parts commonly used for food.

ROOTS

Root Systems The largest part of the plant is often the root system. Roots take up more space in the soil than the top part of the plant that is seen in the air above the ground. In fact, some roots will go down into the soil 6, 8, or even 10 ft. Some plants, such as a squash, might have as many as 2,000 miles of roots.

While many roots are in the soil, there are other types that we see above the ground and may not consider as roots. Some plants, such as poison ivy *(Rhus radicans)* and English ivy *(Hedera helix)*, have roots that help them climb trees, walls, and sides of buildings. These are called adventitious roots. *Adventitious roots* appear where roots are not normally expected.

The mistletoe, a popular Christmas plant, has roots that penetrate the bark of trees in the upper branches or crown. These roots grow into the xylem and phloem tissues of the host plant and extract nutrients that originate in the soil. The dodder *(Cuscuta campestris)* has soil roots that die off as the plant gains a foothold in a plant. The dodder plant forms rootlike attachments that penetrate the stem of the host plant and extract nutrients from that plant.

Roots are divided into two major groups: taproot and fibrous roots (Figure 15-2). Knowledge of these two different types of root systems can be of value in caring for and handling plants. The *taproot* is the main root of a plant and generally grows straight down from the stem. It is a heavy, thick root that does not have many side or lateral branches. Taproots are often used for human and livestock consumption, since they are food storage organs. Carrots *(Daucus carota)* and sugar beets *(Beta vulgaris)* are examples (Figure 15-3). Some plants

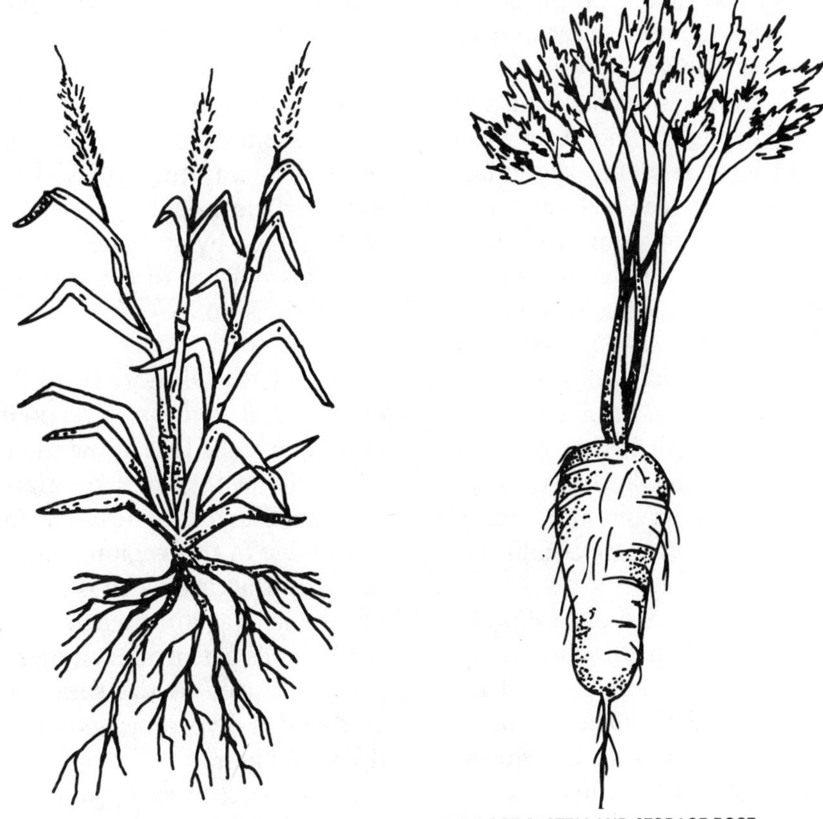

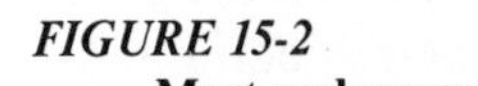

FIGURE 15-2
Most underground roots are parts of either a tap or fibrous root system.

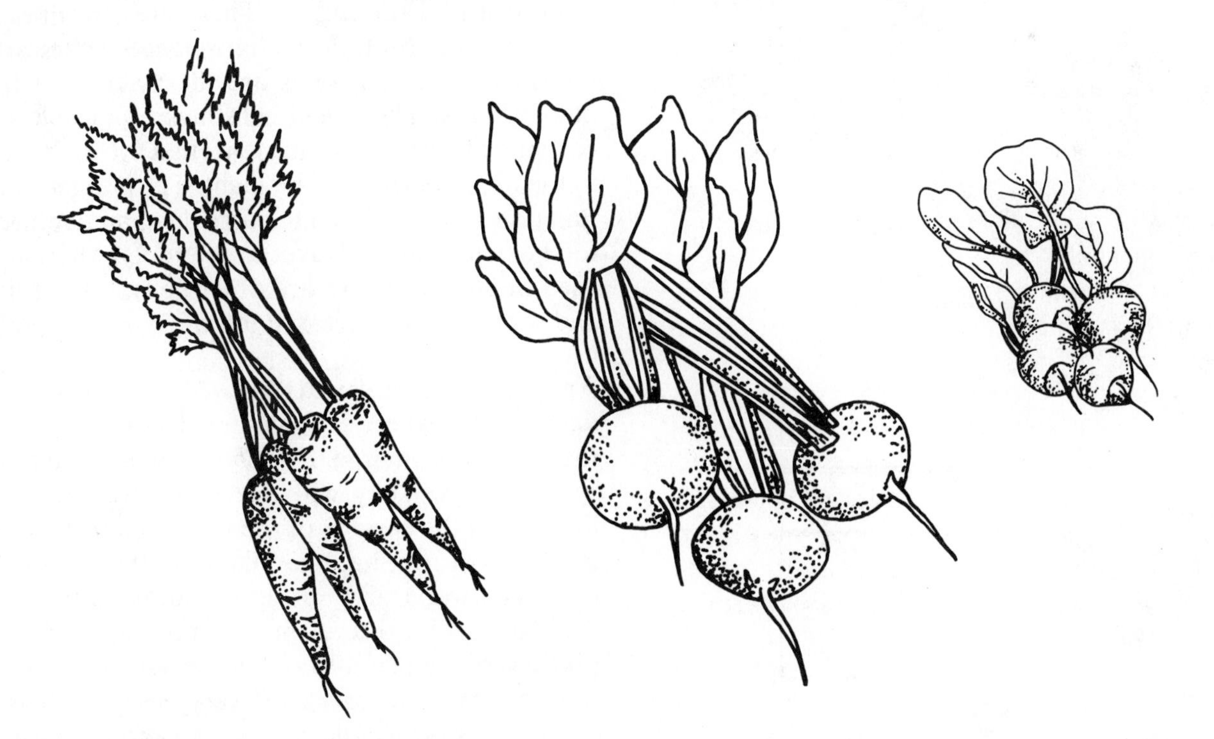

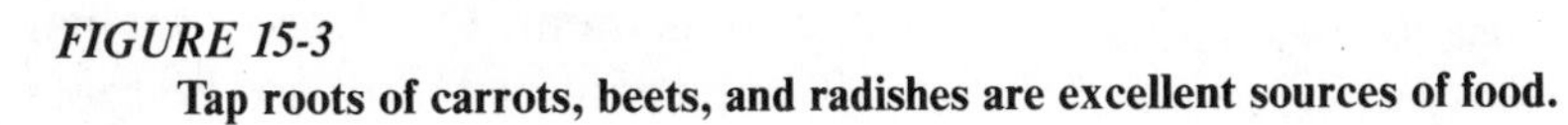

FIGURE 15-3
Tap roots of carrots, beets, and radishes are excellent sources of food.

with taproots are used for ornamental purposes. An *ornamental* plant is used to improve the appearance of a structure or area.

Plants that have a taproot system have the ability to survive periods of drought. Since they grow deep into the soil and have few fine secondary roots, taproots do not stabilize the soil very well. *Fibrous roots* are generally thin, somewhat hairlike, and numerous. The fibrous root system is normally very shallow. Plants such as grasses and corn, and ornamentals such as *Begonia semperflorens*, are good examples of plants with fibrous root systems. There are many small, thin branched roots in this type of system. The result is that they are able to hold soil much better than taproot systems. However, fibrous root systems dry out more easily. Therefore, they cannot tolerate drought conditions.

Root Tissues While there are different systems, all roots look similar when they are examined on the inside (Figure 15-4). The parts of the root have very specific functions in the plant. A knowledge of these parts is helpful in diagnosing diseases and other dysfunctions of plants.

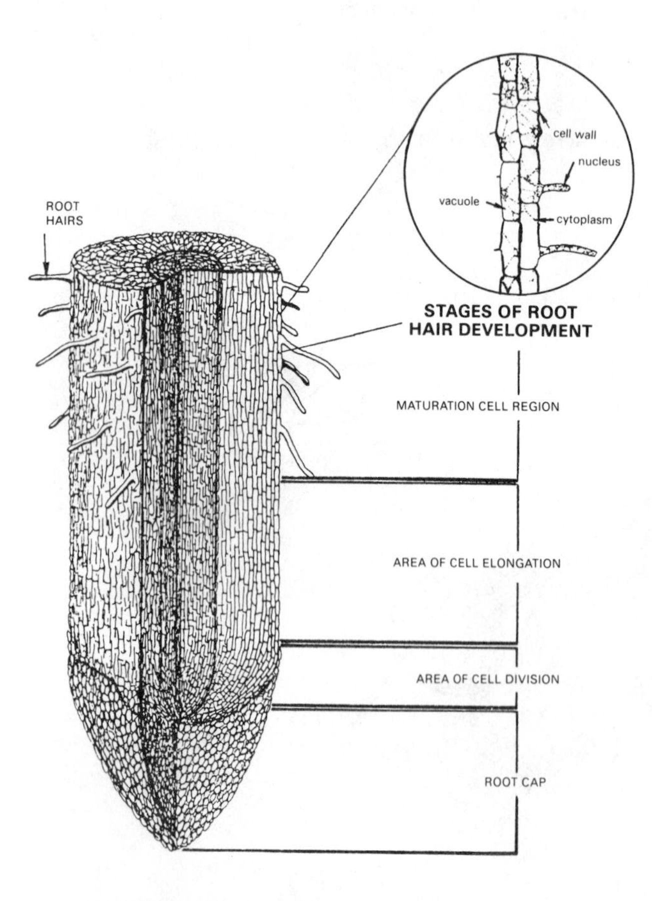

FIGURE 15-4
Cross section of a root showing the major internal parts

The Root Cap

The *root cap* is the outermost part of the root. It protects the tender growing tip as the root penetrates the soil. The root cap is a tough set of cells that are able to withstand the coarse conditions that the root encounters as it pushes its way through soil with rock and small sand particles. As the root cap wears away, the cells are replaced by more cells that develop at the root tip. This portion of the root is known as the area of cell division.

Area of Cell Division and Elongation

The *area of cell division* is responsible for producing more new cells that allow the root to grow longer. The cells in this area multiply in two directions. The small and tougher cells are produced on the front edge of this region. They replace cells of the root cap that are worn off or destroyed as the tip pushes its way through the soil. Small and more tender cells are produced on the back of this area. They are used as the root tip grows longer. The area of cell division is actually very thin, maybe as thick as a hair.

The next area, as you move toward the base of the plant, is the *area of cell elongation*. In this area the cells start to become longer and specialized. They also begin to look like the older cells and will start to do their specific jobs.

Xylem and Phloem

There are many types of cells in the root. To an agriscience technician, the most important ones are the xylem and the phloem cells. The *xylem* cells are responsible for carrying the water and nutrients that are in the soil to the upper portion of the plant. The *phloem* cells function as the pipeline to carry the manufactured food down from the leaves to the roots, where it is used or stored. There are other cells in the root, and some will be discussed later.

Area of Cell Maturation

The *area of cell maturation* is where cells mature. This is also where the root hairs emerge. *Root hairs* are small microscopic roots. They will rise from existing cells located on the surface of the root (Figure 15-4). It is the job of root hairs to take in water and nutrients. The water and nutrients will move into the root hairs, enter the xylem, and move to the upper portions of the plant. Root hairs are small and very tender. They will break off very easily. This is a good reason to handle plants very carefully when transplanting them. Once the root hairs are broken off, they cannot regrow or be replaced.

Although roots are normally hidden from view, they are still important parts of the plants. Roots need the same care and consideration as the other parts in order for plants to grow well.

STEMS Stems are one of the first things the casual observer sees when looking at plants (Figure 15-5). Stems and branches are noticeable in the winter when the leaves are gone. They are easily seen as the plant grows. *Stems* support the leaves, flowers, and fruit.

Types of Stems Aboveground stems are of two types: woody and herbaceous. *Woody* stems are tough and winter hardy. They often have bark around them. *Herbaceous* stems are succulent, often green, and will not overwinter in cold climates.

Not all stems are erect, aboveground structures. Some grow along the ground or even underground. Some stems have a specialized job to perform. These are often referred to as modified stems. Examples of modified stems are bulbs, corms, rhizomes, and tubers. *Bulbs* are shortened stems that are surrounded by modified leaves called scales. Some examples of bulbs are Easter lilies *(Lillium longiflorum)* and onions *(Allium sp.)* (Figure 15-6). *Corms* are thickened, compact fleshy stems. An example of a corm is the gladiola *(Gladiolia sp.)* (Figure 15-7). *Rhizomes* are thick stems

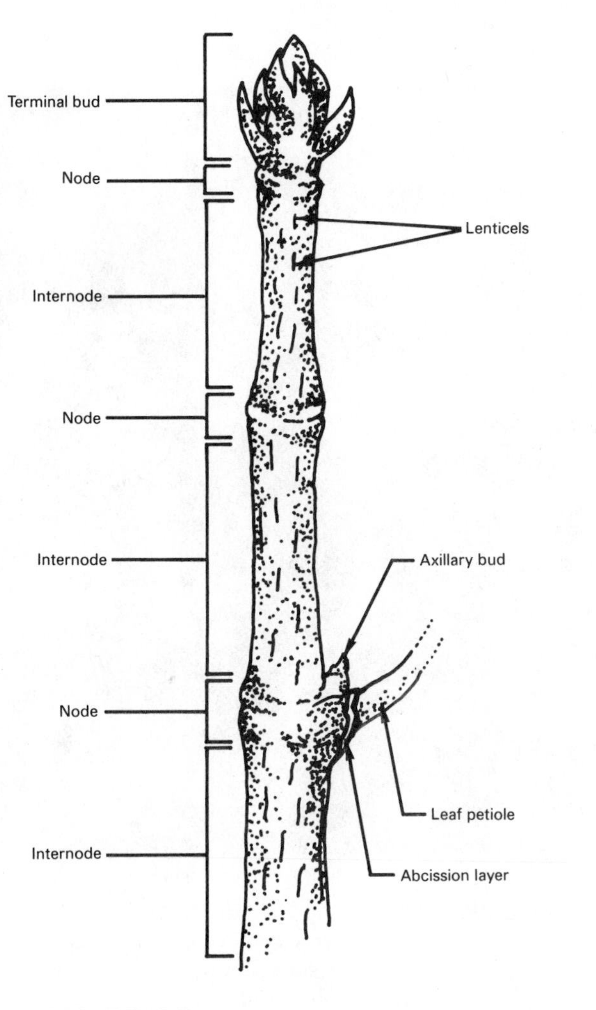

FIGURE 15-5
Important parts of stems

AGRI-PROFILE

CAREER AREAS: BOTANY/BIOLOGY/TAXONOMY

In the last two centuries, plant scientists have studied individual plant specimens and devised a system of classification of plants based on characteristics of the plant parts. ***(Courtesy of the National FFA Organization)***

Everyone relies heavily on our system of plant classification. Without it we could not order, buy, sell or instruct others about a plant unless we had the plant in the presence of both parties at the same time. Taxonomists devote their careers to identifying, classifying, and teaching others about plants. An important and exciting part of their work is that of discovering, studying, and naming new plants in their appropriate places in the classification system. Consider the excitement of discovering plants in far away lands or even at home which have not been observed or recognized even by the most knowledgeable specialists to date.

Botany is a study of plants, and biology is a study of both plants and animals. A knowledge of plant structures is critical to both. Similarly, consumers of plants for food, ornamentation, medicine, shade, wood, and other have some knowledge of plant structures. Plant structures affect plant nutrition, functions, disease susceptibility, adaptation, and use.

that run below the ground. The iris *(Iris germanica)* is an example of a plant with rhizomes (Figure 15-8). *Tubers* are thickened underground stems that store carbohydrates. We often eat an example of this type of stem, the Irish potato *(Solanum tuberosum)* (Figure 15-9).

Parts of Stems Stems have some of the same internal parts as roots. The xylem and the phloem continue to run the length of the stem and into all of the branches of the plant. In a subclass of plants called dicotyledons, the xylem and phloem occur together in tissue called *vascular bundles*. In another important subclass called monocotyledons, the xylem and phloem occur in separate areas (Figure 15-10). This group of tissues is called the *vascular bundle*.

The external parts that are important to the agriscience technician are the node, internode, axillary bud, lenticels, and terminal bud. The *node* is the portion of the stem that is swollen or slightly enlarged as a result of the leaf and a bud rising from it. The *internode* is the area between the nodes. The *axillary bud* grows out of the axil. The *axil* is the angle above a leaf or flower stem, and the stalk. The function of the axillary bud is to develop into a leaf or branch. The *lenticels* are pores in the stem that allow the passage of gases in and out of the plant. The *terminal bud* is located on the tip or top of the stem or its branches. It may be either a vegetative or flowering bud. The *vegetative bud* will produce the stem and leaf growth of the plant. The *flowering bud* will produce flowers.

FIGURE 15-6
Bulbs, such as Easter lilies and onions, are shortened stems surrounded by modified leaves called scales. ***(Courtesy George Rogers Area Vocational-Technical School)***

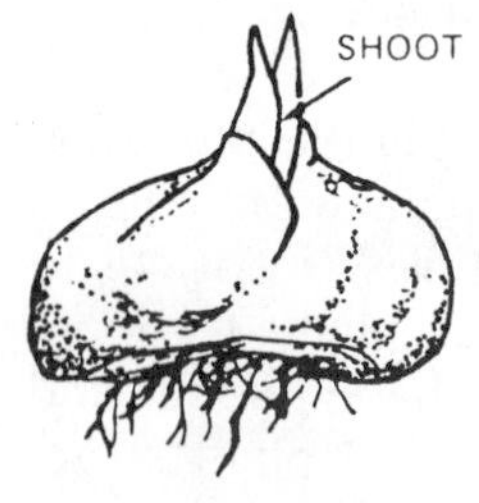

FIGURE 15-7
Corms, such as the gladiolus, are stems that are thick, compact and fleshy. ***(From Ingels/Ornamental Horticulture, copyright 1985 by Delmar Publishers Inc.)***

FIGURE 15-8
Rhizomes are thick stems which grow underground near the surface and give rise to new plants at each node. Johnson grass and iris are examples of plants with rhizome stems. ***(Courtesy George Rogers Area Vocational-Technical School)***

LEAVES The *leaf* of a plant has a very important function. It manufactures food for the plant by using light energy. The leaf exposes itself to as much sunlight as possible.

Leaf Margins

Plants may be identified by the edge, shape, and arrangement of the leaves. The leaf edges are known as *margins*. Leaf margins are named or described according to the toothed pattern on each leaf edge (Figure 15-11).

Leaf Shape and Form

Leaves vary in shape and form according to their species. Therefore, knowledge of the name given to each leaf shape and form is useful in identifying the plant (Figure 15-12).

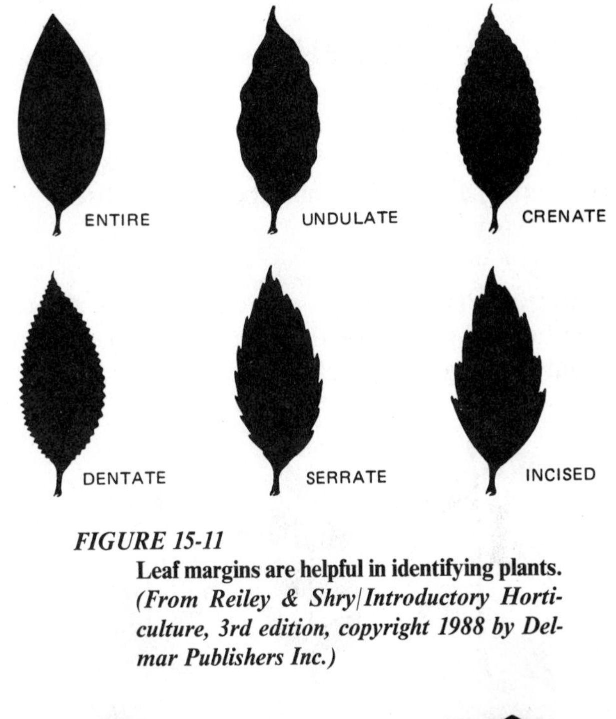

FIGURE 15-11
Leaf margins are helpful in identifying plants. ***(From Reiley & Shry/Introductory Horticulture, 3rd edition, copyright 1988 by Delmar Publishers Inc.)***

FIGURE 15-9
The common Irish potato is really a specialized stem called a tuber. ***(Courtesy George Rogers Area Vocational-Technical School)***

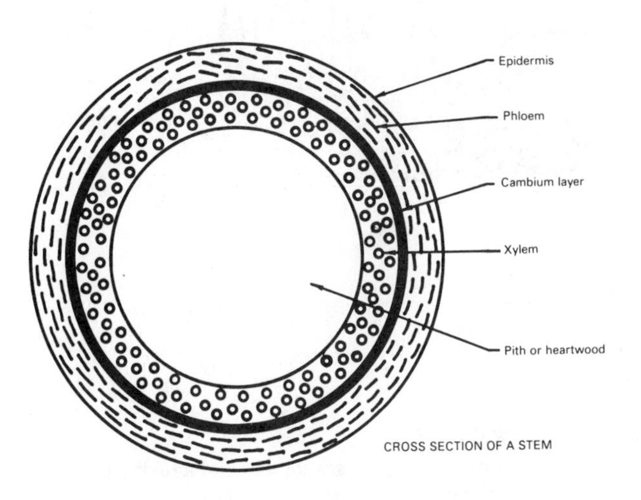

FIGURE 15-10
Xylem and phloem cells make up the vascular system of plants.

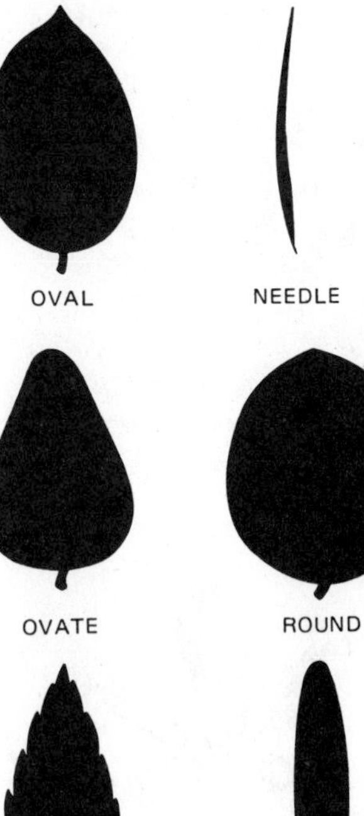

FIGURE 15-12
Names given to various leaf shapes ***(From Reiley & Shry/Introductory Horticulture, 3rd edition, copyright 1988 by Delmar Publishers Inc.)***

Types of Leaves Leaf types vary according to the species. Therefore, leaf type is also used to identify plant species. A single leaf arising from a stem is called a *simple leaf*. Two or more leaves arising from a common point on the stem is referred to as a *compound leaf* (Figure 15-13).

Leaf Parts A leaf consists of a petiole and blade (Figure 15-14). These are the most familiar parts of a leaf. The *leaf blade* is the wide portion. It may be of many shapes and sizes. The *petiole* is the stem of the leaf. It may be almost absent or may be very long.

Internal Structure The leaf is the food-manufacturing unit for the plant. The process of manufacturing food is called *photosynthesis*. The food that is created through photosynthesis in leaves enables the plant to grow. The process of photosynthesis is illustrated in Figure 15-15 and discussed in greater detail in a later unit.

The *cuticle* is the topmost layer of the leaf. It is waxy and functions as a protective covering for the rest of the leaf. The *epidermis* is the surface

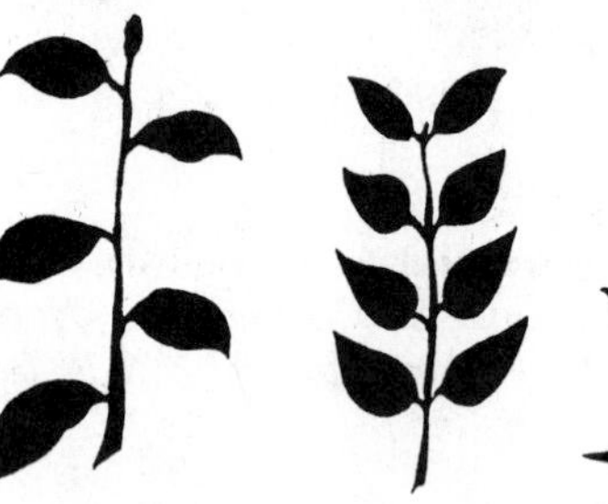

FIGURE 15-13
Examples of various leaf arrangements ***(From Reiley & Shry/Introductory Horticulture, 3rd edition, copyright 1988 by Delmar Publishers Inc.)***

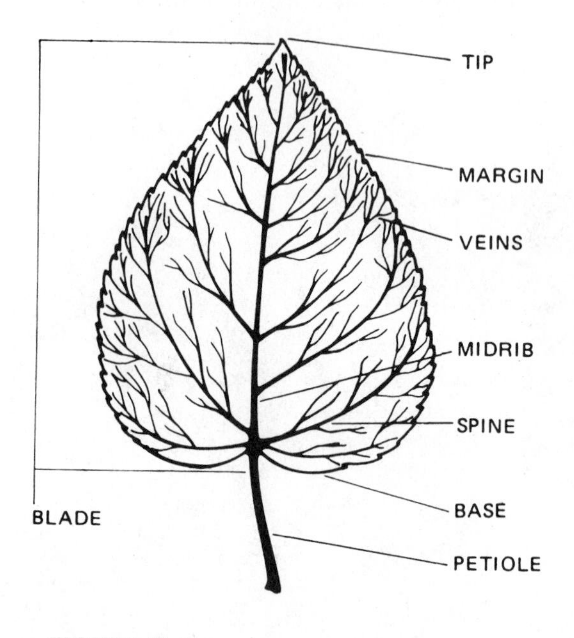

FIGURE 15-14
Parts of a leaf ***(From Reiley & Shry/Introductory Horticulture, 3rd edition, copyright 1988 by Delmar Publishers Inc.)***

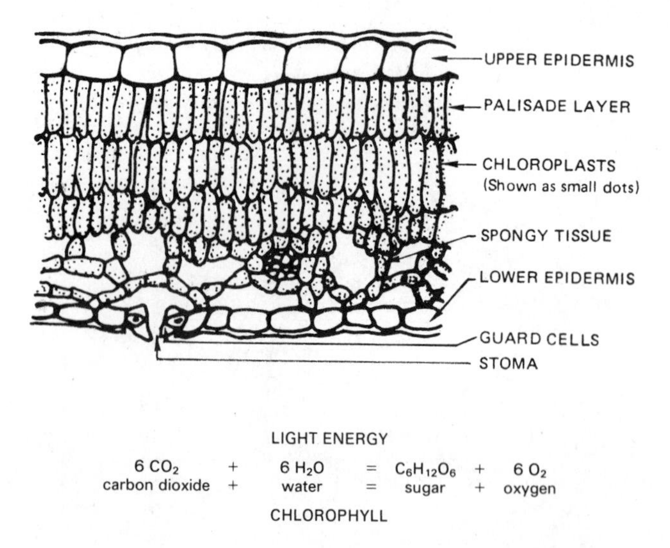

FIGURE 15-15
The leaf is the plant food manufacturing part of plants. ***(From Reiley & Shry/Introductory Horticulture, 3rd edition, copyright 1988 by Delmar Publishers Inc.)***

layer on the lower and upper side of the leaf. The epidermis protects the inner leaf in many ways. The elongated, vertical *palisade cells* give the leaf strength and is the site for the food-manufacturing process. These cells, as well as the lower *spongy layer*, contain chloroplasts. *Chloroplasts* are the parts of the cells that contain chlorophyll and are necessary for photosynthesis to be carried out. The lower layer is irregular and allows the veins, or vascular bundle, to extend into the leaf. The layer of palisades and spongy tissue is often referred to as the *mesophyll*. The vascular bundle contains the xylem and the phloem. These are an extension of the same tissue that is located in the root and runs through the stem to the leaves. The xylem brings the water and minerals from the root. The phloem will remove the manufactured food from the leaf and carry it to the various parts of the plant. There, the food nourishes plant tissue or it will be stored. The lower epidermis contains some special cells called *stoma*. They are openings that allow for the exchange of carbon dioxide and oxygen, as well as some water. The stoma are surrounded by *guard cells* that open and close the stoma. If the plant is stressed by the lack of water or by a low light level, the guard cells will close the stoma. The result is that the plant cannot manufacture food because it will not have all of the ingredients necessary.

FLOWERS Many people see plants for only the beauty of the flower. Others see in the plant only a fruit to eat. Fruit production is only part of the job of the flower. The *flower* has as its primary function the production of seeds needed to continue the species. It is with this structure that the plant scientist will work to produce new and different varieties.

Not all of the beauty that is seen as flowers are actually flowers. The poinsettia *(Euphorbia pulcherrima)* and the flowering dogwood *(Cornus florida)*, for example, have modified leaves called bracts. A *bract* is a modified leaf that is often brightly colored and showy. People often see the red or white bracts and call them flowers. Their function is to protect the flower parts, as well as attract insects for pollination.

Flower Structure Flowers are composed of various parts. These include the filament, anther, pollen, stigma, style, ovary petals, and sepals (Figure 15-16). These are the most important parts of the flowers. Other flower parts are not important in the study of plant structures at this point.

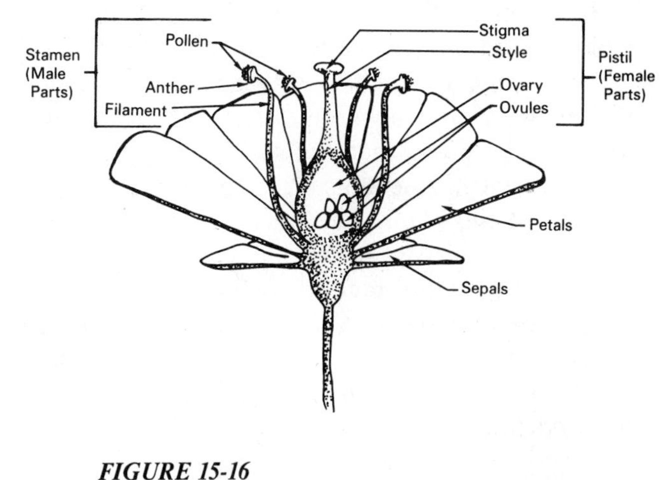

FIGURE 15-16
Major parts of flowers

The male part of the flower is the *stamen*. It consists of the filament, the anther, and the pollen. The *filament* supports the anther. The *anther* manufactures the pollen. The *pollen* is the male sexual reproductive cell. The female part of the flower, the *pistil*, is made up of the stigma, style, and ovary. The *stigma* receives the pollen. The pollen travels down the *style* and into the *ovary*. The ovary contains *ovules*. These are the eggs, which are the female reproductive cells. When the eggs are fertilized by the pollen, they will ripen into seeds.

If a flower contains all of the parts mentioned above, it is a *perfect flower*. If one or more of the parts are missing, it is considered an *imperfect flower*. In some plants, particularly the flowering plants used in horticulture, it is desirable to remove the anther sacs before the pollen ripens. This prevents pollination and stains from the pollen on the petals of the flower. *Pollination* means the union of the pollen with the stigma. In some cases, orchids for example, the unfertilized flower will last for many months.

When plants are bred, it is the flower that is important. The anther sac is removed from the plant to prevent natural pollination. It may be destroyed, or it might be used to pollinate another flower to create a new variety. Many hybrids are created in this way.

The colored *petals* attract insects or other natural pollinators. The flower petals are collectively called the *corolla*. The *sepals* function as a protective device for the developing flower. Collectively, sepals are called the *caylx*.

Fruits, Nuts, and Vegetables After fertilization, the ripening seed develops in the pistil. The pistil then enlarges and becomes the *fruit* (Figure 15-17). The fruit may be of many different shapes and sizes (Figure 15-18). The true fruit are the seeds that carry the male and female genetic characteristics of the plant. However, the fleshy material surrounding the mature seed, and the seed itself, is commonly called the fruit of the plant. The purpose of the fleshy part of the fruit is to attract animals and humans to the seed to help spread it over wide areas. This helps in the reproduction of plants. Entire fruit or just the seed may be moved by wind, water, animals, and humans. Often the fruit is eaten by animals and humans as a source of food. Then the seeds may be discarded, where they can take root and grow into new plants. People assist greatly in spreading seeds and starting new plants when they plant seeds for crops.

There are many kinds of fruits and vegetables. The two terms are sometimes used incorrectly. A *vegetable* can be any part of a plant that is grown for its edible parts. This can be a root, stem, leaf, or ripened flower. However, fruit, a ripened or mature ovary, is a specific plant part. A *nut* is also a type of fruit.

PLANT TAXONOMY

Importance of Classifying Plants

Taxonomy is the science, laws, and principles of classification. In biology, taxonomy provides the means for classifying organisms into established categories according to characteristics. Such classification makes it easier to understand and remember plants and animals by the similarities and differences found in their structure and parts. Living organisms are given Latin names to help scientists and technicians around the world communicate better. Latin is regarded as the universal language for those in professions dealing with the biological sciences. Agriscience has its origin in the biological sciences.

There are about 300,000 species of plants that have been identified and classified. The plant names are based on Latin descriptions and must be approved by a special committee of plant scientists. Carl Linnaeus, a Swedish botanist, developed the present system of plant classification in 1753.

If there were no botanical classification to identify and classify plants, many different species would carry one common name. For example, all clovers would be identified as clover, even though crimson clover is a winter annual and sweet clover is a biennial.

The examples of the complete classification of field corn and the petunia are listed in the table on page 209.

Binomial System Used in Classifying Plants

When identifying a plant by its scientific name, it is not necessary to give its entire classification. Rather, a specific plant can be identified by using the genus and species only, since the genus and species name is not used in combination for any other plant or animal. For example, grain sorghum is *Sorghum* (genus) *vulgare* (species). *Genus* is the taxonomic category between family and species. It is customarily capitalized when written along with a species name. *Species* is the subgroup under genus. Species names are gen-

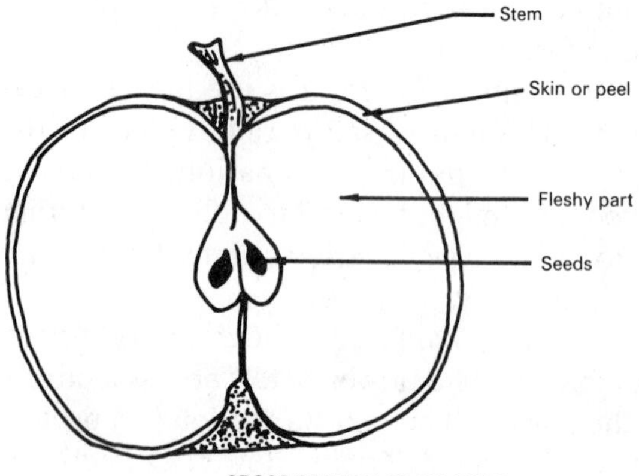

FIGURE 15-17
Seeds develop from the fertilized egg or ovum in the ovary to become part of the fruit of the plant.

FIGURE 15-18
Fruits from different plants vary in size, shape, and taste.

Common Name:	Corn	Petunia
Kingdom:	Plant	Plant
Phylum:	Spermatophyta (seed plants)	Embryophyta
Subphylum:	Angiosperm (seed in fruit)	Angiosperm
Class:	Monocotyledonae (single leaf seed)	Dicotyledonae (two-seed leaf)
Order:	Graminales (grasslike families)	Tubiflorea
Family:	Gramineae (grass family)	Solanaceae
Genus:	Zea (the corns)	Petunia
Species:	Mays (dent corns)	Hybridea
Variety:	Reid's yellow dent	Blue Moon

erally not capitalized when written in combination with their genus. To compare a plant or animal name with that of a human's, the species corresponds to the person's first name and the genus to the person's last name. The system of using genus and species in combination is referred to as a binomial system of classification. *Binomial* means consisting of two names.

Some species are broken down into varieties. A *variety* is a subgroup of plants developed by people, as opposed to species that originate in the wild. Variety is a rank within a species. When writing a plant variety name, it is generally capitalized. An example is Triumph wheat.

Student Activities

1. Observe the plants that are commonly grown in your area. Classify them by the type of root system that they have.
2. Make a chart of the plants listed in item 1 and indicate the common and scientific names, and their response to drought or their water maintenance requirements.
3. Make a collection of different kinds of leaves. Classify each according to its shape and type of margin.
4. Sketch the parts of roots, stems, and leaves. Label all items.
5. Make a bulletin board showing the major parts of plants.
6. Ask your teacher to provide a microscope and slides of plant tissue. Diagram the plant parts and label the cells and other structures that you see.

Self-Evaluation

A. MULTIPLE CHOICE

1. The primary function of the root is to

   a. make sure that the plant will grow.
   b. anchor the plant and to supply water and nutrients.
   c. ensure that the plant can be propagated.
   d. hold up the stem of the plant and provide propagation material.

2. The portion of the root that takes in the water and plant nutrients is the

   a. root cap.
   b. area of root division.
   c. root hair.
   d. area of cell maturation.

3. The major types of root systems are

   a. area of cell division and fibrous.
   b. fibrous and root cap.
   c. cuttings and root hairs.
   d. fibrous and taproot.

4. The area of cell division is

   a. responsible for the production of new cells on the tip of the root.
   b. where the cells will start to specialize.
   c. located in the area where the root hairs start to erupt from the wall of the epidermal cell.
   d. where the roots drop off on special plants like the dodder.

5. The phloem

   a. is the pipeline that carries the water and nutrients from the soil to the leaves.
   b. is the part of the stem that gives support to the node.
   c. is the part of the leaf that holds it to the stem.
   d. carries the manufactured food from the leaves to the roots.

6. Herbaceous stems

   a. are tough and have bark around them.
   b. come from herbs.
   c. are green and are not winter hardy.
   d. are part of the bulb.

7. The node

   a. is the part of the stem that supports the flower.
   b. is the part of the stem where the leaf is attached.
   c. is the part of the stem that carries the nutrients.
   d. will become detached when dry weather sets in.

B. MATCHING

______	**1.** Root cap	**a.** Located on the tip of the stem
______	**2.** Terminal bud	**b.** The wide portion of the leaf
______	**3.** Leaves	**c.** Protects the root tip as it moves in soil
______	**4.** Cuticle	**d.** Manufacture food for the plant
______	**5.** Blade	**e.** The topmost layer on the leaf
______	**6.** Guard cells	**f.** Surround the stoma

C. COMPLETION

1. The roots are responsible for ______ the plant.
2. An ______ plant is a plant used to improve the appearance of an area.
3. The area of ______ ______ is where the cells start to become specialized.
4. ______ are thick stems that run below the ground.
5. ______ are pores in the stem that allow the gases to pass through the stem.
6. ______ ______ are cells that give the leaf strength.
7. The ______ is the male part of the flower.

8. When a flower contains the stamen, pistil, petals, and sepals, it is considered a ______ flower.

9. The ______ is an enlargement that results after fertilization.

10. A ______ can be any part of the plant that is grown for its edible parts.

D. TRUE OR FALSE

______ 1. Plants with a large amount of thin, hairlike roots have a fibrous root system.

______ 2. Plants with taproot systems are more likely to survive in a dry period.

______ 3. The root cap protects the young growing tip of the root.

______ 4. The xylem carries the nutrients and water down the stem of the plant.

______ 5. Herbaceous stems are tough and winter hardy.

______ 6. Bulbs are stems that are thick and compact.

______ 7. The vascular bundle is only in the leaf of the plant.

______ 8. The leaf consists of the petiole and the blade.

______ 9. The stomates allow for the passage of gases only through the leaf surface.

______ 10. The bract is the colored petal located in the flower.

UNIT 16 Plant Physiology

OBJECTIVE To determine how plants make food and describe the relationships among air, soil, water, and essential plant nutrients for good plant growth.

Competencies to Be Developed

After studying this unit, you will be able to

- ☐ explain how plants make food.
- ☐ describe the roles of air, water, light, and media in relation to plant growth.
- ☐ trace the movement of minerals, water, and nutrients in plants.
- ☐ describe the ways that various plants store food for future use.
- ☐ compare the activity in a plant during exposure to light and periods of darkness.
- ☐ explain how plants protect themselves from disease, insects, and predators.

TERMS TO KNOW

Physiology
Photosynthesis
Chlorophyll
Chloroplasts
Glucose
Light intensity
Respiration
Transpiration
Turgor
Osmosis
Semi-permeable membrane
Plant nutrition
Plant fertilization
Macronutrients
Micronutrients
Ion
Anions
Cations
Deficiency
pH
Precipitate
Acid
Alkaline
Leached
Chlorosis

The life of a plant from its beginning to its maturity is a complex process. Many factors influence and directly control how a plant grows and what it produces. Growth, as in all living organisms, occurs by the division of cells and their enlargement as the plant increases in size (Figure 16-1). As the plant grows to maturity, the cells are produced, divide, grow, and become specialized organs. These specialized organs are stems, leaves, roots, flowers, fruits, and seeds (Figure 16-2). The study of how these organs function and the complex chemical processes that permit the plant to live, grow, and reproduce is *physiology*. An understanding of the processes of germination, photosynthesis, respiration, adsorption of water and nutrients, translocation, and transpiration will enable the agriscience technician to maximize production of plants. The technician who works with interior and other ornamental plants must especially understand the environment of the plant, because it is not in its native habitat.

PHOTOSYNTHESIS The most important chemical process in the atmosphere is photosyn-

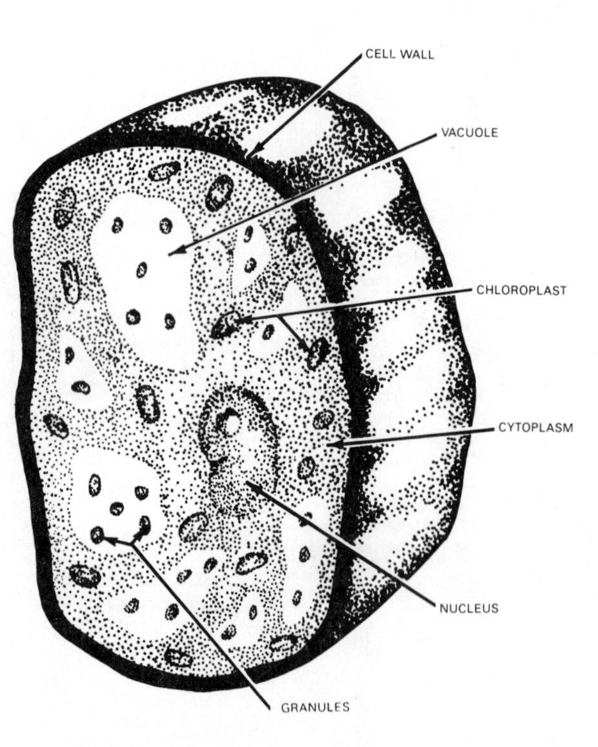

FIGURE 16-1
Major parts of a plant cell

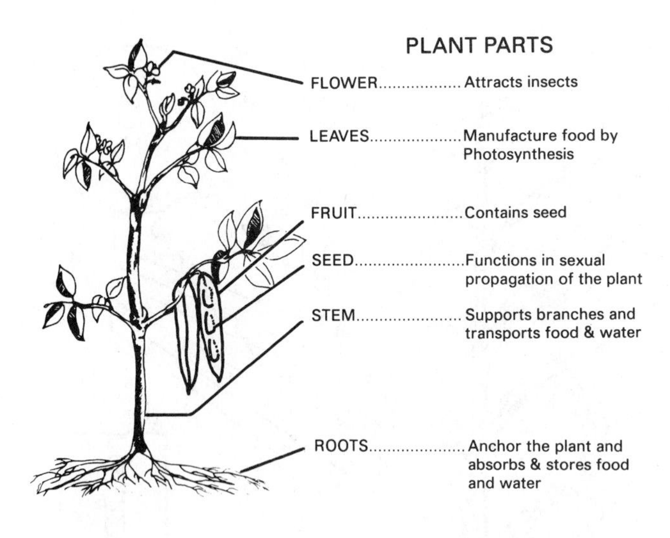

FIGURE 16-2
Major parts of a typical plant ***(From Reiley & Shry/Introductory Horticulture, 3rd edition, copyright 1988 by Delmar Publishers Inc.)***

thesis. Without this chemical process, maintenance of life on this planet would not exist. Plants need carbon dioxide to manufacture food. Animals need oxygen to live. The complex chemical process of photosynthesis permits both to live and to support each other. Although the process is complex, it is not difficult to understand.

Photosynthesis is a series of processes in which light energy is converted to chemical energy in the form of a simple sugar. Chlorophyll and chloroplasts are also essential in this process. *Chlorophyll* is the green material inside the chloroplast. It is the substance that gives the green color to plant leaves. *Chloroplasts* are small membrane-bound bodies inside cells that contain the green chlorophyll pigments. The chloroplasts are located in the mesophyll of the leaf. They are the site of the actual conversion of solar (light) energy into stored energy —simple sugars.

Photosynthesis is the conversion of carbon dioxide and water in the presence of light and chlorophyll into glucose, oxygen, and water. *Glucose* is a simple sugar and contains the building blocks for other nutrients. A simple chemical definition of photosynthesis is:

$$6CO_2 + 12H_2O \xrightarrow[\text{chlorophyll}]{\text{light energy}} C_6H_{12}O_6 + 6O_2 + 6H_2O$$

The rate at which the foodmaking process occurs depends on and varies with the light intensity, temperature, and concentration of carbon dioxide in the atmosphere. *Light intensity* is also known as the quality of light or the brightness of light. Light must be present with sufficient brightness for the process to be successful. Plants have been able to adapt to various levels of light brightness. A knowledge of the level of light required for plants to grow well is essential, particularly for indoor plant production.

Temperature is also an important factor in the process of food manufacturing in the leaf. Photosynthesis occurs best in a temperature range of 65–85°F (18–27°C). Extremes of temperatures slow down or completely stop the process of photosynthesis. A lack of carbon dioxide will affect photosynthesis, too. Carbon dioxide is especially important in the beginning of the process. Under normal outdoor conditions, its availability is not a problem. However, in enclosed conditions such as those found in a greenhouse, a carbon dioxide shortage could be a limiting factor. To correct this problem, a carbon dioxide generator might be used.

RESPIRATION All living cells carry on the process of respiration. *Respiration* is a process by which living cells (plant or animal) take in oxygen and give off carbon dioxide. Unlike photosynthesis, which occurs only in the light, respiration occurs both day and night. It is not easily measured during the day, because the presence of photosynthesis will mask or obscure the occurrence of respiration. Respiration is a breaking-down process. It uses the sugars and starches produced by photosynthesis and converts them into energy. The chemical equation for respiration is:

$$C_6H_{12}O_6 + 6O_2 \longrightarrow 6CO_2 + 6H_2O + \text{heat}$$

A comparison of the activities that occur during photosynthesis and respiration may be helpful in understanding the two processes.

PHOTOSYNTHESIS	RESPIRATION
1. Food is produced.	1. Food is used for plant energy.
2. Energy is stored.	2. Energy is released.
3. It occurs in cells that contain chloroplasts.	3. It occurs in all cells.
4. Oxygen is released.	4. Oxygen is used.
5. Water is used.	5. Water is produced.
6. Carbon dioxide is used.	6. Carbon dioxide is produced.
7. It occurs in sunlight.	7. It occurs in dark as well as light.

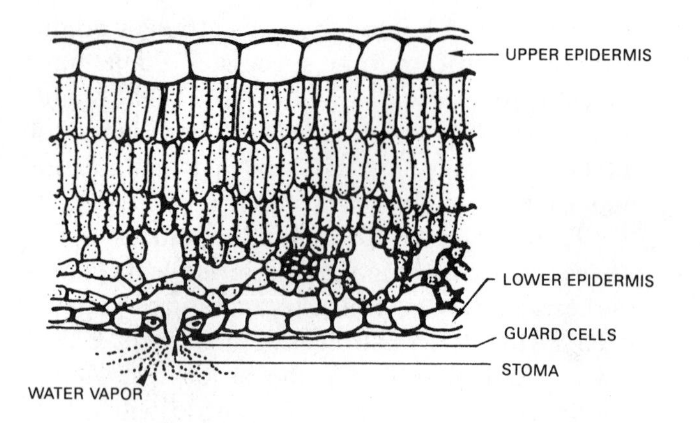

FIGURE 16-3

The process of transpiration ***(Adapted from Reiley & Shry/Introductory Horticulture, 3rd edition, copyright 1988 by Delmar Publishers Inc.)***

TRANSPIRATION *Transpiration* is the process by which a plant loses water vapor (Figure 16-3). The loss takes place primarily through the leaf stoma. The plant transpires about 90% of the water that enters through the roots. Water saturates all of the spaces between the cells throughout the plant. About 10% of the water that enters from the roots is used in chemical processes and in the plant tissues. Functions of this water include transporting minerals throughout the plant, cooling the plant, moving sugars and plant chemicals, and maintaining turgor pressure. *Turgor* means a swollen or stiffened condition as a result of being filled with liquid.

Transpiration is greatly influenced by humidity, wind and other air movement, and temperature (Figure 16-4). As humidity in the air around the plant increases, the rate of transpiration decreases. Conversely, as humidity decreases, the rate of transpiration increases. Increased air movement around the plant increases the rate of transpiration, due to the evaporation caused by air movement. Similarly, as temperatures increase, the rate of transpiration increases.

Often during dry weather or when plants are not watered, transpiration causes the plant to lose water faster than it can be replaced by the root system. When this occurs, the guard cells will close the stoma in the leaves, thus slowing down the rate of transpiration. This mechanism enables the plant to preserve the water it contains. If there is water in the soil, the plant may wilt slightly, but it will recover. However, if there is insufficient moisture in the soil, the plant may not be able to recover.

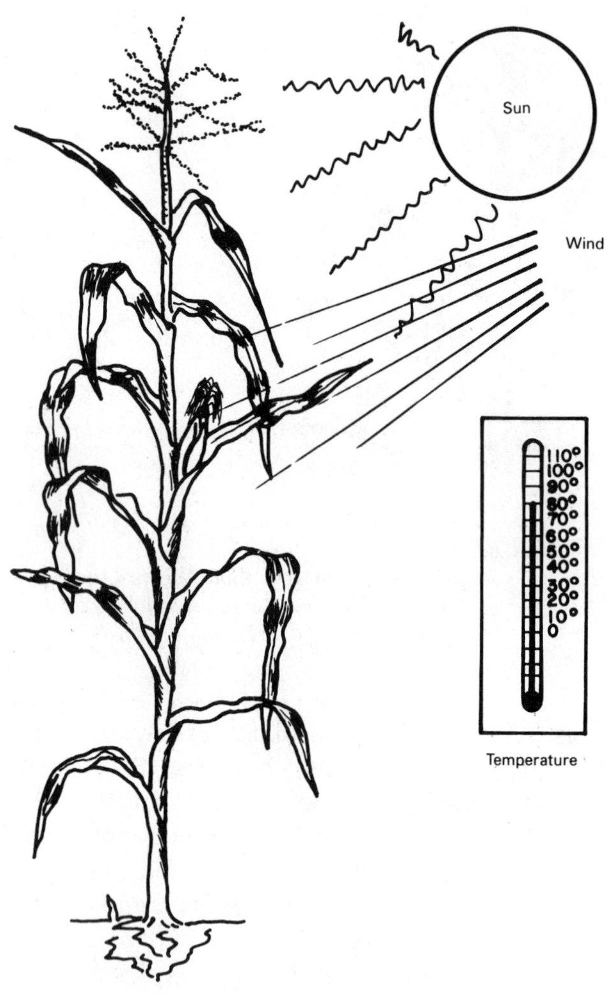

FIGURE 16-4

Factors that influence transpiration

SOIL Productive soil provides a natural environment for the root zone. There are air, water, and nutrients for the plant. Root hairs penetrate the pore spaces in the soil and absorb plant nutrients (Figure 16-5). A process called osmosis is used to get nutrients into root cells so they can be transported into the remainder of the plant. *Osmosis* is a process whereby materials in solution move from areas of high concentration to areas of low concentration through a semi-permeable membrane (Figure 16-6). A *semi-permeable membrane* is a structure that permits a solution to move through it in a direction controlled by the concentration of solutions. The epidermis of a root hair is a semi-permeable membrane. When fertilizer is added to the soil, a high concentration of nutrients occurs in that soil. When the soil moisture dissolves nutrients from the fertilizer, they can move through the semi-permeable membrane of the root hair. They enter the cells of the plant, where nutrient concen-

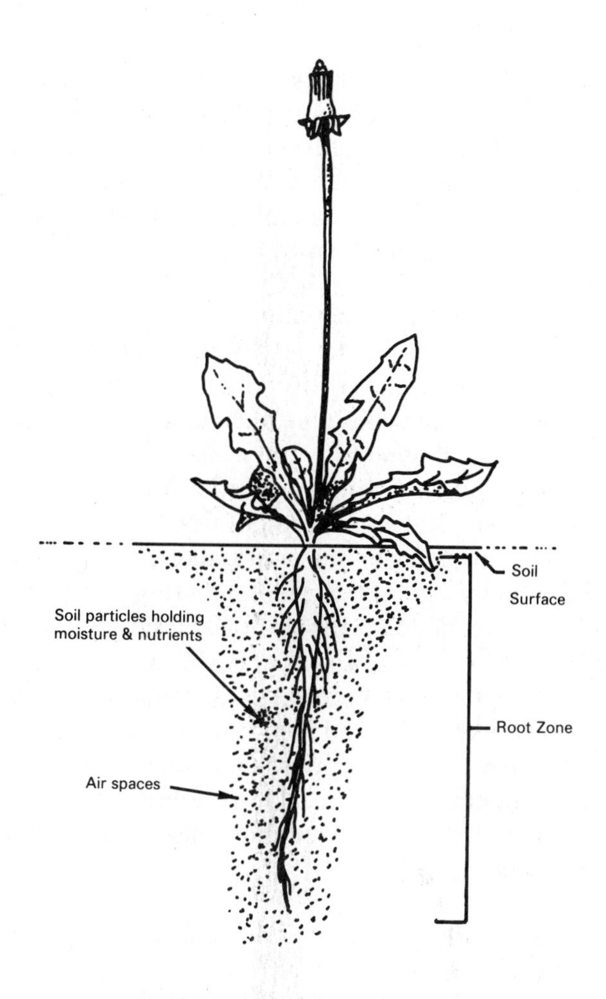

FIGURE 16-5
Root hairs in soil pores

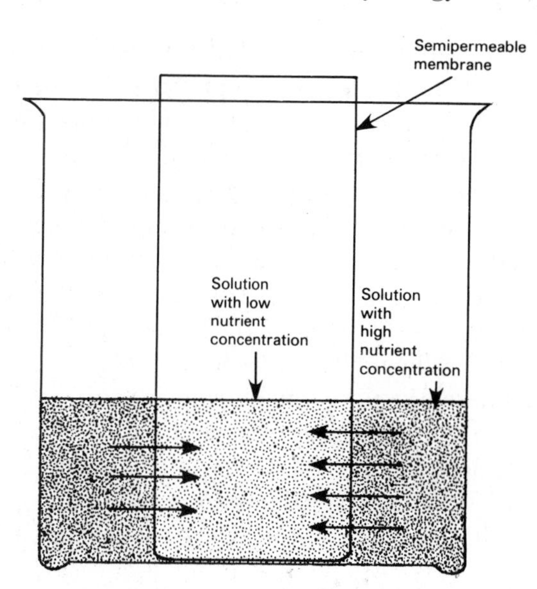

FIGURE 16-6
The process of osmosis

trations are lower. Once inside the root-hair cell, nutrients can be transported to other parts of the plant as needed.

To allow the root hairs to move through the soil, the soil must have spaces between the particles of sand, silt, and clay. Such spaces are called pores. Their job is to store air, water, and nutrients, as well as permit root penetration.

AIR The air or atmosphere that surrounds the aboveground portion of the plant must supply carbon dioxide as well as oxygen. Generally this is no problem when the plants grow outdoors in fields. When plants are transplanted into artificial or unnatural environments, consideration must be given to the quality of the air surrounding the plant. In a greenhouse or other enclosed system, the quality of the atmosphere needs monitoring. The presence and levels of carbon dioxide and pollutants must be understood for maximum production. With certain crops in greenhouses, such as carnations and roses, the addition of some carbon dioxide might be desirable to increase crop production. In areas where crops are growing near industrial plants or cities, or along major highways, the technician must be aware of the many types of pollutants that might cut production or severely damage the plants.

WATER The most essential ingredient for all life is said to be water. Nutrients in the soil must first be dissolved in water before they can be absorbed in the roots. Once inside the plant, water carries the nutrients to the leaves, where these nutrients chemically combine with water in the photosynthetic process. Sugars and other plant foods manufactured in the leaves are then transported throughout the plant by water. Water helps to control the temperatures in and around plants through transpiration. Finally, water gives the plant support by keeping the cells in a turgor state. It is important, therefore, that water used for plant production be of good quality and in adequate supply.

PLANT NUTRITION This is often confused with plant fertilization. There is a difference. *Plant nutrition* refers to availability and type of basic chemical elements in the plant. *Plant fertilization* is the process of adding nutrients to the soil or leaves so these chemicals are added to the growing environment of the plant. Before chemicals supplied as fertilizer can be taken up and used by plants, they generally undergo various changes.

Essential Nutrients There are 16 elements that are essential for normal plant growth. These are required in various amounts by plants and must be available in the relative proportions needed if the plants are to produce well. Three are used in huge amounts and are obtained from the atmosphere and water around the plant. They are carbon (C), hydrogen (H), and oxygen (O). Then there are six elements that are used in relatively large amounts and are called *macronutrients*. The macronutrients are nitrogen (N), phosphorus (P), potassium (K), calcium (Ca), magnesium (Mg), and sulfur (S). These are all obtained from the soil.

An additional seven elements are used in small quantities and are called *micronutrients* (trace elements). The micronutrients are also obtained from the soil. They are boron (B), copper (Cu), Chlorine (Cl), iron (Fe), manganese (Mn), molybdenum (Mo), and zinc (Zn). For a plant to grow at maximum efficiency, it must have all essential plant nutrients. The absence of any one of these nutrients will cause the plant of grow poorly or show some signs of poor health.

AGRI-PROFILE

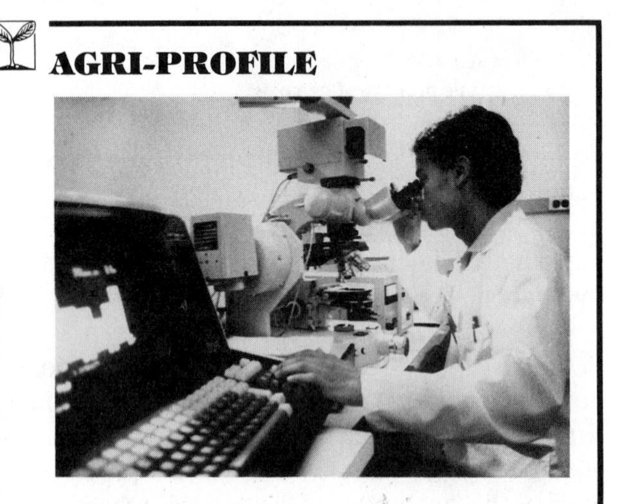

CAREER AREA: PLANT PHYSIOLOGY
Plant physiologists use regular and electronic microscopes, computers, and many specialized analytical tools in their work. ***(Courtesy of the National FFA Organization)***

Physiology refers to the many functions that occur inside of plants. These include familiar activities such as osmosis, nutrient uptake, translocation, respiration, photosynthesis, food movement, and food storage. Other complex functions are known only to specialists in the field.

Plant physiologists work closely with technicians and scientists in other fields of plant science. They may be consultants to or collaborators with specialists in agronomy and horticulture. The work of plant physiologists is typically done as college or university faculty, employees of state or national research institutes, or specialists with agriscience corporations developing and selling seeds and plant materials.

Remembering the 16 Plant Nutrients Various schemes have been devised to help you remember the names of the 16 plant nutrients. One technique is to first learn the chemical symbols. Then use the symbols to make a logical string of words that are easy to remember. One such string of words that uses the symbols of most of the nutrients is, "C. Hopkin's cafe, mighty good." By remembering this phrase, you can recall the symbols of 10 of the 16 nutrients as follows: C HOPKNS CaFe Mg (carbon, hydrogen, oxygen, phosphorus, potassium, nitrogen, sulfur, calcium, iron, and magnesium. The remaining ones are boron, copper, chlorine, manganese, molybdenum, and zinc. Can you devise a string of words to help you remember the symbols of these micronutrients?

Ions Plant nutrients are absorbed from the soil-water solution that surrounds the root hairs of the plant. In fact, 98% of the nutrients obtained from the soil are absorbed in solution, while the other 2% are extracted by the root directly from soil particles. Most of the nutrients are absorbed as charged ions. An *ion* is an atom or a group of atoms that has an electrical charge.

Ions are either negatively charged and called *anions*, or positively charged and called *cations*. The electrical charge in the soil is paired so that the overall effect in the soil is not changed. These ions compete and interact with each other according to their relative charges. For example, nitrogen, in its nitrate form, has a negative charge and chemical formula NO_3-. Therefore, nitrates are anions with negative charges.

On the other hand, potassium has a positive charge and is an example of a cation (K+). Potassium nitrate, $K+NO_3-$, is a combination of potassium and nitrate consisting of one nitrate ion and one potassium ion. Calcium nitrate, $Ca++(NO_3-)_2$, has two nitrate ions and one calcium ion. The reason is that the calcium cation has two positive charges. As you might guess, this could be confusing, but it serves to illustrate the need to understand chemistry to manipulate plant fertility if conditions are not ideal in the natural environment.

The balance of ions is important and needs to be carefully monitored for good plant growth. Opposite charges attract each other, but ions with similar charges compete for chemical reactions and interactions in the soil-water environment. Some ions are more active than others and might be able to compete better in the soil. Further study would be needed to thoroughly understand why soil tests may indicate the presence of a certain element in sufficient amounts for plant growth, yet the plants may show deficiency symptoms (Figure 16-7).

Deficiency means a shortage of a given nutrient available for plant use. A good example of a nutrient deficiency symptom is blossom-end rot of tomato. The end opposite the stem is called the blossom end. This is common in gardens and occurs when there is not enough water to dissolve and carry calcium to the plant in sufficient quantities. The calcium deficiency produces a tomato that looks good from the top but, when picked, the bottom end is rotten.

Soil Acidity and Alkalinity

The chemistry of plant elements in the soil can be affected by pH. Soil *pH* is a measurement of acidity (sourness) and alkalinity (sweetness) (Figure 16-8). Many of the nutrients in soil form very complex combinations and are capable of precipitating out of solution where they are unavailable to the plant. *Precipitate* occurs when a solid is dropped out of solution. If the soil pH is *acid* or extremely low, some micronutrients become too soluble and occur in high-enough concentrations to harm the plants (Figure 16-9).

On the other hand, if soil pH is very high in the *alkaline* range, many of the nutrients can be precipitated out and not be available to the plants. The pH of soils can be determined with low-cost test kits. Fortunately, soil pH can be corrected by adding lime if the pH needs to be raised and sulfur if it needs to be lowered. Such practices are common, since soil pH is seldom perfect for the crop being grown.

Plant Nutrient Functions

The importance of carbon, hydrogen, and oxygen has already been discussed under the topic of photosynthesis. The other nutrients have very specific functions and must be available in the appropriate amounts and form. The effect

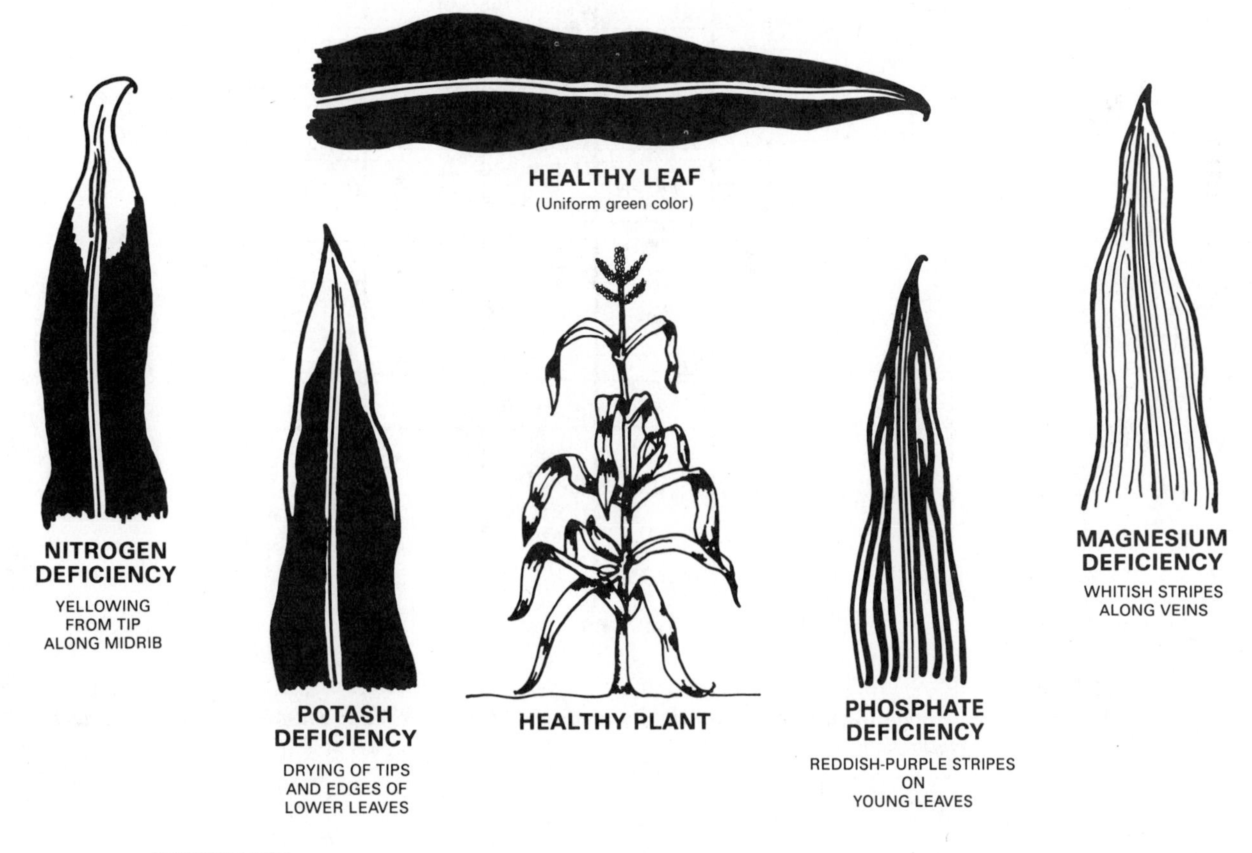

FIGURE 16-7
Nutrient deficiencies decrease plant health, vigor, and growth.

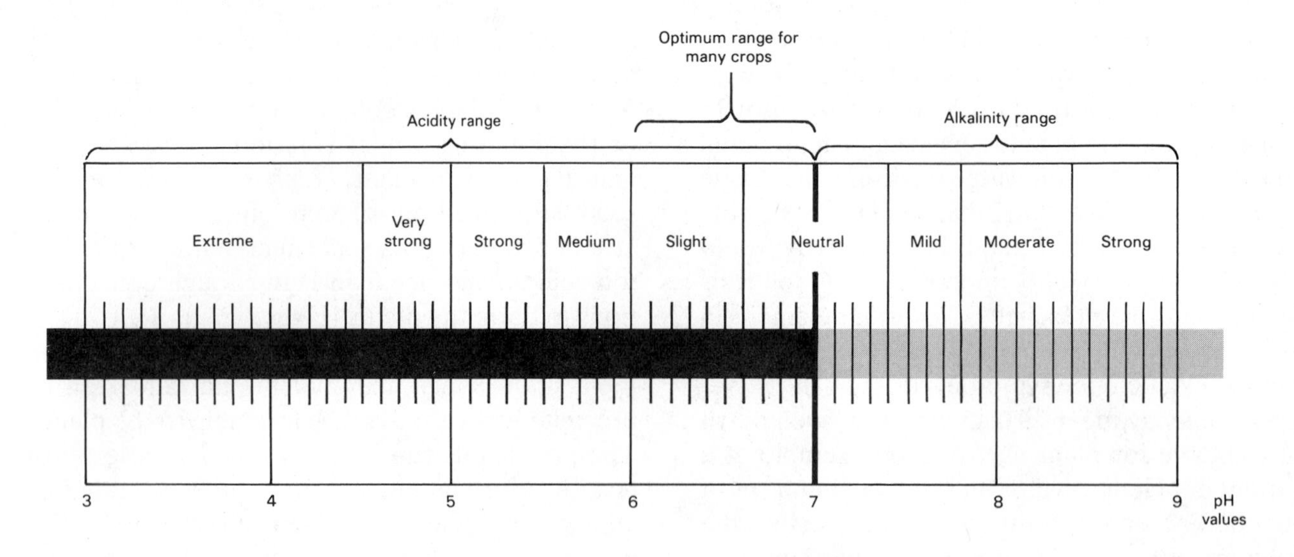

FIGURE 16-8
The pH scale

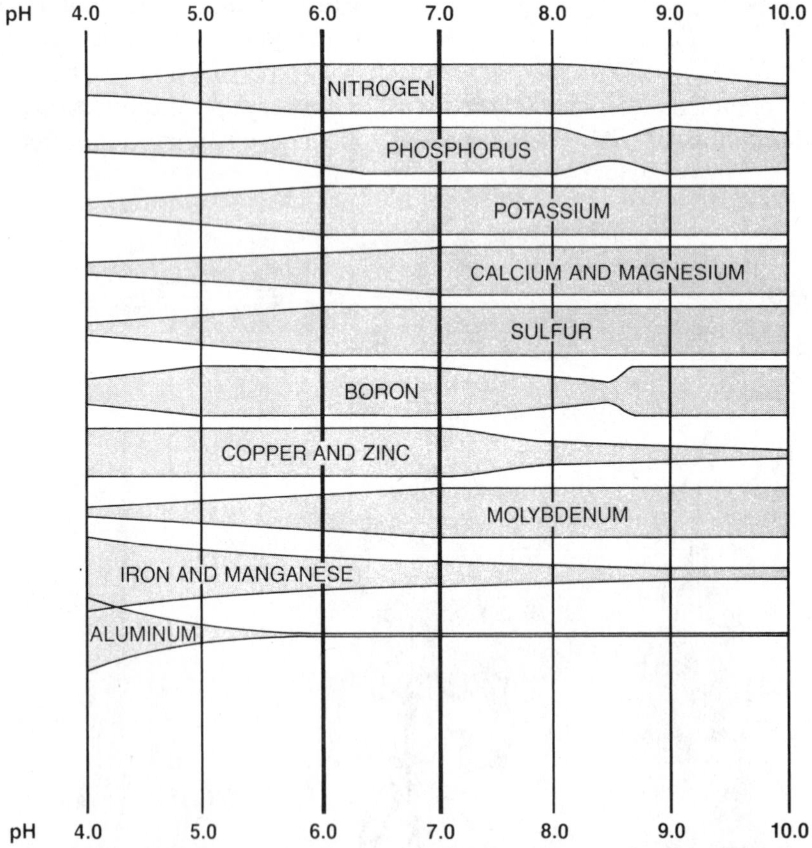

pH 4.0 5.0 6.0 7.0 8.0 9.0 10.0

NITROGEN
PHOSPHORUS
POTASSIUM
CALCIUM AND MAGNESIUM
SULFUR
BORON
COPPER AND ZINC
MOLYBDENUM
IRON AND MANGANESE
ALUMINUM

pH 4.0 5.0 6.0 7.0 8.0 9.0 10.0

FIGURE 16-9
The effect of soil pH on nutrient availability ***(From Plaster/Soil Science and Management, copyright 1987 by Delmar Publishers Inc.)***

of plant nutrients may be likened to a chain—the weakest link will determine how much the chain will pull. Similarly, the nutrient in shortest supply will determine the maximum growth achievable by the plant.

Nitrogen

This element is present in the atmosphere as a gas. It is added to the soil in some fertilizers. Since it exists in nature as a gas, it is easily *leached* (washed out of the soil). Nitrogen is responsible for the vegetative growth of the plant and its dark-green color. When nitrogen is lacking, some deficiency signs are reduced growth and yellowing of the leaves. This yellowing is referred to as *chlorosis*. Excess nitrogen can cause succulent, weak, spindly growth, and a darker-than-normal green color.

Phosphorus

In nature, phosphorus is present as a rock and is not easily leached out of the soil. It is important in seedling and young plant growth. It will help the plant develop a good root system. Some symptoms of a deficiency of phosphorus are reduced growth, poor root systems, and reduced flowering. Thin stems and browning or purpling of the foliage are also signs of poor phosphorus availability.

Potassium

Potassium is mined as a rock and made into fertilizer, but it can be leached from the soil. If too much potassium is present, it can cause a nitrogen deficiency. A lack of potassium will show up as reduced growth or shortened internodes, and sometimes as marginal burn or scorch (brown leaf edges). Dead spots in the leaf, and plants that wilt easily, can also be an indication of a potassium deficiency.

Calcium

This element is often supplied by adding lime to the soil. It can be leached out and does not move very easily throughout the plant. Too much calcium can cause a high pH and reduce the availability of some elements to the plant. A lack of this element can stop bud growth and result in death of root tips, cupping of mature leaves, and blossom-end rot of many fruits. Pits on root vegetables can also be a sign of calcium deficiency.

Magnesium

Magnesium can be added by using high-magnesium lime. It can also be leached from soil. If magnesium is lacking, some reduction of growth and marginal chlorosis can be noticed. In some plants, even interveinal chlorosis can be seen. Cupped leaves and a reduction in seed production can also be symptoms of magnesium deficiency. Foliage plants are commonly lacking this nutrient.

Sulfur

Present in the atmosphere as a result of combustion, sulfur is often an impurity in fertilizers carrying other nutrients. As a result, it is rarely deficient. However, if sulfur is deficient, a yellowing of the entire plant may result.

The following chart summarizes typical symptoms exhibited by plants deficient in various micronutrients.

NUTRIENT	EXCESS	DEFICIENCY
Iron	Rare	Interveinal chlorosis, especially on young growth
Zinc	Might appear as an iron deficiency	Interveinal chlorosis, reduction in leaf size, short internodes
Molybdenum	Not known	Interveinal chlorosis on older leaves; may also affect leaves in the middle of the plant
Boron	A blackening or death of tissue between veins	Failure to set seed; death of tip buds
Copper	Might occur in low pH; will appear as an iron deficiency	New growth small and misshapen, wilted
Manganese	Brown spotting on leaves; reduced growth	Interveinal chlorosis of the leaves and brown spotting; checkered effect possible

FOOD STORAGE

When the plant makes its food through photosynthesis, it often manufactures more than it needs to maintain itself. This excess is often stored in the plant for future use. Such food may be stored in roots, stems, seeds, or fruits.

Roots

The most common type of root that serves as a storage organ is the taproot. Some common examples of plants with extensive storage capacity are sugar beets, carrots, radishes, and turnips. The sugars and carbohydrates are transported down the phloem and into the root cells. It is held here as the root enlarges. Most of the time this type of plant is a short-term crop that does not take long to mature. This type of root system is easy to dig or harvest.

Stems The stems of plants usually contain cells that are necessary for plant support. Some specialized stems, however, are excellent food-storage organs (Figure 16-10). Some are used for propagation and some for food. The most common specialized stem used for food is the tuber. It is an enlarged portion of a stem containing all of the parts of a normal stem. Nodes, internodes, and buds can be identified in tubers. The Irish potato is an example of a tuber.

Corms and bulbs are other examples of specialized stems that contain large amounts of food made by photosynthesis. A rhizome is yet another. Often these are used for propagation purposes.

Seeds As the ovule of a plant matures, it stores food for the young embryo to start its growth when it germinates. Both animals and humans uti-

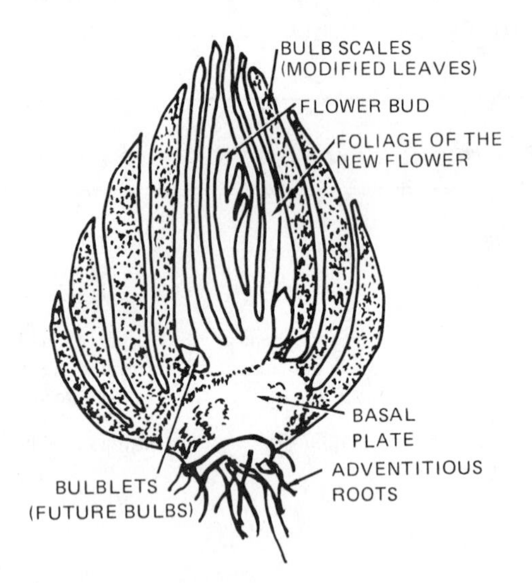

SCALES

A NON-TUNICATE BULB WITH LOOSE SCALES BUT NO OUTER TUNIC.
EXAMPLE: LILY

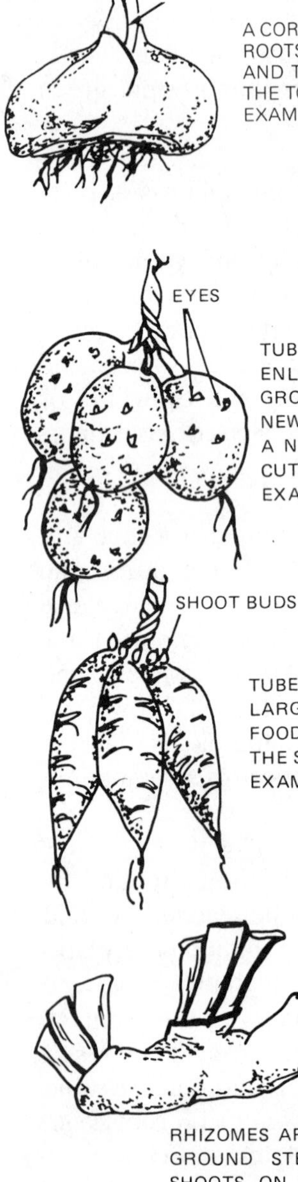

FIGURE 16-10
Examples of stems that are major food storage organs for the plant ***(From Ingels/Ornamental Horticulture, copyright 1985 by Delmar Publishers Inc.)***

lize seeds as major food sources. They help the plant by spreading the seed to new locations, increasing the plant's chances for survival.

SUMMARY This unit has covered some basic principles of plant physiology. Physiology is complex and must be studied in great depth for proficiency. Plant physiologists typically have a master's or a doctor's degree. However, most technicians and scientists in the plant sciences will have some training in plant physiology. A basic knowledge of soils and how plants use nutrients and function in general will help greatly in the successful production and management of plants.

Student Activities

1. Make a bulletin board showing the cross-section of a leaf. Label the various cells and leaf parts. Include the formula for photosynthesis.
2. Write a letter to a person new to plant production and explain the importance of photosynthesis.
3. Collect plants or pictures and make a display that could be used by others to help identify nutrient deficiencies.
4. Select a crop that interests you and conduct research to determine the optimum nutrient requirements.
5. Set up a demonstration to explain osmosis.
6. Make a list of all the chemical symbols of plant nutrients and see if you can write a sentence or story to help you remember them.

Self-Evaluation

A. MULTIPLE CHOICE

1. The study of functions and the complex chemical processes that allow plants to grow is known as

   a. plant taxonomy.
   b. plant physiology.
   c. plant nutrition.
   d. photosynthesis.

2. Chlorophyll is important in plants because it

   a. creates an atmosphere where it can determine the osmotic pressure.
   b. allows the plant to make good xylem tissue.
   c. gives the green color to plants.
   d. is also known as the chloroplasts.

3. The rate at which photosynthesis is carried out depends on

   a. the amount of fertilizer in the water.
   b. the amount of oxygen in the atmosphere.
   c. the amount of respiration carried on during the daylight hours.
   d. the light intensity, temperature, and concentration of carbon dioxide.

4. Photosynthesis will work best in which temperature range?

   a. 50–60°F
   b. 60–70°F
   c. 65–85°F
   d. 85–95°F

5. Respiration

a. uses food for plant energy.
b. stores energy.
c. occurs in cells that contain chlorophyll.
d. uses carbon dioxide.

6. Plant nutrition is

a. plant food added to the plant pot.
b. use of basic chemical elements in the plant.
c. chemical processes providing plants with elements for growth.
d. the measurement of acidity (sourness) and alkalinity (sweetness).

B. MATCHING

_____ **1.** pH
_____ **2.** Osmosis
_____ **3.** Corm
_____ **4.** Root
_____ **5.** Sulfur
_____ **6.** Leaves
_____ **7.** Fertilization

a. Movement through a semi-permeable membrane
b. Addition of nutrients to the plant growing environment
c. Storage organ for excess plant food
d. Site of photosynthesis
e. Measurement of acidity and alkalinity
f. Macronutrient
g. Specialized stem

C. COMPLETION

1. Extremes of temperature will slow down or completely stop _____ .

2. Respiration will occur only in the _____ .

3. When the temperature increases, the rate of transpiration _____ .

4. Soil provides a natural environment for the _____ _____ .

5. The spaces in between the soil particle, where the soil water is found, is called the _____ .

D. TRUE OR FALSE

_____ **1.** Respiration is a building process that uses sunlight to work.

_____ **2.** The process by which the plant loses water is perspiration.

_____ **3.** All plant elements carry on the same function in the plant.

_____ **4.** Humidity refers to the amount of water in the atmosphere.

UNIT 17 Plant Reproduction

OBJECTIVE To determine the methods used by plants to reproduce themselves and explore new propagation technology.

Competencies to Be Developed

After studying this unit, you will be able to

- ☐ distinguish between sexual and asexual reproduction.
- ☐ explain the relationship between reproduction and plant improvement.
- ☐ draw and label the reproductive parts of flowers and seeds.
- ☐ state the primary methods of asexual reproduction and give examples of plants typically propagated by each method.
- ☐ explain the procedures used to propagate plants by tissue culture.

TERMS TO KNOW

Propagation
Reproduction
Sexual reproduction
Asexual reproduction
Clone
Vegetative
Hybrid
Hybrid vigor
Germinate
Dormant
Imbibition
Scarify
Viable
Cuttings
Rooting hormone
Stem tip cuttings
Stem section cuttings
Cane cuttings
Heel cuttings
Single-eye cuttings
Double-eye cuttings
Leaf cutting
Leaf petiole cuttings
Leaf section cuttings
Split-vein cuttings
Root cuttings
Layering
Simple layering
Tip layering
Air layering
Grafting
Tissue culture

MATERIALS LIST

seed catalog
grafting knife
grafting rubber
grafting wax
cutting knife
rooting hormone
stock plants
rooting media
tissue-culture tubes
tissue-culture media
scalpel
razor blade (single edge)
tweezers
50% alcohol solution
sanitary work area

Plant *propagation* or *reproduction* is simply the process of increasing the numbers of a species or perpetuating a species. The two types of plant propagation are sexual and asexual. *Sexual reproduction* is the union of an egg (ovule) and sperm (pollen), resulting in a seed. Two parents creating a third individual is referred to as sexual propagation. In plants, it involves the floral parts. It may involve one or two plants. *Asexual reproduction* utilizes a part or parts of only one parent plant. The purpose is to cause the parent plant to make a duplicate of itself. The new plant is a *clone* (exact duplication) of its parent. Since this type of reproduction uses the vegetative parts of the plant, namely the stems, roots, or leaf, it is often referred to as *vegetative* propagation.

Some advantages of sexual propagation are that it is often less expensive and quicker than some other methods. It is the only way to obtain new varieties and also capture hybrid vigor. A *hybrid* is a plant obtained by crossbreeding. *Hybrid vigor* refers to the tendency for hybrid plants to be stronger and survive better than purebred plants. Sexual propagation is a good way to avoid passing on some diseases. In some plants it is the only way to propagate them.

Asexual propagation has many advantages as well. In some cases it is easier and less expensive to obtain plants. In some species or cultivars, it is the only way to propagate them.

SEXUAL PROPAGATION

A seed is made up of the seed coat, endosperm, and embryo (Figure 17-1). The seed coat functions as a protector for the seed. Sometimes it can be very thin and soft, or it may be very hard and impervious to water or moisture. The endosperm functions as a food reserve. It will supply the new plant with nourishment for the first few days of its life. The embryo is the young plant itself. When a seed is fertilized and matures, it will be dormant. When it is given favorable conditions it will *germinate*, which means that the seed will start to sprout and the plant will grow.

Seed propagation starts with quality seed. Crop production by sexual reproduction means consideration can be given to the type of plant needed or the variety that is best adapted to a particular area or purpose. Hybrid plants are developed by cross-pollinating two different varieties. Many varieties on the market are the result of hybridization or crossbreeding. Seeds of hybrid plants cost more than open-pollinated varieties. But the increased quality of the plants generally offsets the increased cost of the seed. New varieties are being developed for disease and insect resistance. It is natural to expect the seeds of such improved varieties to cost more than standard or regular seeds. Some varieties have unusual cultural or product characteristics.

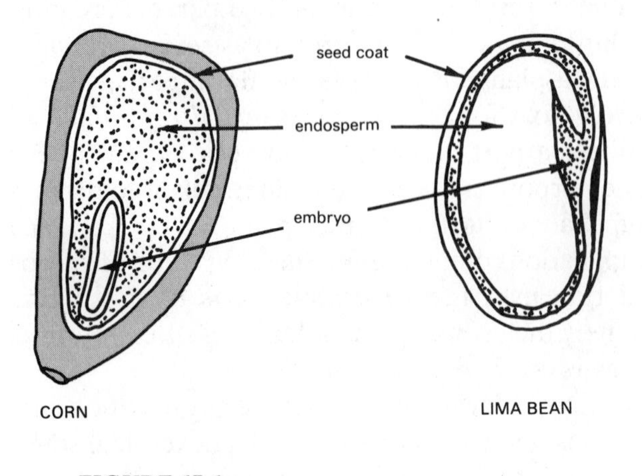

FIGURE 17-1
Parts of a seed

Seeds collected from plants used for production other than seed production generally will not save money in the long run. Seeds from such plants are often small. As a result, they are often poorly managed and improperly handled and stored. It is recommended that seed saved from season to season be stored in a sealed jar at 40°F and in low humidity.

Germination

When seed is harvested or collected, it is normally mature and in a *dormant* or resting state. To germinate or start to grow, it must be placed in certain favorable conditions. The four environmental factors that must be right for effective germination are water, air, light, and temperature.

Water

Imbibition (the absorption of water) is the first step in the germination process. The seed, in its dormant stage, contains very little water. The imbibition process allows the seed to fill all its cells with water. Then if other conditions are favorable, the seed breaks its dormant stage and germinates.

A good germination medium is important. The medium must not be too wet or too dry. An adequate and continuous supply of water must be available. This is often difficult to control with crops directly seeded in the field. It is much easier to control in crops planted for transplanting. A dry period during the germination process will result in the death of the young embryo. Too much water will result in the young seed rotting out. In some species, the seed coat is very hard and water cannot penetrate to the endosperm. In cases like this it is necessary to scarify the seed.

A common way to *scarify* seed is to nick the seed coat with a knife or a file. Another method is to soak the seeds in concentrated sulfuric acid. This requires special care and experience, since sulfuric acid is a dangerous material. Another technique is to place seeds in hot water (180 to 212°F) and allow them to soak as the water cools. This process takes 12 to 24 hours. A warm, moist scarification process may be utilized by simply placing the seeds in warm, damp containers and letting the seed coat decay over a period of time.

Air

Respiration takes place in all viable seed. *Viable* seed is alive and capable of germinating. Oxygen is required. Even in nongerminating seeds, a small amount of oxygen is required even though respiration is low. As germination starts, the respiration rate increases. It is important that the seed be placed in good soil or media that is loose and well drained. If the oxygen supply is limited or reduced during the germination process, germination will be reduced or inhibited.

Light Some seeds are stimulated to grow by light. Some are inhibited by the presence of light. It will be necessary to have some knowledge of the presence or absence of special light requirements. Many of the agronomic crops do not require light for germination. In fact, light will inhibit germination. Ornamental bedding plants are more likely to require light for germination. Some crops with a requirement for light are ageratum, begonia, impatiens, and petunia. Lettuce also requires light for successful germination. Seeds of these plants are often deposited on the surface by nature, and the grower should follow the same procedure for successful germination.

Temperature Heat is another important requirement for germination. The germination rate and percentage of seed that germinates are affected by the availability of heat. Some seeds will germinate over a wide range of temperatures, while others have more narrow limits. In the agronomic crops that are direct seeded in the field, the only way to control heat is to plant when the ground is warm. In horticultural crops, particularly bedding plants and perennials, knowledge of a plant's specific heat requirements for germination will result in more efficient production. The germination requirements are often listed in the seed catalogs. Requirements for popular bedding plants are listed in the table at the bottom of this page.

ASEXUAL PROPAGATION

As stated previously, asexual propagation is using the vegetative parts of the plant to increase the number of plants (Figure 17-2). The primary advantages are economy, time, and plants that are identical to the parents. The primary methods of asexual propagation are cuttings, layering, and grafting.

Stem Cuttings Herbaceous and woody plants are often propagated by *cuttings* (vegetative parts that the parent plant used to regenerate itself). Types of cuttings are named for the part of the plant from which they come. There are stem tip cuttings, stem cuttings, cane cuttings, leaf cuttings, leaf petiole cuttings, and root cuttings.

The procedure for taking cuttings is relatively simple. The equipment needed is a sharp knife or a single-edge razor blade. Sharp equipment will make the job easier and will reduce injury to the parent plant. To prevent the possibility of diseases spreading, it is best to dip the cutting tool in bleach water made with one part bleach to nine parts water. The tool can also be dipped in rubbing alcohol.

The flowers and flower buds should be removed from all cuttings. This allows the cutting to use its energy and food storage for root formation instead

PLANT	TIME TO SEED BEFORE LAST FROST (WEEKS)	GERMINATION TIME (DAYS)	GERMINATION TEMPERATURE REQUIREMENTS (DEGREES F)	GERMINATION LIGHT REQUIREMENTS LIGHT (L) DARK (D)
Ageratum	8	5-10	70	L
Aster	6	5-10	70	—
Begonia	12+	10-15	70	L
Coleus	8	5-10	65	—
Cucumber	4	5-10	85	—
Eggplant	8	5-10	80	—
Marigold	6	5-10	70	—
Pepper	8	5-10	80	—
Portulaca	10	5-10	70	D
Snapdragon	10	5-10	65	L
Tomato	6	5-10	80	—
Watermelon	4	5-10	85	—
Zinnia	6	5-10	70	—

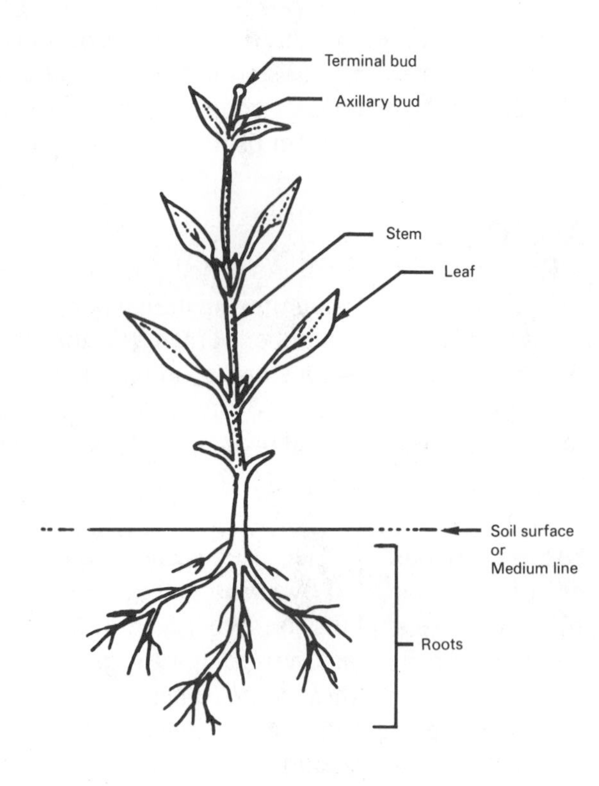

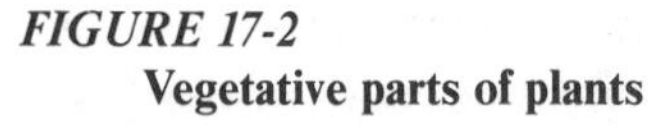
FIGURE 17-2
Vegetative parts of plants

of flower and fruit development. A rooting hormone containing a fungicide is used to stimulate root development. A fungicide is a pesticide that helps prevent diseases.

Rooting hormone is a chemical that will react with the newly formed cells and encourage the plant to make roots faster. The proper way to use a rooting hormone is to put a small amount in a separate container and work from that container. This procedure will ensure that the rooting hormone does not become contaminated with disease organisms. Do not put the unused hormone back in the original container.

Cuttings are normally placed in a medium consisting of coarse sand, perlite, soil, a mixture of peat and perlite, or vermiculite. It is best to use the correct medium for a specific plant in order to obtain the most efficient production in the shortest possible time. The rooting medium should always be sterile and well drained, with moisture retention ability to prevent the medium from drying out. The medium should be moistened before inserting the cuttings. It should then be kept continuously and evenly moist while the cuttings are forming roots and new shoots.

Stem and leaf cuttings do best in bright, but indirect light. However, root cuttings are often kept in the dark until new shoots are formed and start

AGRI-PROFILE

CAREER AREAS: PLANT BREEDING/PLANT PROPAGATION/TISSUE CULTURE

The plant breeder or plant geneticist crossbreeds plants from various sources in an effort to develop new varieties with selected characteristics. ***(Courtesy USDA Agricultural Research Service)***

Plant breeders' objectives might include making plants faster growing, disease resistant, drought tolerant, insect resistant, wind resistant, frost tolerant, more beautiful, or better flavored, depending on the use of the plant. Much plant breeding occurs in greenhouses, with plants having small flowers that require the use of magnifying glasses and tweezers to transfer pollen or remove reproductive parts. Generally, such research is followed by field trials and seed production which gives the plant breeder variety in the settings where work is done.

Asexual reproduction involves rooting, budding, grafting, layering, and other procedures other than pollination. Tissue culture is a procedure developed in biotechnology which permits the production of thousands of new plants identical to a superior plant. The procedure is relatively cheap and easy, and is used extensively to reproduce ornamental plants. Many jobs are available in the area of plant reproduction.

to grow. For most plants, the most popular method of making cuttings is by stem cutting. On herbaceous plants, stem cuttings may be made almost any time of the year. Stem cuttings of many woody plants are normally taken in the fall and/or the dormant season.

Stem Tip Cuttings

Taken from the end of the stem or branch, *stem tip cuttings* normally include the terminal bud. A piece of stem about 2–4″ long is selected and the cut is made just below the node. The lower leaves that would be in contact with the media are removed. The stem is dipped in the rooting hormone and is gently tapped to remove the excess rooting hormone. The cutting is then inserted into the rooting media. The cutting should be inserted deep enough so the plant material will support itself. It is important that at least one node be below the surface of the media (Figure 17-3).

Stem Section Cuttings

Stem section cuttings are prepared by selecting a section of the stem located in the middle or behind the tip cutting. This type of cutting is often used after the tip cuttings are removed from the plant. The cutting should be 2–4″ long, and the lower leaves should be removed. The cutting should be made just above a node on both ends. It is then handled as a tip cutting. Make sure that the cutting is positioned with the right end up. The axial buds are always on the top of the leaves.

Cane Cuttings

Some plants, such as the dumbcane *(Diffenbachia sp.)*, have canelike stems. These stems are cut into sections that have one or two eyes or nodes to make *cane cuttings*. The ends are dusted with activated charcoal or a fungicide. It is best to allow the cane to dry in the open air for an hour or two. The cutting is then placed in a horizontal position with half of the cane above the surface of the medium. The eyes or nodes should be facing upward. This type of cutting is usually potted when the roots and new shoots appear (Figure 17-4).

Heel Cuttings

Heal cuttings are used with woody-stem plants. A shield-shaped cut is made about halfway through the wood around the leaf and the axial bud. Rooting hormone may be used as it is in the other types of cuttings. The cutting is inserted horizontally into the medium (Figure 17-5). If space or stock material is in short supply, propagators might make the single-eye and the double-eye cuttings.

Single-eye Cuttings

When the plant has alternate leaves, *single-eye cuttings* are used. The eye refers to the node. The stem is cut about ½″ above and below the same node (Figure 17-6). The cutting

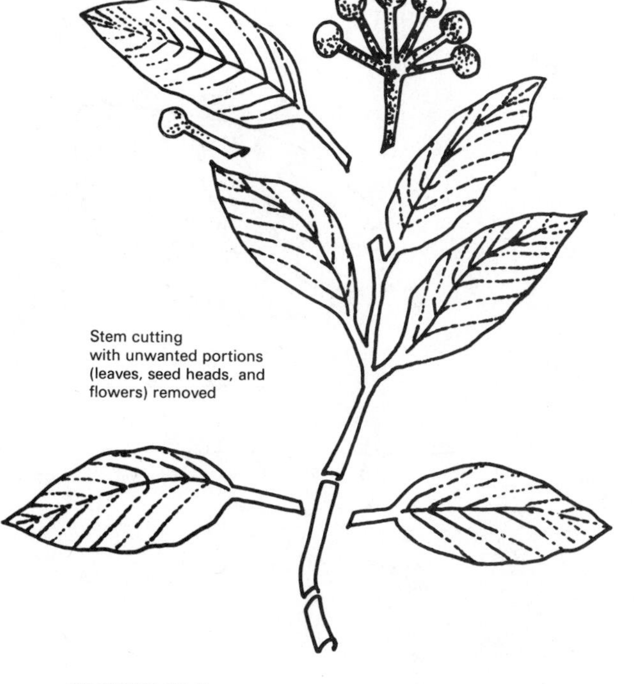

FIGURE 17-3
Stem tip cuttings

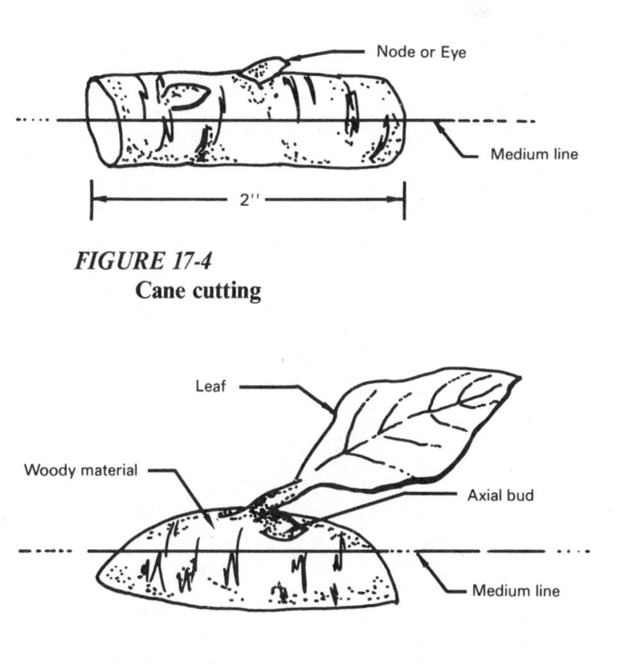

FIGURE 17-4
Cane cutting

FIGURE 17-5
Heel cutting

may be dipped in rooting hormone and then placed either vertically or horizontally in the medium.

Double-eye Cuttings When plants have opposite leaves, the *double-eye cutting* is the preferred type. It is prepared the same way as the single-eye cutting (Figure 17-7).

Leaf-type Cuttings For many of the indoor herbaceous plants, a leaf-type cutting will produce plants quickly and efficiently. This type of cutting will not normally work for woody plants, however.

Leaf Cuttings A cutting made from a leaf without a petiole is referred to as a *leaf cutting*. To prepare a leaf cutting, detach the leaf from the plant with a clean cut and dip the leaf into the rooting hormone. Place the leaf cutting vertically into the medium. New plants will form at the base of the leaf and may be removed when they have formed their own roots (Figure 17-8).

Leaf Petiole Cuttings For *leaf petiole cuttings*, a leaf with a petiole about ½–1½″ is detached from the plant. The lower end of the petiole is dipped into the rooting medium and is then placed into the medium. Several plants will form at the base of the petiole (Figure 17-9). These plants may be removed when they have developed their own roots. The cutting may be left in the medium to form new plants.

Leaf Section Cuttings Fibrous-rooted begonias are frequently propagated using *leaf section cuttings*. The begonia leaves are cut into wedges, each containing at least one vein (Figure 17-10). The sections are then placed into media.

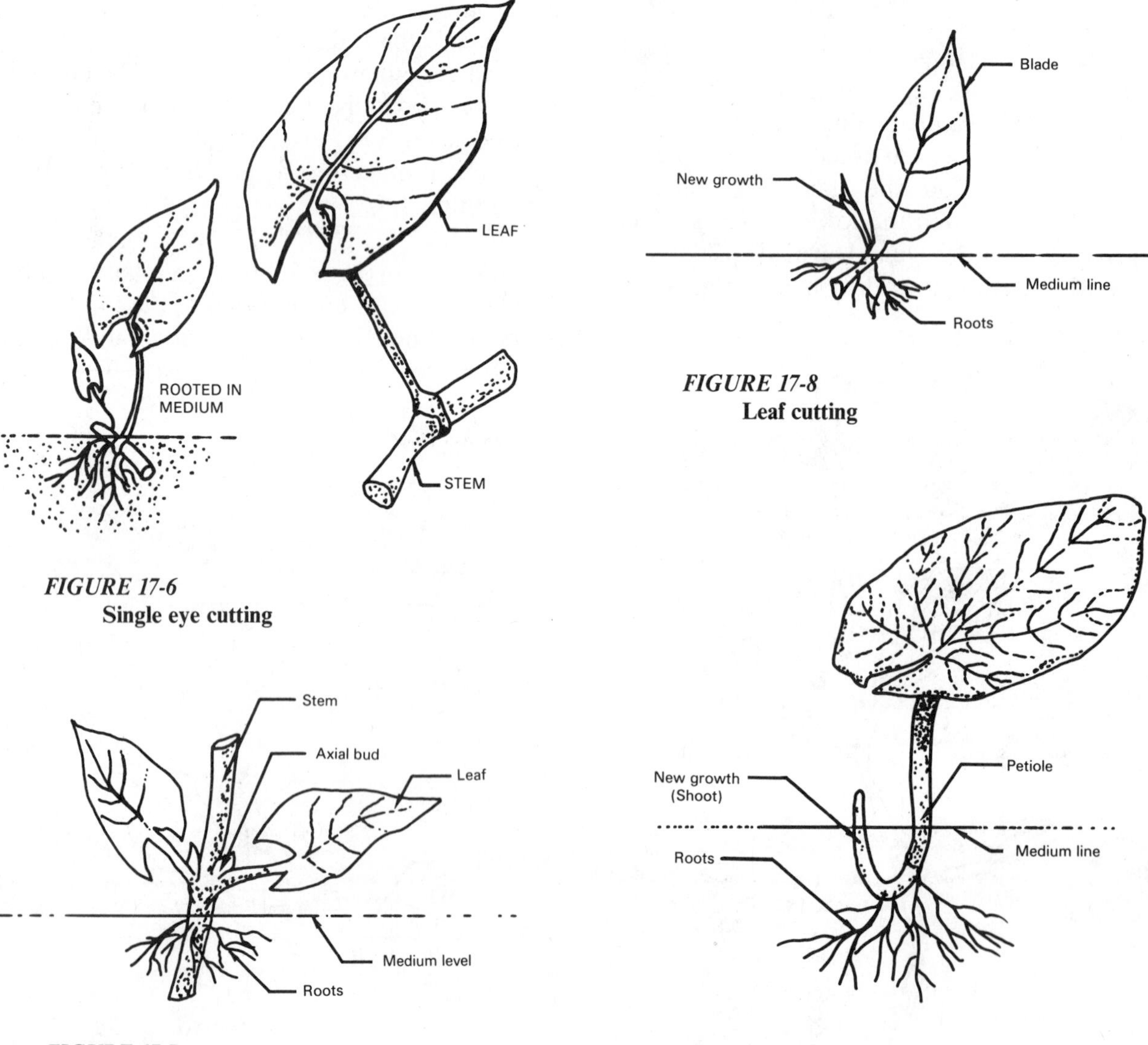

FIGURE 17-6
Single eye cutting

FIGURE 17-8
Leaf cutting

FIGURE 17-7
Double eye cutting

FIGURE 17-9
Leaf petiole cutting

New plants will form at the vein that is in contact with the media. A section-type leaf cutting is made with the snake plant *(Sanseveria sp.)*. The leaf is cut into 2–3″ sections. It is a good practice to make the bottom of the cutting on a slant and the top straight. This is done so you can tell the top from the bottom (Figure 17-11). The sections are placed in the medium vertically. Roots will form reasonably soon and new plants will start to appear. These are to be cut off from the cutting as they develop root systems. The original cutting may be left in the media for more plants to develop.

Split-vein Cuttings *Split-vein cuttings* are often used with large leaf types, such as begonias and other large-leaf plants. With split-vein cuttings the leaf is removed from the stock plant and the veins are slit on the lower surface of the leaf (Figure 17-12). The cutting is then placed on the rooting medium with the lower side down. It might be necessary to secure the leaf to make it lay flat on the surface. A good method is to use small pieces of wire, bending them like hair pins and pushing them through the leaf to hold it in place. The new plants will form at each slit in the leaf.

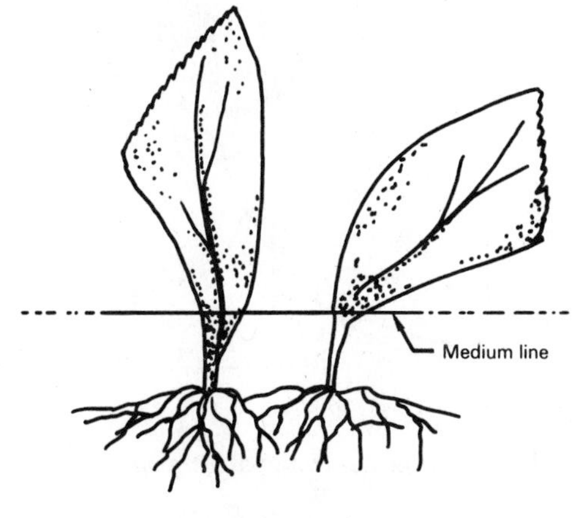

FIGURE 17-10
Leaf section cutting

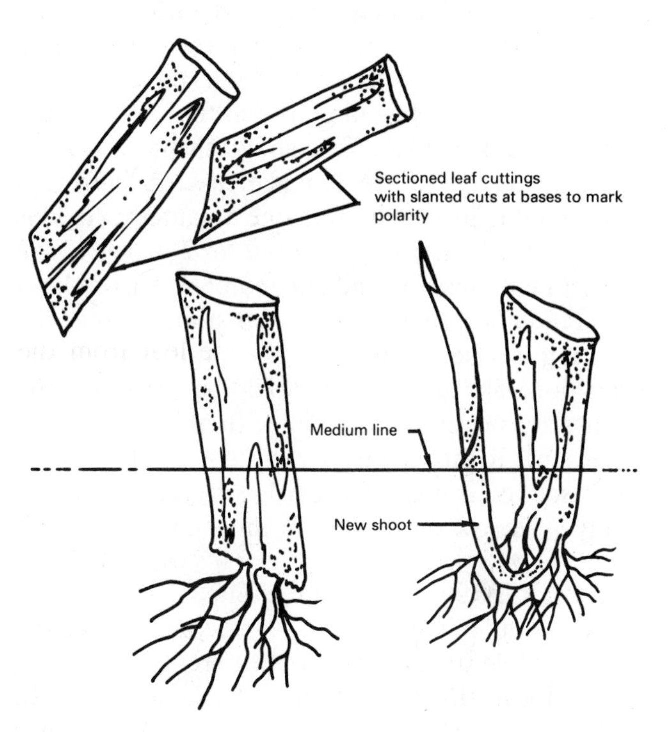

FIGURE 17-11
A leaf section of a snake plant

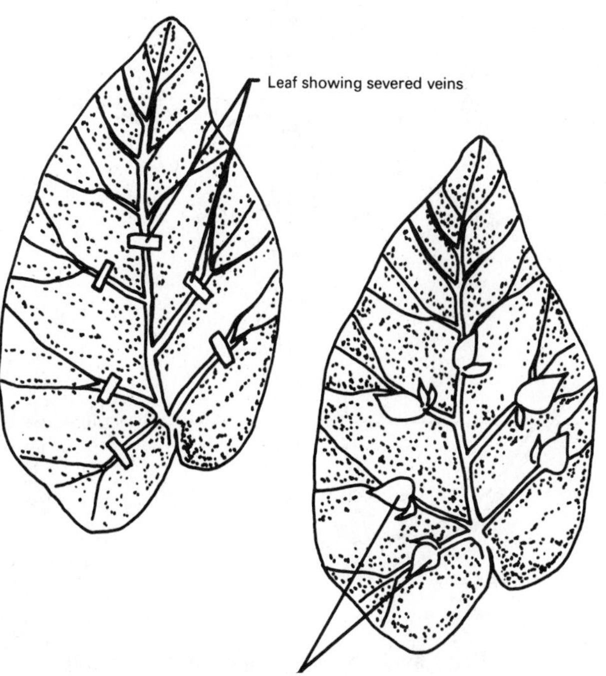

FIGURE 17-12
A split vein cutting

Root Cuttings It is best to use plants that are at least 2 to 3 years old for making *root cuttings*. The cuttings should be made in the dormant season when the roots have a large supply of carbohydrates in reserve. In some species the root cuttings will develop new shoots, which in turn will develop root systems. In others, the root system will be produced before new shoots develop.

If the plant has large roots, the root section should be 4–6″ long. To distinguish the top from the bottom of the root, make the top cutting a straight cut and the bottom one a slanted cut. This type of cutting should be stored for 2 to 3 weeks in moist peat moss or sand at a temperature of about 40°F. When removed from the storage area, the cutting is inserted into the medium in a vertical position. The slanted cut should be down and the top straight cut just level with the top of the medium. If the plant typically has small roots, a 1-2″ section is used. The cutting is placed horizontally ½″ below the surface of the medium.

LAYERING In many plants, stems will develop roots in any area that is in contact with the media while still attached to the parent plant. After roots form, shoots develop at the same point. An advantage of this type of vegetative propagation is that the plant does not experience water stress and sufficient carbohydrates are supplied to the new plant that is forming. A discussion of some of the more common methods of *layering* follows.

Simple Layering *Simple layering* is a very easy method that can be used on azaleas, rhododendrons, and other plants. A stem is bent to the ground and is covered with medium. It is advantageous to wound the lower side of the stem to the cambium layer. The last 6–10″ of the stem is left exposed (Figure 17-13).

Tip Layering Raspberries and blackberries are propagated using *tip layering*. With this method, a hole is made in the medium and the tip of a shoot is placed in the hole and covered. The tip will start to grow downward and will then turn to grow upward. Roots will form at the bend. When the new tip appears above the medium, a new plant is ready to be transplanted. It will be necessary to separate the new plant from the parent by cutting the stem just before it enters the medium (Figure 17-14).

Air Layering Many foliage plants are propagated using *air layering*. Some ornamental trees, such as dogwood, can also be reproduced by this type of layering. The stem is girdled with two cuts

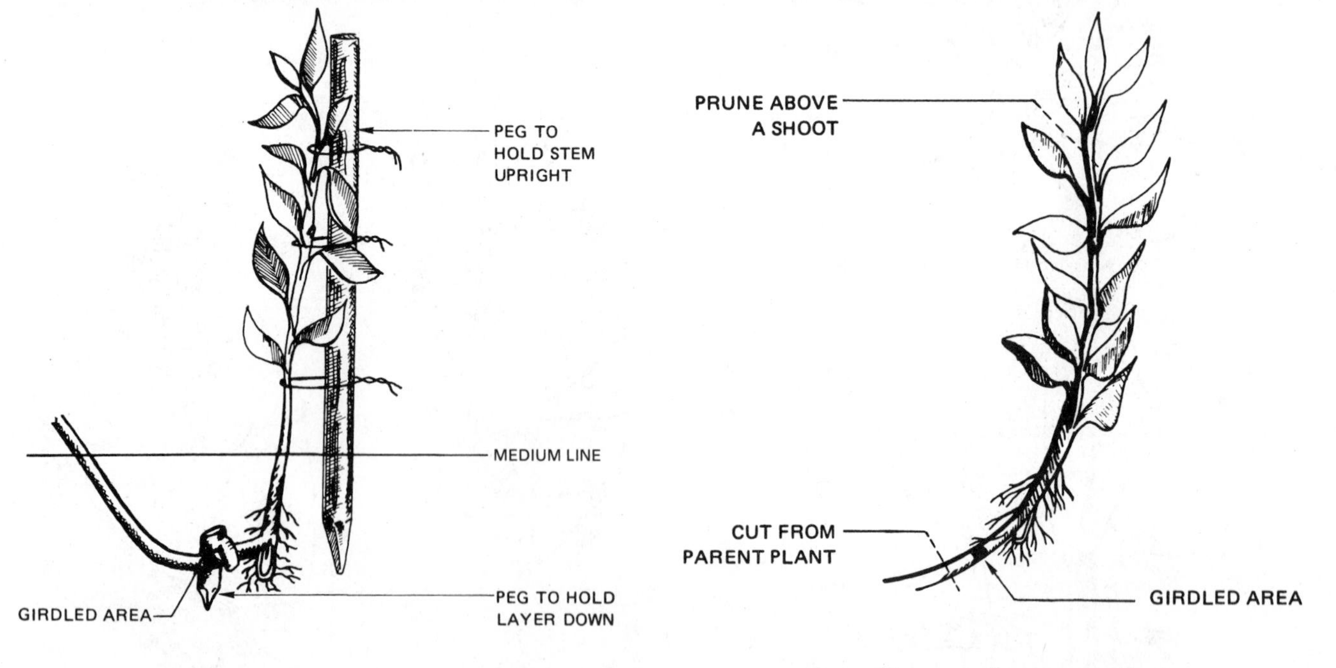

Pegs hold the layer upright and in the ground.

Rooted layer cut from parent plant.

FIGURE 17-13
Simple layering ***(From Reiley & Shry/Introductory Horticulture, 3rd edition, copyright 1988 by Delmar Publishers Inc.)***

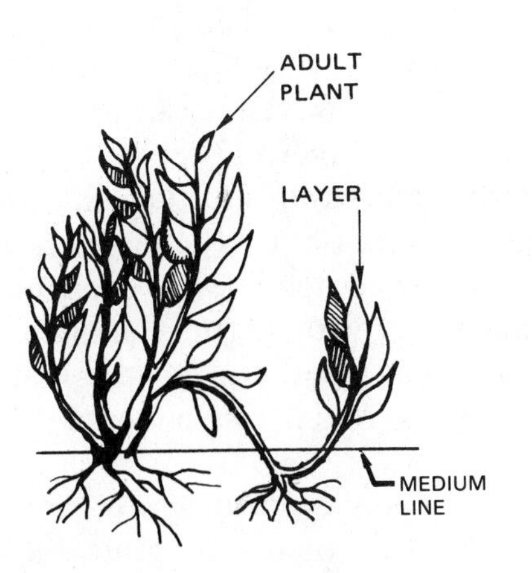

FIGURE 17-14
Tip layering ***(From Reiley & Shry/Introductory Horticulture, 3rd edition, copyright 1988 by Delmar Publishers Inc.)***

about 1″ apart. The bark is removed. The wound is dusted with a rooting hormone and surrounded with damp sphagnum moss (Figure 17-15). Plastic is wrapped around the moss-packed wound and tied at both ends. In a few weeks, depending on the plant, roots will appear throughout the moss. The stem is cut just below the newly formed root ball, and the ball is planted into a well-drained potting medium.

Division Some plants are easily propagated by dividing or separating the main part into smaller parts. If the plant has rooted crowns, they are separated by cutting or pulling them apart. The resulting clumps are planted separately. If the stems are not attached to each other, they are pulled apart. If the crowns are joined together by horizontal stems, they are cut apart with a knife

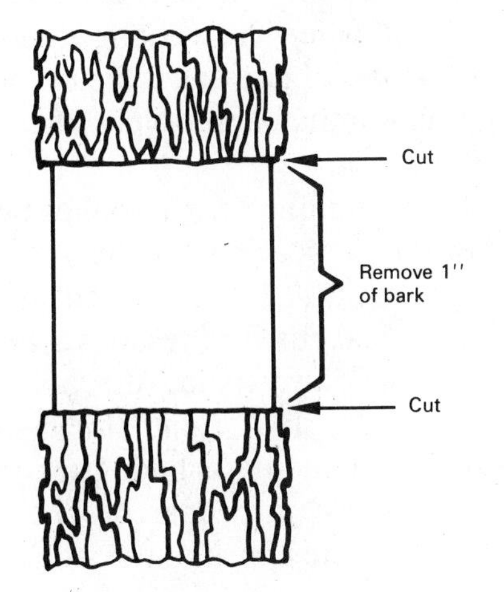

1. Prepare stem.

2. Soak Sphagnum moss and squeeze out excess water.

3. Pack damp moss over girdled area and tie.

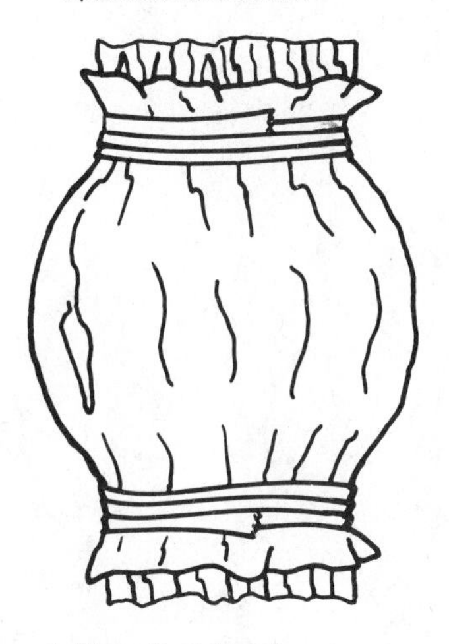

4. Wrap with plastic and tape ends tightly.

FIGURE 17-15
Air layering

(Figure 17-16). It is a good practice to dust the divided plants with a fungicide.

Some plants that grow from bulbs or corms form little bulblets or cormels at their base. To produce more plants from this type, simply separate the newly formed plant part and place it into a good medium (Figure 17-17).

Grafting *Grafting* is a procedure for joining two parts of a plant together so they grow as one. This method of asexual propagation is used when plants do not root well as cuttings or when the root system is inadequate to support the plant for good growth. Grafting will allow the production of some unusual combinations of plants. For instance, several varieties of apples can be grown on one tree. Some nut trees can be made to grow varieties other than their own. Some unusual foliage plants can also be made by grafting. Finally, dwarf fruit trees are created by grafting regular varieties on dwarfing root stock.

Hosta Root Clump Before Division

Hosta Root Divisions

FIGURE 17-16
Division

The top part of the plant that is to be propagated is called the scion. The rootstock or stock will be the new plant's root system and will supply the nutrients and water. The graft union is where the two parts meet (Figure 17-18).

To ensure successful grafting, the following conditions are necessary: (1) the scion and the rootstock must be compatible, (2) each must be at the right stage of growth, (3) the cambium layer of each section must meet, and (4) the graft union must be protected from drying out until the wound has healed.

There are many types of grafts. Some common grafts are whip or tongue graft, bark graft, cleft graft, bridge graft, and bud graft. Each type is used for a special purpose. The most commonly used and the easiest is bud grafting.

Bud Grafting The union of a small piece of bark with a bud and a rootstock is called bud grafting. It is most useful when the scion material is in short supply. This type of grafting is faster and will make a stronger union than other types of grafting.

T-Budding T-budding is a popular type of bud graft, where a vertical cut about ½″ long is made on the rootstock. A horizontal cut is made at the top of the vertical cut. The result is a T-shape. The bark is loosened by twisting the point of a knife at the top of the T. A small shield-shaped piece of the scion, including a bud, bark, and a thin section of the wood, is prepared. The bud is pushed under the loosened bark of the stock plant.

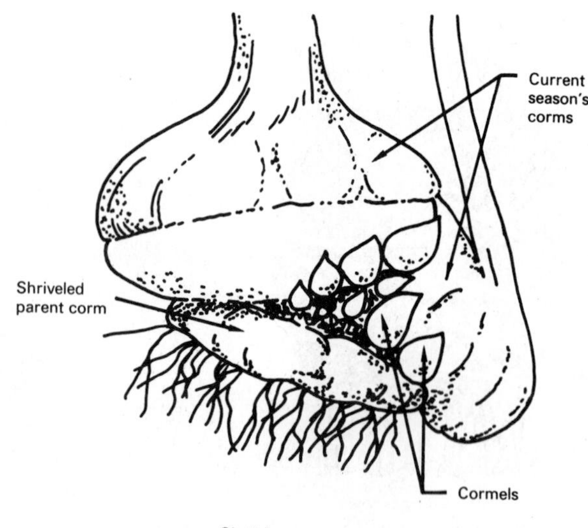

Gladiolus corm system

FIGURE 17-17
Separation

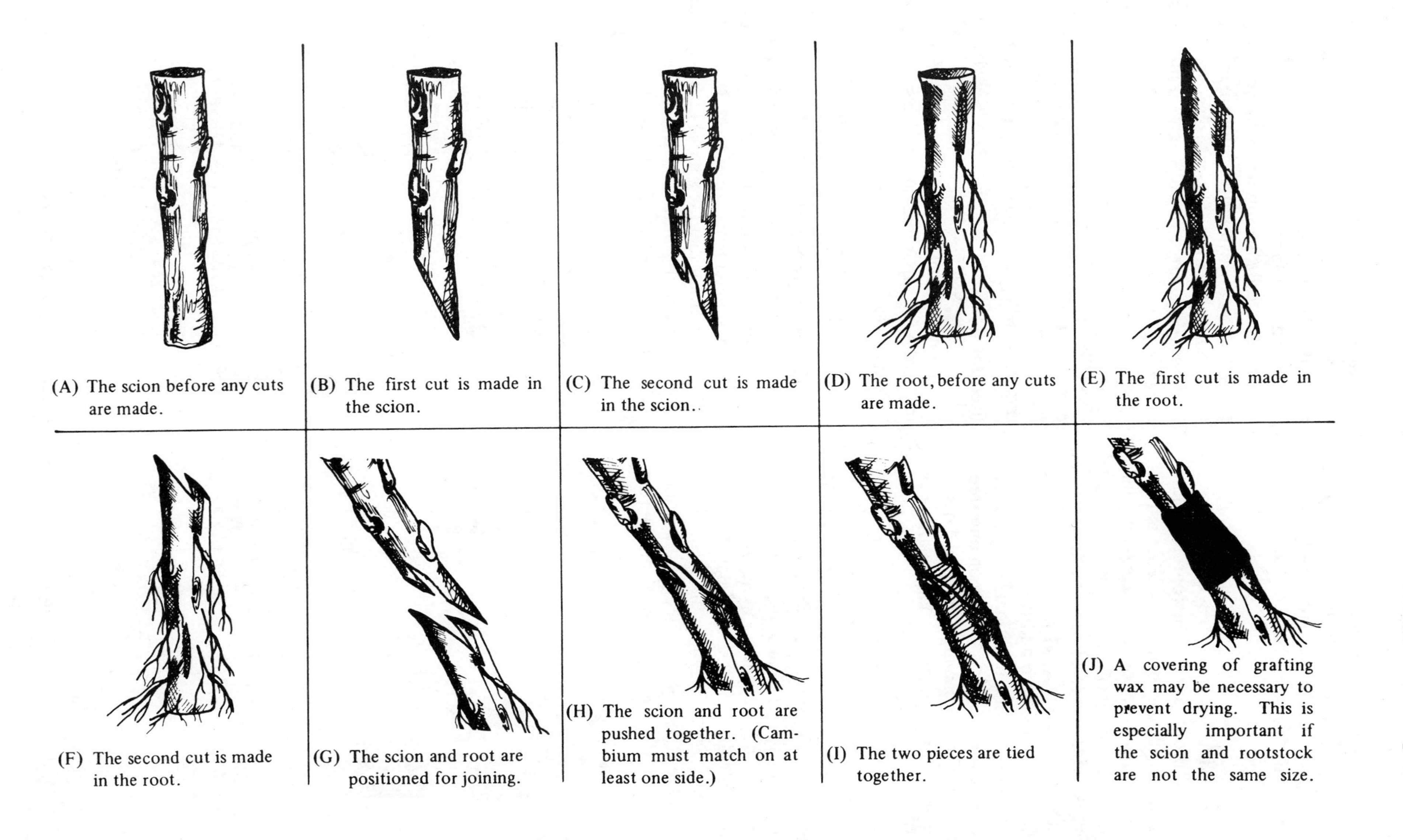

FIGURE 17-18
Whip graft ***(From Reiley & Shry/Introductory Horticulture, 3rd edition, copyright 1988 by Delmar Publishers Inc.)***

The union is wrapped with a piece of rubber band called a budding rubber. The bud is exposed. After the bud starts to grow, the remainder of the rootstock plant is cut off above the bud graft (Figure 17-19).

TISSUE CULTURE

A new and exciting method of plant propagation is micropropagation, or *tissue culture*. Instead of using a large part of the plant as in other types of vegetative or asexual propagation, a very small and actively growing part of the plant is used. The result is many new plantlets from a section of a leaf. The process must be done in a clean atmosphere and is not successful in the greenhouse or other traditional propagation areas. Tissue culture requires the use of very sanitary conditions. There are many commercial tissue-culture laboratories producing a large variety of plants today.

Advantages over Traditional Methods

The greatest advantage of this type of propagation is that numerous plants can be propagated from a single disease-free plant. Plants can be propagated more efficiently and economically than with traditional methods of asexual reproduction. The main disadvantage is that the work area must be clean. All of the equipment must be sterile. In commercial production by tissue culture, there is more expense in equipment and facilities than there is with traditional methods of propagation.

The materials necessary for tissue culture are (1) a clean, sterile area in which to work, (2) clean plant tissue, (3) a multiplication medium, (4) a transplanting medium, (5) sterile glassware, (6) sterile tools, (7) a razor or an X-acto® knife, and (8) tweezers.

Preparing Sterile Media

The first step in preparing for tissue culture is to prepare the medium in which the tissue will grow. The medium is commercially available from many of the scientific supply houses. The Virginia Cooperative Extension Service offers the following formula and procedure for preparing media for experimentation in tissue culture on a small basis.

1. Use a quart jar to mix the following materials:
   - □ ⅛ cup of sugar
   - □ 1 tsp of soluble, all-purpose fertilizer. The label will indicate that all of the major and minor elements are present. It is especially important that the soluble fertilizer contain ammonium nitrate. If it is lacking, add ⅓ tsp of 35–0–0 soluble fertilizer.

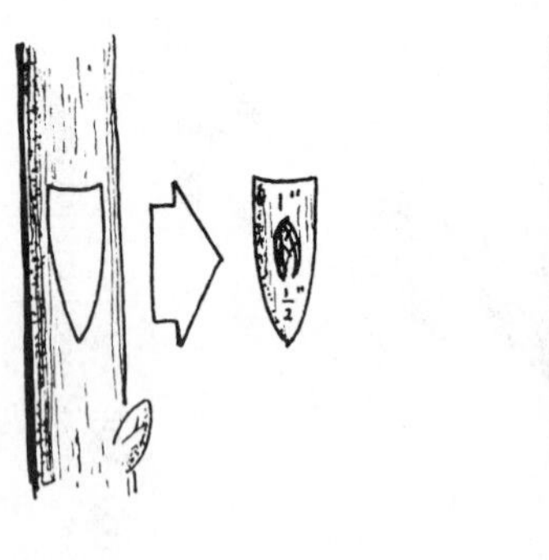

STOCK

A T-CUT IS MADE THROUGH THE BARK.

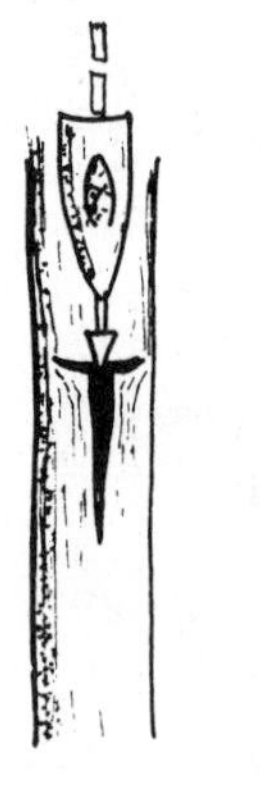

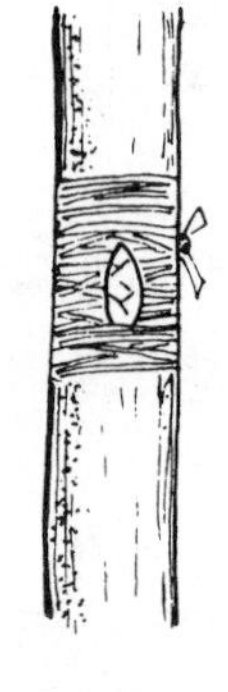

FIGURE 17-19
Making a T-bud ***(From Ingels/Ornamental Horticulture, copyright 1985 by Delmar Publishers Inc.)***

- □ 1 tablet (100 mg) of inositol (myo-inositol). This can be obtained from most health-food stores.
- □ ¼ of a pulverized tablet containing 1 to 2 mg of thiamine
- □ 4 tbsp of coconut milk, the source of cytokinin. This is obtained from a fresh coconut. Freeze the remainder for later use.
- □ 3 to 4 grains of a rooting hormone containing 0.1 active ingredient IBA

2. Fill the jar with purified, distilled, or deionized water.
3. Shake the jar to dissolve all materials.

After the medium is dissolved, prepare the culture tubes using test tubes with lids or other suitable glass. Fill the culture tubes one-quarter of the way with sterile cotton balls. Use one or two per tube. They do not need to be packed tightly. Pour the prepared medium into the culture tubes to just below the top level of the cotton. Place the lid on loosely.

After all medium is placed in culture tubes, it is ready to be sterilized. Sterilization may be done in two ways: (1) heat in a pressure cooker for 30 minutes or (2) heat in an oven for 4½ hours at 320°. After it is removed, place the culture tubes in a clean area and allow them to cool (Figure 17-20). If several days will go by before using all of the tubes, wrap them in small groups in plastic wrap or foil before sterilizing.

Sterilizing Equipment and Work Areas The tools and equipment used for tissue culture must also be sterilized. This can be done as the medium is sterilized by placing the tweezers, razor blade, or knife in the pressure cooker or oven. After the initial sterilization, they may be cleaned by dipping them in alcohol before and after each use.

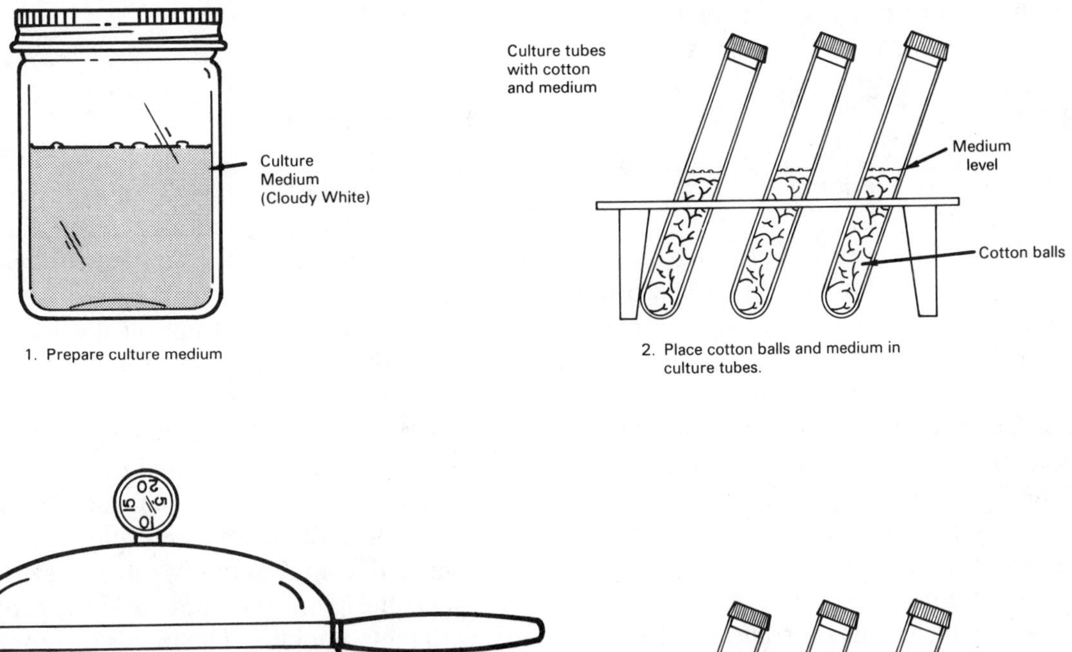

FIGURE 17-20
Summary of tissue culture medium

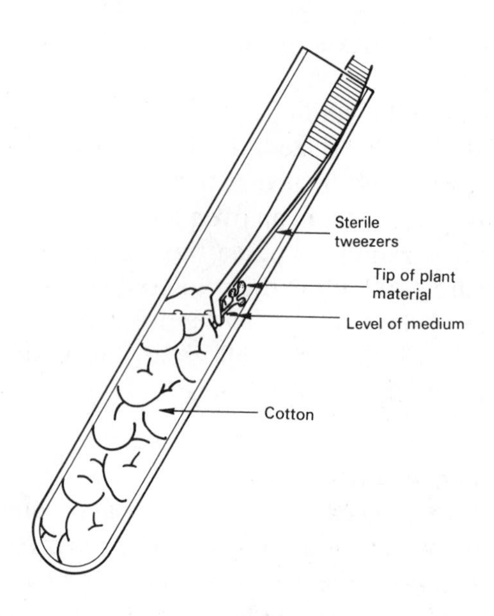

FIGURE 17-21
Plant material in a culture tube

The work area must be thoroughly cleaned and sterilized. Wash the area with a disinfectant. Keep a mist bottle filled with a mixture of 50% alcohol and sterilized water to spray work areas and tools, as well as the hands and arms of the propagator.

Preparing Plant Tissue and Placement in the Culture Tube After the growing medium is properly prepared and cooled and the work area properly cleaned, the next step is to prepare the plant tissue. Various parts of the growing plant may be tissue cultured. For the production of vigorous plantlets, use only actively growing portions. With some species of plants, only a small ¼″ square section of the leaf is used, while for others ½″ of the shoot tip is used. With ferns, ¼″ of the tip of the rhizome is used. Remove the part of the plant to be used. Also remove excess plant material. Submerge the plant part into a solution of one part commercial bleach and nine parts water for about 10 minutes. Remove the tissue with sterile tweezers and rinse the material in sterile water. Once the plant part has been sterilized in the bleach solution, it can be handled only with sterile tweezers and must not touch any nonsterile surface.

When the plant material has been sterilized and rinsed, remove any damaged tissue with a sterilized razor blade. Remove the lid from a properly prepared culture tube or jar and place the plant material on the cotton. Take care that the plant material is not completely submerged (Figure 17-21). Recap quickly to avoid contamination from the air. It is best if the material is placed in front of the propagator so that work is not done over the uncapped culture tube.

Reminder!

IT IS IMPORTANT THAT ALL WORK AND THE TRANSFERRING OF MATERIALS BE DONE QUICKLY AND IN A CLEAN ENVIRONMENT. Scrub all areas with disinfectant and clean all tools with a disinfectant solution. Any contamination may lead to unsuccessful work. Bacteria and fungus will grow in unclean culture tubes and will overtake the new plant growth.

Storing Tissue Cultures After all plant material has been cultured, put the cultures in an evenly warmed (70–75°F) and well-lighted area. The plant tissue will NOT do well in direct sunlight. If any contamination has occurred, it will be evident in 48 to 96 hours as mold or rotting on the medium. If contamination occurs, remove the contaminated tubes and wash them for reuse.

When the plantlets have grown to a satisfactory size, take them out and transplant them into a good growing medium. The plantlets are very fragile. Handle them very carefully. As each plant is removed from the culture tube, wash it thoroughly. Place the plant in a pot inside a protected area with high humidity. The plants are coming out of a well-protected environment with plenty of humidity and light. After they have adapted to the pot and are growing well, they may be treated like any other growing plant. This process will take about 3 to 6 weeks from the beginning to a successfully growing plant.

Student Activities

1. List the crops that are grown commercially in your area. Visit a site or sites where plants are propagated and ask the grower to discuss the propagation methods used.
2. List the prevalent crops that are grown from seed in your area and the popular varieties of each. Study a seed catalog and determine the requirements for germination of each variety and why the varieties are popular in your community.
3. Plant a seed in a jar filled with medium. Place the seed near the edge of the jar so that you can see what is happening. Keep a journal of daily observations as the seed or plants change. You may want to do this with several seeds in different jars and vary the amount of water, light, air, or temperature in each jar. Note your observations and the conditions regarding each seed. Write your conclusions about what is best for maximum germination results.
4. Experiment with different kinds of plants and various kinds of cuttings. Keep a journal to determine the best type of cutting for specific plants. Keep notes on different media, temperatures, light, and rooting hormones.
5. Practice making each type of graft discussed in this unit under the supervision of your instructor.
6. Research additional types of grafting methods.
7. Conduct an experiment with the tissue-culture method of propagation. Keep notes on the different kinds of plants used, the time required for root formation, and observations regarding the benefits of propagation by tissue culture.
8. Prepare a statement about the propagation methods best suited to your purpose.

Self-Evaluation

A. MULTIPLE CHOICE

1. Propagation is defined as

   a. the union of an egg and sperm.
   b. the process of increasing the numbers of a species.
   c. a cheaper method of propagation than with seeds.
   d. the only way to propagate some species and cultivars.

2. A seed consists of

   a. a root, stem, and flower.
   b. a root, seed coat, and endosperm.
   c. a seed coat, endosperm, and embryo.
   d. an embryo, cotyledons, and new plant.

3. A type of stem cutting used where stock material is limited and has alternate leaves is a

   a. stem tip cutting.
   b. cane cutting.
   c. simple layering.
   d. single-eye cutting.

4. A cutting that is usually made from a large-leaf plant with the veins split is

   a. split vein cutting.
   b. leaf petiole cutting.
   c. terminal tip cutting.
   d. tissue propagation.

5. Grafting is

   a. a type of sexual propagation.
   b. a type of hybridization.
   c. a method by which two plants are propagated.
   d. a method of joining two parts of two different plants.

6. The most common type of grafting is

   a. T-budding.
   b. simple layering.
   c. stem-tip propagation.
   d. scion cut out of the stock plant.

B. MATCHING

______ **1.** Outer seed coat
______ **2.** Germinate
______ **3.** Imbibition
______ **4.** Tip cutting
______ **5.** Root cutting
______ **6.** Division

a. A cutting from an end of a branch containing a terminal bud
b. Best taken when the plant is dormant
c. The absorption of water into the young seed
d. Functions as a protector for the seed
e. When plants are separated and then replanted
f. When a seed starts to sprout

C. COMPLETION

1. ______ propagation utilizes a part or parts of one parent plant.
2. The ______ will supply food to the young seedling until it is able to make its own.
3. A ______ might contain fungicide and is used to help plants produce roots more quickly.
4. ______ cuttings are normally made of a section containing one or two nodes.
5. Tissue culture is also known as ______ .
6. A major aspect of tissue culture is that the area to be worked in must be ______ and ______ .

D. TRUE OR FALSE

______ **1.** A clone is almost like the parents.

______ **2.** The embryo is actually the young plant.

______ **3.** All seeds need light to germinate.

______ **4.** An advantage of tissue culture is that only one plant can be made from a disease-free plant.

______ **5.** Double-eye cuttings are often used when plants have opposite leaves.

______ **6.** Herbaceous plants are propagated by cuttings.

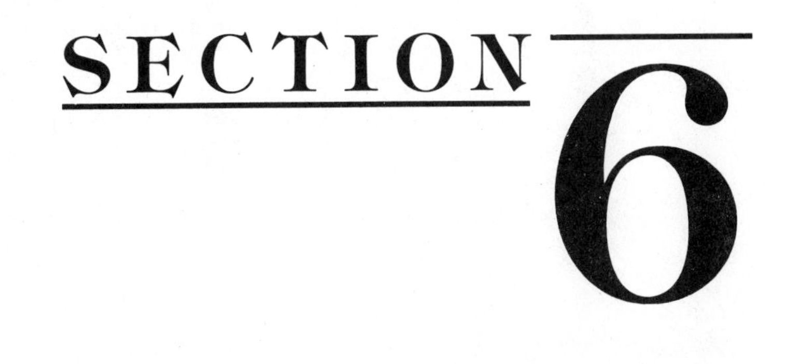

SECTION 6

Crop Production

UNIT 18 Home Gardening

OBJECTIVE To plan, plant, and manage a home garden.

Competencies to Be Developed

After studying this unit, you will be able to

- ☐ analyze family needs for homegrown fruits, vegetables, and flowers.
- ☐ determine the best location for a garden.
- ☐ plan a garden to meet family needs.
- ☐ establish perennial garden crops.
- ☐ prepare soil and plant annual garden crops.
- ☐ list recommended cultural practices for selected garden crops.
- ☐ protect the garden from excessive damage from drought and pests.
- ☐ harvest and store garden produce.
- ☐ describe the use of cold frames, hotbeds, and greenhouses for home production.

TERMS TO KNOW

Seasonal
Square foot
Successive
Variety
Loamy
Peat moss
Clod
Furrows
Climate
Cultivation
Herbicides

MATERIALS LIST

grid-type paper
pencil and eraser
seed catalogs

Gardening is an activity that can be enjoyed by all members of the family. It provides fresh fruits and vegetables for immediate use or to sell for profit. Gardening is both an art and a science. It is demanding of the gardener, both in skill and creativity. A garden is alive and changing every day, presenting new challenges to the gardener (Figure 18-1).

ANALYZING A FAMILY'S GARDENING NEEDS First, you must decide what vegetables and flowers the family likes. It would not make sense to plant sweet potatoes, green beans, and marigolds if no one in the family cared for these vegetables and flowers. The home gardener should provide a seasonal and continuous variety of vegetables and flowers. *Seasonal* means pertaining to a certain season of the year. Fresh vegetables are important to everyone's diet. Plan to have plenty available during the growing season and to store some for future use. Choose only those vegetables that will produce high yields.

When it has been decided what vegetables and flowers the family likes, it is time to figure out how much ground will be needed. It is better to have a smaller garden that is well cared for than to have a garden that goes to waste because it becomes too much to handle. A good rule of thumb for four grown people is to start with a plot 10 ft. wide and 26 ft. long, or 260 ft.^{2} A *square foot* is an area equal to 12″ × 12″.

FIGURE 18-1
While some people seem to have a natural talent for gardening, the growth of plants is based on scientific principles and procedures. ***(Courtesy Denmark High School Agriculture Education Department)***

THE GARDEN PLAN

A prospective gardener should make a sketch on paper detailing the amount and placement of the various crops. At this stage, it is important to consider *successive* plantings. These crops follow each other in the season so that the ground is occupied throughout the growing season. Fall crops can follow spring and summer crops in many areas. This can be easily done by planting perennial crops and different varieties of specific annual crops. *Variety* is a category within a species of a plant. Allow adequate space between the rows for cultivation.

LOCATING THE GARDEN

Depending on where you live—in the city, suburbs, or a rural area—the location of the garden is an important consideration. Your garden should be convenient to the house. And it should be accessible to a water supply; on loamy, well-drained soil; in a sunny spot; and visible from your home, if possible. *Loamy* means a granular soil with a balance of sand, silt, and clay particles.

Next, you must visualize where to locate the garden. What happens to the proposed spot when there is a heavy rain? Or, what happens when it is very dry? From the chosen spot, look up to determine whether trees or branches will cause problems by excessively shading the garden. When planting flower beds around the house, remember that along the south and west sides, the heat will be reflected onto these beds, and they may require extra water. Select the best site you can for both your vegetable and flower gardens.

PREPARING THE SOIL

Conditioning the Soil

Garden soil should be loose and well drained. The ideal soil type should be granular—like coffee grounds—so that water will soak in rapidly. Few soils are originally found this way. Soil-building practices can improve the soil over time, and the gardener must prepare it to achieve the best results.

If the proposed site has not been previously planted, it is a good idea to add organic matter and plow, spade, or rototill it in the fall. The decayed organic matter and the freezing and thawing of the soil during the winter months will help improve its physical condition. The application of fine organic matter in the spring can also improve the soil condition and fertility.

Materials such as composted leaves and grasses, peat moss, composted sawdust, and sterilized weed-free manure are good soil conditioners. *Peat moss* is a soil conditioner made from sphagnum moss. A good rate of organic matter to apply is one pound of dry material per square foot.

Preparing to Plant

If the site selected has never been planted, it is best to remove the sod before digging up the soil. Then spread the organic matter over the soil. When moisture conditions are favorable, turn the soil with a shovel, spade, plow, or rototiller. If you are turning the soil over by hand, break up any large clods. A *clod* is a lump or mass of soil.

After the spading is completed, make the planting beds. A good procedure is to heap the soil to make raised rows with grooves or *furrows* between them. Another method is to prepare raised beds that are 4–8′ wide with sunken walks between them (Figure 18-2). The furrows or sunken walkways can drain off excess rain.

Next, level the raised beds with a rake, but do not push the soil back into the furrows. The next step is to walk over the beds or tamp them with the head of a garden rake to firm them and to eliminate air pockets. Finally, use the hoe and back of the garden rake to push and pull the soil to level it and to break the clods into fine particles (Figure 18-3). The soil is now ready for planting.

COMMON GARDEN CROPS AND VARIETIES

Consider the Climate

When choosing varieties of vegetables and flowers to plant in the home garden, consider the climate. *Climate* refers to the weather conditions for a specific region. Do not plant crops outdoors until all danger of frost is gone. Except for a few mountainous areas and the northern tier of states, almost all areas of the United States are frost-free from June through August (Figure 18-4).

Plants grow rapidly in the three to four months of frost-free weather. However, this is not enough time for long-season vegetables such as eggplant, cantaloupe, and watermelon to mature reliably. Much of the country has an intermediate growing season, with five to six months of frost-free weather. Long-season crops can be grown in these areas. Growing seasons of seven to ten months are common across the mid-South, South, and low-elevation regions of the Southwest and West. In these regions, gardeners can grow both spring and fall crops of many varieties. However, these same regions are susceptible to summers that are so hot or dry that only a few crop varieties can survive the intense summer heat.

Consider the Variety There are both warm-season and cool-season crops. Certain flowers and vegetables must have continuous cool weather to do well. Heat will quickly make the plants dry up or go to seed. Where the growing season is five months or longer, outdoor seeding in the late summer usually results in an excellent harvest during the cool fall season. Most of the cool-season plants can withstand light frosts with little damage.

There are two types of warm-season flowers and vegetables: those that mature quickly and those that require four months or more from planting to harvest. The quick-maturing types are almost always started in the garden when the soil is warm. The late maturing types are usually started indoors and are transplanted to the garden after all danger of frost is past.

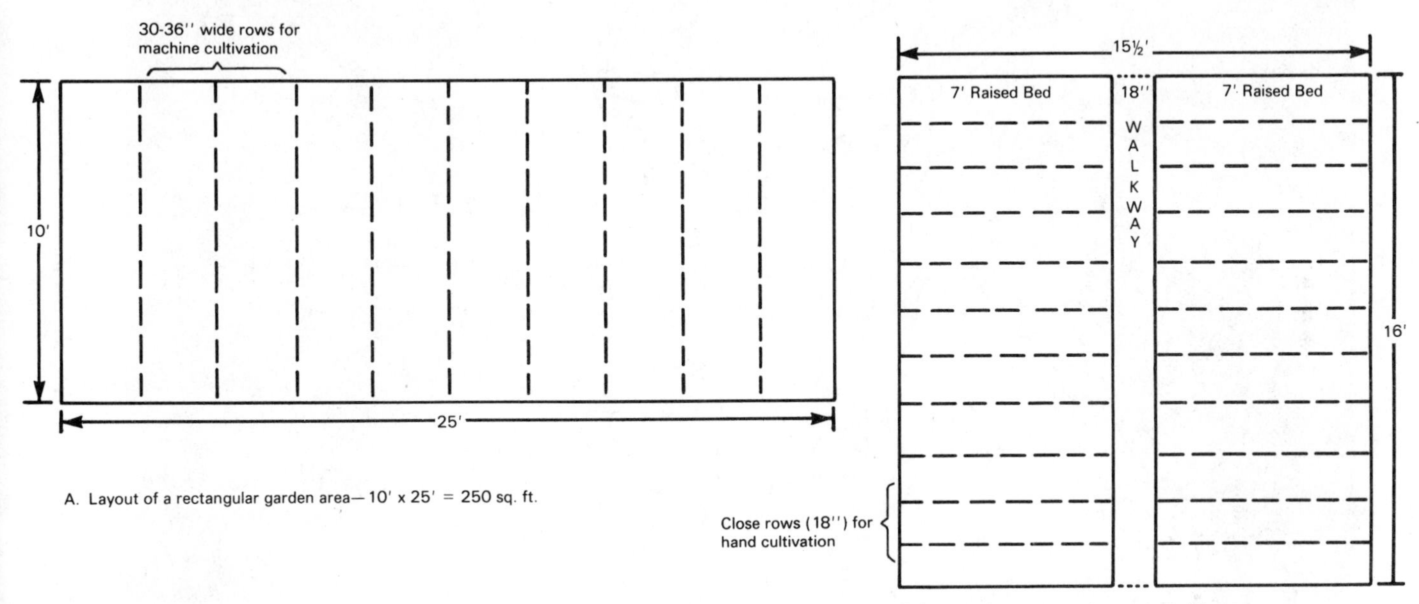

FIGURE 18-2
Layout for rectangular and square gardens

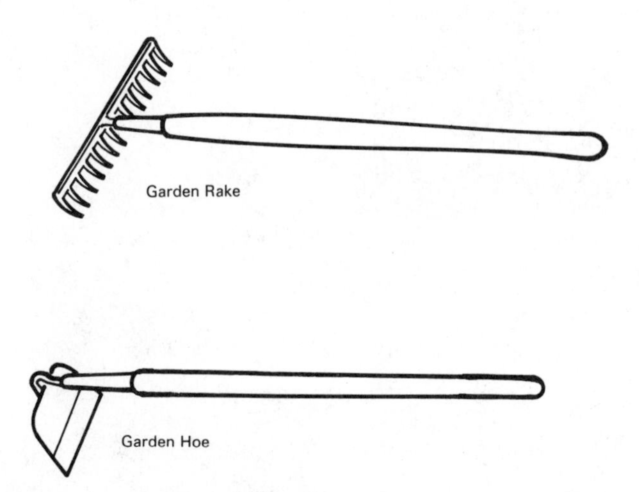

FIGURE 18-3
The garden rake and hoe are used to break up clods and loosen, smooth, and level the soil.

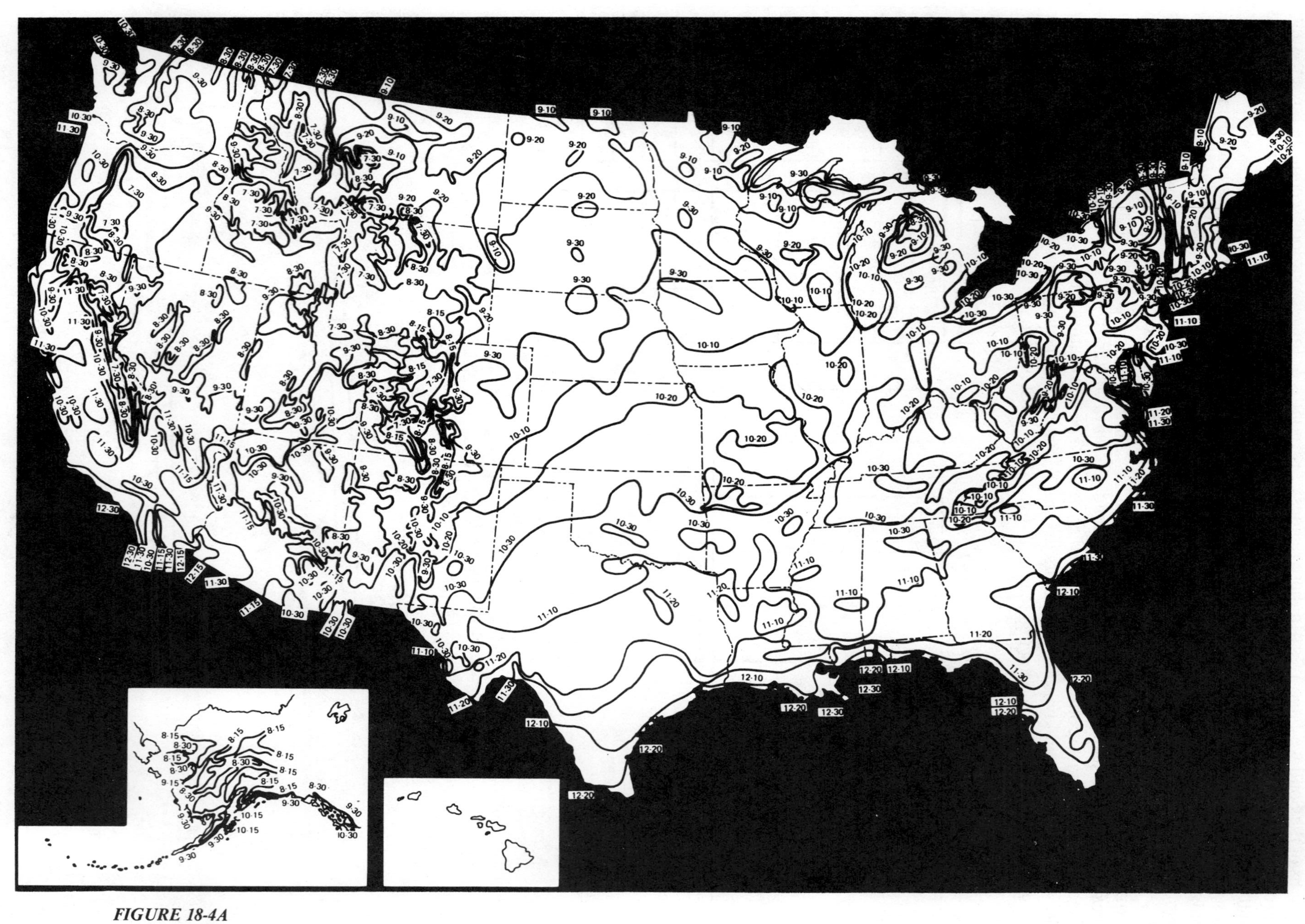

FIGURE 18-4A
Average last freeze dates in the United States *(From Reiley & Shry/Introductory Horticulture, 3rd edition, copyright 1988 by Delmar Publishers Inc.)*

Crop	Planting dates for localities in which average date of last freeze is—						
	Apr. 10	Apr. 20	Apr. 30	May 10	May 20	May 30	June 10
Asparagus [1]	Mar. 10–Apr. 10	Mar. 15–Apr. 15	Mar. 20–Apr. 15	Mar. 10–Apr. 30	Apr. 20–May 15	May 1–June 1	May 15–June 1.
Beans, lima	Apr. 1–June 30	May 1–June 20	May 15–June 15	May 25–June 15			
Beans, snap	Apr. 10–June 30	Apr. 25–June 30	May 10–June 30	May 10–June 30	May–15–June 30	May 25-June 15	
Beet	Mar. 10–June 1	Mar. 20–June 1	Apr. 1–June 15	Apr. 15–June 15	Apr. 25–June 15	May 1–June 15	May 15–June 15.
Broccoli, sprouting [1]	Mar. 15–Apr. 15	Mar. 25–Apr. 20	Apr. 1–May 1	Apr. 15–June 1	May 1–June 15	May 10–June 10	May 20–June 10.
Brussels sprouts [1]	Mar. 15–Apr. 15	Mar. 25–Apr. 20	Apr. 1–May 1	Apr. 15–June 1	May 1–June 15	May 10–June 10	May 20–June 10.
Cabbage [1]	Mar. 1–Apr. 1	Mar. 10–Apr. 1	Mar. 15–Apr. 10	Apr. 1–May 15	May 1–June 15	May 10–June 15	May 20–June 1.
Cabbage, Chinese	(2)	(2)	(2)	Apr. 1–May 15	May 1–June 15	May 10–June 15	May 20–June 1.
Carrot	Mar. 10–Apr. 20	Apr. 1–May 15	Apr. 10–June 1	Apr. 20–June 15	May 1–June 1	May 10–June 1	May 20–June 1.
Cauliflower [1]	Mar. 1–Mar. 20	Mar. 15–Apr. 20	Apr. 10–May 10	Apr. 15–May 15	May 10–June 15	May 20–June 1	June 1-June 15.
Celery and celeriac	Apr. 1–Apr. 20	Apr. 10–May 1	Apr. 15–May 1	Apr. 20–June 15	May 10–June 15	May 20–June 1	June 1–June 15.
Chard	Mar. 15–June 15	Apr. 1–June 15	Apr. 15–June 15	Apr. 20–June 15	May 10–June 15	May 20–June 1	June 1–June 15.
Chervil and chives	Mar. 1–Apr. 1	Mar. 10–Apr. 10	Mar. 20–Apr. 20	Apr. 1–May 1	Apr. 15–May 15	May 1–June 1	May 15–June 1.
Chicory, witloof	June 10–July 1	June 15–July 1	June 15–July 1	June 1–20	June 1–15	June 1–15	June 1–15.
Collards [1]	Mar. 1–June 1	Mar. 10–June 1	Apr. 1–June 1	Apr. 15–June 1	May 1–June 1	May 10–June 1	May 20–June 1.
Cornsalad	Feb. 1–Apr. 1	Feb. 15–Apr. 15	Mar. 1–May 1	Apr. 1–June 1	Apr. 15–June 1	May 1–June 15	May 15–June 15.
Corn, sweet	Apr. 10–June 1	Apr. 25–June 15	May 10–June 15	May 10–June 1	May 15–June 1	May 20–June 1	
Cress, upland	Mar. 10–Apr. 15	Mar. 20–May 1	Apr. 10–May 10	Apr. 20–May 20	May 1–June 1	May 15–June 1	May 15–June 15.
Cucumber	Apr. 20–June 1	May 1–June 15	May 15–June 15	May 20–June 15	June 1–15		
Eggplant [1]	May 1–June 1	May 10–June 1	May 15–June 10	May 20–June 15	June 1–15		
Endive	Mar. 15–Apr. 15	Mar. 25–Apr. 15	Apr. 1–May 1	Apr. 15–May 15	May 1–30	May 1–30	May 15–June 1.
Fennel, Florence	Mar. 15–Apr. 15	Mar. 25–Apr. 15	Apr. 1–May 1	Apr. 15–May 15	May 1–30	May 1–30	May 15–June 1.
Garlic	Feb. 20–Mar. 20	Mar. 10–Apr. 1	Mar. 15–Apr. 15	Apr. 1–May 1	Apr. 15–May 15	May 1–30	May 15–June 1.
Horseradish [1]	Mar. 10–Apr. 10	Mar. 20–Apr. 20	Apr. 1–30	Apr. 15–May 15	Apr. 20–May 20	May 1–30	May 15–June 1.
Kale	Mar. 10–Apr. 1	Mar. 20–Apr. 10	Apr. 1–20	Apr. 10–May 1	Apr. 20–May 10	May 1–30	May 15–June 1.
Kohlrabi	Mar. 10–Apr. 10	Mar. 20–May 1	Apr. 1–May 10	Apr. 10–May 15	Apr. 20–May 20	May 1–30	May 15–June 1.
Leek	Mar. 1–Apr. 1	Mar. 15–Apr. 15	Apr. 1–May 1	Apr. 15–May 15	May 1–May 20	May 1–15	May 1–15.
Lettuce, head [1]	Mar. 10–Apr. 1	Mar. 20–Apr. 15	Apr. 1–May 1	Apr. 15–May 15	May 1–June 30	May 10–June 30	May 20–June 30.
Lettuce, leaf	Mar. 15–May 15	Mar. 20–May 15	Apr. 1–June 1	Apr. 15–June 15	May 1–June 30	May 10–June 30	May 20–June 30.
Muskmelon	Apr. 20–June 1	May 1–June 15	May 15–June 15	June 1–June 15			
Mustard	Mar. 10–Apr. 20	Mar. 20–May 1	Apr. 1–May 10	Apr. 15–June 1	May 1–June 30	May 10–June 30	May 20–June 30.
Okra	Apr. 20–June 15	May 1–June 1	May 10–June 1	May 20–June 10	June 1–20		
Onion [1]	Mar. 1–Apr. 1	Mar. 15–Apr. 10	Apr. 1–May 1	Apr. 10–May 1	Apr. 20–May 15	May 1–30	May 10–June 10.
Onion, seed	Mar. 1–Apr. 1	Mar. 15–Apr. 1	Mar. 15–Apr. 15	Apr. 1–May 1	Apr. 20–May 15	May 1–30	May 10–June 10.
Onion, sets	Mar. 1–Apr. 1	Mar. 10–Apr. 1	Mar. 10–Apr. 10	Apr. 10–May 1	Apr. 20–May 15	May 1–30	May 10–June 10.
Parsley	Mar. 10–Apr. 10	Mar. 20–Apr. 20	Apr. 1–May 1	Apr. 15–May 15	May 1–20	May 10–June 1	May 20–June 10.
Parsnip	Mar. 10–Apr. 10	Mar. 20–Apr. 20	Apr. 1–May 1	Apr. 15-May 15	May 1–20	May 10–June 1	May 20–June 10.
Peas, garden	Feb. 20–Mar. 20	Mar. 10–Apr. 10	Mar. 20–May 1	Apr. 1–May 15	Apr. 15–June 1	May 1–June 15	May 10–June 15.
Peas, black-eye	May 1–July 1	May 10–June 15	May 15–June 1				
Pepper [1]	May 1–June 1	May 10–June 1	May 15–June 10	May 20–June 10	May 25–June 15	June 1–15	
Potato	Mar. 10–Apr. 1	Mar. 15–Apr. 10	Mar. 20–May 10	Apr. 1–June 1	Apr. 15–June 15	May 1–June 15	May 15–June 1.
Radish	Mar. 1–May 1	Mar. 10–May 10	Mar. 20–May 10	Apr. 1–June 1	Apr. 15–June 15	May 1–June 15	May 15–June 1.
Rhubarb [1]	Mar. 1–Apr. 1	Mar. 10–Apr. 10	Mar. 20–Apr. 15	Apr. 1–May 1	Apr. 15–May 10	May 1–20	May 15–June 1.
Rutabaga			May 1–June 1	May 1–June 1	May 1–20	May 10–20	May 20–June 1.
Salsify	Mar. 10–Apr. 15	Mar. 20–May 1	Apr. 1–May 15	Apr. 15–June 1	May 1–June 1	May 10–June 1	May 20–June 1.
Shallot	Mar. 1–Apr. 1	Mar. 15–Apr. 15	Apr. 1–May 1	Apr. 10–May 1	Apr. 20–May 10	May 1–June 1	May 10–June 1.
Sorrel	Mar. 1–Apr. 15	Mar. 15–May 1	Apr. 1–May 15	Apr. 15–June 1	May 1–June 1	May 10–June 10	May 20–June 10.
Soybean	May 1–June 30	May 10–June 20	May 15–June 15	May 25–June 10			
Spinach	Feb. 15–Apr. 1	Mar. 1–Apr. 15	Mar. 20–Apr. 20	Apr. 1–June 15	Apr. 10–June 15	Apr. 20–June 15	May 1–June 15.
Spinach, New Zealand	Apr. 20–June 1	May 1–June 15	May 1–June 15	May 10–June 15	May 20–June 15	June 1–15	
Squash, summer	Apr. 20–June 1	May 1–June 15	May 1–30	May 10–June 10	May 20–June 15	June 1–20	June 10–20.
Sweetpotato	May 1–June 1	May 10–June 10	May 20–June 10				
Tomato	Apr. 20–June 1	May 5–June 10	May 10–June 15	May 15–June 10	May 25–June 15	June 5–20	June 15–30.
Turnip	Mar. 1–Apr. 1	Mar. 10–Apr. 1	Mar. 20–May 1	Apr. 1–June 1	Apr. 15–June 1	May 1–June 15	May 15–June 15.
Watermelon	Apr. 20–June 1	May 1–June 15	May 15–June 15	June 1–June 15	June 15–July 1		

[1] Plants.
[2] Generally fall-planted (table 5).

FIGURE 18-4B
Average last freeze dates in the United States ***(Courtesy USDA)***

AGRI-PROFILE

CAREER AREA: GARDENER/CARETAKER/HOMEOWNER

Carol Reese of Starkville, Mississippi, examines blueberries on the family pick-your-own operation in northeastern Mississippi. ***(Courtesy USDA Agricultural Research Center)***

Fruit and vegetable gardening has long been an important enterprise for providing fresh and wholesome food for rural and suburban families. Similarly, flower gardening provides a rewarding pastime and greatly increases the beauty of homes in urban as well as rural and suburban settings. Career opportunities exist for gardeners on estates, institutions, colonial farms, truck farms, and residential neighborhoods.

Nationally, there is a resurgence of small farms and truck gardens where roadside stands and farmers' markets have created new interest in farm-fresh fruits and vegetables. Pick-your-own operations provide opportunities for people to harvest their own field-fresh produce and save the producer on the cost of labor.

People have a new awareness of the benefits of fresh food and pay premium prices where freshness is assured. This has created new markets for garden produce and opened new career opportunities in production, processing, and marketing.

Annuals, Biennials, and Perennials Flowers and vegetables are classified as either annuals, biennials, or perennials. An annual is a plant whose life cycle is completed in one growing season. Growth is very rapid. Practically all vegetables, except asparagus, rhubarb, and parsley, are annuals.

A biennial is a plant that takes two growing seasons or two years from seed to complete its life cycle. Some biennials bloom very little or not at all the first year, but they come into full bloom the second year and then go to seed.

A perennial is a plant that lives on from year to year. A gardener commonly treats some plants that are true perennials as either annuals or biennials. This is done because, when planted each year from seed, perennials produce higher quality blooms than do older plants that remain from year to year. Perennial crops such as asparagus, artichokes, rhubarb, and some herbs and flowers should be planted in one section of the garden, separate from the annuals. By such separation, the perennials do not interfere with the cultivation of the annuals.

Vegetable or Flower The kinds of vegetables and flowers to plant depend upon the individual tastes of the members of your family. There are certain top-ranking vegetables that are popular everywhere. Certain ones are popular only within certain regions of the country. The most popular vegetables are tomatoes, snap beans, onions, cucumbers, peppers, radishes, lettuce, carrots, corn, beets, cabbage, squash, and peas. Favorites of specific regions are artichokes in Louisiana and California, southern peas in warm southern climates, and melons in warm summer regions.

The gardener has a wide selection of flowers from which to choose. Generally, flower varieties are chosen because of their characteristics. Various varieties may survive well in poor soil, have a wonderful fragrance, or grow well in the shade. A few of the flowers that do well in poor soils are alyssum, cactus, cosmos, marigolds, petunias, and phlox. Flowers that are particularly fragrant are alyssum, carnations, petunias, sweet peas, and sweet william. Those that do well in partial shade are ageratums, begonias, coleus, impatiens, and pansies.

CULTURAL PRACTICES FOR GARDENS

Cultivating Never put off cultivation. *Cultivation* is the act of preparing and working the soil.

It is easier to do a little each day than to let weeds get ahead or the soil get too hard. The main purpose of cultivation is to control weeds and to loosen and aerate the soil. A sharp hoe is a necessary tool for any gardener. Buy one that can fit in between plants for easy cultivation. Begin cultivation as soon as weeds or grass break through the soil. Do not wait until the weeds and grasses threaten to take over the garden. When cultivating, take care not to damage the roots of the vegetables and flowers. Use short shallow scraping motions instead of chopping deeply into the soil. Deep hoeing or cultivation will destroy desirable plant roots.

Weeding Begin weeding as soon as the weeds appear. When weeds grow near the plants, their roots can become intertwined with the crop. Therefore take care to avoid pulling out both the weed and the plant. When weeding, select a time of day when the soil and plants are fairly dry so the weeds wither when pulled. Be careful to shake the soil from weed roots so the weeds will die quickly. Rake persistent weeds such as purslane, crabgrasses, and Bermuda grasses out of the garden to prevent them from taking root again.

Weed-control chemicals called *herbicides* can be used in the home garden, with care. However, the use of herbicides is best suited to larger gardens. Chemicals are very specific in their actions in certain types of plants, so they must be used with caution to prevent injury to the vegetables and flowers being grown. Before using any herbicides, obtain detailed information from an extension agent or an informed garden center salesperson. Always read the label on the container and use the chemical strictly according to the label instructions.

Watering During a growing season, there will probably be some days or weeks of dry weather when the garden must be watered. Most soils require at least 1″ of water per week, either through rain or irrigation. If water is needed, make it a practice to soak the soil to a depth of 6″. Frequent, light waterings tend to promote shallow root development.

When watering, run the water on the soil near the plants. A good sprinkler head, soaker hose, or water breaker is needed to assure even distribution of water. Let the garden hose run for 15 to 30 minutes to water 100 $ft.^2$ at a depth of several inches. You should water whenever the soil lacks moisture 1–2″ below the surface.

Protection from Pests Your garden can be damaged from a variety of insects and diseases. To help prevent severe damage, follow these measures:

1. Rotate crops so that the same or a related crop does not occupy the same area every year. This helps control soilborne diseases.
2. Watch closely for insects. Pick them off by hand or knock them off with a hard stream of water.
3. When using sprinklers, water early in the day so the foliage can dry before nightfall. Diseases of leaves and fruit prosper in damp conditions.
4. Keep weeds out of the garden. Weeds may harbor diseases or insects and will interfere with spraying and dusting of crops.
5. Use enough fertilizer and lime to promote vigorous growth.

HARVEST AND STORAGE OF GARDEN PRODUCE

Harvesting Harvest the vegetables and flowers at the peak of quality or at the stage of maturity that you and your family members prefer (Figure 18-5). Some vegetables will lose their quality within a day or two, while others hold it for a week or more.

For the best results for flowers, pick them in the early morning or late afternoon. Immediately place the cut flowers in warm water. Display the flowers in a cool place.

Storage Many vegetables that are not canned or frozen for later use can be stored successfully if proper temperature and moisture conditions are met (Figure 18-6). Only vegetables that are of good quality and at the proper stage of maturity should be stored.

Warm Storage Vegetables that require higher storage temperatures than others are squash, pumpkins, and sweet potatoes. These vegetables may be stored on shelves in a furnace room or in an upstairs storage area. Damp basement areas are not recommended. Squash and pumpkins should be kept in a heated room, with a temperature between 75 and 85°F for two weeks to harden the shells.

Cool Storage Most vegetables require cool temperatures and relatively high humidity for suc-

Asparagus–From seeds—3rd year.
From roots—after 1st year.
Cut when spears are 6″ to 10″ high, in spring.

Beans, Pole–When pods are nearly full size.

Beans, Bush–When pods are nearly full size.

Beans, Bush Lima–When tender, pods are nearly full size.

Beets–When bulbs are 1¼″ to 2″ in diameter.

Swiss Chard–Outer leaves can be harvested anytime.

Broccoli–Before green clusters begin to open.

Brussels Sprouts–When sprouts are firm, pick from bottom up on stalks.

Cabbage–When heads are solid, before splitting.

Cauliflower–After blanching, when firm curds are 2″ to 3″ in diameter.

Carrots–When top of root is 1″ to 1½″ in diameter.

Celery–When about 2/3 mature, harvest as needed.

Collards–When leaves are large but tender. Can harvest up to winter.

Corn–When kernels are filled out and milky. The silk at the tip of the ear should be dry and brown.

Cucumbers–When slender and dark green.

Eggplant–When half grown, glossy and bright.

Endive–When leaves are tender and desired size. About 15″ in diameter.

Kale–When young and tender.

Kohlrabi–When 2″ to 3″ in diameter.

Lettuce–Leaf—When tender and desired size.
Head—When round and firm.

Muskmelons–When stem separates easily from fruit.

Mustard Greens–When large leaves are still tender.

Okra–Cut pods when about 2 or 3 inches long.

Onions–For Fresh use—When ¼″ to 1½″ in diameter.
For Storage—When tops shrivel at the bulb and fall over.

Parsley–Any time the outer leaves are desired size.

Parsnips–After hard frost. Can leave in ground all winter for spring use.

Peas–When pods are well-filled, before seeds are largest.

Peppers–When solid and nearly full size.

Pumpkins–When skin is hard and not easily punctured. Cut with some stem on.

Radishes–When desired size.

Rhubarb–When stems are of desired size, twist off near base of plant and discard leaves. Do not pick more than 1/3 of plant during a season.

Spinach–When outer leaves are large enough.

Squash–Summer—When skin is soft.
Winter—When skin is hard, cut with part of stem on, before frost.

Tomatoes–When uniformly red and firm.

Turnips–When 2″ to 3″ in diameter.

Watermelons–When underside is yellow and thumping produces a muffled sound.

FIGURE 18-5
Recommended times to harvest vegetables

cessful storage. The storage area should be cool, dark, and ventilated. The room should be protected from frost, heat from a furnace, or high outdoor temperatures.

The suggested temperature and relative humidity for storage of crops is shown in Figure 18-6. This list will provide you with a general idea of the type of storage area that each vegetable requires.

COLD FRAMES, HOTBEDS, AND GREENHOUSES FOR HOME PRODUCTION

Cold Frames This is an extremely convenient aid for the gardener. A cold frame is a bottomless wooden box with a sloping glass roof (Figure 18-7). It can be constructed by using plywood or common lumber and a window sash. The size of the sash should determine the dimensions of the box. The length and width should be 2″ smaller than the overall dimensions of the sash.

To construct a cold frame, the following procedures should be helpful.

1. Make the front of the box approximately 8″ high and the back of the box about 12″ high.
2. Cut the two sides on an angle, from 12″ on one end to 8″ on the other.
3. Nail the four sides together.
4. Try the sash or top for size.
5. Hinge the sash to the back of the frame for easy use. The hinge allows you to prop the top open slightly in warm weather for ventilation.
6. Place the cold frame in a southern exposure, with good protection from the wind and proximity to a water supply.

Once the cold frame is built, it can serve three purposes:

1. It can be used for a protective home for seedlings that have been started indoors. The seedlings can continue to grow inside the cold frame and become "hardened off" before they are transplanted to the garden.
2. You can begin plants in the cold frames. Seeds can be planted directly into the soil in the cold frame. After they are 4-6″ tall, they can be transplanted straight into the garden.
3. You can grow vegetables such as lettuce and endive into the fall. Sow seeds in early autumn. Cover the cold frame with a blanket or tarp during extremely cold nights.

Crop	Temperature °F	Relative Humidity Percent
Asparagus	32	85-90
Beans, snap	45-50	85-90
Beans, lima	32	85-90
Beets	32	90-95
Broccoli	32	90-95
Brussels sprouts	32	90-95
Cabbage	32	90-95
Carrots	32	90-95
Cauliflower	32	85-90
Corn	31-32	85-90
Cucumbers	45-50	85-95
Eggplants	45-50	85-90
Lettuce	32	90-95
Cantaloupes	40-45	85-90
Onions	32	70-75
Parsnips	32	90-95
Peas, green	32	85-90
Peppers, sweet	45-50	85-90
Potatoes	38-40	85-90
Pumpkins	50-55	70-75
Rhubarb	32	90-95
Rutabagas	32	90-95
Spinach	32	90-95
Squash, summer	32-40	85-95
Squash, winter	50-55	70-75
Sweet potatoes	55-60	85-90
Tomatoes, ripe	50	85-90
Tomatoes, mature green	55-70	85-90
Turnips	32	90-95

FIGURE 18-6
Recommended temperature and relative humidity for storage of fresh vegetables *(Courtesy University of Maryland Cooperative Extension Service)*

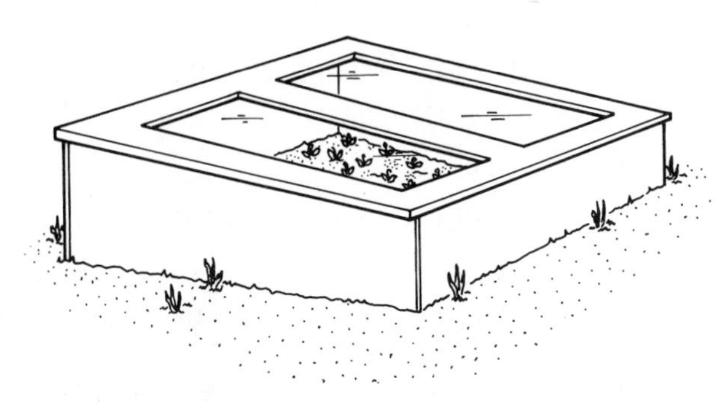

FIGURE 18-7
Cold frames are structures with glass or plastic covers for admission of heat and light, used for starting garden plants.

Hotbeds A hotbed is simply a cold frame with a heat source. In many areas, a cold frame is usually adequate. However, in colder areas, some type of artificial heat is necessary. The hotbed should be located on well-drained land.

Electricity is a convenient means of heating hotbeds. Use either lead- or plastic-coated heating cables or 25 W frosted light bulbs. You can control the temperature with a thermostat. If you are using light bulbs, attach the electrical fixtures to strips of lumber and suspend the bulbs 10–12″ above the soil surface. Allow one 25 W bulb per 2 $ft.^2$ of space. When using a heating cable, a standard 60′ length will heat a 6 × 6′ or a 6 × 8′ bed. Lay cable loops about 8″ apart for uniform heating. Lay the cable directly on the floor of the bed unless drainage material is needed.

FIGURE 18-8
A home greenhouse can be attached to the house for convenience and efficiency.

Greenhouses Growing flowers and vegetables in a greenhouse can be enjoyable as well as profitable. The conventional greenhouse is designed primarily to capture light and control temperatures. It can be free standing, but most often it is attached to a building to provide convenient access, simplified construction, and a potential source of supplemental heat if needed (Figure 18-8). The greenhouse can provide an environment for starting plants, hardening them off, or completely growing the plants.

In summary, gardening is both an art and a science. You can keep a garden for fun, reduce the cost of food for the family, generate income, or improve the beauty of the surroundings. If gardening is to be profitable, the operator must read extensively, get help occasionally with disease and insect problems, and perform garden chores in a timely fashion. For many people, the garden is a real source of satisfaction and creates a feeling of self-sufficiency and achievement.

Student Activities

1. Discuss with family members the types of vegetables and flowers that they want in the home garden.
2. Select vegetable and flower varieties for the home garden from seed catalogs.
3. Sketch a garden plot containing both flowers and vegetables that are to be grown.
4. Make a calendar indicating the dates to plant and harvest the various crops in the garden.
5. Start and manage your own garden area. Use the flowers and vegetables at home or sell them for profit.
6. Learn to calculate area in square feet. A square foot is an area that is 1′ long and 1′ wide. An area that is 1′ long and 2′ wide contains 2 $ft.^2$ Area (A) in square feet is calculated by multiplying length (L) in feet times width (W) in feet. Therefore, $A = L \times W$ or $A = LW$.
   - **a.** How many square feet are in a garden that is 10′ long × 5′ wide?
     A = ______ sq. ft.
   - **b.** How many square feet are there in a lawn that is 40 × 70′?
     A = ______ sq. ft.

c. What is the area of a building lot that is 109 × 150′?
A = ______ sq. ft.

d. Suppose the lot in item "c" above is covered with lawn except for the house, which is 28 × 42′. What is the area of the house?
A = ______ sq. ft.

e. In item "c" above, what is the area of the lawn?
A = ______ sq. ft.

Self-Evaluation

A. MULTIPLE CHOICE

1. Gardening is both a

a. science and art.
b. chore and hard work.
c. science and hobby.
d. hobby and job.

2. A good rule of thumb for planning a garden for four people is to start with a plot that is

a. 40 × 60′.
b. 15 × 25′.
c. 10 × 26′.
d. 3 × 7′.

3. The south and west sides of a house may not be the best locations for a flower garden because they

a. are not warm enough.
b. reflect heat.
c. have poor exposure to the sun.
d. collect rainwater.

4. The furrow in a garden is a

a. sunken walkway.
b. hole.
c. place to plant seeds.
d. pile of weeds.

5. An annual is a plant whose life cycle is completed in

a. two growing seasons.
b. one growing season.
c. four growing seasons.
d. three growing seasons.

6. When cultivating, a gardener should use a ______ for best results.

a. rake
b. rototiller
c. shovel
d. hoe

7. The technical name for weed control chemicals is

a. pesticide.
b. fungicide.
c. herbicide.
d. weed killer.

8. After cutting flowers from the garden, it is best to immediately put them in

a. hot water.
b. cold water.
c. salty water.
d. warm water.

9. To heat a hotbed, frosted light bulbs or ______ are commonly used for heat sources.

a. a furnace
b. lead- or plastic-coated heating cables
c. a fire
d. a small stove

10. The conventional greenhouse is designed to

a. add more living space to the home.
b. keep plants warm.
c. capture light and control temperatures.
d. protect plants from pests and diseases.

B. MATCHING

______	**1.** Perennial	**a.** 12 x 12″
______	**2.** Peat moss	**b.** Fragrant flower
______	**3.** Square foot	**c.** More than two growing seasons
______	**4.** Loamy	**d.** Long-season vegetable
______	**5.** Watermelon	**e.** Persistent weed
______	**6.** Sweet pea	**f.** Soil type
______	**7.** Bermuda grass	**g.** Form of organic matter

C. COMPLETION

1. A cold frame is a bottomless wooden box with a sloping ______ ______ .

2. You should store only the vegetables that are of good quality and at the proper stage of ______ .

3. You should plant only the vegetables and flowers that are liked by ______ ______ .

4. Fertilizer and lime should be used to promote ______ ______ .

5. Watering should be done to a depth of ______″.

UNIT 19 Fruit and Nut Production

OBJECTIVE To determine the opportunities in and identify basic principles of fruit and nut production.

Competencies to Be Developed

After studying this unit, you will be able to

- ☐ determine the benefits of fruit and/or nut production as a personal enterprise or career opportunity.
- ☐ identify fruit and nut crops.
- ☐ plan a fruit or nut enterprise and prepare a site for planting.
- ☐ describe how to plant fruit and nut trees and utilize appropriate cultural practices in fruit and nut production.
- ☐ list the appropriate procedures for harvesting and storing at least one commercial fruit or nut crop.

TERMS TO KNOW

Pomologist
Semi-dwarf
Dwarf
Pome
Drupe
Nursery
Cultivars
Propagated
Rootstock
Susceptible
pH
Pollination
Vegetative
Terminal
Lime

MATERIALS LIST

paper, pencil, and eraser
old seed catalogs
pruning shears
scissors, glue, and index cards

Home fruit operations can provide high-quality and bountiful varieties of fruit and nuts. The fruits grown at home can be enjoyed at their peak of ripeness. This quality is difficult to obtain from supermarket fruits. Home fruit and nut operations can be enjoyed as hobbies or turned into profitable commercial enterprises.

CAREER OPPORTUNITIES IN FRUIT AND/OR NUT PRODUCTION

Like other agricultural enterprises today, the production of fruit and nut crops is more a business than a way of life. The orchardist or small-fruit grower has to be a good financial manager, as well as a scientist and farmer. Producing food, however, can be a very enjoyable and rewarding profession.

Career Descriptions The production, harvest, and marketing of fruits is a large industry. The United States accounts for approximately 25% of the combined world crops of apples, pears, peaches, plums, prunes, oranges, grapefruit, limes, lemons, and other citrus fruits. There is a shortage of trained pomologists today in the fruit industry and profession. A *pomologist* is a fruit grower or scientist.

Career opportunities exist at all levels of the industry. You can be a grower, orchardist, plant manager, foreman, or technician. People who work in this field must be able to propagate fruit and nut trees and vines, as well as plant and transplant, prune, thin and train and fertilize them. They must also be able to provide protection and control of diseases and insects.

IDENTIFICATION OF FRUITS AND NUTS

Types of Fruits There are three broad categories of fruit—tree, small bush and cane, and vine. These categories describe the growth habit of the fruit. Each category includes numerous kinds of fruit.

Tree Fruits Tree fruits are popular with home gardeners as well as large producers. One drawback, however, is that some types of trees can take several years before producing the first harvest. There are three types of fruit trees based on growth habit: standard, semi-dwarf, and dwarf. These three terms refer to the size of the tree and the length of time it takes from planting it to producing the first crop.

A standard tree is one that has its original rootstock and reaches normal size. A standard apple tree can grow to be 30′ high. However, *semi-dwarf* and *dwarf* trees are smaller. They are standard varieties of trees, but they are grafted onto dwarfing rootstocks. Such rootstocks cause the tree to produce less annual growth. Thus, it remains smaller throughout its life. The semi-dwarf tree averages 10–15′ in height, while dwarf trees average 4–10′. Dwarf trees are better suited to the home garden because they bear earlier and take up less space than semi-dwarf or standard trees (Figure 19-1).

Small Bush Fruits

These fruits are excellent selections for home gardens and small farms. The best choices are strawberries, blueberries, raspberries, and thornless blackberries. Small bush fruits grow either low to the ground or only 3–4′ high. They require less maintenance than tree fruits or vine fruits. They tend to bear quickly—usually between 9 months and one year after planting.

Vine Fruits

The best known vine fruit is the grape. Grapes will occupy the land for many years, so site selection is important. The vines need to be trained and pruned for adequate production and yield. Grapes require a growing season that has at least 140 frost-free days.

Common Tree Fruits

Tree fruits commonly grown in the home garden are apples, cherries, peaches, pears, plums, and quince. There are two types of tree fruits: pome and drupe. The common pome fruits are apples and pears. A *pome* is a fruit with a core and embedded seeds. A *drupe* has a large, hard seed called a stone. Stone fruits include peaches, plums, quince, and cherries.

When growing tree fruits, particular attention must be paid to the climate of the area. Peaches, plums, and cherries are especially sensitive to climatic conditions.

Common Bush Fruits

Small bush and cane fruits that are popular with gardeners and commercial operations are strawberries, blueberries, red raspberries, currants, gooseberries, and thornless blackberries. A factor to consider in growing bush fruits is to buy plants from a reliable *nursery*. Some of these fruits will bear a harvest in the summer and again in the fall.

Common Vine Fruits

Grapes are the most prevalent vine fruit. They are easily grown and have a wide range of flavors. The disadvantage is the three- to four-year wait for the vines to reach maturity before they produce fruit.

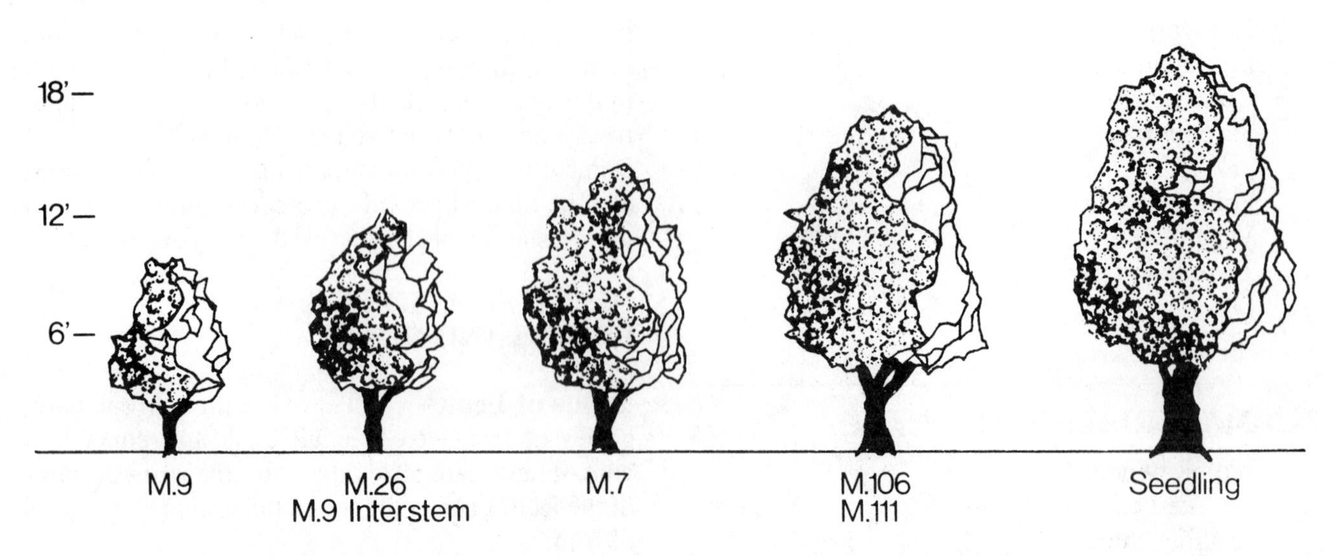

FIGURE 19-1
Rootstocks control size of fruit trees. *(Courtesy University of Maryland Cooperative Extension Service)*

Types of Nuts Nut trees are planted for numerous purposes. They can be useful for shade, landscaping, food, and wood. Wherever trees grow, a nut tree can also be grown. Popular nut trees grown in the United States are pecan, black walnut, filbert, hickory, almond, English walnut, and chestnut.

PLANNING FRUIT AND NUT ENTERPRISES

Selecting Fruit or Nuts to Plant

Climatic Considerations Certain varieties and *cultivars* of fruits or nuts are more suited to one type of climate than another. For example, citrus fruits can only be grown in southern-type climates such as Florida, Texas, and California. This information is a key factor when deciding which fruit variety to select. Even though you may absolutely love nectarines, they may not grow successfully in your area. Recommendations for each fruit crop for a given area can be obtained from your local county agricultural extension agent.

Rootstock Selection Since many fruit crops are not *propagated* by seed, selection of a *rootstock* is important. Select one that is true to its variety and is hardy enough to support the fruit. Most nurseries will readily give this information about the rootstock. The rootstock will determine whether the tree is a standard, semi-dwarf, or dwarf tree.

Frost Susceptibility Locate crops in an area that is protected from frost. You can also select varieties that are late bloomers if your area is *susceptible* to late frosts. Planting areas on hillsides, near ponds, or near cities will remain warmer on clear, frosty spring nights. Such locations decrease the amount of damage due to frosts. Apricots and sweet cherries are especially susceptible to late frost damage. The home gardener can protect some fruit crops from frost by remulching or irrigating strawberries and by placing cloth or plastic sheets over trees. Commercial growers use irrigation sprinklers, heaters, and fans to help reduce frost damage.

Soil Conditions Soil requirements vary from one type and variety to the next. A good rule of thumb, however, is to select soils that are well drained and loamy. Consult the county agricultural extension agent or a reputable nursery to determine if your soil is suitable for the fruit or nut species that you plan to use.

Fertility The production of high-quality fruit is dependent upon the fertility of the soil. On the average, soil should have a pH (acidity level) of 5.5 to 6.0. However, for growing blueberries, the soil

AGRI-PROFILE

CAREER AREAS: POMOLOGY/FRUIT GROWER/NUT GROWER/PACKER

Near Albany, Georgia, a farm manager examines pecan trees for pecan aphids and mites, but hopes to also find sufficient numbers of predatory lacewings and lady beetles to eliminate the need to spray. ***(Courtesy USDA Agricultural Research Service)***

Pomology is the scientific study and cultivation of fruit, including nuts. Careers in pomology include work on farms, and in nurseries, orchards, and groves. Fruit and nut specialists may be plant breeders, propagators, educators, food scientists, and consultants. Technical level jobs are available for graders, packers, supervisors, and managers.

In the past, fruit production meant hard labor because of hand picking. However, the need for mechanical harvesting has opened new opportunities for agricultural engineers, electronics technicians, plant breeders, systems managers, food scientists, and inventors. Mechanical harvesters range from hydraulically controlled buckets, to tree shakers, to sensor-directed robots.

should be more acidic—a pH of 5.2 or less. Fruit crops should be fertilized annually after they are planted. However, too much fertilizer can damage roots and create other problems. Apply one pound of 10-10-10 fertilizer per 100 ft.^{2} of surface soil for good results.

Planning for Pollination

Most fruit crops require pollination. *Pollination* is defined as the transfer of pollen from the male stamen to the stigma or female part of the plant. Pollination is necessary for the fruit to set or form. The gardener should choose crops that are known for their high-quality and abundant pollen. The pollen is distributed by wind or insects. Sometimes it is necessary to have several varieties that bloom at the same time for adequate cross-pollination. Similarly, it may be necessary to plant a tree just for its pollen quality and availability. Many fruit growers keep honeybees to help assure good pollination.

When favorable conditions exist, just two varieties of a particular fruit crop are necessary for cross-pollination. Choose varieties that bloom at the same time or, at least, overlap for a period of time. Pears, plums, and cherries all tend to bloom at approximately the same time. However, apple varieties differ greatly in their time of bloom. Careful selection of apple varieties is very important to ensure cross-pollination.

Growing and Fruiting Habits

The various types of fruit and nut crops have different kinds of growth patterns and fruiting dates. This fact is especially true of fruit trees. There are some varieties of apples that require as long as eight years before they bear fruit. Newly planted peach and sour cherry trees can bear fruit in as little as three years. It pays to understand what to expect from the variety you have chosen. Most fruit trees go through two stages of growth, a vegetative period and a productive period. The *vegetative* period consists of rapid, vigorous growth of the tree. In younger trees, 4–5′ of growth may occur in a single season. When the tree enters the productive period or fruit bud formation, a change is noticeable. Fruit buds are larger and more plump. On apple, pear, cherry, and plum trees, short, spurlike growths occur (Figure 19-2). On the peach tree, the peach buds become larger and occur in clusters, with two large ones surrounding a center one. The vigorous *terminal* growth slows down to only 18–20″ in a season.

SOIL AND SITE PREPARATION

Selecting the Site Since many fruits require full sunlight for maximum production and yield, choose a site that allows for maximum exposure to the sun. A southern exposure is best. To help pre-

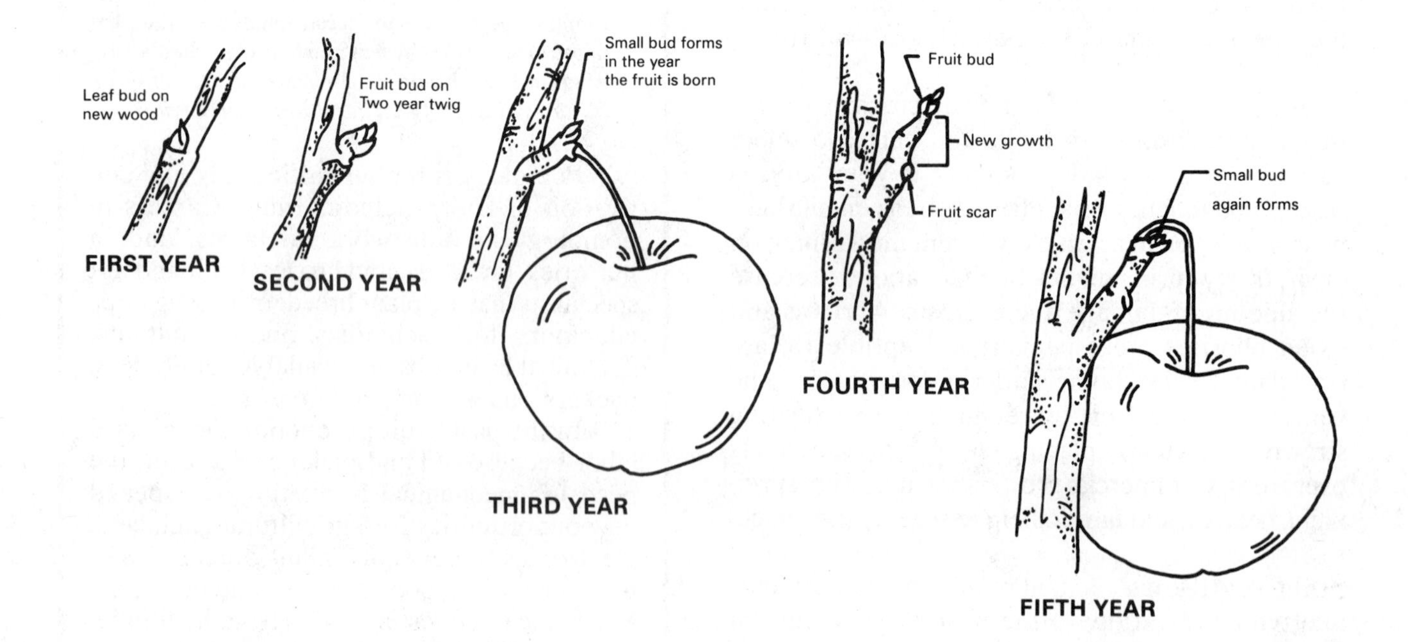

FIGURE 19-2
Growth of an apple fruiting spur

vent frost damage, select land with a slight slope and good air circulation. The soil for the fruit crops should be well drained and textured medium.

Soil Preparation The soil for fruit and nut trees should first be deeply plowed. These root systems grow deep into the soil. *Lime* and fertilizer should be added at this time. Deep plowing can be accomplished the year before you plant the trees. This permits the freezing and thawing action of the winter months to improve the texture of the soil. In addition, a soil sample should be taken a year in advance. Since many fruit and nut crops have specific pH requirements, a soil sample tested by a reputable testing laboratory will provide important information.

Any amendments such as organic matter, lime, or fertilizer should be added and disced into the soil prior to planting. Care should be taken when planting small bush and cane fruits in soils that have been planted with vegetable crops such as peppers, tomatoes, and potatoes. These vegetables may leave residues in the soil that cause diseases.

PLANTING ORCHARDS OR SMALL FRUIT AND NUT GARDENS

When to Plant Tree stock from reputable nurseries can be planted either in the spring or fall. There are advantages to planting in both seasons. Fall planting is advisable if the soil will be less than desirable or wet in the spring. Planting in the fall also allows the crop to get its root system established before rapid top growth occurs in the spring. Finally, weather conditions may be more stable in the fall than in the spring.

The advantages of planting in the spring are:

1. The crop can make considerable top and root growth before the stress of the winter season.
2. The threat of damage by rodents is reduced.
3. The threat of winter kill is reduced.

Laying Out the Orchard Each fruit or nut species and type will have its own planting distance. It is important to follow the specific planting distance for each variety, too. When planting small trees, the distance seems enormous. However, the trees will grow and fill the spaces in time. It is important to plant the trees in a straight line. A crooked row of trees will be obvious for years to come.

Planting the Fruit or Nut Crop Each fruit or nut variety has its own specifications for planting. With trees, make the hole just wide enough to accommodate the roots without crowding them. The hole should be deep enough to allow the tree to be set 2–3″ deeper than it was in the nursery. The two sides of the hole should be parallel and the bottom should be as wide as the top. By making the hole to these specifications, the soil can be packed uniformly around the roots to avoid leaving air spaces. Such spaces permit the roots to dry out.

When planting the tree, add some prepared soil to the hole. Then spread the roots, adding more soil until all roots are covered. Move the tree up and down several times, while firming the soil around the roots. This procedure allows all roots to be in contact with the soil and avoids air pockets.

Before the hole is completely filled, add two gallons of water. After the water soaks in, add the final soil. Most trees will need to be staked for good support. An 8′ metal pole can be driven 8″ from the tree. Tie the tree to the pole by using wire strung through a piece of rubber hose (Figure 19-3). After planting, keep a 12″ circular weed-free and grass-free area around the tree.

CULTURAL PRACTICES IN FRUIT AND NUT PRODUCTION

Fertilization A complete fertilizer, such as a 5-10-10 or a 10-10-10, should be used on fruit and

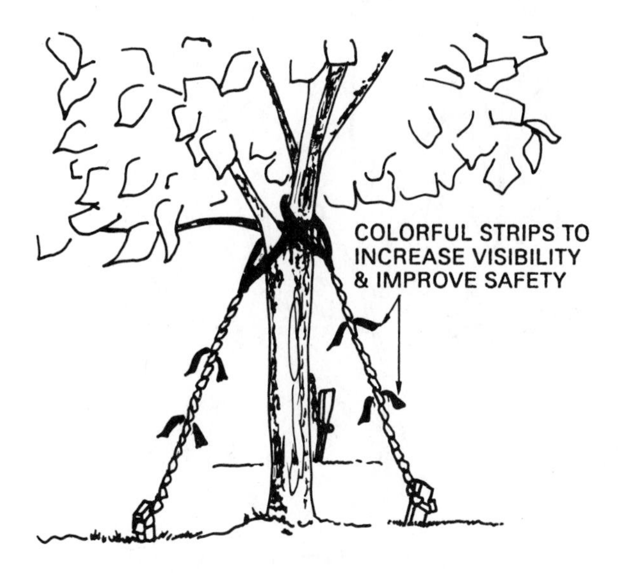

FIGURE 19-3
One method of staking a tree ***(From Ingels/ Ornamental Horticulture, copyright 1985 by Delmar Publishers Inc.)***

nut trees. After filling the hole of a newly planted tree, apply one pound of the 5–10–10 or one-half pound of 10–10–10 fertilizer around each tree. Do not place the fertilizer closer than 8–10″ from the trunk. Excessive fertilization can result in damage or death to young trees. After the first year, apply fertilizer at the rate of one pound times the age of the tree. After the tree is six to eight years old, do not increase the rate of application. As the tree becomes larger, fertilize all of the ground area under the limb spread.

The above rates apply to all tree fruits except pears. Pears should not be fertilized, since increased susceptibility to fire blight more than offsets the benefit from fertilizing.

Fertilization requirements for the various bush and cane fruits vary considerably. Therefore, it is advisable to consult a grower's guide for each individual variety.

Pruning The purpose of pruning and training young fruit trees is to establish a strong framework of branches that will support the fruit. This framework will resemble the shape best suited for the tree and for the fruit it will bear. Each tree fruit has a specific way that it should be pruned. For example, peach trees are pruned for an open center, V-shape (Figure 19-4). However, apple trees should have a strong central leader and be pruned into a Christmas-tree scaffold (Figure 19-5).

FIGURE 19-4
Peach tree pruned to the open center or V-shape pattern

Pruning a grapevine depends on its fruiting habit. Next year's grape clusters are on shoots produced during the current growing season. Therefore, this new growth should be carefully pruned. Once the shoots have fruited and the leaves have fallen, the shoots are called canes. The canes will produce the new shoots and next year's crop.

A common method of training grapevines is the Four-Arm Kniffin System. In this system, the vine is pruned to form a double T on a two-wire trellis system. Two arms originating at the two wires are trained to grow in opposite directions from the trunk. The vines are pruned to remove all canes except those that are saved for fruiting and renewal growth (Figures 19-6 and 19-7).

Pruning and thinning small bush and cane fruits should be accomplished according to the specific standards for each fruit. Pruning, thinning, and training these smaller fruits is an essential practice for high-quality yields. These practices also help prevent pests and diseases.

Disease and Pest Control A great deal of the time and expense in fruit and nut production is spent controlling diseases, insects, and other pests. Control of diseases and insects is a major component of fruit and nut production. The type of pesticide needed and how often it is applied varies tremendously among all the varieties of fruits and nuts.

Each tree fruit, nut, and small fruit must be treated individually. What may work on one variety could be disastrous when applied to another. Most orchardists will have a spray schedule worked out for each of their crops. An example of a schedule is shown in Figure 19-8. Tree fruits require more pesticide than any other fruit. Before applying any pesticide, however, read the label thoroughly.

HARVEST AND STORAGE

Harvesting Harvesting fruits can be a daily practice. The advantage of homegrown fruit is that you can pick it at its peak of ripeness. However, fruits for commercial use must be picked several days ahead so they will withstand shipping and handling.

For the best quality of fruit, harvest apples, pears, and quince when they begin to drop, soften, and become fully colored. Some varieties will ripen over a two-week period, requiring picking every day. Other varieties will ripen all at once.

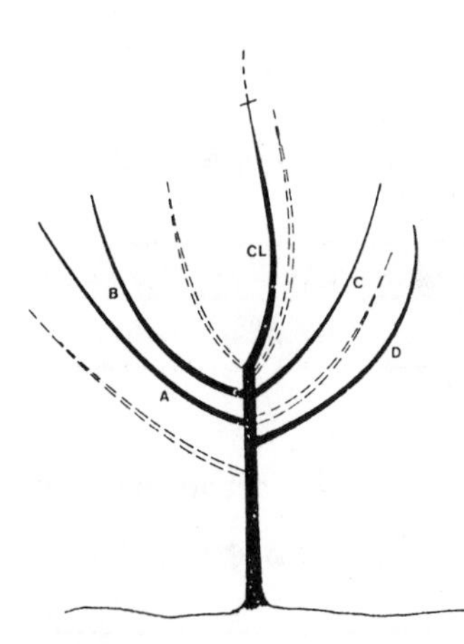

Sketch of a young apple tree after one year's growth. All limbs with broken lines should be removed. The central leader (CL) should be tipped if it is more than two feet long.

1

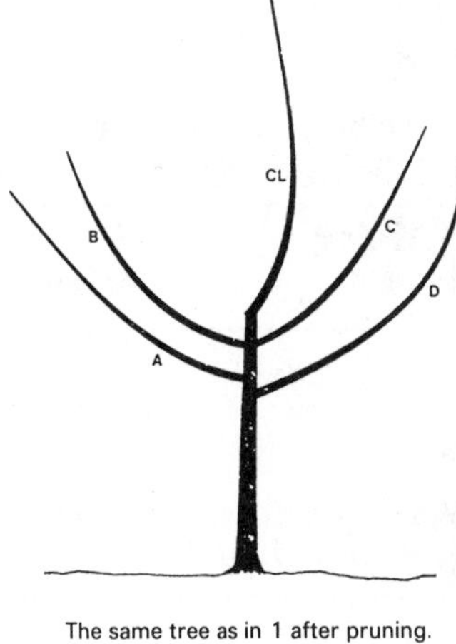

The same tree as in 1 after pruning.

2

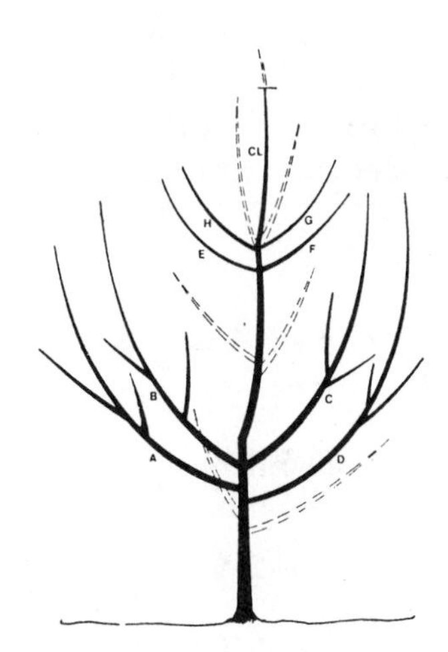

Sketch of how the tree in 2 may look after the second year. Branches with broken lines should be removed. The central leader (CL) should be tipped if it grew more than two feet.

3

The same tree as in 3 after pruning and spreading of branches. All limbs in the first tier of branches (A, B, C & D) have been spread with wooden spreaders, with a sharp-pointed nail in each end, to illustrate the beginning of the Christmas-tree shape. Limbs E and F only, in the second tier of branches, have been spread with wire with sharpened ends.

4

FIGURE 19-5
Apple trees showing the pruning cuts to be made on them.

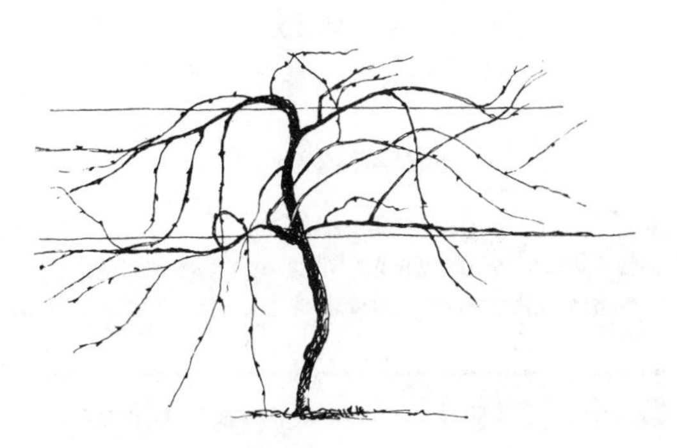

FIGURE 19-6
A vine before pruning, which has been trained to the four-arm Kniffin system *(Courtesy University of Maryland Cooperative Extension Service)*

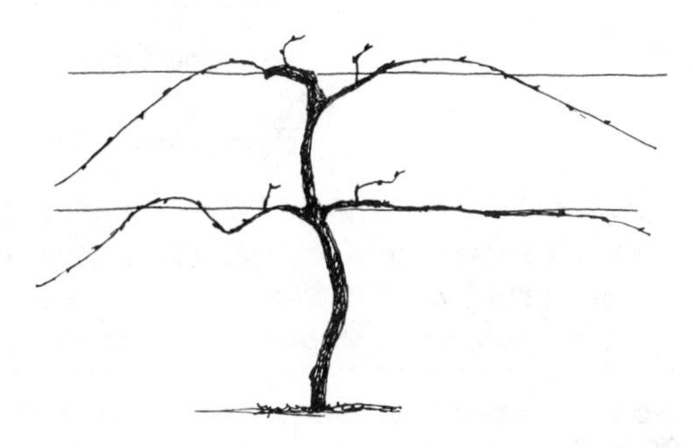

FIGURE 19-7
The same vine as seen in figure 19-6 after pruning *(Courtesy University of Maryland Cooperative Extension Service)*

Cherries

Young, nonbearing cherry trees, especially those near or among established orchards, should be protected from serious defoliation by cherry leafspot.

Timing and Major Pests Involved	Materials	Concentration (Amount Per 100 Gallons)*	
		Dilute	LV Min–Max
Dormant			
Scale insects	Superior Oil 70 sec	2 gal	Not recommended
Comments: Apply before buds swell only if scale insects are present. ***Do not apply oil if temperatures are likely to drop below 40°F within 24 hours after the spray.***			
Prebloom to Bloom			
Brown rot	Benomyl 50W	4 oz	¾–1¼ lb
	+Captan 50W	1 lb	3–5 lb
	or		
	Rovral	8 oz	1½–2½ lb
Comments: Apply when first blooms open and repeat at intervals of 4 to 5 days throughout the bloom period, especially if weather is warm and wet. ***Do not apply insecticides during bloom.***			
Petal Fall to Shuck Fall			
Brown rot	Benomyl 50W	4 oz	¾–1¼ lb
Leaf spot	+Captan 50W	1 lb	3–5 lb
Mildew	or		
	Ferbam 76W	1 lb	3–5 lb
	+Sulfur 95 mfw	3 lb	9–15 lb
Leafrollers	Guthion 50W	8 oz	1½–2½ lb
Aphids	or		
	Parathion	1½ lb	4½–7½ lb
Comments: Apply when half the petals have fallen. Repeat when most of the shucks have fallen. Neither captan nor ferbam will control mildew.			
Cover Sprays			
Brown rot	Captan 50W	1½ lb	4½–7½ lb
Leaf spot	or		
	Ferbam 76W	1½ lb	4½–7½ lb
Curculio	Guthion 50W	8 oz	1½–2½ lb
	or		
	Parathion 15W	1½ lb	4½–7½ lb
Comments: Usually three to four sprays applied at intervals of 10 to 14 days beginning 10 to 12 days after shuck fall. See comments on peach tree borer control. **Caution:** Usually three or four sprays are applied in this "cover spray" period. Not more than three sprays of captan, however, should be used on sensitive varieties like *Schmidt, Emperor Francis* and *Giant*.			

* Where maximum pesticide LV concentration is used, do not exceed 80 gal per acre; spray volumes of less than 25 gallons per acre are not recommended.

FIGURE 19-8
Example of a pest control spray schedule for cherries

Commodity	Freezing point	Place to store	Storage conditions		Length of storage period
			Temperature	Humidity	
	° F.		° F.		
Fruits:					
Apples	29.0	Fruit storage cellar	Near 32° as possible	Moderately moist.	Through fall and winter.
Grapefruit	29.8	do.	do.	do.	4 to 6 weeks.
Grapes	28.1	do.	do.	do.	1 to 2 months.
Oranges	30.5	do.	do.	do.	4 to 6 weeks.
Pears	29.2	do.	do.	do.	See text.

FIGURE 19-9

Storage recommendations for fruits *(Courtesy USDA)*

Peaches, plums, apricots, and cherries should be harvested when the green disappears from the surface skin of the fruit. A yellow undercolor should be developed by this time. The fruit should be soft when pressed lightly in a cupped hand.

The nut varieties ripen from August to November. Harvest nuts immediately after they fall from the tree. Most of the nuts that do not fall can be knocked off the tree with poles.

The nut varieties of pecan, hickory, chestnut, and Persian walnut will lose their husks when ripe. However, the husk of black walnuts and other types of walnuts will have to be removed by hand.

Grapes should only be harvested when fully ripe. The best way to judge when the grape is at full maturity is to sample an occasional grape.

Small bush and cane fruits such as strawberries and raspberries should be harvested when they are fully ripe. The best indicator is when the fruit is fully colored. Once these fruits and berries reach maturity, they should be picked every day. When harvesting, pick when the fruit is dry to avoid mildew and mold damage.

Storage Storing fruits is limited to those varieties that mature late in the fall or those that can be purchased at the market during the winter months. The length of time that fruits can be stored depends on the variety, stage of maturity, and soundness of the fruit at harvest.

For long-term storage of apples, the temperature should be as close to 32°F as possible. Apples can be stored in many ways, but they should be protected from freezing. A cellar, or another area below ground level that is cooled by night air, is a good place for apple storage. There should be moderate humidity in the storage area. Pears may be stored with the apples (Figure 19-9).

Nuts should be air dried before they are stored. Nuts, especially pecans, keep longer if they are left in the shell and refrigerated at 35°F. They can also be frozen if they are kept in their shells. Chestnuts have special requirements. They should be stored at 35–40°F with high humidity, shortly after harvest.

Small bush and cane fruits and grapes are not suitable for storage. These fruits are much too perishable to keep for even a day or two. They are best utilized as soon as they are harvested.

Student Activities

1. Define the "Terms To Know."
2. Develop a bulletin board display explaining opportunities in fruit- and nut-related occupations. Include those that can be found on the local, state, and national levels.
3. Bring a specimen of some type of unusual fruit or nut to class. Discuss its origin and the techniques for growing the fruit or nut.
4. Prepare fruit and nut flash cards using pictures from seed catalogs.
5. Prepare a planting plan for fruits and nuts, drawn to scale. Indicate the types, quantity, and location of plants to be grown.

Self-Evaluation

A. MULTIPLE CHOICE

1. Most fruits and nuts require ______ to grow properly.

a. full sunlight
b. partial sunlight
c. partial shade
d. no sunlight

2. The stock for a fruit or nut crop can be planted in the

a. fall or spring.
b. spring only.
c. summer only.
d. fall only.

3. When planting a fruit tree, you should make the hole deep enough to allow the tree to be set ______ deeper than in the nursery.

a. 3–5″
b. 2–3″
c. 1-2″
d. 6–8″

4. Nuts can be stored in the refrigerator at the optimum temperature of

a. 40°F.
b. 32°F.
c. 38°F.
d. 35°F.

5. The pear is an example of a ______ fruit.

a. drupe
b. pome
c. bramble
d. aggregate

B. TRUE OR FALSE

1. A standard tree is one that has its original rootstock and grows to normal size.

2. A semi-dwarf tree averages a mature height of 30–40′.

3. Small bush and cane fruits tend to bear quickly, usually between 9 months and one year after planting.

4. Rootstock selection is important to fruit and nut crops, since most of these crops are not propagated by seed.

5. Frost susceptibility is not a factor to fruit crops.

6. Too much fertilizer can be a hazard to fruit trees.

7. Control of diseases and insects is a major component of fruit and nut production.

8. An apple is an example of a drupe fruit.

UNIT 20 Vegetable Production

OBJECTIVE To determine the opportunities in and identify the basic principles of vegetable production.

Competencies to Be Developed

After studying this unit, you will be able to

- ☐ determine the benefits of vegetable production as a personal enterprise or career opportunity.
- ☐ identify vegetable crops.
- ☐ plan a vegetable production enterprise and prepare a site for planting.
- ☐ describe how to plant vegetable crops and utilize appropriate cultural practices.
- ☐ list appropriate procedures for harvesting and storing at least one commercial vegetable crop.

TERMS TO KNOW

Olericulture
Vegetable
Herbaceous
Home gardening
Market gardening
Truck cropping
Olericulturalist
Angiosperms
Monocotyledons (monocots)
Dicotyledons (dicots)
Aeration
Green manure
Cover crop
Transplants
Broadcasting
Germinate
Arid
Semi-arid
Precooling
Hydrocooling

MATERIALS LIST

seed catalogs
scissors, index cards, and glue
pen, paper, eraser, and ruler
lima bean seeds, containers, and soil mix
hand trowels
tomato seeds, soil mix, and flats

Olericulture is the study of vegetable production. A *vegetable* is the edible portion of an herbaceous plant. *Herbaceous* means vegetables that are used either fresh or processed. The production of vegetables can be classified into three categories: home garden, market garden, or truck crop. *Home gardening* usually refers to the vegetable production for one family and doesn't involve any major selling of the crops. *Market gardening* refers to growing a wide variety of vegetables for local or roadside markets. *Truck cropping* refers to large-scale production of a few selected vegetable crops for wholesale markets and shipping.

VEGETABLE PRODUCTION FOR HOME PROFIT OR AS A CAREER

Home Enterprise Growing vegetables in a home garden is enjoyed by millions of Americans. It has become a part of the life style of most families with access to a little bit of ground. The vegetables produced can be used for fresh table consumption or stored for later use. Vegetable gardening not only produces nutritious food, but also provides outdoor exercise from spring until fall.

The gardener who enjoys this type of activity can plant enough to provide for the family and harvest vegetables for sale on a small scale for some tax-free income. Fresh, homegrown vegetables are usually superior in quality to those found in the supermarkets. In addition, the gardener can grow those vegetables that may be too expensive to buy.

Career Opportunities The vegetable industry is a very large and complex portion of the horticultural industry today. Even though there are millions of homeowners who garden, the majority of vegetables consumed by the public are grown commercially. The commercial vegetable industry is fast moving, intensive, and competitive. It is a business that is continually changing as the demands for certain vegetables fluctuate with the tastes of the American public.

There are numerous and varied career opportunities in the vegetable industry. These include being the owner of a small market gardening business, a member of a larger truck-crop business, a vegetable wholesaler or retailer, or a worker in a vegetable processing plant. With adequate education and training, a person can become an *olericulturist* who develops pest-resistant strains and new varieties of vegetables, and does other specialized work. The opportunities are endless in the vegetable industry.

IDENTIFYING VEGETABLE CROPS

Vegetables can be identified in various ways—by their botanical classification, according to their edible parts, or by the growing season required.

Botanical Classification

All vegetables belong to the division of plants known as *angiosperms*. These are plants with ovules and an ovary. From this division, the vegetables can be grouped into either Class I, *monocotyledons* (having only one seed leaf) or Class II, *dicotyledons* (having two seed leaves) (Figure 20-1). The vegetables can be further grouped into a family, a genus, a species, and sometimes a variety. Examples of a few of the more popular vegetable families include Cruciferae—the mustard family, which contains brussels sprouts, cabbage, cauliflower, collards, cress, kale, turnips, mustard, watercress, and radish. Another is Leguminosae—the pea family, which contains bush beans, lima beans, cow peas, kidney beans, peas, peanuts, soybeans, and scarlet runner beans. Cucurbitaceae—the gourd or melon family—contains pumpkins, cucumbers, cantaloupe, casaba melons, and watermelons. Solanaceae—the nightshade family—contains eggplants, ground cherries, peppers, and tomatoes.

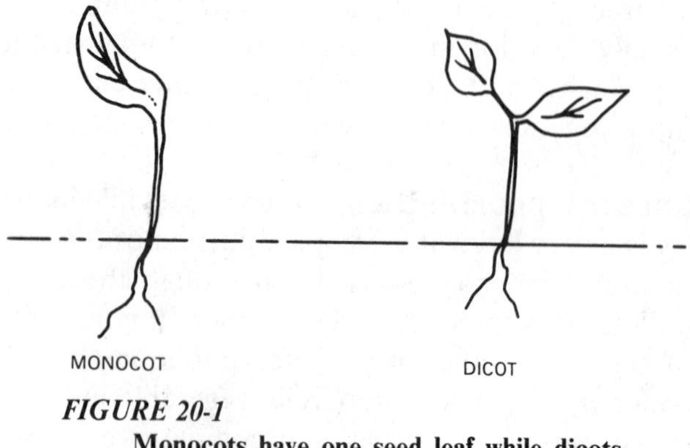

FIGURE 20-1
Monocots have one seed leaf while dicots have two.

Edible Parts

Vegetables can also be classified by the part of the vegetable that is eaten. There are three groupings of vegetables according to their use: (1) vegetables of which leaves, flower parts, or stems are used; (2) vegetables of which the underground parts are used; and (3) vegetables of which the fruits or seeds are used. See Figure 20-2 for a list of these categories and the vegetables in each one.

Growing Seasons

There are basically two growing seasons—warm and cool. Cool-season vegetable crops grow best in cool air and can withstand a frost or two. Some of these crops, such as asparagus and rhubarb, can even endure winter freezing. This group of crops is planted early in the spring and late in the season for fall and winter harvest. This category contains mostly leaf and root crops. See Figure 20-3 for a list of cool-season crops.

The warm-season or warm-weather crops are those that cannot withstand cold temperatures, especially frosts. These vegetables require soil warmth to germinate and long days to grow to maturity. They must have very warm temperatures to produce their edible parts. The edible portions of these crops are basically what can be picked off the standing plant or the fruit. A list of warm-season crops can be found in Figure 20-4.

PLANNING A VEGETABLE-PRODUCTION ENTERPRISE

Planning the Garden

Prior to planting a vegetable garden or truck crop, the operator needs to plan. The gardener must decide where to plant as well as what to plant. Without some advance planning, the vegetable garden or truck farm could turn into a total disaster.

Selecting the Site

Choose a site that is convenient to a water supply. The site should also be exposed to the sun a minimum of 50% during the day. In other words, a minimum of 8 to 10 hours of direct sun is needed. You should also consider the type of trees that are around the proposed site. Trees can provide too much shade. They can also

PLANTS OF WHICH THE FRUITS OR SEEDS ARE EATEN		PLANTS OF WHICH THE LEAVES, FLOWER PARTS OR STEMS ARE EATEN		PLANTS OF WHICH THE UNDERGROUND PARTS ARE EATEN	
Family	**Vegetable**	**Family**	**Vegetable**	**Family**	**Vegetable**
Grass, *Gramineae*	Sweet corn, *Zea mays*	Lily, *Liliaceae*	Asparagus, *Asparagus officinalis* var. *altilis* Chives, *Allium schoenoprasum*	Lily, *Liliaceae*	Garlic, *Allium satvium* Leek, *Allium porrum* Onion, *Allium cepa* Shallot, *Allium ascalonicum* Welsh onion, *Allium fistulosum*
Mallow, *Malvaceae*	Okra (gumbo), *Hibiscus esculentus*	Goosefoot, *Chenopodiaceae*	Beet, *Beta vulgaris* Chard, *Beta vulgaris* var. *cicla*	Yam, *Dioscoreaceae*	Yam (true), *Dioscorea batatas*
Pea, *Leguminosae*	Asparagus or Yardling bean, *Vigna sequipedalis* Broad bean, *Vicia faba* Bush bean, *Phaseolus vulgaris* Bush Lima bean, *Phaseolus limensis* Cowpea, *Vigna sinensis* Edible podded pea, *Pisum sativum* var. *macrocarpon* Kidney bean, *Phaseolus vulgaris* Lima bean, *Phaseolus limensis* Pea (English pea), *Pisum sativum* Peanut (underground fruits), *Arachis hypogaea* Scarlet runner bean, *Phaseolus coccineus* Sieva bean, *Phaseolus lunatus* Soybean, *Glycine max* White Dutch runner bean, *Phaseolus coccineus*	Orach, *Atriplex hortensis*	Spinach, *Spinacia oleracea*	Goosefoot, *Chenopodiaceae*	Beet, *Beta vulguris*
Parsley, *Umbelliferae*	Caraway, *Carum carvi* Dill, *Anethum graveolens*	Parsley, *Umbelliferae*	Celery, *Apium graveolens* Chervil, *Anthriscus cerefolium* Fennel, *Foeniculum vulgare* Parsley, *Petroselinum crispum*	Mustard, *Cruciferae*	Horseradish, *Armoracia rusticana* Radish, *Raphanus sativus* Rutabaga, *Brassica campestris* var. *napobrassica* Turnip, *Brassica rapa*
Martynia, *Martyniaceae*	Martynia, *Proboscidea louisiana*	Sunflower, *Compositae*	Artichoke, *Cynara scolymus* Cardoon, *Cynara cardunculus* Chicory, witloof, *Chichorium intybus* Dandelion, *Taraxacum officinale* Endive, *Chichorium endivia* Lettuce, *Lactuca sativa*	Morning Glory, *Convolvulaceae*	Sweet Potato, *Ipomoea batatas*
Nightshade, *Solanaceae*	Eggplant, *Solanum melongena* Groundcherry (husk tomato), *Physalis pubescens* Pepper (bell or sweet), *Capsicum frutescens* var. *grossum* Tomato, *Lycopersicon esculentum*	Mustard, *Cruciferae*	Brussels sprouts, *Brassica oleracea* var. *gemmifera* Cabbage, *Brassica oleracea* var. *capitata* Cauliflower, *Brassica oleracea* var. *botrytis* Collard, *Brassica oleracea* var. *viridis* Cress, *Lepidium sativum* Kale, Borecole, *Brassica oleracea* var. *viridis* Kholrabi, *Brassica oleraceae* var. *gongylodes* Mustard leaf, *Brassica juncea* Mustard, Southern Curled, *Brassica juncea* Pak-Choe, Chinese Cabbage, *Brassica chinensis* var. *crispifolia* Pe-tsai, Chinese cabbage, *Brassica pekinensis* Seakale, *Crambe maritima* Sprouting Broccoli, *Brassica oleracea* var. *italica* Turnip, Seven Top, *Brassica rapa* Upland cress, *Barbarea verna* Watercress, *Rorippa nasturtium-aquaticum*	Parsley, *Umbelliferae*	Carrot, *caucus carota* var. *sativa* Celeriac, *Apium graveolens* var. *rapaceum* Hamburg parsley, *Petroselinum crispum* var. *radicosum* Parsnip, *Pastinaca sativa*
Gourd or Melon, *Cucurbitaceae*	Chayote, *Sechium edule* Cucumber, *Cucumis sativus* Cushaw, *Cucurbita moschata* Gherkin, *Cucumis anguria* Cantalope (Muskmelon), *Cucumis melo* Pumpkin, *Cucurbita pepo* Summer squash (bush pumpkin), *Cucurbita pepo* Squash, *Cucurbita maxima* Watermelon, *Citrullus lunatus* Winter melon, *Cucumis melo* var. *inodorus*			Nightshade, *Solanaceae*	Potato, *Solanum tuberosum*
				Sunflower, *Compositae*	Black salsify, *Scorzonera hispanica* Chicory, *Chicorium intybus* Jerusalem artichoke, *Helianthus tuberosus* Salsify, *Tragopogon porrifolius* Spanish salsify, *Scolymus hispanicus*

FIGURE 20-2

Classification of vegetable crops by botanical families and crop use

Asparagus
Beet
Broad bean
Broccoli
Brussels sprouts
Cabbage
Carrot
Cauliflower
Celery
Chard
Chicory
Chinese cabbage
Chive
Collard
Endive
Garlic
Globe artichoke
Horseradish
Kale
Kohlrabi
Leek
Lettuce
Mustard
Onion
Parsley
Parsnip
Pea
Potato
Radish
Rhubarb
Salsify
Spinach
Turnip

FIGURE 20-3
Cool-season crops

Cowpea
Cucumber
Eggplant
Lima bean
Muskmelon
New Zealand spinach
Okra
Pepper, hot
Pepper, sweet
Pumpkin
Snap bean
Soybean
Squash
Sweet corn
Sweet potato
Tomato
Watermelon

FIGURE 20-4
Warm-season crops

take away nutrients from the soil that are needed by the vegetable plants. Some kinds of trees can also produce toxins, which are harmful to certain vegetables. For example, the walnut tree is toxic to the tomato. Buildings and structures can also cast shade that can slow down or prevent growth of a vegetable crop. Avoid trying to grow vegetables closer than 6–8′ from the northern side of a one-story structure—farther for higher structures. Both the south and west sides of a building have good light and often radiate heat late in the day (Figure 20-5).

The type of soil in the area selected is important as well. The majority of vegetables grow best in a well-drained, loamy soil. Avoid ground that develops puddles after it rains. This is a sign of poor drainage. To test the drainage of a site, dig a trench 12″ wide and 18″ deep. Fill it with water and

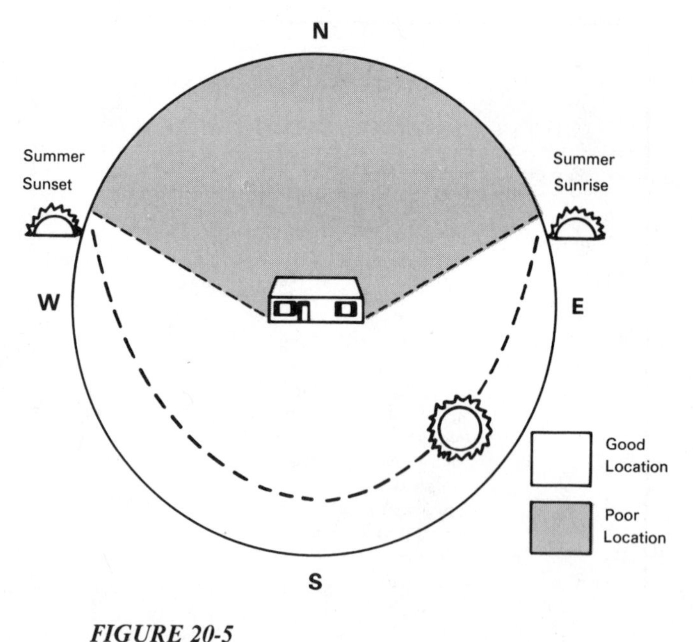

FIGURE 20-5
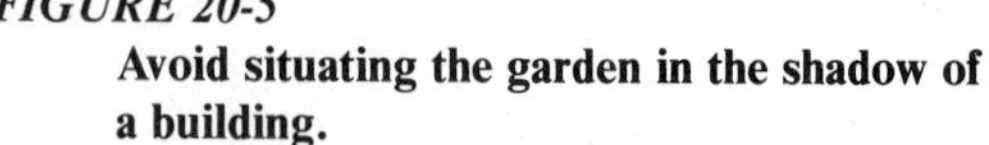
Avoid situating the garden in the shadow of a building.

observe how long it takes for the water to drain away. If it takes one hour or less, the soil can be considered well drained. If the area selected has supported vegetation before, even if it has only been weeds, it will probably support a vegetable crop. If the site selected needs some alterations to its structure, adding organic matter can help. The organic matter can improve the drainage and allow air to move readily through the pores of the soil. If possible, the garden soil should be about 25% organic matter. To accomplish this, put a layer of organic matter 2″ thick over the soil and work it in to a depth of at least 4″. If necessary, repeat this procedure until the final mix contains approximately 25% organic matter (Figure 20-6).

Deciding on the Garden Size The size of the garden will depend largely on the amount of ground that is available and the number of people for which the garden will provide. A large garden 100′ × 50′ will produce enough vegetables for the annual needs of a large family. Figure 20-7 shows a suggested planting plan for a home garden 100′ × 50′.

If this much land is not available, an arrangement to meet the fresh vegetable needs for a medium-sized family is a plot 40′ × 50′. A planting guide for such a garden can be found in Figure 20-8. When deciding on the size of the garden, it would be wise to remember that the larger the garden, the more care it will take. The gardening will be more productive and you will get better crops if you plant

Amounts of Organic Matter Needed to Cover 100 Square Feet at Various Depths

Depth (inches)	To cover 100 sq. ft.
6	2 cubic yards
4	35 cubic feet
3	1 cubic yard
2	18 cubic feet
1	9 cubic feet

1 cubic yard = 27 cubic feet

FIGURE 20-6
Amounts of organic matter to apply per 100 square feet of garden soil

and manage a smaller garden than if you take on a bigger garden that is neglected and full of weeds.

Deciding What to Plant There are several items that a potential vegetable gardener or truck cropper must consider when deciding which vegetables to plant. If the garden will be used for the family only, the first thing to determine is which vegetables the family likes and how often each vegetable will be served. Select vegetable varieties that the family enjoys, and concentrate on the kinds that are definitely better when they are freshly harvested from the garden. Examples are tomatoes, carrots, corn, snap beans, and cantaloupe. Second, consider the maintenance that some vegetables require. The easier they are to grow and maintain, the better.

A third factor to consider is the length of the harvest time for each vegetable. If the object is fresh consumption, then vegetables that can be harvested over a long period of time are best. For instance, carrots can be harvested for months, whereas cabbage should be picked at maturity. The last item to consider is the use for which the vegetables are planted. If the vegetables are going to be for fresh use, plant only for that purpose. Only so many surplus vegetables can be stored in a root cellar or a freezer.

The truck cropper has a different set of questions to consider. First, which vegetables are in high demand in the area? Which ones will produce the highest return for the cost and labor involved? The truck-cropping enterprise is a business and must make a reasonable profit to be considered a good

AGRI-PROFILE

CAREER AREAS: VEGETABLE PRODUCER/ PROCESSOR/DISTRIBUTOR/ PRODUCE MANAGER

Vegetable growers and field supervisors know that vegetables must be harvested at just the right time and rushed to fresh markets to obtain top value for the crop. ***(Courtesy USDA Agricultural Research Service)***

Career opportunities in vegetable production are similar to those in pomology. Vegetables and fruits are frequently grown, processed, and marketed by the same people. In addition to the careers mentioned in the "Fruits and Nuts" unit, many people find rewarding careers that pay well in the area of distributing and marketing produce. Here truckers, wholesalers, and retailers move the product from producer to consumer.

Excellent jobs are available in super markets for produce stockers and foremen. Products include fruits, vegetables, and, possibly, ornamental plants. Tasks may include inventory, ordering, handling, stocking, and displaying produce to keep it fresh and attractive. In large cities as well as small towns, street vendors are frequently seen selling premium fruit and, occasionally, vegetables. Roadside stands provide opportunities for younger members of families to develop business skills while earning money for present and future needs.

NORTH

Row			Feet between rows
1. Sweet corn	1st planting	Sweet corn (2nd planting)	3′
2. Sweet corn	1st planting	Sweet corn (2nd planting)	3′
3. Sweet corn	1st planting	Sweet corn (2nd planting)	3′
4. Sweet corn	1st planting	Sweet corn (2nd planting)	3′
5. Sweet corn	3rd planting		3′
6. Sweet corn	3rd planting		3′
7. Tomatoes (staked)	Plant pole beans near tomato stakes in early July without disturbing the tomato plants		4′
8. Tomatoes (staked)			4′
9. Tomatoes (staked)			4′
10. Early Potatoes			3′
11. Early Potatoes			3′
12. Pepper	Eggplant	Chard (Swiss)	3′
13. Lima bean (bush)			3′
14. Lima bean (bush)			3′
15. Lima bean (bush)			3′
16. Snapbeans (bush)	This entire area may be replanted after harvest with such crops as: endive, cauliflower, Brussels sprouts, spinach, kale, beets, cabbage, broccoli, turnips, lettuce, carrots, and late potatoes. (rows 16–31)		3′
17. Snapbeans (bush)			3′
18. Broccoli			3′
19. Early cabbage			3′
20. Onion sets			3′
21. Onion sets			2′
22. Carrots			2′
23. Carrots			2′
24. Beets			2′
25. Beets			2′
26. Kale			2′
27. Spinach			2′
28. Peas			2′
29. Peas			2′
30. Lettuce (1st planting)	Lettuce (2nd planting)	Lettuce (3rd planting)	2′
31. Radish (1st planting)	Radish (2nd planting)	Radish (3rd planting)	2′
32. Strawberries			2′
33. Strawberries			3′
34. Asparagus	Rhubarb		3′
35. Asparagus			3′
			3′

WEST — EAST: 100′ — width: 50′

SOUTH

NOTE: The time and number of plantings will vary according to locality and family needs.

FIGURE 20-7

Plan for a home garden 100 × 150 feet

ROW					FEET BETWEEN ROWS
					1½
1	EARLY CABBAGE	(TURNIPS)			2
2	EARLY POTATOES	(LATE CABBAGE)			2
3	PEAS	(BEANS)	PEAS	(KALE)	2
4	KALE	(CARROTS)	TURNIPS	(BEANS)	2
5	PARSLEY		PARSNIPS		2
6	ONIONS AND RADISHES (SAME ROW)		CARROTS	(SPINACH)	2
7	SWISS CHARD	(SPINACH)	SPINACH	(BEETS)	2
8	LETTUCE	(BEANS)	BEETS	(KALE)	2
9	BEANS	(SPINACH)	BEANS	(LETTUCE)	2
10	PEPPERS		EGGPLANT		2
11	LIMA BEANS, BUSH				2
12	TOMATOES, LATE				2
13	TOMATOES, EARLY				2½
14	SQUASH		CUCUMBERS		2½
15					
16	CORN, SWEET interplanted with POLE LIMA BEANS				2½
17	CORN, SWEET interplanted with WINTER SQUASH				2½
18					2

50′ (width) × 40′ (length)

NOTE: Items in parentheses are succession crops planted after the first crop is harvested or planted between the rows of mature plants to permit germination and early growth before the first crop is removed.

FIGURE 20-8
Planting guide for a garden serving a medium-sized family

business. Second, what vegetables can be raised in the soil and climate? Truck cropping should be a high-income business, so high quality and quantity of produce is very important. Third, what combination of vegetables can be grown to keep the land occupied and provide cash flow from sales throughout the season?

Other questions to be explored are: (1) can the diseases and insect problems be managed? (2) will continuous use of the land for the same crops create pest buildup? (3) will the soil stand continuous cropping? and (4) what perennials may be included to reduce the labor involved?

PREPARING A SITE FOR PLANTING

Preparing the Soil Before planting vegetables, the soil must be properly prepared. This will generally mean adding organic matter, lime, and fertilizer. The land should be plowed if it is in sod.

Plowing

Land should be plowed or spaded to a depth of 6–8″. The soil can be plowed either in the spring or the fall. Fall plowing has advantages. First, the freezing and thawing of the winter months can improve the physical condition of the soil. Second, the exposure of the soil to the weather can result in a reduced population of insects. Spring plowing should be done only a few weeks prior to planting. Take care not to plow the soil when it is too wet. Doing so can destroy the physical structure of the soil, which results in hard clumps. The moisture content of the soil may be judged by pressing a handful of soil into a ball. If the ball crumbles easily, the soil is ready to plow. If the soil sticks together, it is too wet to plow.

Maintaining Organic Matter

Organic matter increases the water-holding and absorption capacity of the soil. Furthermore, it helps prevent erosion and promotes aeration in the soil. *Aeration* refers to the movement of air through soil. If it is available, and can be secured cheaply, animal manure is the best material for maintaining the organic content of the soil. It is also a good source of nutrients. Manure from animals other than sheep and poultry can be turned under at the rate of 15 to 20 tons per acre or 20–30 bushels per 1,000 ft.^{2}. Sheep and poultry manures can be used, but at half this rate. One note of caution: animal manures are low in phosphorus. Therefore, it may be necessary to use high-phosphorus fertilizer to balance the nutrients. For best results, apply fresh manure in the fall and well-rotted manure in the spring.

Another type of organic matter to use is green manure in the form of a cover crop, especially if it is a legume. *Green manure* is an active, growing crop that is then turned under to help build the soil. Green vegetation incorporated into the soil rots more quickly than dry material does. A cover crop should be planted in early fall. A *cover crop* is a close-growing crop planted to prevent erosion.

Liming The need for lime should be determined through the results of a soil test. A soil test determines the pH of the soil. A pH greater than 7.0 indicates low acidity, a pH of 7.0 is neutral, and a pH below 7.0 indicates acidity.

If the test indicates a pH of 6.0 or less, lime should be applied. The lime should be mixed into the top 3–4″ of the soil. The amount of lime to add depends upon the type of soil, the level of the pH desired, and the form of lime used. Some plants, such as blueberries, require acid soils. For such plants, special materials are available to make the soil acidic.

Vegetable crops grow best at specific pH levels. Figure 20-9 lists the vegetables and their optimum pH level for maximum growth.

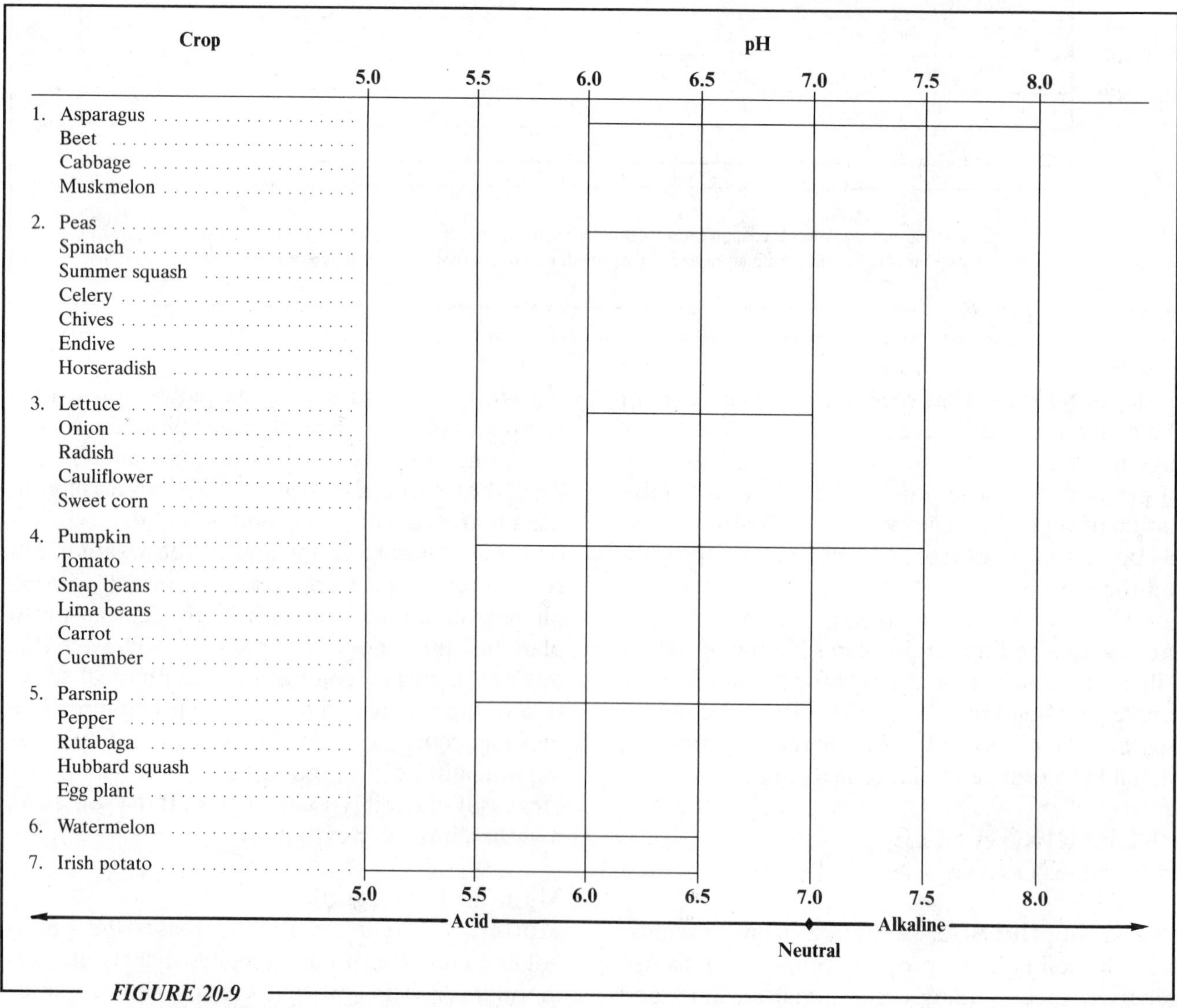

FIGURE 20-9
Optimum pH range for vegetable crops

Fertilizing The amount and type of fertilizer to add to the soil is also determined by a soil test. The test will help you avoid applying too much or too little fertilizer. The application of commercial fertilizer is done to directly increase the amounts of nutrients available to the plants. Soil without additives cannot provide the best combination of conditions for all vegetables.

Commercial fertilizers usually contain nitrogen, phosphoric acid, and potash. Each vegetable crop needs varying amounts of each of these three elements. In general, fertilizer ratios used for home gardens are 5–10–10, 10–10–10, and 5–10–5 (Figure 20-10). The first number in a fertilizer ratio or grade represents the percentage of nitrogen, the second shows the percentage of phosphoric acid, and the third indicates the percentage of potash in the fertilizer. A general guide for home gardeners is that most vegetables will need about 3–4 lbs. of fertilizer per 100 ft.[2] Consult a county extension agent for exact fertilizer recommendations.

PLANTING VEGETABLE CROPS

Vegetable crops are initially grown from seeds. Some seeds can be sown where the plants will grow. Others will need to be grown into transplants and the seedlings then transplanted to where the crop will grow. *Transplants* are plants grown from seeds in a special environment, such as cold frames, hotbeds, or greenhouses.

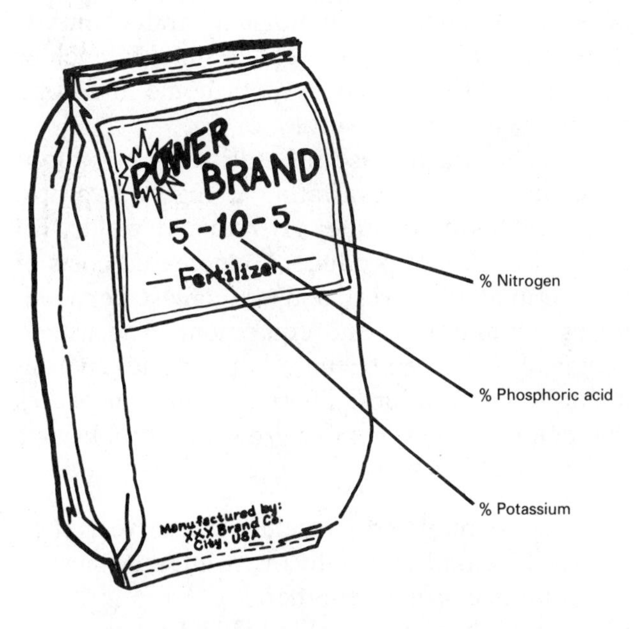

FIGURE 20-10
The fertilizer analysis or ratio on a bag tells the percentage of nitrogen, phosphoric acid and potassium in the bag.

The method used depends upon the vegetable's climatic requirements and the germinating characteristics of the seed.

Planting Seed

Depending upon the size and scope of the garden and the quantity of seed to be planted, vegetable seeds are planted by one of several methods. These are: (1) by hand in hills or rows, (2) through broadcasting by hand or machine, (3) with one-row hand seeders, and (4) with single- or multiple-row, tractor-drawn seeders.

Broadcasting means to scatter around. Regardless of the type of planting method used, the seeds should be planted at the proper depth. The soil should also be left smooth and firm over the seed. For most vegetables, make sure the seed is fresh. Seed more than one year old will probably not germinate at the rate necessary to produce good yields.

Most vegetable seeds should be planted at a moderately shallow depth. It is best to plant in loamy-textured soil, with adequate moisture and no danger of frost. The seed will germinate best when it is planted at a depth no more than four times the diameter of the seed. *Germinate* means to produce a plant and roots from a seed. Regardless of how deep the seeds are planted, the soil surface should be level and firmly pressed after the seed is planted. This will help prevent the seed from washing away or water from puddling above the seed after a rain.

Transplanting Vegetable Seedlings

There are three methods used in transplanting, depending on the number of plants that are being transplanted. The methods are: (1) hand setting, (2) hand-machine setting, and (3) riding-machine setting. In home gardening situations, hand setting is most appropriate.

Hand setting involves six steps: (1) dig a hole slightly bigger and deeper than the root ball of the plant being transplanted, (2) add some fresh soil to the hole, and (3) place the plant in the soil a little deeper than it grew in its original container. Try to keep as much of the original soil as possible around the roots, (4) pull soil in and around the plant and firm it slightly, (5) add about one-half pint of water and let it soak into the soil, and (6) pull in some dry soil to level off and cover the wet area. This helps to prevent loss of moisture and baking of the soil.

The vegetable transplants should be set out on a cloudy day or late in the afternoon. Further, it is best to set plants just before or just after a rain.

CULTURAL PRACTICES

Cultivating Cultivation, or intertillage of crops, is an old agricultural practice. The benefits of cultivation are: (1) weed control, (2) conservation of moisture, and (3) increased aeration. Cultivation increases the yield of most vegetable crops, mainly because the weeds are controlled.

All types of equipment are used to cultivate crops, from small hand tools to large tractors. The cultivation should be shallow and done only when there are weeds to be killed. Excessive cultivation is not beneficial and may even be damaging.

Controlling Weeds Weeds are easier to control when they are small, using shallow cultivation. Weeds that are older and more established are more difficult to control. For them, deeper cultivation is required to rid the area of weeds. This can result in injury to the roots of vegetable crops.

Using herbicides is recommended for large home gardens and commercial vegetable production plots. The use of any herbicide depends on the registration of the herbicide by federal and state Environmental Protection Agencies (EPA). Do not use any herbicide unless it is clearly stated on the label that it will treat a particular crop. Always seek the latest information for use of herbicides on vegetables.

Irrigation Nearly all commercial vegetable operations irrigate their crops. Irrigation is especially important in California and other *arid* and *semi-arid* or dry regions of the West. Irrigation is important because nature's rainfall is rarely uniform and adequate for high yields.

Water for irrigation can be obtained from a stream, lake, well, spring, or stored storm water. Definite laws and regulations guide the practice of irrigation in each state. There are several types of irrigation systems. They include sprinkler, surface or furrow, and subirrigation.

Sprinkler systems are very versatile. They can be used in almost any situation. They deliver shallow and light irrigation, which is good in promoting seed germination and application of fertilizers.

Surface irrigation has the advantage of requiring a relatively low investment. The topography of the land is important for surface irrigation. The land must be gently sloping and uniform. This method is well adapted to irrigating vast areas of land.

Subirrigation requires lots of water. With this system, the water is added to the soil so that it permeates the soil from below. This method is expensive and suited only to commercial enterprises.

Trickle irrigation, a form of subirrigation, differs in that the water is delivered under low pressure near the plants. The water seeps through porous hose under the rows. This system is expensive, but gives excellent results with increased yields and with less water consumed than with other methods.

Mulching A mulch is created whenever the surface of the soil is modified artificially by using straw, leaves, paper, polyethylene film, or a special layer of soil. Mulching helps control weeds, regulates soil temperatures, conserves soil moisture, and provides clean vegetables. The mulching material should be spread around the plants and between the rows.

Plastic film is frequently used with cantaloupes, cucumbers, eggplants, summer squash, and tomatoes. It is important that the plastic be laid over moist soil that has been freshly prepared. It is advisable to lay plastic mulch one to two weeks before planting the vegetables. This allows the soil to warm up underneath the mulch.

Pest Control Good pest control practices are essential for quality vegetables. A combination of cultural, mechanical, biological, and chemical methods can be used. Insecticides and fungicides can be effectively used by both home and commercial vegetable growers.

Insect and disease infestations bring about losses that can be devastating to vegetable crops. Types of losses include: (1) reduced yields, (2) lowered quality of produce, (3) increased costs of production and harvest, and (4) increased expenditures for materials and equipment. The use of chemicals is limited because of toxic and environmental considerations. There are many measures that can help the vegetable grower control insects and diseases. They are:

1. Dispose of old crop residues that serve as shelter to enable pests to overwinter or survive from one crop to another.
2. Rotate the location of individual crops.
3. Choose insect- and disease-resistant vegetable varieties.
4. Treat seeds with fungicides and insecticides.

5. Control weeds.
6. Use adequate amounts of fertilizer.
7. Follow recommended planting dates.

HARVESTING AND STORING VEGETABLES

Harvesting For the best possible quality, vegetables should be harvested at the peak of their maturity. This is possible when the vegetables will be used for home consumption, sold at a local market, or processed. The harvesting quality varies with each vegetable crop. Some will hold their quality for only a few days, while others can maintain quality over a period of several weeks. Frequent and timely harvest of crops is needed if the vegetable grower wishes to supply the market or the kitchen table with high-quality produce over a period of time. Specifications for harvesting each kind of vegetable crop should be consulted for the best results.

Harvesting is done by hand and through the use of machines. Mechanical harvesting is especially useful for the commercial grower. When harvesting a crop, care must be taken to prevent injury to the crop by machinery and handling.

Storing When storing vegetables, be sure they are at their proper stage of maturity. There are specific temperatures and humidity levels for storage of vegetable crops. Most crops need 90–95% humidity. Most homeowners find it difficult to reproduce the exact temperatures and humidity levels, but commercial growers can accomplish storage through specialized facilities (Figures 20-11A and 20-11B).

Vegetables differ in the amount of time that they can be stored. Some may be stored for several months, others for only a few days. For the vegetables that do store well, storage is essential to prolong the marketing period.

Fresh vegetables utilized for storage should be free of skin breaks, bruises, decay, or diseases. Such damage or diseases will decrease the life of the vegetable in storage.

Refrigeration storage is recommended. This type of storage reduces (1) respiration and other metabolic activities; (2) ripening; (3) moisture loss and wilting; (4) spoilage from bacteria, fungus, or yeast and (5) undesirable growths, such as potato sprouts.

One activity that can increase the successful storage of vegetables is precooling. *Precooling* is the process of rapid removal of the heat from the crop before storage or shipment. One method of precooling is termed *hydrocooling*. This is a method wherein vegetables are immersed in cold water and stay under water long enough to get rid of the heat.

STORAGE CONDITIONS

Vegetable	Temperature (°F)	Relative Humidity (%)	Storage Life
Pea, English	32	95-98	1-2 weeks
Pea, southern	40-41	95	6-8 days
Pepper, chili (dry)	32-50	60-70	6 months
Pepper, sweet	45-55	90-95	2-3 weeks
Potato, early	—[1]	90-95	—[1]
Potato, late	—[2]	90-95	5-10 months
Pumpkin	50-55	50-70	2-3 months
Radish, spring	32	95-100	3-4 weeks
Radish, winter	32	95-100	2-4 months
Rhubarb	32	95-100	2-4 weeks
Rutabaga	32	98-100	4-6 months
Salsify	32	95-98	2-4 months
Spinach	32	95-100	10-14 days
Squash, summer	41-50	95	1-2 weeks
Squash, winter	50	50-70	—[4]
Strawberry	32	90-95	5-7 days
Sweet corn	32	95-98	5-8 days
Sweet potato	55-60[3]	85-90	4-7 months
Tamarillo	37-40	85-95	10 weeks
Taro	45-50	85-90	4-5 months
Tomato, mature green	55-70	90-95	1-3 weeks
Tomato, firm ripe	46-50	90-95	4-7 days
Turnip	32	95	4-5 months
Turnip greens	32	95-100	10-14 days
Water chestnut	32-36	98-100	1-2 months
Watercress	32	95-100	2-3 weeks
Yam	61	70-80	6-7 months

Adapted from R. E. Hardenburg, A. E. Watada, and C. Y. Wang, *The Commercial Storage of Fruits, Vegetables, and Florist and Nursery Stocks*, USDA Agriculture Handbook 66 (1986).

[1]Spring- or summer-harvested potatoes are usually not stored. However, they can be held 4-5 months at 40°F if cured 4 or more days at 60–70°F before storage. Potatoes for chips should be held at 70°F or conditioned for best chip quality.

[2]Fall-harvested potatoes should be cured at 50–60°F and high relative humidity for 10–14 days. Storage temperatures for table stock or seed should be lowered gradually to 38–40°F. Potatoes intended for processing should be stored at 50–55°F; those stored at lower temperatures or with a high reducing sugar content should be conditioned at 70°F for 1-4 weeks, or until cooking tests are satisfactory.

[3]Sweet potatoes should be cured immediately after harvest by holding at 85°F and 90–95% relative humidity for 4–7 days.
Winter squash varieties differ in storage life.

FIGURE 20-11A
Temperatures and humidity for storing vegetables

Crop	Temperature °F	Relative Humidity Percent
Asparagus	32	85-90
Beans, snap	45-50	85-90
Beans, lima	32	85-90
Beets	32	90-95
Broccoli	32	90-95
Brussels sprouts	32	90-95
Cabbage	32	90-95
Carrots	32	90-95
Cauliflower	32	85-90
Corn	31-32	85-90
Cucumbers	45-50	85-95
Eggplants	45-50	85-90
Lettuce	32	90-95
Cantaloupes	40-45	85-90
Onions	32	70-75
Parsnips	32	90-95
Peas, green	32	85-90
Peppers, sweet	45-50	85-90
Potatoes	38-40	85-90
Pumpkins	50-55	70-75
Rhubarb	32	90-95
Rutabagas	32	90-95
Spinach	32	90-95
Squash, summer	32-40	85-95
Squash, winter	50-55	70-75
Sweet potatoes	55-60	85-90
Tomatoes, ripe	50	85-90
Tomatoes, mature green	55-70	85-90
Turnips	32	90-95

FIGURE 20-11B
Temperatures and humidity for storing vegetables. ***(Courtesy of the University of Maryland Cooperative Extension Service)***

Harvesting and Storage of Tomatoes

Harvesting Almost all of the tomatoes grown commercially are harvested by machines specifically designed for tomato harvest. All tomatoes grown for processing are harvested by machines, whereas tomatoes for the fresh market may still be picked by hand.

The tomato's degree of ripeness before it is harvested depends upon the purpose of harvest. Tomatoes that will be canned or processed are fully ripened on the vine before harvesting. However, the tomato being picked for a local market will be picked when it is pink. If tomatoes are to be shipped long distances, they are picked at the mature green stage.

Storing Tomatoes can be stored for only a short period of time. The best temperature range for short-term storage is between 54° and 59°F. Tomatoes that are picked when they are three-fourths ripe and put in well-ventilated storage with low humidity, and a temperature between 34° and 36°F, will keep for about three weeks.

Student Activities

1. Define the "Terms To Know."
2. Using outdated seed catalogs, prepare flash cards of warm- and cool-season vegetables.
3. Bring in an unusual vegetable to show to the class. Be prepared to tell the class which part is edible.
4. Select a site for planting a vegetable crop and prepare a planting plan to scale.
5. Take a soil sample from the site selected in the previous student activity. Submit the sample to the local county extension service for analysis and lime and fertilizer recommendations for the vegetables planned.
6. Plant lima bean seeds in containers to observe the growth of the stem and leaves. Determine if the plant is a monocotyledon or a dicotyledon.
7. Plant tomato seeds in the greenhouse for later transplanting.

8. Develop a chart with pictures illustrating the various insects that attack vegetable plants. Find pictures of insects in vegetable catalogs or extension service bulletins.

9. Store vegetables under different conditions (light/dark, dry/humid, warm/cool) in the classroom or at home. Record what happened under each set of conditions. Note which conditions are best for each vegetable.

Self-Evaluation

A. MULTIPLE CHOICE

1. The part of horticulture that has to do with the production of vegetables is called

   a. viticulture.
   b. olericulture.
   c. floriculture.
   d. pomology.

2. Vegetable gardening produces fresh, nutritious food and provides the gardener with

   a. exercise.
   b. weeds.
   c. pests.
   d. diseases.

3. The majority of vegetables grow best in

   a. well-drained, clay soil.
   b. well-drained, sandy soil.
   c. well-drained, silt soil.
   d. well-drained, loam soil.

4. For vegetable gardening, the soil should be plowed to a depth of

   a. 1–2″.
   b. 3–5″.
   c. 6–8″.
   d. 10–12″.

5. The first number in the analysis of a fertilizer represents the percentage of ______ in the bag.

   a. nitrogen
   b. calcium
   c. phosphoric acid
   d. potash

6. The best time to transplant vegetable plants is on

   a. a bright, sunny day.
   b. a rainy day.
   c. a cloudy day.
   d. a snowy day.

B. MATCHING

______ 1. Cultivation
______ 2. Herbicides
______ 3. Polyethylenefilm
______ 4. Olericulturalist
______ 5. Monocotyledon
______ 6. Precooling

a. Weed-control chemicals
b. Rapid removal of heat from vegetables
c. A person involved in the study of vegetable production
d. Intertillage of crops
e. Having only one seed leaf
f. A type of mulching material

C. COMPLETION

1. The scientific name for the mustard family of vegetables is ______ .

2. A site for the vegetable garden should be one that is exposed to the sun a minimum of ______ % of the day.

3. To correct the acidity level in the soil, _____ is commonly added.
4. The type of irrigation that allows water to permeate the soil from below is called _____ irrigation.
5. Whenever the surface of the soil is artificially modified through the use of straw, leaves, plastic film, or paper, a _____ is created.
6. Tomatoes for use in the local market are best harvested in the _____ stage of maturity.

UNIT 21 Grain, Oil, and Specialty Field-Crop Production

OBJECTIVE To determine the nature of and approved practices recommended for grain, oil, and specialty field-crop production.

Competencies to Be Developed

After studying this unit, you will be able to

- ☐ define important terms used in crop production.
- ☐ identify major crops grown for grain, oil, and special purposes.
- ☐ classify field crops according to use and thermo requirements.
- ☐ describe how to select field crops, varieties, and seed.
- ☐ prepare proper seedbeds for grain, oil, and specialty crops.
- ☐ plant field crops.
- ☐ describe current irrigation practices for field crops to meet their water needs.
- ☐ control pests in field crops.
- ☐ harvest and store field crops.

TERMS TO KNOW

Field crops
Grain crops
Malting
Forage
Cover crops
Green manure crops
Oilseed crops
Linen
Linseed oil
Ginning
Seed pieces
Cash crop
Thermo
Cereal crops
Seed legume crops
Root crops
Sugar crops
Tuber crops
Stimulant crops
Conventional tillage
Minimum tillage
No-till
Row crop planters
Drill planters
Broadcast planters
Irrigation
Sprinklers
Surface irrigation
Subsurface irrigation
Mechanical pest control
Cultural control
Biological control
Genetic control
Chemical control
Threshing

MATERIALS LIST

samples of field crops and products made from field crops
bulletin board materials
crop magazines
reference materials on crops

The cultivation of the land and the growing of crops began about 10,000 years ago. The need to produce food for the animals that humans had captured and begun to domesticate caused early humans to change from hunters to farmers. There were no guidelines to follow in selecting plants. Early agriculturists had to rely on observing what the animals were eating to decide which plants to grow. Trial and error and thousands of years of selection have led to the crops being grown today. New types, varieties, and uses of plants continue to be developed in response to current needs and in anticipation of future demands for food and plant fiber by an ever-increasing world population.

In the United States, the production of grain, oil, and specialty crops occupies more than 450 million acres. These crops are normally called *field crops*. This acreage represents nearly 20% of the land mass in the United States. American agriculturists are the most efficient in the world, producing enough food for themselves and about 80 other people. As a result, fewer than 3% of American workers are engaged in the production of food and fiber for the other 97%. The efficiency of the American agriculturist allows the U.S. population to spend only about 16% of their income on food. It also allows for sizable exports of food crops all over the world. This has resulted in a favorable balance of trade for the United States for many years.

MAJOR FIELD CROPS IN THE UNITED STATES

Grain Crops There are seven major grain crops in the United States. *Grain crops* are grasses that are grown for their edible seeds. These crops are corn, wheat, barley, oats, rye, rice, and grain sorghum.

Corn Corn is, by far, the most important of the field crops grown in the United States. Production of corn, in amount and in value, exceeds the combined production of all other grain crops. Thirty-five to forty percent of the corn produced annually in this country is grown in the midwestern states commonly called the Corn Belt. These states include Iowa, Illinois, Nebraska, Indiana, Minnesota, and Ohio. The United States accounts for nearly 50% of the corn produced in the world.

Originating in Central America, corn served as a major part of the diet of the Indians of the area for 3,000 years before the settlement of the New World by Europeans. Corn was unknown to the rest of the world until explorers found Indians growing corn.

Less than 10% of the corn grown in the United States is for human consumption. The rest is used for livestock feed, alcohol production, and hundreds of other products.

The major types or classifications of corn are dent corn, flint corn, popcorn, sweet corn, flour or soft corn, and pod corn (Figure 21-1). About 95% of the corn grown in the United States is dent corn.

Wheat The most important grain crop in the world is wheat (Figure 21-2). In the United States, wheat ranks second to corn in bushels produced. Leading states in the production of wheat are Kansas, Oklahoma, Washington, Texas, and Montana.

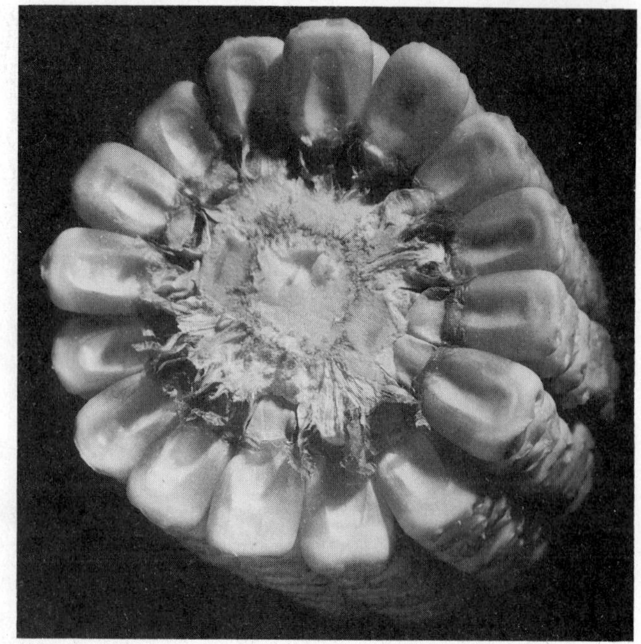

FIGURE 21-1
Corn production in the United States exceeds all other grains combined. *(Courtesy American Society of Agronomy)*

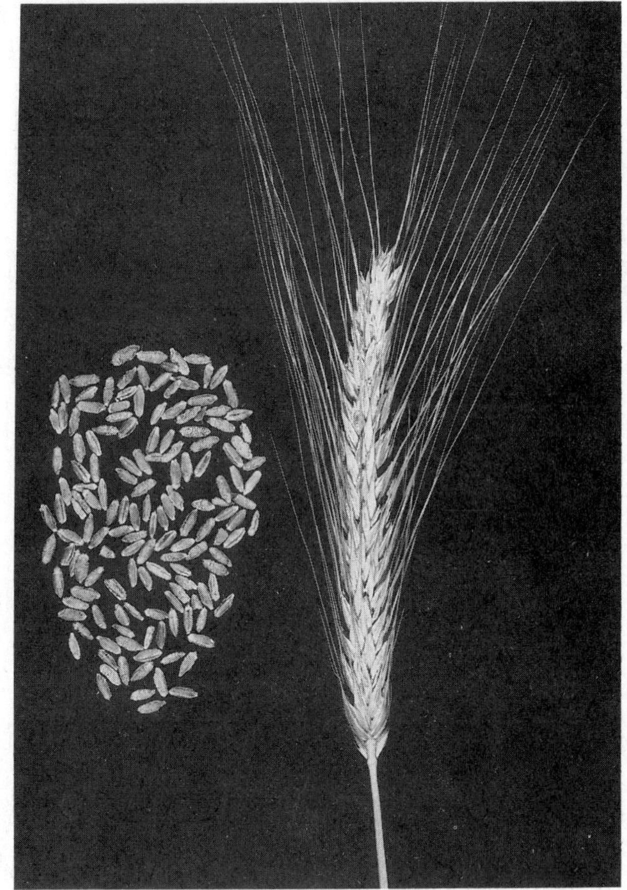

FIGURE 21-2
Wheat is the most important grain crop grown in the world. *(Courtesy American Society of Agronomy)*

Wheat is used primarily for human consumption. The wheat is ground into flour, which is then made into products such as bread, cakes, cereal, macaroni, and noodles. Other uses of wheat include the manufacture of alcohol and livestock feed.

Types of wheat grown in this country include common, durum, club, Poulard, Polish, Emmer, and spelt. Most of this wheat is of the common type. Classes of common wheat include hard red spring, hard red winter, soft red winter, and white.

Barley The leading states in the production of barley in the United States include North Dakota, Montana, Idaho, Minnesota, and California (Figure 21-3). Barley production ranks fifth among the grain crops produced in the United States.

Most barley is used for livestock feed, and it has about the same food value as corn. The production of barley for malting is also important. *Malting* is the process of preparing grain for the production of beer and other alcoholic beverages.

Oats Oats as a grain crop are fourth in acres produced in the United States (Figure 21-4). Major oat-producing states are South Dakota, Minnesota, Iowa, North Dakota, and Wisconsin.

FIGURE 21-3
Most of the barley grown in the United States is used for livestock feed. *(Courtesy American Society of Agronomy)*

The value of oats in adding bulk and protein to the diets of livestock is well documented. However, about 5% of the oats produced in the United

FIGURE 21-4
Oats provide bulk and protein to the diets of animals as well as cereal and cookies for humans. *(Courtesy American Society of Agronomy)*

States are made into oatmeal and cookies. Oats are also used in the production of plastics, pesticides, and preservatives. In addition, they are important in the paper and brewing industries.

Rye Although rye is grown in nearly every state in the United States, it is the least economically important grain crop (Figure 21-5). Most rye is grown in South Dakota, North Dakota, Minnesota, Georgia, and Nebraska.

About 25 to 35% of this rye is used for grain. The rest is used for forage, as a cover crop, or as a green manure crop. A *forage* is hay or grass grown for animals. *Cover crops* are planted to protect the soil from erosion. *Green manure crops* are grown to be plowed under to add organic matter to the soil. The rye grown for grain is used for livestock feed, flour, whiskey, and alcohol production.

Rice Rice is the major grain crop grown for food for almost half of the people in the world.

FIGURE 21-5
Most rye is grown for forages and cover crops. *(Courtesy American Society of Agronomy)*

FIGURE 21-6
Rice is the only commercially grown grain crop that will grow in standing water. *(Courtesy American Society of Agronomy)*

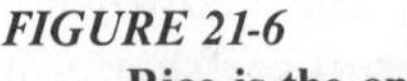

Most of the rice in the United States is grown in California, Louisiana, Texas, Mississippi, and Arkansas. Rice is the only commercially grown grain crop that can grow and thrive in standing water (Figure 21-6). The types of rice grown in the United States are short grain, medium grain, and long grain.

The majority of the rice grown in this country is used for human consumption. The excess that is produced is exported to other countries of the world.

Sorghum Sorghum grown in the United States is used primarily for livestock feed (Figure 21-7). It is about equal to corn in food value. Other uses of sorghum include forage, the manufacture of syrup or sugar, and the making of brooms. Leading states in the production of sorghum include Texas, Kansas, and Nebraska. Irrigated sorghum is also grown in California and Arizona.

The five types of sorghum are grouped according to use. They are grain, forage, syrup, grass, and broomcorn.

Oilseed Crops Crops that are grown for the production of oil from their seeds are called *oilseed crops*. These crops are growing in importance each year as Americans rely more and more heavily on vegetable oils and less on animal fats in their diets. Important crops grown for the oil contained in their seeds are soybeans, peanuts, corn, cottonseed, safflower, flax, and sunflowers. Corn and cottonseed will be discussed in other sections of this unit.

Soybeans There are approximately 70 million acres of soybeans grown in the United States each year. With an average yield of about 30 bushels per acre, the production of soybeans grosses more than $12 billion each year.

FIGURE 21-7
Sorghum is a grain crop grown for syrup and livestock feed. ***(Courtesy American Society of Agronomy)***

AGRI-PROFILE

CAREER AREAS: BROKER; ELEVATOR MANAGER; GRAIN HANDLER/PRODUCER
A Mexican farmer, an entomologist from Mexico's Instituto Nacional de Investigaciones Agricolas, and a USDA scientist collaborate to reduce the impact of corn-infesting insects that migrate from south to north as the season progresses. ***(Courtesy USDA Agricultural Research Service)***

Corn, wheat and other small grains, sugar cane, soybeans, sugar beets, and other specialty crops are grown on large acreages in the United States, Canada, and many nations of the world. Grain, oil, and specialty crops account for large amounts of the Americas' exports and do much to help maintain our balance of payments in foreign trade. Grain brokers, futures brokers, market reporters, grain elevator operators, crop forecasters, farmers, and others owe their jobs to these crop enterprises.

Custom combine operators, haulers, maintenance crews, cooks, and other service workers follow the grain harvest from Mexico to Canada. At season's end, the crews return south and prepare for the next season, when the cycle is repeated.

Along with crop enterprises are jobs in building and storage construction, systems engineering, machinery sales and service, welding, irrigation, custom spraying, hardware sales, agricultural finance, chemical sales, and seed distribution.

Oil and grain products are the major uses of soybeans. The meal resulting from the extraction of oil from soybeans is an important source of protein in livestock feeds. Soybeans are also used for hay, pasture, and other forage crops. Research has led to the development of hundreds of other uses for soybeans (Figure 21-8).

Large centers of production include the midwestern states of the Corn Belt. Major soybean-producing states include Illinois, Iowa, Missouri, Arkansas, and Minnesota.

Peanuts

The peanut is actually a pea rather than a nut, in spite of its nutlike taste and shell. It is grown primarily in the South, where warm temperatures and a long growing season are keys to success. Leading peanut-producing states are North Carolina, Georgia, Alabama, and Florida.

One ton of peanuts in the shell will yield about 500 lbs. of peanut oil and 800 lbs. of peanut-oil meal. The remaining 700 lbs. is mostly shells. The oil meal is used for livestock feed and as a good

FIGURE 21-8
There are many uses for soybeans. ***(Courtesy of the American Soybean Association)***

protein source in human diets. Other food products produced from peanuts include peanut butter and dry roasted peanuts.

Safflower

The production of safflower for oil occurs mainly in California. Safflower plants grow to 2–5′ in height and have flower heads that resemble Canadian thistles. The oil comes from the wedge-shaped seeds that the plant produces. The seeds may contain from 20 to 35% oil.

Safflower oil is used in the production of paint and other industrial products. It is also used for cooking oil and for low-cholesterol diets.

Flax

Originally the production of flax was mostly for fiber. Flax fibers contained in the stems of plants are used to produce *linen* (a fabric made from flax fibers).

The oil produced from the seed of the flax plant is called *linseed oil*. It is an important part of many types of paint and has hundreds of uses in industry. The linseed-oil meal that is left after the extraction of the oil is an excellent source of protein for animal feeds.

Most flax is grown in North Dakota, South Dakota, Minnesota, and Wisconsin.

Sunflowers

The production of oil-type sunflower seed has been important in the United States during the last 20 years (Figure 21-9). Most of the sunflower production is located in North Dakota, South Dakota, Minnesota, and Texas.

FIGURE 21-9
The sunflower has emerged as an important oil producing crop in recent years. ***(Courtesy American Society of Agronomy)***

There are two types of sunflowers grown commercially in the United States—oil-type and non-oil-type. About 90% of sunflower production is of the oil-type. Oil-type sunflower seeds contain 49 to 53% oil. The meal left after the oil has been removed is high in bulk and contains 14 to 19% protein. It is used for livestock feed. The oil is used for margarine and cooking oil. Sunflower oil can also be used as a substitute for diesel fuel in tractors.

Specialty Crops

Fiber crops, sugar crops, and stimulant crops are grouped into the category called specialty crops. Some specialty crops grown in the United States are cotton, sugar beets, sugarcane, and tobacco.

Cotton

Cotton originated in Central and South America (Figure 21-10). It has been an important crop in the South since colonial days. The cotton plant requires warm temperatures and a long growing season in order to reach maturity. These requirements are best met in the states of Texas, Arizona, New Mexico, California, Mississippi, Arkansas, and Louisiana.

In 1984–85, 13 million bales of cotton were harvested in the United States. Since only about 9

FIGURE 21-10
Cotton has been an important crop since colonial times. ***(Courtesy American Society of Agronomy)***

million were needed by the textile industry, the rest was exported to other countries of the world.

The seed must be removed from cotton after it is harvested. This process is called *ginning*. The cotton seed is then processed to remove its oil, which is a major contributor to the vegetable-oil needs of the United States. The seed, after the extraction of the oil, is ground into a high-protein animal feed.

Sugar Beets

The production of sugar beets for sugar accounts for about 35% of the refined sugar produced in the United States. This crop is grown for its thick, fleshy storage root in which sugars are accumulated. Centers of production of sugar beets are the western states and the upper Midwest.

Sugarcane

Sugarcane production in the United States is concentrated in subtropical areas of Florida and the Gulf Coast states. It is also grown in Hawaii. Sugarcane accounts for about 65% of the sugar refined in the United States.

This crop, which is a grass, is grown from sections of stalk called *seed pieces* rather than from seed. It takes about 2 years for sugarcane to reach the harvesting stage in Hawaii. In the South, sugarcane is harvested about 7 months after planting, with a corresponding loss of yield. The same field of sugarcane can be harvested several times before needing to be replanted.

Tobacco

Tobacco is a native American product that was used by the Indians in religious rites (Figure 21-11). It is produced in the southeastern states of North Carolina, Virginia, Kentucky, Florida, Georgia, South Carolina, Maryland, and Ohio as a cash crop. A *cash crop* is grown strictly to sell for cash. Production of tobacco is steadily declining in the United States because cigarette smoking and other uses of tobacco are also on the decline, as Americans become more health conscious.

Tobacco production requires large amounts of labor and is therefore best adapted to small farming operations. Warm temperatures and plenty of rainfall are required for optimum production of high-quality tobacco.

FIGURE 21-11
Tobacco is a native American crop. ***(Courtesy American Society of Agronomy)***

CLASSIFICATION OF FIELD CROPS

There are a number of ways of classifying field crops. Three of these are by use, by thermo requirements, and by life span. *Thermo* refers to heat requirements or growing season.

The classification of field crops according to use is as follows:

1. *Cereal crops* are grasses grown for their edible seeds. They include corn, wheat, barley, oats, rice, rye, and sorghum.
2. *Seed legume crops* are nitrogen-fixing crops that produce edible seeds. Included in this class are soybeans, peanuts, field peas, field beans, and cowpeas.
3. *Root crops* are grown for their thick, fleshy storage roots. Beets, turnips, sweet potatoes, and rutabagas are root crops.
4. Forage crops are grown for hay, silage, or pastures for livestock feed. Examples of forage crops include alfalfa, clover, timothy, orchardgrass, and many other crops used for their stems and leaves.
5. *Sugar crops* are grown for their ability to store sugars in their stems or roots. They include sugarcane, sugar beets, and sorghum. Corn is also used to produce sugar.
6. Oil crops are produced for the oil content of their seeds. Examples are soybeans, peanuts, cottonseed, flax, rapeseed, and castor beans.
7. *Tuber crops* are grown for their thickened, underground storage stems. Potatoes and Jerusalem artichokes are examples of tuber crops.
8. *Stimulant crops* are grown for their ability to stimulate the senses of the user. Examples of stimulant crops are tobacco, coffee, and tea.

Crops can also be classified according to their thermo requirements. The two major thermo groups are warm season and cool season.

Warm-season crops must have warm temperatures in order to live and grow. They are adapted best to areas where freezing or frost seldom occur. They also normally require longer growing seasons than cool-season crops. Examples of warm-season crops are cotton, tobacco, field peas, and citrus.

Cool-season crops are normally grown in the northern half of the United States, where freezing temperatures are normal. These crops often need a period of cool weather in order to attain maximum production. Most of the grains, tubers, and apples are cool-season crops.

Crops can also be classified according to their life spans. They may be annual, biennial, or perennial.

SELECTION OF FIELD CROPS

There are a number of factors to consider when selecting which field crops to grow. Some of these factors are:

1. Crops that will grow and produce the desired yields under the type of climate available. Be sure to consider length of growing season, average yearly rainfall, average temperatures, humidity, and prevailing winds.
2. Crops that are adapted to the type of soil available. Consider soil pH, soil type, soil depth, and soil response to fertilizers.
3. Demand or markets available for the crop to be produced.
4. Labor requirements and availability of labor for the crop.
5. Machinery and equipment necessary to grow the crop.
6. Availability of enough land to justify production of the crop.
7. Pest-control problems.
8. Expected yields.
9. Anticipated production costs. Can a reasonable profit be expected?

SEEDBED PREPARATION

The purpose of seedbed preparation for field crops is to provide conditions that are favorable for the germination and growth of the seed to be planted (Figure 21-12). Not only does the seedbed need to be prepared for seed germination, but the area under the seedbed must also be prepared for the root growth of the crop.

SOIL IS:	
Fine, pulverized	Seed in direct contact with soil particles
Firm	Prevents drying. Good root hair medium
Loose and mellow	Permits air circulation and seedling growth
Free of trash and weeds	Trash and weeds shredded and plowed under
Fertile	Adequate plant nutrients
Free of insects	Approved practices observed
Free of plant diseases	Approved practices observed
Moist	Absorbs and holds water

FIGURE 21-12
Characteristics of a good seedbed

Eliminating competition from weeds and crop residues is a consideration when preparing a seedbed for planting. Proper seedbed preparation can also increase the availability of soil nutrients to plants.

Seedbeds should not be overworked. The texture of the soil should be porous and allow for free movement of air and water. Small seed requires a seedbed with a finer texture than larger seeds require. The seedbed should contain enough fertility to encourage germination and growth until additional fertilizer can be applied. The control and elimination of weeds, insects, and diseases is also an important consideration when preparing seedbeds for field crops.

There are several methods of properly preparing seedbeds for field crops. They can be divided into three general categories: conventional tillage, reduced or minimum tillage, and no-tillage.

In *conventional tillage*, the land is plowed normally, turning under all of the residue from the previous crop (Figure 21-13). The soil is then disked or harrowed to smooth and further pulverize the soil for the seedbed.

Reduced or *minimum tillage* is a system of seedbed preparation that works the soil only enough so that the seed can make contact with the soil and germinate. Minimum tillage systems usually combine several operations into one pass across the field. This method reduces the amount of soil compaction, conserves soil moisture, and usually provides less opportunity for soil erosion.

FIGURE 21-13
The moldboard plow is the primary tillage machine used in conventional tillage systems. ***(Courtesy of Deere & Company)***

No-till preparation of seedbeds involves planting seeds directly into the residue of the previous crop, without exposing the soil. Seed is usually planted in a narrow track opened by the seed planter. It is extremely important that good management practices be employed when using the no-till method of seedbed preparation. Such practices will ensure control of insects and diseases and eliminate competition from previous crop residues.

PLANTING FIELD CROPS

The invention of the seed planter was one of the most important events for American agriculture. There are three general types of planters used in planting field crops today. They are row crop planters, drill planters, and broadcast planters.

Row crop planters plant seeds in precise rows and with even spacing within the rows. Three types of row crop planters are drill planters, hill-drop planters, and checkrow planters. Drill planters drop seeds individually in a row at set distances apart. Hill-drop planters drop two or three seeds together in rows. Checkrow planters plant several seeds together in a checkered pattern in the field to permit cross-cultivation. Row crop planters are used to plant corn, soybeans, sorghum, and cotton.

Drill planters plant seeds in narrow rows at high population rates. Drills are available in many row spacings and planter widths. Seeding rates are less accurate than with row crop planters. Some field crops that are planted with drill planters are wheat, oats, barley, rye, and many grasses. Fertilizer and pesticides are usually also applied at the same time the seed is planted with drill planters.

Broadcast planters scatter the seed in a random pattern on top of the seedbed (Figure 21-14). The accuracy of seeding using this type of planter is the poorest of the planting methods. Broadcast seeders cover wide areas and usually plant seeds much faster than other methods. They are sometimes employed when weather conditions make it difficult to get machinery into the fields being planted. Airplanes are often used as broadcast planters. Knapsack seeders and spinners are other types of broadcast seeders.

One disadvantage of using broadcast seeders is that some seed needs to be covered in order to germinate and to protect it from loss. This means that a second trip often must be made over the field to cover the seed. Small grains, grasses, and legumes are often planted by broadcasting.

There are other considerations in planting field crops. They include the date to plant, germination rate of seeds, uniformity of seed, weather conditions, and insect- and disease-control problems.

MEETING WATER NEEDS OF FIELD CROPS

The soil that plants grow in acts as a storage vat for the water needed by the plant. Under ideal soil conditions, approximately one-half of the pore space is filled with water. About one-half of this water is available for use by plants. Unfortunately, this ideal condition seldom exists and often much less water is available for plant use than is needed. Factors that affect the water available for crops include the type of soil, natural rainfall, water-table levels, and prevailing winds.

When conditions are such that sufficient water is not available for the crop being grown, irrigation may be the answer to obtaining profitable yields. *Irrigation* is the artificial supplying of water to crops.

The irrigation of crops has been practiced for more than 5,000 years. The Nile River was used by the Egyptians to irrigate crops grown in the fertile deserts of the area. The Chinese diverted the water from many rivers to irrigate their rice fields. Even

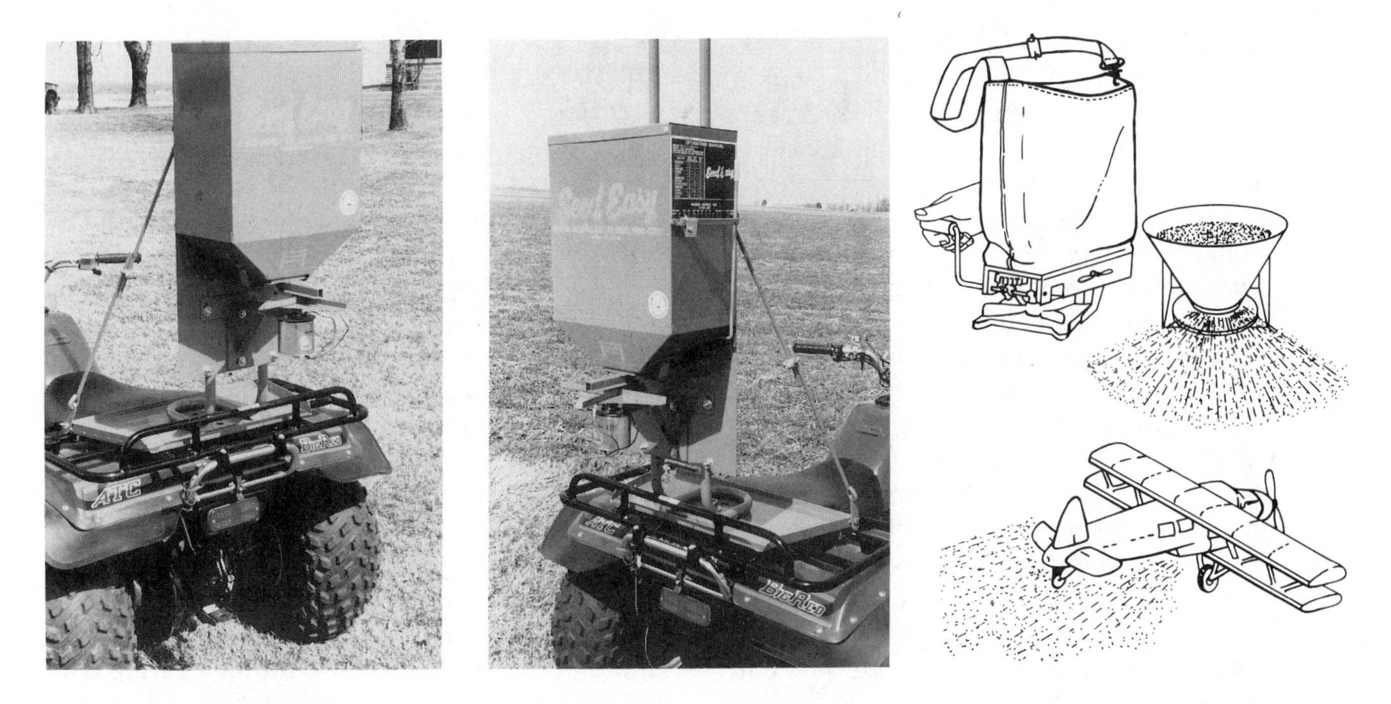

FIGURE 21-14
Three types of broadcast seeders are knapsack, spinner, and airplane. ***(Photos courtesy of Riburn Industries Inc.)***

the Indians of the West used irrigation to allow for the production of corn in arid areas.

The major methods of supplying irrigation water to crops are sprinklers, surface irrigation, and subsurface irrigation.

Sprinklers spray water through the air, much like rainfall. This is the least efficient means of using supplemental water. Much of the water evaporates before it can be used by the crop being irrigated. Some of the water is also placed where it is not available to crop roots.

In *surface irrigation*, water gets to the crop by gravity, flowing over the surface of the soil or in ditches or furrows. It is an inexpensive means of providing water to crops. However, the cropland may need to be leveled before it can be irrigated. Containment borders may also interfere with other farming operations.

Subsurface irrigation supplies water to the roots of crops underground. Use of pipes under the ground to deliver the water to the crops makes subsurface irrigation expensive to set up. It has the advantage of low operating costs once in operation, and permits the most efficient use of water.

PEST CONTROL IN FIELD CROPS

The control of pests in field crops is often the factor that determines whether or not a profit is made. Pests of field crops may include diseases, weeds, insects, and animals. They may destroy the seed before it germinates; attack the growing crop; or render the harvested crop unsalable, unusable, or even consumed. Economics losses from plant pests total billions of dollars each year.

There are three main categories of losses from plant pests. They are reduced yields, reduced quality, and spoilage.

Reduced yields occur when weeds germinate and grow faster than the crop being grown. Weeds compete successfully with the crop for moisture and nutrients, often causing the crop plant to be unhealthy. Parasitic plants actually feed on the host crop and may cause it to break off and die.

Damage from insects also causes the crop to yield less than expected. The damage may occur during the insect's feeding or while it lays eggs in and on the crop. Reduction of yield may also occur as insects spread disease from plant to plant.

Diseases can reduce yields by interfering with the plant's ability to manufacture food. They can also cause other plant processes to be changed, affecting the health and welfare of the plant.

The reduction of quality may result from such things as weed seeds or rodent hairs and droppings intermixed with the crop. Foreign materials may cause flavors that are objectionable to users of the crop. Most foreign materials must be removed before the crop can be used.

Damage from insects and diseases can make the crop less desirable in appearance and can increase processing costs tremendously. Food crops may be deemed unsuitable for human consumption and a total loss if they are too severely infested with insects and diseases.

Spoilage of crops results when weeds hinder the drying process. Insects may also cause stored crops to overheat and mold.

Methods of controlling pests in field crops include mechanical control, cultural practices, biological control, genetic control, and chemical control.

Mechanical pest control refers to anything that affects the environment of the pest or the pest itself. Cultivation is the normal mechanical control of weeds. Cultivation of the soil may also expose insects and soilborne diseases to the air. The effect of the suddenly raised or lowered temperatures as a result of exposure to the sun and air often proves fatal to many pests. Other types of mechanical control of pests include pulling or mowing weeds, screens, barriers, traps, and electricity.

Cultural control refers to adapting farming practices to control pests. Some cultural controls include timing farming operations to eliminate pests, rotating crops, planting resistant varieties, and planting trap crops that are more attractive to insects than is the primary crop.

Biological control of plant pests involves the use of predators or diseases as the control mechanism. The release of sterile male insects and the use of baits and repellents are also examples of this type of pest control. When using insects or diseases to control crop pests, it is important that the control be specific to the intended pest.

The development of varieties of crops that are resistant to pests is called *genetic control.* This may involve making the crop less attractive to the pest because of its taste, shape, or blooming time. Developing more rapidly growing crops that crowd out weeds is also an example of genetic pest control. Crops with resistance to diseases also fall into this category.

Chemical control of plant pests involves the use of pesticides to control pests of field crops. Excellent management practices must be exercised when using chemicals to control pests. Care should be taken to correctly identify the pest to be controlled and the chemical to be used. Dosage, runoff, and pesticide residues need to be carefully monitored.

HARVESTING AND STORING FIELD CROPS

Harvesting field crops at the proper stage of maturity is a key to maximizing profits. The harvest of the crop is the culmination of a growing season of work and anticipation of the rewards of a job well done.

The development of mechanical harvesting equipment allowed field-crop producers to harvest thousands of bushels of grain daily and with less labor than previously required. This allowed for tremendous increases in the amount of food available for people and animals and a greatly improved standard of living.

The primary harvesting machine for field crops is the combine (Figure 21-15). It performs the tasks of cutting the crop, threshing it, separating it from debris, and cleaning it. *Threshing* refers to the separation of grain from the rest of the plant materials. There are many types of combines adapted to the harvest of specific crops.

The proper storage of crops after harvesting is also important. The threats to the quality of stored crops include heat, moisture, fungi, insects, and rodents.

Drying grain to reduce moisture and heat is important for successful long-term storage. Much grain is harvested with a moisture content that is much too high. If stored without drying, the grain may heat up and encourage the growth of fungi, causing the grain to spoil. Foreign materials, such as weed seeds, may also cause stored crops to spoil.

Stored crops must also be protected from insects and rodents if quality is to be maintained. Rodent droppings, hair, and urine, as well as insect parts, may render stored crops unfit for human consumption. Reduction of food value and

FIGURE 21-15
The combine is the primary harvesting machine for field crops. ***(Courtesy Deere & Company)***

spoilage are other hazards of stored crops when insects and rodents are not controlled.

The production of field crops generates more income for American agriculturists than any other production enterprise. With more than 20% of the land in the United States currently being used for growing crops, and with the excess production contributing to a more favorable balance of trade, crop production is likely to remain very important in the future.

Student Activities

1. Define the "Terms To Know."
2. Compile a list of the field crops grown in your area.
3. Write a report on a field crop of interest to you.
4. Select a field crop and determine as many uses for it as possible.
5. Visit a local crop farm and talk to the operator about the advantages and disadvantages of growing a particular crop.
6. Prepare an advertisement to promote a field crop.
7. Construct a bulletin board about field crops and products made from them.
8. Visit a farm-machinery dealer. Make a list of all the equipment sold there that is used in the production and harvesting of field crops.
9. Make a collection of as many different field crops as you can.
10. Do a germination test on the seeds of several types of field crops. Observe the number of days required for germination to take place and the percentage of the seeds that germinate. Also conduct germination tests under a variety of environmental conditions, such as warm and cool, wet and dry, or with and without light. Compare the results to determine optimum conditions for the germination of crop seeds.

Self-Evaluation

A. MULTIPLE CHOICE

1. Flax is an example of a/an ______ crop.

   a. oilseed
   b. grain
   c. sugar
   d. fiber

2. The most important grain crop in the world is

   a. corn.
   b. wheat.
   c. rice.
   d. barley.

3. The moldboard plow is the primary tillage machine for the ______ tillage system.

   a. no-till
   b. minimum
   c. conventional
   d. none of the above

4. About 65% of the refined sugar produced in the United States comes from

   a. sugarcane.
   b. sugar beets.
   c. corn.
   d. sorghum.

5. The use of airplanes is an example of ______ seeding.

   a. row crop
   b. drill
   c. aerial
   d. broadcast

6. Grasses grown for their edible seeds are ______ crops.

   a. grain
   b. legumes
   c. oil
   d. sugar

7. Ginning is the process used to

   a. prepare a seedbed for planting.
   b. prepare crops for storage.
   c. remove seeds from cotton.
   d. prepare grain for use as alcohol.

8. An example of genetic control of pests is

   a. planting a crop when insects are not present.
   b. releasing sterile male insects.
   c. releasing an insect that feeds only on a certain weed.
   d. planting a variety of a crop that grows more rapidly than weeds.

9. Tobacco falls into the category of crops called

   a. fiber.
   b. biennial.
   c. thermo.
   d. stimulant.

10. Soybeans are grown for

    a. oil.
    b. hay.
    c. grain.
    d. all of the above.

B. MATCHING

______ **1.** Most important U.S. crop
______ **2.** Only crop to grow in standing water
______ **3.** Irrigation
______ **4.** Preparing barley for alcohol production
______ **5.** Primary harvesting machine
______ **6.** Most important world grain crop
______ **7.** Oil crop
______ **8.** Cloth made from flax
______ **9.** Warm-season crop
______ **10.** Separation of grain from plant

a. Malting
b. Linen
c. Combine
d. Sugarcane
e. Corn
f. Sprinkler
g. Peanut
h. Rice
i. Threshing
j. Wheat

C. COMPLETION

1. Heat, moisture, fungi, rodents, and insects are all problems to be dealt with in the ______ of crops.
2. ______ crops protect the soil from erosion.
3. The primary use for wheat is for ______ ______.
4. Cotton is grown for fiber and ______.
5. Mowing is a means of ______ control for weeds.

Forage and Pasture Management

OBJECTIVE To determine the nature of and approved practices recommended for forage and pasture production and management.

Competencies to be Developed

After studying this unit, you will be able to

- ☐ define important terms used in forage and pasture production and management.
- ☐ identify major crops grown for forage and pasture.
- ☐ select varieties for forage and pasture.
- ☐ prepare proper seedbeds for forage and pasture crops.
- ☐ plant forage crops and renovate pastures.
- ☐ control pests in forage crops and pastures.
- ☐ harvest and store forage crops.

TERMS TO KNOW

Forages
Hay
Silage
Pasture
Rhizomes
Noxious weed
Nurse crop
Overseeding
Carrying capacity
Haylage
Silo

Forage and pasture crops are the most important category of crops grown in the United States, if acres in production is the criterion used in rating. There are more than 475 million acres of land in pasture and range land. Another 60 million acres are used to produce hay. The importance of forages is further emphasized when you consider that half of the pasture and range land in the United States is not suited for the production of cultivated field crops. *Forages* are crop plants that are produced for their vegetative growth.

Forage production can be divided into three general categories: hay, silage, and pasture. *Hay* is forage that has been cut and dried so it contains a low level of moisture. *Silage* is green, chopped forage that has been allowed to ferment in the absence of air. *Pasture* is forage that is harvested by livestock itself. Forages are generally planted with one of these uses specifically in mind.

FORAGE AND PASTURE CROPS

Legumes

Alfalfa The most important forage crop in the United States is alfalfa (Figure 22-1). It is often called the "queen of the forages." Alfalfa is a legume that adds nitrogen to the soil. It is high in protein and other nutrients and very productive on fertile soils. It is also one of the oldest cultivated forage crops, and was mentioned in the Bible and in other early writings.

When properly harvested and stored, alfalfa is the most economical source of nutrients for ruminant animals. With production accounting for nearly 60% of the hay produced in the United States, alfalfa production ranks third to corn and soybeans in dollar value of product produced.

FIGURE 22-1
Alfalfa is the most important forage crop grown in the United States.

The North Central U.S. states account for about two-thirds of the yearly alfalfa production. However, alfalfa can be grown in nearly every state. Wisconsin ranks number one in alfalfa production, followed by South Dakota, Minnesota, Nebraska, Iowa, North Dakota, and Montana.

True Clovers These crops include about 300 species. However, only about 25 have agricultural importance. Clovers of economic importance in the United States include red clover, white clover, crimson clover, ladino clover, and alsike clover (Figure 22-2).

Clover accounts for about 20% of the hay grown in this country. It is a legume with the ability to add nitrogen to the soil. Clover has the disadvantage of lower yields than alfalfa under similar growing conditions. It is popular as hay, pasture, and silage and grows well in combination with many forage grasses.

The Northeast and North Central states produce most of the clover grown in the United States.

FIGURE 22-2
Clover is an important hay and pasture crop in the United States. ***(Courtesy American Society of Agronomy)***

Sweet Clover Sweet clover is used most often in areas that are hot and/or drought stricken, where its ability to survive and produce a crop is unsurpassed. It is used as hay, pasture, and sometimes as a green manure crop. Sweet clover is also used in Texas in a rotation with cotton to help control a cotton-root disease. It is also an excellent source of nectar, which honeybees make into honey.

There are three species of sweet clover grown in the United States—biennial yellow, biennial white, and annual. Biennial white sweet clover yields more than the other species, although biennial yellow sweet clover is usually of higher quality.

Sweet clover will grow in all areas of the United States. Areas of large production include the Dakotas, Minnesota, Wisconsin, Michigan, and the central states to Texas.

Bird's-foot Trefoil This is a comparatively new crop in the United States. It originated in Europe, where it has been a forage crop for 300 years. There are approximately 2 million acres in production in the United States. Bird's-foot trefoil is used as pasture in most cases. Some is also grown for hay.

The food value of bird's-foot trefoil is about equal to that of alfalfa. Because of smaller yields, bird's-foot trefoil is unlikely to seriously challenge alfalfa in importance. It does have the advantage over true clovers in being much longer lived.

States that report large acreages of bird's-foot trefoil include California, Ohio, Iowa, New York, and Pennsylvania.

Lespedeza This legume is grown primarily in the South, where 1 to 2 million acres are harvested as pasture and hay each year. It can grow and thrive on soils that are low in fertility. Because lespedeza has a lower nutritive value than true clovers and alfalfa, it is recommended for feed for beef cattle but not for dairy cattle.

Types of lespedeza include annual and perennial. Most of the lespedeza grown in the United States is of the annual type (Figure 22-3).

Grasses

Bromegrass Bromegrass is an important forage grass throughout the northern half of the United States. It is extremely hardy and grows to a height of 2-3′ (Figure 22-4). Because bromegrass produces many rhizomes, thin stands rapidly thicken with age. *Rhizomes* are horizontal underground stems

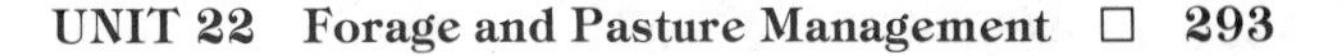

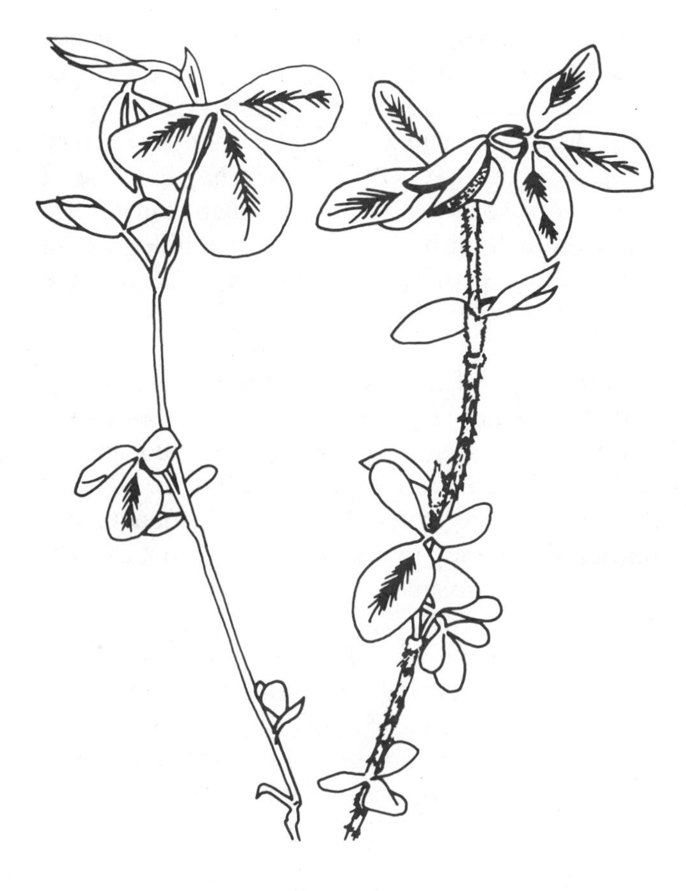

FIGURE 22-3
Most of the lespedeza grown in the United States is of the annual type and falls into one of two classes: Korean (left) and common, or Kobe (right). ***(Courtesy of the American Society of Agronomy & the University of Mississippi)***

from which new plants arise. Fertile soil is required for the production of bromegrass.

Orchard Grass This grass is known for its rapid germination and early spring growth (Figure 22-5). It also recovers quickly after being harvested. Timing of the harvest of orchard grass is very important because quality drops rapidly after the plant reaches maturity. Orchard grass makes excellent pasture and is often used for hay in combination with legumes.

Timothy Timothy is a cool-season grass that grows best when temperatures are between 65 and 72° F (Figure 22-6). The use of timothy as a forage grass has declined in recent years because bromegrass and orchard grass outyield it where all three species are adapted. Most timothy is grown in the northeastern part of the United States.

Reed Canarygrass In areas that are wet or poorly drained, reed canarygrass is often the only answer to producing a forage crop (Figure 22-7). It can produce more than four tons of forage per acre in the cool, damp areas where it thrives. This grass grows as much as 7′ tall and can produce high-

FIGURE 22-4
Bromegrass is grown throughout the Northern half of the United States.

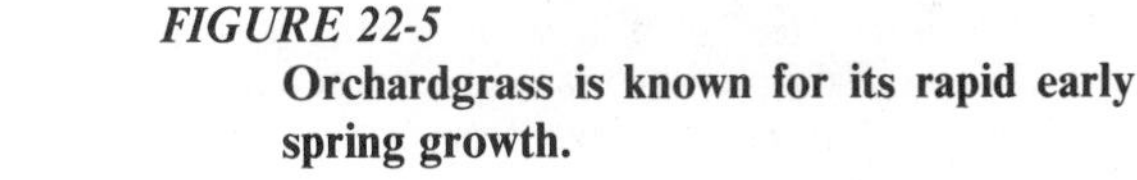
FIGURE 22-5
Orchardgrass is known for its rapid early spring growth.

quality feed if harvested before reaching maturity. Most reed canarygrass is grown in Washington, Oregon, California, Iowa, and Minnesota.

Kentucky Bluegrass

Kentucky bluegrass is grown over much of the United States, even though it is best adapted to the Northeast. It is the major grass of most pastures, lawns, and even golf courses in areas where summers remain fairly cool (Figure 22-8). Kentucky bluegrass is best adapted for pastures. It is not practical to use for hay because of low production. It also goes dormant during hot weather, making it necessary to use bluegrass in combination with other grasses in order to have forages available all summer.

FIGURE 22-6
Timothy grows best when temperatures are between 65 and 72 degrees. ***(Courtesy American Society of Agronomy)***

FIGURE 22-7
Reed canarygrass grows in areas too wet for the production of other grasses.

Fescue

Tall and meadow fescues are perennial grasses that are used for pasture and hay, usually in combination with other grasses and legumes. It is best adapted to the Southeast, where 10 to 20 million acres are grown each year. The quality and palatability of the fescues is lower than that of most other forage grasses.

Bermuda Grass

This is a warm-season grass that is dormant during cool weather. It is adapted to past.ıres and lawns because it grows only 6–12″ tall. Common Bermuda grass is considered to be a weed in some areas, because it tends to crowd out other grasses.

Dallis Grass

Dallis grass grows 2–4′ tall in the warm areas of the South. It cannot stand continuous use as pasture since it needs to be able to recover from close grazing. It is productive earlier in the spring than other warm-season grasses.

Johnson Grass

This grass is so aggressive that it has been declared a noxious weed in many

FIGURE 22-8
Kentucky bluegrass is the major type of grass in the pastures and lawns of the Northeast. ***(Courtesy American Society of Agronomy)***

states of the country. A *noxious weed* is a plant that is prohibited by state law. It is very coarse and grows to a height of 6′ or more. It spreads quickly by rhizomes and often crowds out most other plant species.

SELECTION OF FORAGES

There are many considerations to be made when selecting forages for hay, silage, and pasture. Some of them are:

1. the intended use of the forage.
2. the expected or desired yield.
3. the nutrient value of the crop.
4. the climatic conditions under which the forage will be grown—warm season versus cool season, summer and winter temperatures, humidity, soil type, anticipated rainfall, nutrient level, length of growing season.
5. the pest-control measures required.
6. the methods of establishment required.
7. their compatibility with other forages when grown in mixtures.
8. the expected and desired life of the crop.
9. the care and maintenance required.
10. the equipment and labor necessary for growing, harvesting, and storing the crop.

Forage crops that are adapted to the production of hay include alfalfa, clover, bromegrass, orchard grass, timothy, and fescue. Forages used for pastures are clover, lespedeza, Kentucky bluegrass, and Bermuda grass. Almost any legume or grass crop can be used for silage. In addition, corn and most small-grain crops make excellent silage when harvested at an immature stage.

SEEDBED PREPARATION

In general, preparing a seedbed for forages is much the same as it is for the production of field crops. Residues from previous crops must be dealt with. The soil must be prepared for the planting of seed or the vegetative

AGRI-PROFILE

CAREER AREAS: AGRONOMIST/PLANT PHYSIOLOGIST/FORAGE SPECIALIST

Plant physiologist William Sisson studies water-use efficiency of grasses to determine their adaptability to arid regions where annual rainfall is 10 inches or less. *(Courtesy USDA Agricultural Research Service)*

Forage and pasture provides a variety of career options ranging from farm and ranch to plant breeding and physiology. Grains, such as corn, cut for silage provide enormous volumes of feed for dairy and beef cattle. Crops cut for silage as well as those cut for hay are numbered among the forages. Similarly, pasture and range plants are forages. Those specializing in the science of forage growth and utilization are called agronomists.

Forage specialists are hired by universities and agricultural research centers as well as by large farms and businesses. In certain parts of the United States hay businesses are on the increase as more people do part-time farming, especially those with pleasure horses. Additionally, race tracks need hay and straw, which creates a strong market for these forages in many areas.

Hay business managers, truckers, dealers, and the like also conduct thriving businesses hauling straw and manure from horse barns at race tracks to mushroom farms, where the materials are used as the media for growing mushrooms. Recent droughts in sections of the United States created markets for cross-country shipping of hay, a practice once regarded as too expensive to be worth while.

pieces used in the propagation of some grasses. Necessary seedbed pest-control measures must be practiced. The soil must be amended so that its pH and fertility are suitable for the germination and growth of the intended forage crop. And with some crops, the moisture level of the seedbed may need to be regulated to ensure germination.

One difference between preparing a seedbed for most forages and one for field crops is that the texture of the soil in the seedbed must be finer and the seedbed somewhat firmer than is necessary for most field crops. This is because most forages have very small seeds that have difficulty making firm contact with the soil in seedbeds that are not finely textured.

Conventional tillage is generally used when preparing seedbeds for the growth of forages. The residues from previous crops are plowed down. The soil is then pulverized and leveled by disking and/or harrowing. For very small seeds, further preparation may be necessary. Final preparation of the seedbed should occur immediately before planting the seed.

PLANTING FORAGES AND RENOVATING PASTURES

Forage crops are usually planted by drilling or by broadcasting. The same grain drills that are used for planting small grains are also used to plant forages. Usually there is a separate seedbox on the drill. The forage seed is dropped on the seedbed at the same time that the grain is planted. Many grasses are planted in this way. The grain crop germinates faster than the forage and acts as a nurse corp until the forage crop can become established. A *nurse crop* is used to protect another crop until it can get established.

Many forage crops are also planted alone, using either drills or broadcast planters. With many forages, the seed must be covered after it is broadcast onto the soil by broadcast planters. This is done to protect the seed from pests and the drying effects of sun and wind. Covering the seed may be accomplished by a light disking or by the use of a cultipacker, which has corrugated wheels to press the seed into the seedbed.

Some types of forage seeds are broadcast into growing crops. They must be able to germinate fairly easily, because contact with the seedbed is often minimal. Red clover is often overseeded into stands of forage grasses to make mixed hay. *Overseeding* is the practice of seeding a second crop into one that is already growing. This is usually done during late winter and early spring when freezing and thawing of the soil helps provide contact between seedbed and seed.

Some forages are being planted with no-till planters in live or killed crop residues. The no-till planter opens a narrow furrow in which the seed is planted. Advantages include fewer trips across the field with heavy tractors and tillage equipment. This means that there is less compaction of the soil. There is also little exposure of the seedbed to erosion. Care must be taken to control pests in no-till planting. Another concern is to keep competition for water and nutrients from other plants growing in the seedbed to a minimum.

There are two general methods of renovating pastures in the United States. The existing pasture can be killed with a herbicide, the soil disked heavily to prepare a fine seedbed, and the pasture then reseeded with the desired types of grasses and/or legumes. The pasture is usually treated for insects and diseases at this time if necessary. It is also an ideal time to fertilize and lime the pasture.

The other method of renovating a pasture involves using selective herbicides to kill unwanted species of plants. The pasture is then disked to break up the existing sod. This allows easier entry of moisture and nutrients into the soil. The pasture may or may not be overseeded with desirable species to improve its yield. It is also fertilized and limed at this time according to soil test results.

PEST CONTROL IN FORAGE CROPS

The control of weeds, insects, diseases, and rodents in forage crops helps result in optimum yields. Because many forage crops grow for more than one growing season, pest control is usually an ongoing part of forage crop management.

The proper identification of the pest or pests affecting the crop is very important. To that end, personal experience and the use of trained professionals is often necessary.

The actual methods of pest control are many, and they were discussed in previous units. Chemicals, cultural practices, biological control, genetic control, and the timing of crops are all tools to be used in controlling the pests of forage crops. Not to be overlooked in pest-control management is the control of pests in stored forages.

When chemicals are used to control pests, care must be taken to be sure that the pesticide is properly applied in the recommended dosages. Timing the application of the pesticide for maximum effect is also essential. Care must also be taken to ensure that overspray and runoff of pesticide materials do not adversely affect other than the intended pest. Concern that the residue of pesticides does not end up in food sources is of utmost importance.

HARVESTING AND STORAGE OF FORAGE CROPS

The proper harvesting and storage of forage crops is extremely important in the management of the forage enterprise for maximum profits. Most of the forages grown on American farms and ranches are fed to livestock grown on the same farm. Forages are usually the least expensive source of nutrients available for cattle and sheep.

Pastures The harvesting of pastures involves several factors. One of the most important is that the carrying capacity of the pasture must be determined in order to figure out how many animals can be fed by the pasture available. The *carrying capacity* of a pasture refers to the number of animals for which it will provide feed.

Pastures also need to have time to recover from the ravages of the animals that crop them. The rotation of pastures, or actually the rotation of the animals using the pastures, is important if the pastures are to yield their potential. Pasture rotation also helps to control parasites of the animals grazing on them by breaking the pest's life cycle.

Hay Harvesting hay involves several operations that must be timed fairly accurately if a quality product is to be harvested. The hay must be cut at the optimum time, with an eye on the weather forecast for the next several days. Hay is normally dried in the field by the sun. Nothing ruins hay faster than an unexpected rainstorm. The maturity of the forage being harvested for hay is also critical (Figure 22-9). The more mature the forage becomes, the lower the quality and food value of the hay produced.

There are several types of machines that are designed especially for cutting forages intended for hay. They include the sickle-bar-type mower, rotary mower, flail-type mower, and variations of these types.

The sickle-bar-type mower cuts with the same action as a pair of scissors. These types of mowers are best adapted for forages that are standing up straight.

The rotary mower cuts with blades that move in a circular motion parallel to the ground. This type of mower will cut any type of forage, although higher amounts of horsepower are required when the stands of forage are extremely thick and heavy.

The flail-type mower is most commonly used when harvesting forages for silage, although it will cut crops intended for hay. This type of mower features a cutter head that rotates in a circular motion vertically to the ground. It has the disadvantage of chopping up the hay more than is usually desirable.

TYPE OF HAY	WHEN TO HARVEST
1. Alfalfa	Pre-bud to 1/10 bloom stage
2. Clover	¼ to ½ bloom stage
3. Birdsfoot trefoil	¼ bloom stage
4. Sweetclover	Start of the bloom stage
5. Bromegrass	Medium head stage
6. Timothy	Boot stage to early bloom stage
7. Lespedeza	Early bloom stage
8. Orchardgrass	Full head but before blooming
9. Reed canarygrass	When the first head appears

FIGURE 22-9
Maturity levels of harvesting selected forages for hay.

FIGURE 22-10
Hay is usually harvested with a baler. ***(Courtesy Deere & Company)***

Often included in the cutting operation is crushing or mashing the stems of the forages to hasten their drying. Windrowers are also a part of some mowing operations. They leave the cut forages in rows to make it convenient for gathering by the harvesting machines.

In areas of high humidity, chemicals to speed up the drying of the forage are sometimes sprayed on the forage as it is cut.

The hay is usually raked at least once before it is harvested to allow it to dry more evenly and more rapidly. Legume-type hay should be raked during the part of the day when humidity is highest to keep the loss of leaves to a minimum. Raking hay also puts it into windrows so that it can be gathered more easily.

Once the hay has dried to the desired level of moisture, it is harvested by one of three basic methods. It may be baled, harvested loose with stackers, or cubed.

Hay may be harvested into square or rectangular bales or into round bales of various sizes (Figures 22-10 and 22-11). The baler gathers the hay from the ground, compresses it into a bale, and ties twine or wire around the bale to hold it together and allow for handling it. The bale is then expelled from the baler. It may be allowed to drop on the ground for later removal to storage or, in the case of square bales, ejected directly into a wagon.

There are a number of machines that are designed to move baled hay from the field to the storage site. They include wagons, bale loaders, bale handlers, and bale stackers.

Hay may also be harvested with machines called stack wagons. These machines gather forages from the field and form a dense stack or loaf of weather-resistant hay. Less labor is required for the production of hay stacks than for baled hay. With special handling equipment, stacks can be moved with few problems.

Hay cubes were developed to allow for the full mechanization of the feeding of hay (Figure 22-12). The cuber gathers the dry forage and compresses it into a cube about 1¼″ × 1¼″ × 2–3″. Normally, only legumes are cubed because of difficulty in getting grass cubes to bind together. Artificial binders are sometimes used.

Hay may be stored in buildings or in the open field. Regardless of the method of storage, care must be taken to ensure that quality is maintained. Hay that is stored in buildings must be kept free from pests, particularly rodents. Care must also be taken to ensure that the moisture level of stored hay is low enough that spoilage does not occur.

When hay is stored outdoors, it should be placed on land that is well drained and sloped to drain water away from the hay. The stacks or bales of hay stored outside must be dense enough so that little moisture can penetrate it and cause spoilage. Hay stored outside is often covered with plastic or other materials to protect it from adverse weather conditions.

Silage Harvesting forages for silage is far more simple than making hay. Less equipment and labor is required, and the entire harvesting operation is usually completed in one pass over the field. A forage harvester cuts the green forage, chops it into small pieces, and deposits it into a wagon or truck to be hauled to the storage facility. Cutter heads on the forage harvester may be of either the cutter type or flail type.

Flail-type forage harvesters are used to harvest low-moisture silage called *haylage*. Forages

FIGURE 22-11
Round balers have become popular in recent years because they allow for more efficient use of labor. *(Courtesy Deere & Company)*

FIGURE 22-12
Hay handling is now fully mechanized. *(Courtesy Deere & Company)*

used for haylage are cut by mowers and allowed to dry to a moisture content of 40 to 55% before they are placed in a silo. A *silo* is an airtight storage facility for silage or haylage.

Silage may be stored in silos or in trenches or bunkers. It is also sometimes stored in piles on top of the ground and sealed with plastic or other materials to make it airtight. The production of silage is dependent on storage in an absence of air or oxygen. The fermentation that occurs in the absence of air preserves the silage. Spoilage or rot occurs when green forages are stored in the presence of air.

The production of forages in the United States accounts for about half of all the land in agricultural use. Much of this land is unsuitable for production of other crops. With the production of hay ranking third only to corn and soybeans, forages are extremely important to American agriculture. Because forages are the least expensive sources of nutrients for cattle, sheep, and horses, they are likely to remain essential for years to come.

Student Activities

1. Define the "Terms To Know."
2. Put fresh-cut forage in each of two quart jars. Pack the forage tightly into the jars. Seal one jar and leave the other open to the air. Compare the contents of the two jars after 10 days.
3. Make a collection of as many different forage seeds as you can.
4. Write a report on a forage of economic importance in your area.
5. Make a collection of local forages.
6. Have a forage producer speak to the class about production of forage crops.
7. Visit a machinery dealership to learn about the types of seedbed preparation, planting, and harvesting equipment for forages.
8. Make a bulletin board about forages.

Self-Evaluation

A. MULTIPLE CHOICE

1. The type of mower that cuts with a scissors-type action is a ______ type mower.

   a. rotary
   b. flail
   c. sickle-bar
   d. vibrating

2. A silo is used to store

   a. silage.
   b. hay.
   c. pasture.
   d. cubes.

3. The most important legume hay in the United States is

   a. clover.
   b. lespedeza.
   c. bird's-foot trefoil.
   d. alfalfa.

4. A forage that grows in wet or poorly drained soil is

   a. reed canarygrass.
   b. timothy.
   c. fescue.
   d. bird's-foot trefoil.

5. ______ is suitable for only pasture and lawns because it grows only 6–12″ high.

   a. Dallis grass
   b. Sweet clover
   c. Kentucky bluegrass
   d. Orchard grass

6. Forages that are cut and allowed to dry to 40 to 55% moisture before storing is called

   a. silage.
   b. haylage.
   c. hay.
   d. cubes.

7. The harvesting of ______ requires the most labor of any of the forages.

   a. pasture
   b. hay
   c. silage
   d. haylage

8. The production of forages in the United States equals ______ percent of the land used for crops.

   a. 25
   b. 35
   c. 50
   d. 65

9. A warm-season grass is

   a. sweet clover.
   b. bromegrass.
   c. reed canarygrass.
   d. Bermuda grass.

10. Hay is raked to

   a. allow even drying.
   b. allow the hay to ferment.
   c. reduce its maturity.
   d. remove leaves and weeds.

B. MATCHING

______ **1.** Timothy
______ **2.** Alfalfa
______ **3.** Bromegrass
______ **4.** Sweet clover
______ **5.** Orchard grass

a. Number one grass hay crop
b. Excellent source of nectar
c. Grows very early in the spring
d. Number one forage crop
e. Popular in mixes with clover

C. COMPLETION

1. A disadvantage of ______ type mowers is that they tend to chop up the forage.
2. Sweet clover types include annual and ______ .
3. Seedbeds for forages should be ______ than those for most other crops.
4. ______ is forage harvested by livestock.
5. Horizontal underground stems are called ______ .
6. A ______ weed is prohibited or banned by state law.
7. Crops that are harvested for their vegetative growth are called ______ .
8. Two methods of planting forages are ______ and broadcasting.
9. A ______ gathers dry forage, compresses it, and ties it with twine and wire.
10. The more mature the forage, the ______ its quality.

SECTION 7

Ornamental Use of Plants

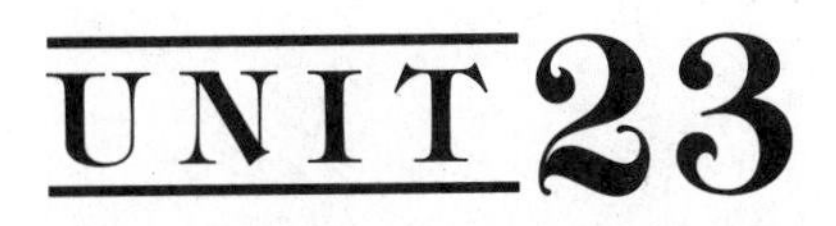

UNIT 23 Indoor Plants

OBJECTIVE To use plants indoors for beautification, and for air quality and pollution control.

Competencies to Be Developed

After studying this unit, you will be able to

- ☐ identify plants that grow well indoors.
- ☐ select plants for various indoor uses.
- ☐ grow indoor foliage plants.
- ☐ grow indoor flowering plants.
- ☐ describe elements of design for indoor plantscapes.
- ☐ describe career opportunities in indoor plantscaping.

TERMS TO KNOW

Floriculture
Succulent
Foliage
Variegated
Herb
Relative humidity
Dormant
Rootbound
Plantscaping
Basic color
Accent color
Texture
Accent
Sequence
Balance
Formal
Symmetrical
Informal
Asymmetrical
Scale
Circulation
Development limitations

MATERIALS LIST

indoor plant reference books
seed catalogs
pencil, grid paper, and eraser
indoor plants
pots of various sizes

The world of indoor plants is a fascinating one. There are plants that have magnificent blooms, unusual shapes, fancy foliage, and fragrant smells. If selected and cared for properly, they can last for years and may even be passed from generation to generation. The industry of indoor plants is the floriculture part of the ornamental horticulture field. *Floriculture* involves the production and distribution of cut flowers, potted plants, greenery, and flowering herbaceous plants. Indoor plants present a challenge to the homeowner who wishes to surround the rooms of the home with living color. They also challenge those who decorate public areas.

PLANTS THAT GROW INDOORS

Almost all plants can be grown indoors. There are plants, however, that favor indoor conditions. Many of the trees and shrubs do better when grown outdoors. Smaller and more succulent plants are best for indoor use. *Succulent* means having thick, fleshy leaves or stems that store moisture. There are a wide variety of shapes and sizes of indoor plants. These plants can be divided into two major groups—those that flower and those that are grown only for their foliage. *Foliage* means stems and leaves.

Popular and Common Flowering Indoor Plants

African Violets One of the most popular and common indoor flowering plants is the African violet (*Saintpaulia ionantha*). This plant can be recognized by its small size and hairy leaves. The leaves are oval-shaped, dark green, and covered with soft, short hairs. The flowers contain four to five petals arranged in a clover pattern. They vary in color from deep purple to brilliant white (Figure 23-1).

Fuchsias Fuchsias (*Fuchsia triphylla*) are plants with very colorful flowers that cascade from the plant. Most have flowers that are two-tone pinks

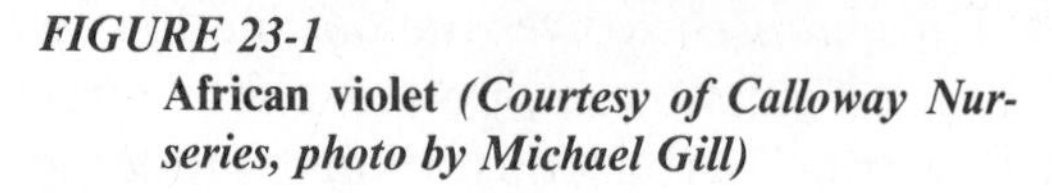

FIGURE 23-1
African violet ***(Courtesy of Calloway Nurseries, photo by Michael Gill)***

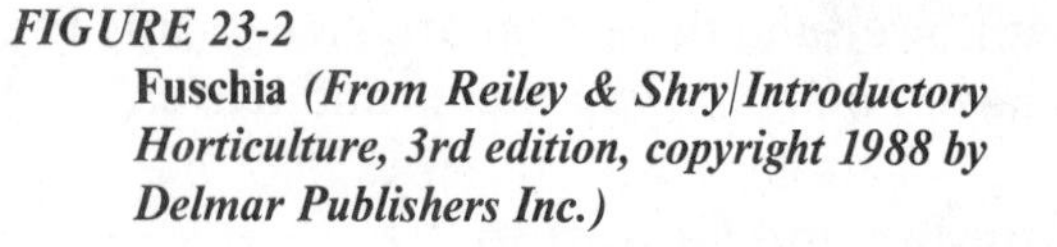

FIGURE 23-2
Fuschia ***(From Reiley & Shry/Introductory Horticulture, 3rd edition, copyright 1988 by Delmar Publishers Inc.)***

FIGURE 23-3
Gardenia ***(Courtesy of Calloway Nurseries, photo by Michael Gill)***

FIGURE 23-4
Geranium ***(From Reiley & Shry/Introductory Horticulture, 3rd edition, copyright 1988 by Delmar Publishers Inc.)***

and reds. The foliage is dark green and the leaves tend to be long, oval shapes with a bronzy hint of color (Figure 23-2).

Gardenias Particularly fragrant, flowering indoor plants, gardenias (*Gardenia jasminoides*) have deep-green shiny foliage and pure white flowers. The leaves are in clusters of three and are pointed (Figure 23-3).

Geraniums One of the most versatile flowering indoor plants is the geranium (*Pelargonium zonale*). These indoor plants are one of the oldest. The leaves are a rounded, yellowish green with scalloped edges. The flower is borne on a stem and consists of many petals in a cluster shaped like a ball. The flower color ranges from the most popular red, to white and pink (Figure 23-4).

Impatiens If an indoor plant with many blooms is desired, the impatiens (*Impatiens wallerana sultanii*) is a good choice. The flowers are small and rounded, with five petals. One petal is shaped like a tube that protrudes from the underside of the flower. Flower colors include white, pink, salmon, coral, lavender, purple, and red. The leaves are

FIGURE 23-5
Impatiens ***(Courtesy of Calloway Nurseries, photo by Michael Gill)***

lance-shaped and have succulent stems (Figure 23-5).

Others There are many other varieties of indoor plants that can be grown indoors for their flowers. Some flowering plants are both indoor and outdoor, such as wax begonias, ageratums, verbenas, and petunias.

Foliage Plants Indoor foliage plants can be divided into five groups for easy identification. The groups are ferns, indoor trees, vines, cacti and succulents, and others. A foliage plant is grown for the appearance of the leaves and stems.

Ferns These foliage plants come in a variety of types. Some of the most popular indoor plant-type ferns are Boston fern (*Nephrolepis exaltata*) Figure 23-6A; asparagus fern (*Asparagus sprengeri*) Figure 23-6B; maidenhair fern (*Adiantum capillus-veneris*), sword fern (*Nephrolepsis cordifolia*), rabbit's-foot fern (*Davallia canariensis*), and staghorn fern (*Platycerium biforcatum*). Ferns are categorized by their long and often multicut leaves. Most ferns are feathery in appearance.

Indoor Trees An indoor tree can be an excellent accent to a room or a terrific addition to a patio or hallway. They can grow to be 6 to 7 feet tall. Some of the more popular indoor trees are Norfolk Island pine (*Araucaria excelsa*) Figure 23-7A; fiddleleaf fig (*Ficus lyrata*), umbrella plant (*Schefflera actinophylla*) Figure 23-7B; rubber plant (*Ficus elastica*) Figure 23-7C; fragrant dracaena (*Dracaena fragrans*), weeping fig (*Ficus benjamina*) Figure 23-7D; and croton (*Codiaeum variegatum*).

FIGURE 23-6
Fern-type plants ***(Courtesy of Calloway Nurseries, photos by Michael Gill)***

Indoor trees vary in the type and size of their foliage, but all have woody-type stems.

Vines The vines are characterized by their habit of climbing or draping from the sides of the pot. Some of the more widely recognized vine-type indoor plants are philodendrons (*Philodendron species*) Figure 23-8A, Wandering Jew (*Tradescantia fluminensis "Variegata"*) Figure 23-8B, grape ivy (*Rhoicissus rhomboidea*), and English ivies (*Hedera species*). These plants can be trained to climb up a piece of wood, around a planter box, or up a wall.

Cacti and Succulents This is a unique group of indoor plants. They are among the easiest plants to grow indoors. This group of plants tend to hold water within their stems and leaves. Cacti and suc-

FIGURE 23-7
Indoor trees ***(Courtesy of Calloway Nurseries, photos by Michael Gill)***

FIGURE 23-8
Vine-type indoor plants ***(Courtesy of Calloway Nurseries, photos by Michael Gill)***

culents survive dry heat, low humidity, and varying temperatures. Cacti usually have some type of prickly needles, whereas succulents do not. One of the most popular succulents is the jade plant (*Crassula arborescens*) (Figure 23-9).

Others or Specimens There are numerous other types of indoor plants that do not fit into any of the categories of ferns, indoor trees, vines, and cacti and succulents. These indoor plants tend to have a special characteristic or feature that people like to display—a specimen plant. A few examples

FIGURE 23-9
Jade plant ***(From Reiley & Shry/Introductory Horticulture, 3rd edition, copyright 1988 by Delmar Publishers Inc.)***

of more popular specimens are the spider plant (*Chlorophytum elatum vittatum*), which shoots off baby plants or "spiders"; peperomia (*Peperomia caperata*), which has pale pink to red stems with deeply grooved heart-shaped leaves (Figure 23-10A); purple passion plant (*Gynura aurantica*), which has rich royal-purple leaves that are covered with velvet-type hair; bromeliads, one variety of which (Figure 23-10B) has a deep pink color at the center, and snake plant (*Sansevieria trifasciata*), which has long, spikelike, thick leaves that are variegated with gold. *Variegated* means having streaks, marks, or patches of color.

SELECTING PLANTS FOR INDOOR USE

There are two rules of thumb when selecting plants for use indoors. The first is to be selective when you purchase the plant. The second is to choose the right plant for the growing conditions available in the place you want the plant to live. Careful attention to these two rules will greatly increase the chance for success with indoor plants.

Purchasing an Indoor Plant

When buying an indoor plant, look for one that appears to be healthy. Look closely at the plant for insects, being careful to check the undersides of the leaves where many insects hide or lay their eggs. Select plants with even green color, no yellowing leaves, and no spots or blotches on the leaves. Avoid plants that have spindly growth or appear wilted. If possible, pur-

A

B

C

FIGURE 23-10
Specimen plants ***(Courtesy of Calloway Nurseries, photos by Michael Gill)***

chase the plant during its growing season. Finally, look for new growth, such as leaf or flower buds.

Choosing the Right Plant

Before selecting an indoor plant, decide where the plant will be placed. Note particularly the light intensity and duration, as well as the temperature of the area where the plant will be placed. Each species of indoor plants has specific conditions for optimum growth. The plant should match the location for best results.

Light The amount of light available for a plant is a factor that is difficult to control. A shadow test can help determine this. To conduct this test, hold a piece of paper up to the light (window or lamp) and note the shadow it makes. A sharp shadow means you have bright or good light. However, if there is barely a shadow visible, the light is dim or poor. It is important to know how much light the plant needs. If a plant needs direct or full sun, then exposure of the sun is needed for at least half of the daylight hours. When a plant needs indirect or

FIGURE 23-11
These plants need indirect or filtered sun. *(Courtesy of Calloway Nurseries, photos by Michael Gill)*

partial sun, the light should be filtered through a curtain or slats. Figure 23-11 shows two plants that fit into this category: Chinese evergreen (A) and dracena (B). Even for plants preferring no direct sunlight, the room should be bright and well lighted. Plants that need shade should be kept in a well-shaded part of the room. In summary, too much or too little light can greatly affect the health of the plant.

Temperature There are three temperature categories in which indoor plants are grouped. They are cool, moderate, and warm. The cool temperature range is 50 to 60°F, with temperatures not falling below 45°. The moderate temperature range is from 60 to 70°F, with a minimum of 50°. The warm temperature range is from 70 to 80°F, with a bottom limit of 60°.

USES OF INDOOR PLANTS

The use of plants indoors is varied. They can be used as room dividers or to brighten up a dull spot in the kitchen or a bathroom. Plants can be used to divide a room into specific living areas. They may be placed in containers with special watering devices, or several plants may be grouped together to form a natural barrier. Containers are available in a number of sizes and shapes, with castors on the bottom so they can be moved easily. Indoor trees and climbing plants are best used as living-area dividers.

Specimen or showy plants, such as Boston fern or date palm (*Phoenix dactylifera*), are good for brightening up dull areas. Plants may be placed on ornate plant stands or in decorative pots. A series of wall shelves may also be used to display interior plants. Plants on shelves should be compact, and perhaps trailing, for a more dramatic effect.

Hanging baskets filled with indoor plants are useful when space is at a premium. Since baskets can easily be suspended from the ceiling, they can utilize the unused space above head level. Care should be used when selecting plant containers so that water does not drip on the furniture or the floor. Select baskets that are designed with a drip tray attached to the base.

Some interesting areas for hanging baskets are stairwells, offices, and foyers. One other element that is unique to hanging baskets is the hook securing the basket to the ceiling. It should be firm enough to hold when the basket's soil is wet and when the plant is being cared for.

Bathrooms are excellent places for indoor plants. These rooms are humid and warm, which provides a good atmosphere for plants. They can be placed on windowsills or around the tub. Shelves can also be installed in a bathroom to display indoor plants. Plants will add a touch of color to a bathroom, especially today as the bathroom is expanding to hold a hot tub or jacuzzi. Philodendrons are excellent in bathrooms.

The kitchen is another room of the home where plants can add a beautiful touch. Refrigerators, other appliances, and kitchen windowsills all make good places for plants that prefer a cooler temperature. Most of the plants in the kitchen should be relatively small, as space is usually at a premium. Several plants suitable for the kitchen are aloes (*Aloe verrucosa*), maidenhair ferns, peperomias, and even a variety of herbs. An *herb* is a plant kept for aroma, medicine, or seasoning.

AGRI-PROFILE

CAREER AREAS: INTERIOR OR INDOOR PLANTSCAPER

Interior plants are frequently grown in greenhouses, shade houses, and outdoors, and then rotated to indoor areas where lower light conditions may restrict plant growth and duration. ***(Courtesy United States Department of Agriculture)***

Foliage plants including small trees have become popular decor for interior public areas such as shopping malls, office buildings, institutions, and housing complexes. Flowering and foliage plants have long been part of the interiors of attractive homes and have provided excellent hobbies for many. However, their extensive use in commercial settings is a relatively recent development which has stimulated many new job opportunities in ornamental horticulture.

Youth in suburban settings may obtain jobs raising and caring for plants, installing interior plantscapes, rotating plants between growing areas and display areas, and contracting to maintain interior plantscapes. Early experiences frequently lead to life-long careers.

With more experience and education, one may become a plantscape designer, business owner-operator, grower, wholesaler, plant doctor, extension specialist, and other. Ornamental horticulture specialists are hired by universities, research institutes, and business firms to develop improved plants and horticultural practices.

GROWING INDOOR PLANTS

Growing these plants is much like growing outdoor plants. Consideration needs to be given to the environment in both cases. Tomatoes will not grow and produce in the desert, nor will sweet corn grow in the shade of a tree.

Both need full sun, but cannot survive in extreme temperatures. Aspects of the environment that must be considered are light, temperature, water, drainage, and feeding conditions. While indoor plants are not harvested like tomatoes or sweet corn, they need to be maintained as well. Some aspects of maintenance that need to be addressed are grooming, repotting, and propagation.

Light

Light is one of the most crucial factors to consider when growing indoor plants. Two aspects of light that need to be determined are intensity and duration. The intensity of light varies at different times of the day and through different windows. A plant near a window with a southern exposure receives more intense sun, and for a longer period of time, than does a plant in a window with a northern exposure. Pulling shades across a window can reduce the intensity of light, and a tree outside a window can reduce both the intensity and duration of light that enters the window. The season of the year also affects the light's intensity and duration. The summer sun is much more intense than winter sun, since it is closer to the Earth during the summer.

Too little light will cause some indoor plants to grow tall and leggy and lose leaves because the long, thin, weak stem can no longer support the plant due to limited photosynthesis. On the other hand, too much light causes indoor plants to wilt and lose their vibrant, green colors. The youngest leaves on plants are affected first by unfavorable conditions. Another item concerning light is the tendency of plants to grow toward the strongest source of light. If you notice that a plant is leaning toward the light, you may need to rotate the plant periodically so that it maintains a balanced shape.

There are five basic light categories for indoor plants. These are full sun, some direct sun, bright indirect light, partial shade, and shade. Full sun is a location that receives at least 5 hours of direct sun a day. This amount of sun can be found in areas that have a southern exposure. Some direct sun occurs in an area that is brightly lit but receives less than 5 hours of direct sun a day. These areas are windows facing East and West.

Bright indirect light describes areas that receive a considerable amount of light but no direct sun.

1

Today's agriscience student must be as comfortable in the laboratory as in the field; they are on their way to becoming the **AGRISCIENTISTS** *of our future.*

2

3

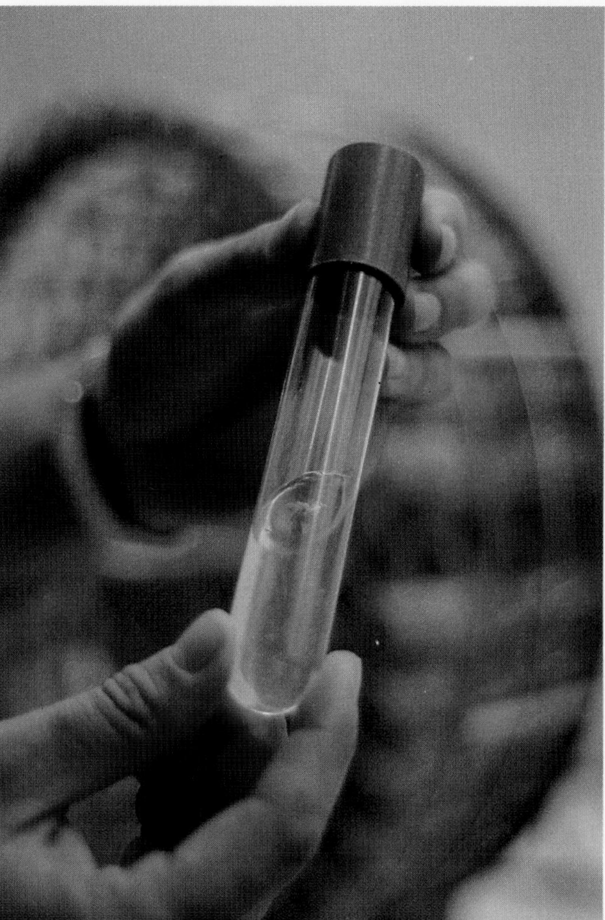

4

The computer is a key component of both agriscience and agribusiness. Today's student learns to use the computer as an aid in all areas of agriculture.

5

6

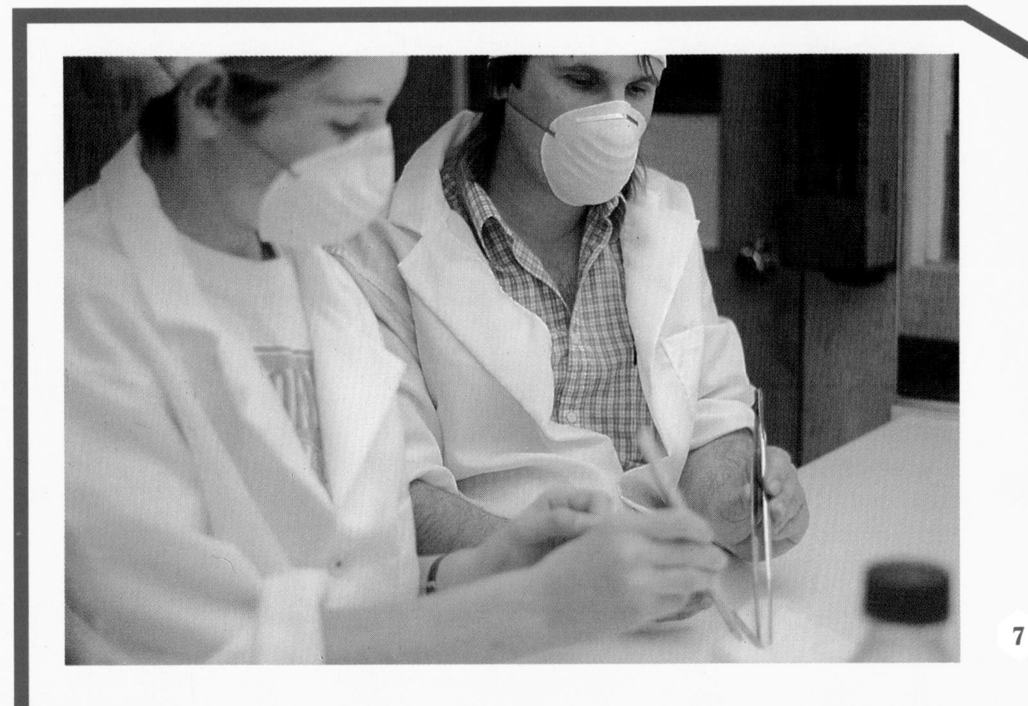

7

Many careers are open to the agriculture student of today and tomorrow. From laboratory research assistant to agriculture engineer — the possibilities are endless.

8

Environmental science is one of the fastest growing agriculture concerns. Soil conservation techniques can be both decorative and functional (9). Water resources are becoming more and more important as we plan for future prosperity (10).

9

10

Students learn plant science through both classroom study and hands-on experience.

11

12

13

Landscaping and ornamental horticulture are "fast-growing" career paths for agriscience students.

14

15

Greenhouse and nursery operations require careful planning and construction. This operation exemplifies both.

16

Careful pruning and training of certain plants will ensure good growth habits.

17

18

19

The study of animal science insures that our livestock is raised as efficiently and humanely as possible. The polled Hereford (Fig. 19), the Santa Gertrudes (Fig. 20), the Charolais (Fig. 21), and the Limousin (Fig. 22) are just a few of the many excellent breeds of cattle developed through animal science techniques.

20

21

22

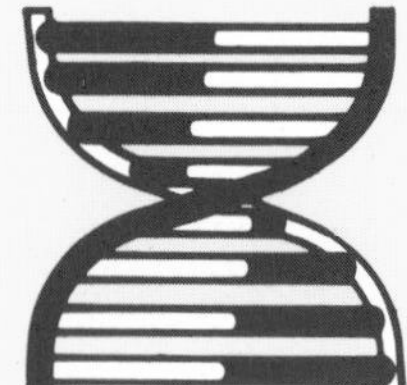

Computers are essential for monitoring biotechnology practices such as genetic engineering, hormonal implants, and embryo transfer as well as for production practices such as milk output and feed calculations.

An example is an area 5′ away from a window that receives full sunlight. Partial shade refers to areas that receive indirect light of various intensities and durations. Areas 5–8′ away from windows that receive direct sun are partial-shade areas. Shade refers to poorly lit sections away from windows that receive direct sun. Very few plants can survive this low-intensity light.

Temperature Plants survive best at constant temperatures. Temperatures that go up and down are not ideal for plant growth. Temperature interacts with light, humidity, and air circulation. It is best to maintain temperatures in a range from 60 to 68°F for optimum indoor plant growth. As temperatures increase beyond this range, the air gets hotter and the available moisture in the air decreases. Although the thermostat in a house reads one temperature, each room usually varies as much as five degrees in one direction or the other. Since plants have various temperature preferences, it is advisable to place plants in rooms that match their specific temperature requirements.

Water Water is an essential ingredient for growth of any living organism. The amount of water needed by indoor plants is usually not as much as one would think. The most likely problem with unhealthy indoor plants is too much water. More plants die from overwatering than from any other cause. As with other environmental factors such as light and temperature, each plant varies in its requirements for water. Unless you know the particular needs of the plant, it is best to water when the soil around the plant is a little on the dry side. Most plants do best if allowed to dry out between waterings. When indoor plants are watered properly, the roots remain more active than if the soil becomes waterlogged or excessively wet.

There are numerous ways to water plants. The basic rule about watering is to wet the soil in the pot until the excess water drains from the bottom. One way to water is to soak the pot in a bucket of warm water for half an hour, remove the pot, and drain it. A second way is to pour the water on top of the soil slowly, filling the pot to the top with water. Allow it to absorb the water until the excess drains from the hole in the bottom of the pot. Then do not rewater until the soil first becomes dry to the touch.

Besides water in the media around their roots, plants also need moisture in the air. The moisture content in the air is referred to as humidity. It is expressed as relative humidity (a relationship between the amount of water vapor in the air compared to the maximum moisture the air will hold at a given temperature). Almost all indoor plants prefer 50% relative humidity. The simplest method for humidifying the air around plants is to set pots in trays filled with gravel and add water to just cover the gravel. Misting is a good shortcut way to add humidity, but it should be done several times a day to be effective. The humidity around plants can also be increased by grouping them together. However, plants should have enough space around them to allow for adequate air circulation.

Drainage Good drainage is achieved by using pots with porous materials in the bottom and drainage holes in the pot. Good drainage is essential for indoor plants. Adding coarse material such as sand or perlite to the soil will improve soil drainage. The addition of gravel or bits of broken clay-pot material to the bottom of a pot before adding the soil is a substantial aid to good drainage.

Fertilizing Plants need nutrients on a regular basis for good health. A balanced fertilizer, such as a 5–5–5, should be applied at regular intervals. Fertilizers with a higher proportion of nitrogen than of phosphorus and potash are often used to keep foliage plants green and healthy looking. Plants should be fertilized at 2- to 6-week intervals, depending on the type of fertilizing material. Slow-release-type fertilizers dissolve slowly and release nutrients evenly over weeks or months. On the other hand, liquid fertilizers suitable for foliar application are used by the plant within a few days. They may also be applied at weekly or biweekly intervals.

Plants should be fertilized while actively growing. Fertilization should be discontinued when the plant is *dormant* or in a resting stage. Flowering plants need more fertilizer. Addition of fertilizer once every 2 weeks is recommended from the time flower buds first appear until the plants stop blooming. Avoid fertilizing plants when the soil is excessively dry. Under such conditions, the fertilizer solution is likely to be too highly concentrated and may cause burned leaves or roots.

Grooming Grooming plants is important even with the best combination of light, temperature, humidity, water, and drainage. Grooming should be done weekly. This task includes removing wilted or withering leaves, flowers, and stems with sharp scissors or shears. The plant should be observed to determine if it is getting leggy or thin. If so, pinch out new growth to force the plant to branch out.

To avoid crooked stems, stake plants when they are young. Climbing or trailing plants need a stake made of bark to enable them to climb and cling to the stake.

Once a week, the plant should be dusted. Dust accumulates on the leaves and blocks the stoma so that the plant cannot breathe or transpire as well as it should. The soil around the base of the plant should be loosened with a fork or small spade to allow air to enter the soil and water to percolate through. To help control insects, mist infested plants with a diluted solution of mild dishwashing soap and water.

Repotting In time, plants develop root systems that are restricted by the pot or container. When this happens the plant is said to be *rootbound*. There is no place for the roots to continue growth, so repotting is necessary for plant health. In general, repotting is done in the spring or the fall. A good rule of thumb for determining pot size is to use one whose diameter at the top of the rim is equal to one-third to one-half the height of the plant. This rule does not apply to plants whose growth habit is tall and slender as opposed to a balanced top growth. When repotting, it is not desirable to move an established plant to a new pot that is more than 2″ wider than the original pot. Excessively large pots result in wasted soil, water, and nutrients.

Flowering plants are best repotted after the flowers have faded. During repotting, check the roots for insects and root damage. Remove any roots that look or feel unhealthy.

Propagation Propagating most indoor plants is relatively easy. The method chosen depends upon the type of plant (whether it is herbaceous or woody, flowering or foliar). Both sexual and asexual methods of propagation are utilized. In sexual propagation seeds may be started in containers with a good potting soil, plenty of moisture, and adequate air circulation. Pots kept in warmer areas of a house increase the speed of seed germination. Covering the pots with glass or plastic held up by stakes will also aid in the germination process.

Some popular methods of asexual propagation of indoor plants are leaf and stem cuttings, removal of plantlets from parent plants, and air layering. Each of these procedures results in the multiplication of the plant.

Flowering Plants Indoor flowering plants may take some extra, special care to ensure blossoming. These plants are more sensitive to the availability of light, so some artificial lighting may be needed. They are also more susceptible to temperature changes. Flowering plants have a particular season in which they flower. It is important to know when to expect the plants to bloom. Some examples of plants with specific bloom times are the Christmas cactus (*Schlumbergera x buckleyi*), which is expected to bloom between October and late January, and the Easter cactus (*Rhipsalidopsis gaertneri*), which blooms in April or May. These two plants are very similar in their leaf types and flowers, but they bloom in opposite seasons of the year. Indoor plants may flower in the winter, spring, summer, or year-round.

Foliage Plants There is such a wide variety of foliage plants grown in homes and offices that specific instructions for their care comprise entire books. However, the correct management of light is probably the greatest single factor for growing foliage plants. Light is the source of energy for the process of photosynthesis, wherein the leaves produce sugars and starches to feed all parts of the plant. Different foliage plants have different light requirements. Therefore, it is recommended that a good reference book on indoor plants be used to determine the particular requirements for any given plant. Regardless of the amount and duration of light, it is desirable to rotate plants so that each side receives the same amount of light over time.

INTERIOR LANDSCAPING OR PLANTSCAPING

Plantscaping is the design and arrangement of plants and structures in indoor areas. This design and arrangement is an art. Interior plantscaping is an activity that is fun. However, it should be approached seriously just as the arrangement of furniture or other interior decorations must be done carefully. The indoor plants should be used to complement people-oriented spaces. The more creative you are in designing, the more distinctive indoor plantscapes will be. There is no right or wrong way to design with indoor plants, but there are elements that enhance the interior plantscaping process.

Design Before beginning a plantscape design, the designer must know the purpose or intent of the plants. Are the plants to be used as a space divider? Are they being used to accent existing furnishings? Will they fill empty space? An answer to these questions will result in an organized design, rather than a happenstance.

Another question to consider is the function or functions of the plants. Are they to create a specific shape, emphasize a specific area, or support a specific architectural feature in the room? If the function of the plantscape is not considered, the end result is not likely to be successful.

When selecting plants for a plantscape, consider the physical characteristics of color, form, and texture. These are determined in concert with the perceptual characteristics of accent, sequence, balance, and scale. These characteristics are the basic tools a designer uses to create the interior plantscape.

Color This is the most important physical characteristic. It can influence emotions, create specific feelings, and add beauty and harmony to the environment. There are two types of color that must be considered by the designer in creating an interior plantscape. First is the basic or background color. *Basic color* is the color of the walls, ceiling, and floor. These colors should influence the selection of the flowering characteristic and foliage of plant material. The second type of color is accent. *Accent color* is the color of the plants or other attention-getting objects.

Form Form refers to shape. There are different forms that plants possess naturally. The most common shapes are round (Figure 23-12A), oval (Figure 23-12B), weeping or drooping (Figure 23-12C), upright, spiky (Figure 23-12D) and spreading or horizontal. These shapes can be used individually or grouped to form an artificially sculptured shape and form. Each plant shape or cluster shape adds its own particular feature to the design.

Texture Plant *texture* refers to the visual or surface quality of the plant or plants. Texture is influenced by the arrangement and size of leaves, stems, and branches. It is described in terms of coarseness or fineness, roughness or smoothness, heaviness or lightness, and thickness or thinness. Coarse-textured plants, such as fiddleleaf fig or prickly pear cactus, should be used in large spaces, whereas a maidenhair fern should be placed in small spaces such as on shelves.

Accent and Sequence *Accent* means a distinctive feature or quality. An accent captures the attention of the viewer. It has a dramatic effect on the visual appearance of the room or part of the room. You can create an accent with the use of color, form, or texture. It can also be created through the use of a sequence. *Sequence* refers to a related or continuous series. Sequence is created with plant material by repeating the same color, texture, or form. The overuse of accents, however, will detract from their function to capture the attention of the viewer.

Balance and Scale *Balance* is the state of equality and calm between items in a design. In a design, two types of balance are utilized—formal or symmetrical and informal or asymmetrical. *Formal* or *symmetrical* balance occurs when the items are equal in number, size, or texture on both sides of the center of the design. *Informal* or *asymmetrical* balance occurs when the items are not equal in number, size, or texture on the two sides of the center. *Scale* refers to the size of items. A large patio with African violets as accents would be out of scale. Plant materials need to complement the size and therefore the scale of the room. A better selection for a patio would be Norfolk Island pines and dracaenas.

A Design Process There are various types of design processes. One that is good to use has three phases. The three-phase process emphasizes the how and why of using plants for the particular space.

Phase One The first phase of the process is preplanning, which includes three steps. The first step is to develop the design objectives by determining the purpose of the plantscape. The second step is to determine the space capacities, which include the habitat and the circulation of people. It is best to know how people will move in the area, referred to as *circulation*, and the amounts of light, temperature, and humidity in the room, referred to as habitat. The third step is to determine the development limitations. *Development limitations* include the amount of money available, room and space characteristics, and habitat limitations created by light, temperature, people, pets, and others.

Phase Two The second phase is developing the plan. The plan will include the basic design and arrangement of plants and materials. The physical and perceptual characteristics of the plants are very important in this phase. Tentative selection of plants with visual placement in the room is part of this step. The planning is done on paper, and many alternatives are examined. The final plan is drawn up with all the plants identified.

Phase Three Implementation is the third phase of the interior plantscaping process, where the design comes to life. There are three steps to this phase. The first step is the preparation of the documents. If there is any construction involved, such as platforms, decks, or planter boxes, drawings for these objects must be done. Drawings and specifications for the installation and maintenance of water lines and fixtures, electrical devices, and the plants should also be listed at this point. This step is very important if you are designing for someone other than yourself. The second step in this phase is the installation of all physical modifications. Selection of plant containers and the planting of the indoor plants happens in this phase. The

A

C

B

D

FIGURE 23-12
Physical characteristics of form *(Courtesy of Calloway Nurseries, photos by Michael Gill)*

plants are actually put in place. The final step is evaluation of the project. A thorough look at the plants, their containers, and their position in the room or area must be done. If all looks balanced, no changes need to be made. If not, minor adjustments may be made. The interior plantscape is now complete.

Plantscape Maintenance Maintenance of the plantscape with attention to light, moisture, humidity, and grooming is important. These factors are discussed in preceding parts of this unit.

CAREERS IN INDOOR PLANTSCAPING Interior plantscaping is a career field within the large industry of horticulture. Horticulture is the study of plants, especially garden crops, both indoor and outdoor, for human consumption, for aesthetic purposes, or for medicinal purposes. Interior plantscaping is a part of floriculture, which is a division of horticulture. Plantscaping is both a science and an art. It is a career area that allows for creativity and use of scientific knowledge and technology.

The opportunities for a career in indoor plantscaping are numerous. An individual may be involved in growing indoor plants, designing interior plantscapes, installing plantscapes, maintaining plantscapes, or selling or servicing plantscape materials. The plantscapes can be small-scale within a residence or home, or large-scale in public areas such as shopping malls or office buildings. This area of horticulture offers many individuals the chance to be entrepreneurs with a relatively low investment. The education required for a career in indoor plantscaping varies. Persons with a high school diploma, a two-year or four-year college degree, or a four-year degree can be successful in this area.

Student Activities

1. Take an inventory of species, types, and numbers of indoor plants in your home.
2. Select indoor plants from reference books and seed or plant catalogs that are suitable for your home.
3. Develop a plan for an interior plantscape of a room in your home, the school office, or another location.
4. Repot an indoor plant.
5. Propagate indoor plants using one or more methods.
6. Perform a shadow test at several windows in your home or classroom to determine the intensity of the light available.
7. Prepare hanging baskets for use in the home or office.
8. Take a trip to a local nursery, florist shop, or grocery store and identify the indoor plants.

Self-Evaluation

A. MULTIPLE CHOICE

1. One of the most popular flowering indoor plants is the

   a. spider plant.
   b. petunia.
   c. African violet.
   d. Norfolk Island pine.

2. An example of an indoor tree is a

   a. philodendron.
   b. cactus.
   c. fiddleleaf fig.
   d. zebra plant.

3. When purchasing an indoor plant, check the underside of the leaves for

a. powdery mildew.
b. insects.
c. price tags.
d. brown spots.

4. The moderate temperature range for indoor plants is

a. 60 to 70°F.
b. 45 to 55°F.
c. 90 to 100°F.
d. 35 to 45°F.

5. If an indoor plant needs direct or full sun, then the plant will need sun for at least

a. the entire day.
b. one hour.
c. one-fourth of the daylight hours.
d. one-half of the daylight hours.

6. The place in a home not usually considered an ideal location for an indoor plant is the

a. bathroom.
b. basement.
c. kitchen.
d. living room.

7. Which of the following is not considered an environmental factor when growing indoor plants?

a. weather
b. light
c. temperature
d. humidity

8. When watering plants, the ideal temperature for the water is

a. ice cold.
b. hot.
c. tepid.
d. cold.

9. Which of the following is not a physical characteristic of an indoor plant?

a. texture
b. form
c. color
d. balance

10. Interior plantscaping is a career area in which area of horticulture?

a. agronomy
b. floriculture
c. forestry
d. arboriculture

B. MATCHING

_____ 1. *Ficus lyrata*
_____ 2. Gardenia
_____ 3. Croton
_____ 4. Wandering Jew
_____ 5. jade plant
_____ 6. *Asparagus sprengeri*

a. A fern-type foliage plant
b. Has unusually colorful leaves
c. A succulent-type foliage plant
d. A fragrant flowering indoor plant
e. A tree-type indoor plant
f. A vine-type foliage plant

C. COMPLETION

1. The two environmental factors that are most important to the survival of indoor plants are _____ and _____ .

2. An indoor plant that is a tree can be identified by its usually _____ stem.

3. The bathroom is a great place for indoor plants because of the _____ usually found in a bathroom.

4. Indoor plants survive best at a ______ temperature.
5. More indoor plants die from ______ than any other cause.
6. Adding a coarse material such as sand to the soil will improve an indoor plant's ______ .
7. A 5-5-5 fertilizer is an example of a ______ fertilizer.
8. When repotting an indoor plant, do not move a plant to a pot that is more than ______ wider than its original pot.

UNIT 24 Turfgrass Use and Maintenance

OBJECTIVE To understand growth and development of turfgrasses and the establishment and cultural practices involved in maintaining these plants.

Competencies to Be Developed

After studying this unit, you will be able to

- ☐ identify and describe careers available in the turfgrass industry.
- ☐ identify turfgrass plant parts.
- ☐ select turfgrass species for various purposes and locations.
- ☐ state the basic cultural practices for turfgrass production and maintenance.
- ☐ list the basic steps for turfgrass establishment.

TERMS TO KNOW

Turfgrasses
Playability level
Putting greens
Seminal roots
Adventitious roots
Warm-season turfgrass
Cool-season turfgrass
Crown
Rhizome
Stolon
Extravaginal growth
Vegetative spreading
Tillers
Intravaginal growth
Seed culm
Inflorescence
Induction
Sheath
Blade
Ligule
Collar
Recuperative potential
Texture
Fine texture
Coarse texture
Cultivar
Bunch-type grass
Utility-type grass
Fungal endophytes
Transition zone
Thatch
Complete fertilizer
Fertilizer analysis
Syringing
Seed blend
Seed mixture
Seed certification programs
Sodding
Turf
Sprigging
Stolonizing
Plugging

THE TURFGRASS INDUSTRY It has been estimated that the value of the turfgrass industry in the United States is $25 billion. This industry is large and diverse. Career opportunities in this field expanded rapidly in the 1980s and are projected to continue growing in the years ahead. *Turfgrasses* are grasses that are mowed frequently to maintain a short and even appearance.

Golf course superintendents and other athletic-field managers maintain turfgrass at a certain level of playability. The *playability level* means its suitability for the intended use. It is determined by the type of sporting or recreational event. For example, *putting greens* are areas used for playing golf, and the turfgrass is very short. These areas are maintained to provide a surface for consistent, yet adequate putting speeds. A football field may be managed so it offers secure footing and sufficient turfgrass resiliency for player safety (Figure 24-1).

Lawn-care services offer many job positions and are one of the largest employers within the industry. Job opportunities include lawn-care specialist, branch manager, owner-operator, and others (Figure 24-2).

The maintenance of turfgrasses at large government, apartment, university, commercial, and private complexes requires personnel trained in turfgrass management.

Sod-production farms and landscaping businesses are involved in the establishment and installation of turfgrass for lawns. Turfgrass specialists are also needed for these segments of the industry (Figure 24-3).

The turfgrass industry supports a substantial sales force. Companies that produce or distribute seed, fertilizer, pesticides, and turfgrass equipment require an extensive support and sales staff (Figure 24-4).

Federal and state governments and private companies hire turf specialists and scientists with advanced degrees. Career opportunities in these areas offer challenging positions in research, teaching, and extension.

TURFGRASS GROWTH AND DEVELOPMENT

Turfgrasses are plants grouped into the *Poaceae* family.

FIGURE 24-1
Sports turf will require maintenance practices that provide an acceptable level of playability. ***(Courtesy Denmark High School Agriculture Education Department)***

FIGURE 24-3
Sod production is an important part of the turfgrass industry. ***(Courtesy Brouwer Equipment, Ltd.)***

FIGURE 24-2
Lawn care services offer many jobs and are one of the largest employers within the turfgrass industry. ***(Courtesy ChemLawn Services Corporation)***

FIGURE 24-4
Equipment demonstration by a sales representative ***(From Emmons/Turfgrass Science & Management, copyright 1984 by Delmar Publishers Inc.)***

These grasses differ from other grass plants because they can withstand mowing at low heights. They can also tolerate vehicle and foot traffic. These traits have made turfgrasses the most widely used ornamental crop in the United States. To properly maintain turfgrasses, an understanding of their growth and development is required.

THE TURFGRASS PLANT

The grass plant can be divided into two broad areas known as the root and shoot systems. The root system consists of adventitious and seminal roots. The shoot system includes the stem and leaves of the plant.

Root System

The *seminal roots* develop from the seed during seed germination. They initially anchor the seed into the soil. The seminal root system will be active for 6 to 8 weeks. The *adventitious roots* develop from the nodes of stem tissue. They usually comprise the entire root system of a mature turfgrass stand.

Turfgrass roots are multibranching and fibrous. They are responsible for nutrient and water absorption. They also prevent soil erosion, since they effectively stabilize and anchor the plant into the soil.

Seasonal change in root growth is dependent upon soil temperature and moisture. Active root growth for warm-season turfgrasses occurs in the summer. A *warm-season turfgrass* is one of a group of grasses adapted to the southern region of the United States. A *cool-season turfgrass* is a plant adapted to the northern region of the United States.

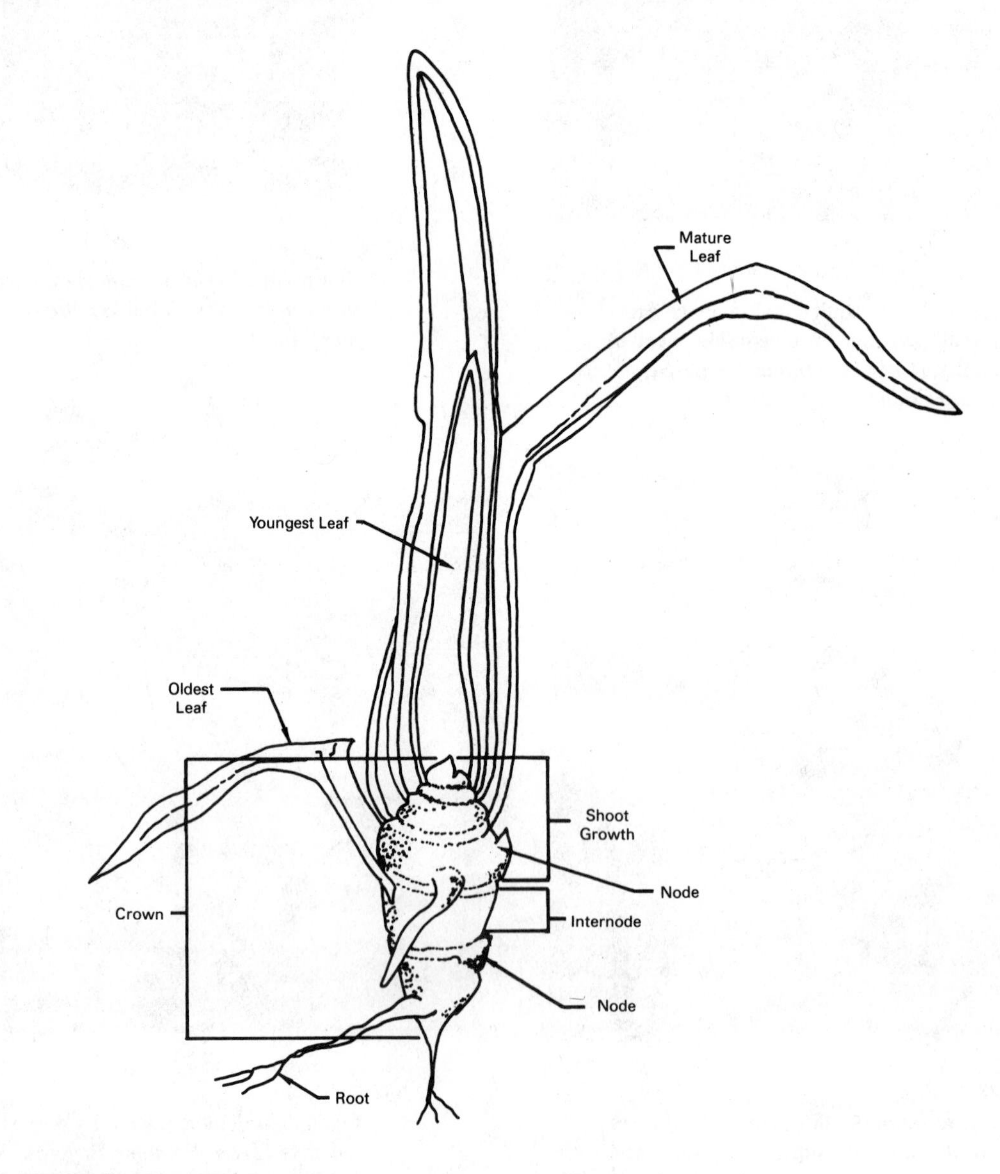

FIGURE 24-5
The crown of a turfgrass plant

These turfgrasses have active root growth in the fall and early spring. Optimum growth occurs at temperatures of 60–75°F.

Rooting depth is affected by plant species, soil factors, and cultural or maintenance practices. Average rooting depth for turfgrasses is 6–12″. The warm-season grasses will have deeper root systems than the cool-season grasses. Well-drained, sandy-loam soils with neutral soil pH provide an optimum growing medium for turfgrass roots.

Cultural practices that influence rooting depth and growth are mowing, fertilization, and irrigation. Frequent and low mowing heights will reduce rooting depth. In addition, fertilization programs that emphasize only shoot growth will impair root growth. Light and frequent irrigation will cause shallow-root grasses. Heavy, but infrequent irrigation will be more conducive to deep root growth.

Turfgrass maintenance practices should attempt to optimize the rooting potential of turfgrass plants. An extensive root system allows a plant to recover from drought and other stress conditions more rapidly.

Shoot System The shoot system consists of stems, leaves, and seed head or inflorescence.

Stems

Turfgrass stems include the crown, tillers, rhizomes, stolons, and seed culms. The crown is the major growth or meristematic tissue of the grass plant. The *crown* is a stem with the nodes stacked on top of each other (Figure 24-5). All root, leaf, and other shoot growth originates from this area. The crown is located at the base of the grass plant in the soil surface area.

Rhizomes and stolons are horizontal stems. A *rhizome* is a creeping underground stem, whereas a *stolon* is an aboveground stem (Figure 24-6). They both originate from an axillary bud on the crown and will penetrate through the lower leaf sheath. This type of growth is referred to as *extravaginal growth*. Rhizome and stolon growth allows for vegetative spreading of turfgrasses. *Vegetative spreading* means reproduction by plant parts other than seed.

Tillers

Tillers are new shoots of a grass plant that develop at the axillary bud of the crown. They form within the lower leaf sheath of the plant (Figure 24-7). This type of growth is referred to as *intravaginal growth*. Increased tillering will enhance turfgrass density. All turfgrasses produce tillers. Optimum tillering for the cool-season grasses occurs in the spring and fall months. Warm-season grasses have optimum tillering during the summer. Under low-moisture and high-temperature stress conditions, tiller, rhizome, and stolon development is reduced.

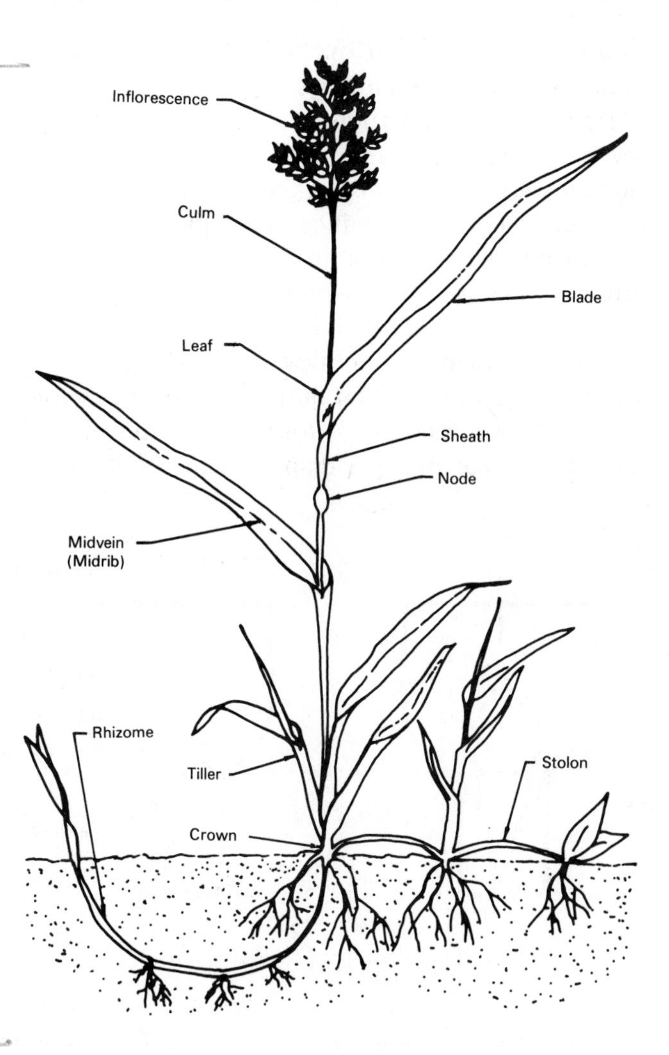

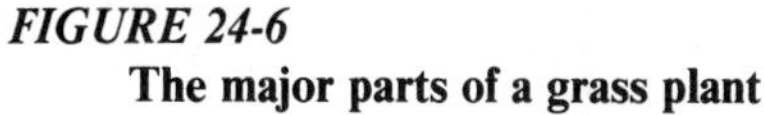
FIGURE 24-6
The major parts of a grass plant

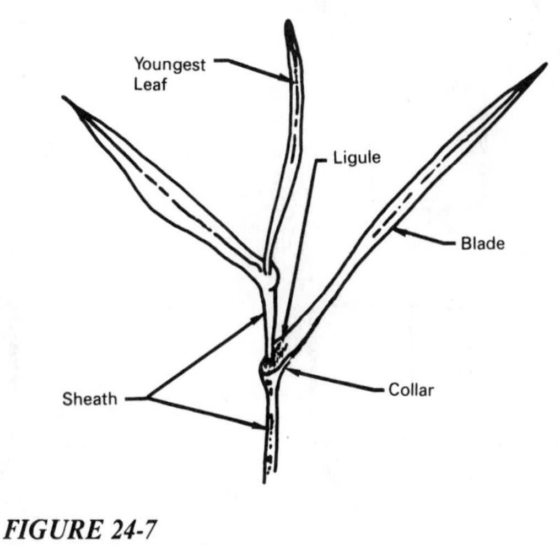

FIGURE 24-7
The leaf blade and leaf sheath of a grass plant

Seed Culm and Inflorescence The *seed culm* or seed stem supports the inflorescence of the plant. The seed culm originates at the top of the crown (Figure 24-6). *Inflorescence* is the arrangement of and flowering parts of a grass plant. Cool-season grasses produce their inflorescence in the spring. Warm-season turfgrasses produce their inflorescence in the late summer.

Flower *induction* or initiation is caused by several environmental conditions. Temperature and photoperiod are the two induction processes for grasses. When seed heads form, they will cause a decrease in playability and appearance of the turfgrass stand. Mowing will remove the seed head and thus improve turfgrass quality.

Leaf The turfgrass leaf consists of the sheath and blade (Figure 24-7). The *sheath* is the lower portion of the leaf and may be rolled or folded over the shoot system. The *blade* is the upper portion of the leaf. At the junction of the blade and sheath is the collar and ligule (Figure 24-8). The *ligule* is located on the inside of the leaf and is a membranous or hairy structure. The *collar* can be found on the outside of the leaf and is a light-green or white-banded area (Figure 24-8). These two fea-

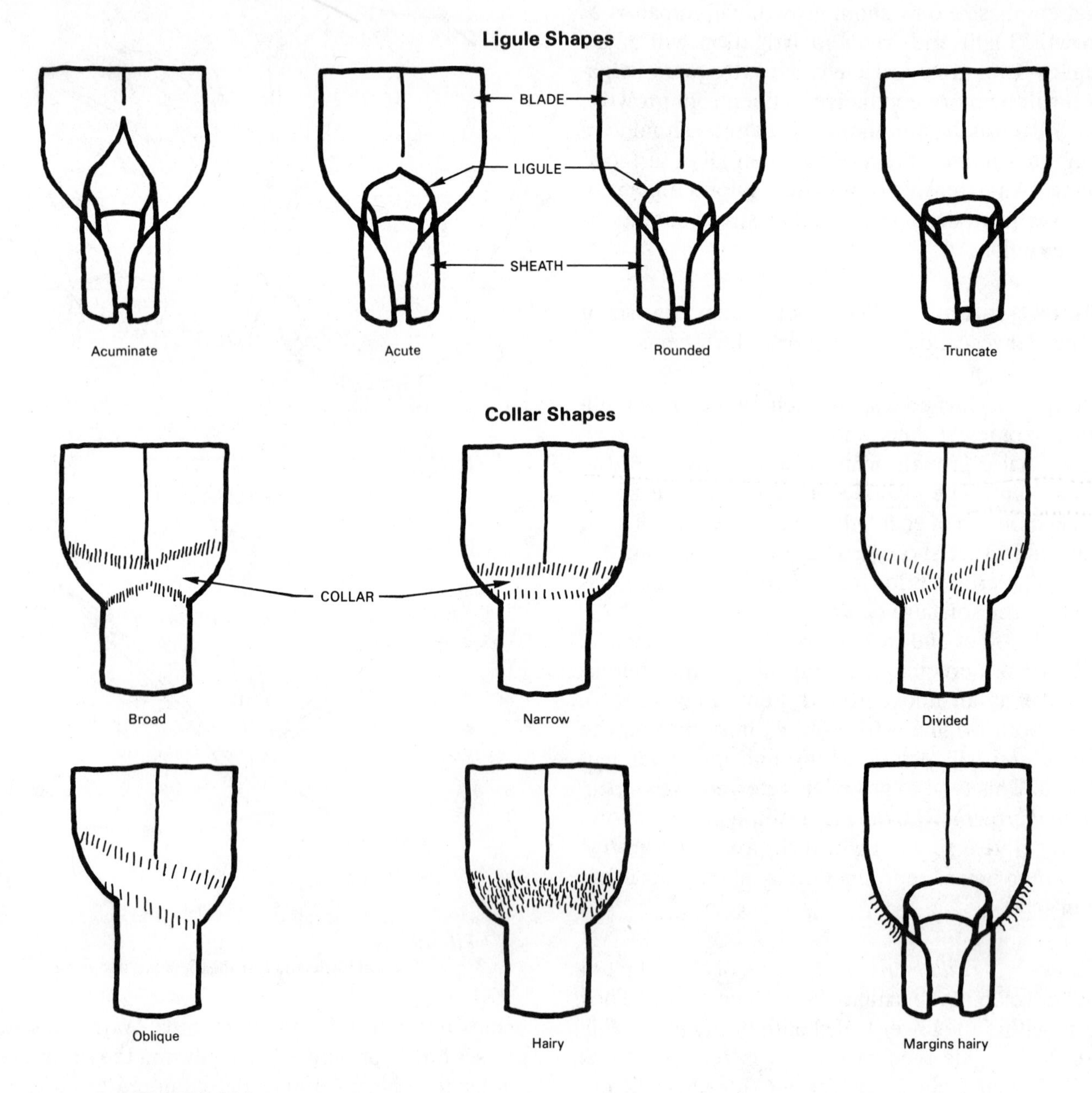

FIGURE 24-8
Ligules and collars of a grass

tures are important vegetative traits for turfgrass identification.

Turfgrass growth is dependent upon the production and utilization of carbohydrates. The turfgrass leaf is responsible for photosynthesis and, ultimately, carbohydrate production. Reserve carbohydrates will be stored in crown, rhizome, and stolon tissue.

During unfavorable growth conditions, the plant will go into dormancy. All leaf tissue will die back. However, when favorable environmental conditions occur, carbohydrate reserves from the crown and other stem tissue will be used for plant regrowth.

Turfgrass maintenance programs attempt to optimize root growth and carbohydrate accumulation. Greater recuperative potential and plant persistency occur when these two basic concepts of growth are understood and managed. *Recuperative potential* is the ability of a plant to recover after being damaged.

TURFGRASS SPECIES

There are some 7,500 plants classified as grasses. Only a few dozen are considered for turfgrass use. Turfgrasses are divided into two major groups based on climatic adaptation. The two groups are cool-season and warm-season grasses (Figure 24-9).

Cool-Season Turfgrasses

These turfgrasses originated in Europe and Asia and have optimum growth at temperatures from 60 to 75°F. They predominate in the northern and central regions of the United States. Species adaptation within the cool-season group is determined by rainfall, soil fertility, and turf use.

COOL SEASON	WARM SEASON
Kentucky bluegrass	Bermudagrass
Red fescue	St. Augustinegrass
Tall fescue	Bahiagrass
Perennial ryegrass	Centipedegrass
Creeping bentgrass	Zoysiagrass
Crested wheatgrass	Carpetgrass

FIGURE 24-9
A list of the cool and warm season turfgrasses present in the United States

Kentucky Bluegrass (*Poa pratensis* L) This is used extensively in residential and commercial lawns, in recreational facilities, and along highway

AGRI-PROFILE

CAREER AREAS: GOLF COURSE SUPERINTENDENT/TURFGRASS GROWER/GROUNDSKEEPER

Palettes of turf are inspected by this grower before delivery to the job site where attractive turfgrass will add to the beauty, function, and value of the area. ***(Courtesy of the National FFA Organization)***

In the last three decades, turfgrass production and management has become big business in America. In one eastern state, turfgrass recently became the number one crop based on acres covered. Starting in the fifties, growth and development of golf courses stimulated the turfgrass industry with high-paying salaries for golf course superintendents and other turfgrass specialists.

Today, career opportunities in turfgrass production, management, service, supervision, research, and consultation are extensive. Turfgrass technicians and specialists generally work in attractive and appealing surroundings. Many work outdoors in sunny weather and indoors when the weather is bad. In many localities, salaries have become quite attractive even for laborers.

Educational programs in turfgrass are available in high schools, technical schools, colleges, and universities. With the movement towards urbanization, interest in open spaces, concern for the environment, and increasing population, the outlook for careers in turfgrass production and management is excellent.

rights-of-way. It performs well with moderate levels of maintenance. The plant has a medium leaf texture and an extensive rhizome system. *Texture* refers to leaf width. *Fine texture* turf contains grasses with narrow blades, whereas *coarse texture* is turf with wide-blade grasses.

Kentucky bluegrass grows best with full sun, moist and fertile soil, and a mowing height of 1½–2½″. More than 100 cultivars of Kentucky bluegrass have been developed for certain geographic areas and specific maintenance conditions. A *cultivar* is a plant of the same species that has been discovered and propagated because of its unique characteristics. Cultivar differences exist with respect to disease tolerance, leaf width, color, and other traits.

Tall Fescue (*Festuca arundinacea* Screb) Tall fescue is a coarse-textured, bunch-type grass used in home lawns or as a utility-type turf. A *bunch-type grass* grows in clumps rather than spreading evenly over the ground. However, the bunching tendency of tall fescue can be overcome by heavy seeding, which produces thick stands of grass. A *utility-type grass* refers to a turfgrass adapted to lower maintenance levels. Tall fescues have an extensive root system and are one of the most drought-tolerant, cool-season species. However, they are prone to winter injury in the northern range of the cool-season zone.

Recent breeding work has introduced many new and improved cultivars of tall fescue. These new cultivars have medium leaf texture, with more aggressive rhizome development. This turfgrass will not tolerate low mowing heights and should be mowed at 2½–3″. Insect resistance and plant persistency is excellent for tall fescues infected with fungal endophytes. *Fungal endophytes* are microscopic plants growing within a plant.

Red Fescue (*Festuca rubra* L) Red fescue is a fine-textured turfgrass well adapted to shady, dry locations. It has excellent drought tolerance and can persist on rather infertile soils and under acid soil conditions (pH 5.5–6.0).

Red fescue is often seeded in mixtures with Kentucky bluegrass for lawn turf. It is not used on athletic fields as a permanent turf since it has a poor recuperative potential. However, in the South it may be overseeded into dormant Bermuda grass greens to provide winter play endurance and color.

Red fescue functions satisfactorily if mowed at 1.5–2.5″ and provided with minimal levels of nitrogen fertilizer and water. The major pest problem of red fescue is leaf spot disease.

Perennial Ryegrass (*Lolium perenne* L) This grass has a medium leaf texture and is most often used in seed mixes with other turfgrasses. It has rapid germination and excellent seedling vigor. It is often used in seed mixes to provide soil stabilization during the establishment period.

The perennial ryegrasses are used extensively for recreational turf. They have good wear tolerance and can be rapidly established if the turfgrass stand is damaged. They are also utilized in winter overseeding programs for warm-season turfgrasses.

Perennial ryegrasses require a moderate level of maintenance to form an attractive turf. As a group, they have poor disease resistance. Improved plant persistence and insect resistance occur if they are infected with fungal endophyte.

Creeping Bentgrass (*Agrostis palustris* Huds) Used in close-cut, high-maintenance areas, this is an extremely attractive turfgrass and fits well when used on putting greens or bowling greens mowed at 5/6″. It can be maintained at higher mowing heights (½–¾″) for use on football fields, golf course fairways and tees, lawns, and formal gardens.

The plant has extensive stolon growth and is a fine-textured turfgrass. It is best adapted to slightly acid soils with a pH of 5.5–6.0 that have good internal drainage and are not prone to compaction. Creeping bentgrass has excellent cold tolerance, but poor heat tolerance.

Creeping bentgrass requires a high level of maintenance to produce a quality turf. It requires proper irrigation, disease control, mowing, and cultivation practices when grown as a sports turf. Because of these high-maintenance requirements, creeping bentgrass is not recommended as a lawn turf.

Warm-Season Turfgrasses These turfgrasses originated in Africa, North and South America, and southeastern Asia, but they are adapted to the southern United States. They make optimum growth at temperatures of 80–95°F. These grasses go into winter dormancy as temperatures drop below 50°F. There are some 14 species of warm-season turfgrasses found throughout the world.

Bermuda Grass (*Cynodon dactylon* L) This grass is considered the most important and widely

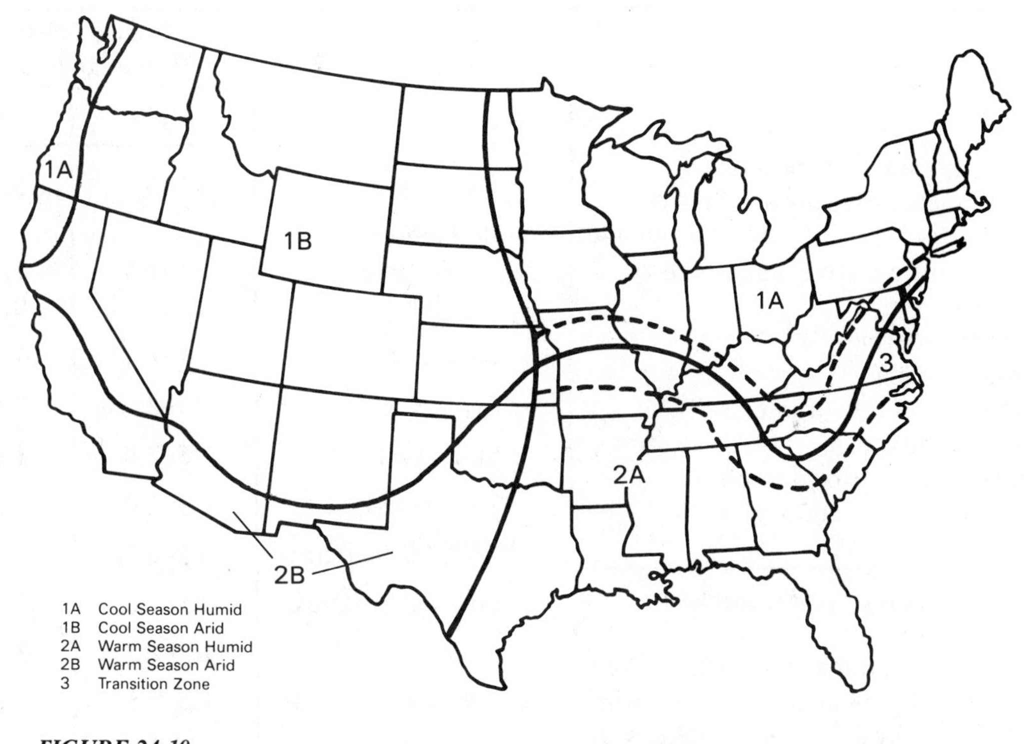

FIGURE 24-10
The major turfgrass adaption zones

used warm-season turfgrass in the United States. It is principally used as a lawn turf and sports turf. Improved breeding lines have provided fine-textured Bermuda grasses capable of being used on putting greens and fairways. The common type of Bermuda grass is a medium-textured turfgrass used for airport runways, rights-of-way, and other low-maintenance areas.

The plant spreads by both stolon and rhizome growth. It has excellent wear resistance, recuperative potential, and drought tolerance. Bermuda grass can persist on a wide range of soil types and soil pH. It has poor shade tolerance and low winter hardiness.

The improved types of Bermuda grass require a high level of maintenance and must be vegetatively established. Recommended mowing heights for these grasses range from ¼ to 1″. The common type of Bermuda grass is established by seed, requires a higher mowing height than the improved cultivars, and is often considered a weed in certain situations.

Zoysia Grass (*Zoysia japonica* Steud.) Zoysia grass has excellent low-temperature hardiness and is found in home lawns as far north as New Jersey. It can also be found on golf course fairways and tees within the transition zone. The *transition zone* is a geographic area of the United States where the warm-season and cool-season adaptation zones overlap (Figure 24-10).

Zoysia grass has good drought and shade tolerance. It can survive in a wide range of soil types. However, this plant will not perform well under poorly drained soils that remain waterlogged. Though zoysia grass has excellent wear resistance, it has a low recuperative potential. It is not as aggressive as Bermuda grass.

Zoysia grass requires a low to moderate level of maintenance. It is vegetatively established, since seed germination is poor. New techniques to increase seed germination are presently being evaluated. This grass should be mowed between 1 and 2″. Two major pest problems of zoysia are nematodes and billbugs.

TURFGRASS CULTURAL PRACTICES Mowing, irrigation, and fertilization are the most common and most important cultural practices performed to maintain turfgrass stands. These practices have tremendous effects on turfgrass quality and persistency. Improper mowing, fertilization, and irrigation are major causes of poor lawns (Figure 24-11).

Mowing Mowing will influence the functional use, persistency, and aesthetic value of a turfgrass.

1. Using the wrong turfgrass species or cultivars.
2. Using poor quality seed.
3. Mowing the lawn too closely.
4. Permitting excessive growth between mowings.
5. Using too little or too much lime or fertilizer.
6. Improper watering.
7. Too much shade.
8. Droughty or poorly drained soils.
9. Too much traffic.
10. Damage by insects or disease.
11. Improper use of chemicals.

FIGURE 24-11
Major causes of poor quality lawns

Species	Mowing height range in inches and centimeters	
	(in)	(cm)
Bahiagrass	2-4	5.0-10.2
Bermudagrass		
Common	0.5-1.5	1.3-3.8
Hybrids	0.25-1	0.6-2.5
Carpetgrass	1-2	2.5-5.0
Centipedegrass	1-2	2.5-5.0
St. Augustinegrass	1.5-3.0	3.8-7.6
Zoysiagrass	0.5-2.0	1.3-5.0
Creeping bentgrass	0.2-0.5	0.5-1.3
Colonial bentgrass	0.5-1.0	1.3-2.5
Fine fescue	1.5-2.5	3.8-6.4
Kentucky bluegrass	1.5-2.5	3.8-6.4*
Perennial ryegrass	1.5-2.5	3.8-6.4*
Tall fescue	1.5-3.0	3.8-7.6
Crested wheatgrass	1.5-2.5	3.8-6.4
Buffalograss	0.7-2.0	1.8-5.0
Blue grama	2.0-2.5	5.0-6.4

*Some cultivars will tolerate closer mowing.

FIGURE 24-12
Recommended mowing heights for different turfgrasses

Grasses used for recreational purposes must be playable. The playability of an athletic-field turf is principally determined by its mowing height. A uniform turf surface, fine leaf texture, and freedom from weed encroachment require proper mowing height and frequency.

Turfgrasses are capable of being mowed because their growing point, the crown, is located just below or at the soil surface. However, mowing does have several adverse effects on the grass plant. Reduced rooting depth and decreased carbohydrate reserves occur in mowed turfs. Improper mowing practices only accentuate these and many other adverse effects, causing a decline in turfgrass quality.

Mowing Height

The recommended range for mowing heights is listed in Figure 24-12. If one mows below or above this range various problems will occur. Mowing below the desired range will reduce photosynthesis, thus preventing carbohydrate production within the plant. This will cause a reduction in rooting and decrease the recuperative potential of the plant.

Mowing above the recommended height will increase thatch buildup and leaf texture. It will also decrease turf density and appearance. *Thatch* is the buildup of organic matter on the soil around the turfgrass plants. Excessive thatch will cause poor water infiltration, increased disease activity, and decreased rooting depth.

Mowing Frequency

This practice is determined by the mowing height and the growth rate of the plant. No more than one-third of the top growth should be removed per mowing (Figure 24-13). This will allow sufficient leaf area for photosynthesis after mowing. The lower the mowing height, the more frequent the mowing if the one-third rule is observed. For example, creeping bentgrass mowed at 1/4″ will be mowed 5 to 6 times per week. However, tall fescue cut at 3″ will require only one mowing per week. Under ideal growing conditions, grasses will require more frequent mowing than they will under poor growing conditions.

Fertilization

Fertility programs attempt to supply adequate levels of plant nutrients to allow for favorable plant growth. Proper fertilization should increase turfgrass density and color. Lush or succulent growth caused by excessive fertilization should be avoided.

A complete fertilizer is often recommended, with the rate of application based on the amount of nitrogen required per 1,000 ft^2 (Figure 24-14). A *complete fertilizer* consists of nitrogen, phosphorus, and potassium. The *fertilizer analysis* states the percentage of nutrients by weight in the fertilizer. A 10-6-4 fertilizer consists of 10% nitrogen, 6% phosphorus, and 4% potassium. A formula can be used to determine fertilizer amounts based on nitrogen-rate recommendations (Figure 24-15). Nitrogen is the most important element in turfgrass fertilization.

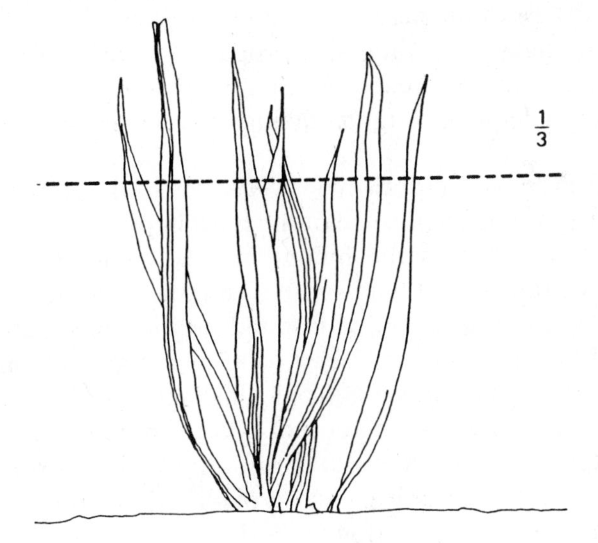

FIGURE 24-13
No more than 1/3 of the top growth should be removed per mowing.

Time to Apply	Nitrogen Applied Per 1,000 Square Feet	Example: Equivalent Fertilizer Per 1,000 Square Feet of Area
September	1 lb	8 to 10 lb 12-4-8 **or** 10 lb 10-6-4 **or** 5 lb 20-10-10
October	1 to 1½ lb	8 to 12 lb 12-4-8 **or** 10 to 15 lb 10-6-4 **or** 5 to 8 lb 20-10-10
November/December or February/March	1 to 1½ lb	8 to 12 lb 12-4-8 **or** 10 to 15 lb 10-6-4 **or** 5 to 8 lb 20-10-10
May-June 20	0 to ½ lb	0 to 4 lb 12-4-8 **or** 0 to 5 lb 10-6-4 **or** 0 to 3 lb 20-10-10

FIGURE 24-14
Fertility requirements and types of fertilizers for a Kentucky bluegrass

Proper timing and application rate are the keys to a successful fertilization program. Timing of fertilization will vary depending upon turfgrass species and geographic location. For cool-season turfgrasses, the fall is the desired time period. Optimum root growth occurs during this time. On the other hand, fertilization of warm-season turfgrasses should be done during late spring or the summer months (Figure 24-16).

Improper fertilization will reduce turfgrass quality. Turfgrasses receiving insufficient amounts of fertilizer will lack color, density, and recuperative potential. Excessive fertilization will reduce heat and drought tolerance, increase disease and insect damage, and cause excessive top growth. Appropriate nitrogen rates range from ¼ to 1 lb. of nitrogen per 1,000 ft^2. The rate is dependent on growth and environmental conditions. Phosphorus and potassium rates should be determined by soil testing.

Irrigation The application of water to turf may accomplish several different objectives. First, sufficient soil moisture will allow for optimum plant growth. Irrigation is also used to establish turfgrass, reduce plant surface temperatures, and rinse in fertilizer and pesticide applications. *Syringing* is a light application of water to a turfgrass. It may be used to reduce plant temperatures or to remove dew and frost from the turfgrass leaf.

The amount of irrigation needed to maintain optimum plant growth is dependent upon many factors. Some factors are turfgrass species, geographic location, soil type, weather conditions, and turfgrass use. General recommendations are to apply 1″ of water per week during the summer months. It takes approximately 620 gallons of water per 1,000 ft^2 to provide 1″ of water. This amount of water will produce a green and actively growing turfgrass stand.

$$\text{Percent in fertilizer analysis} \times \text{Weight of fertilizer} = \text{Pounds of nutrient}$$

FIGURE 24-15
Formula for calculating fertilizer amounts

Heavy, but infrequent irrigation will force root growth deep into the soil. Light and frequent irrigation will keep the surface soil moist, thus encouraging shallow rooting.

The best time to irrigate is at night, when evaporation and wind are low. However, an increase in disease activity will occur at this time. Therefore, early morning watering is often selected since evaporation, wind, and disease activity are at a minimum.

Time to Apply	Nitrogen Applied Per 1,000 Square Feet	Example: Equivalent Fertilizer Per 1,000 Square Feet of Area
April or May	1 lb	8 lb 12-4-8 **or** 10 lb 10-6-4 **or** 5 lb 20-10-10
June	1 lb	8 lb 12-4-8 **or** 10 lb 10-6-4 **or** 5 lb 20-10-10
July	1 lb	8 lb 12-4-8 **or** 10 lb 10-6-4 **or** 5 lb 20-10-10
August	1 lb	8 lb 12-4-8 **or** 10 lb 10-6-4 **or** 5 lb 20-10-10

FIGURE 24-16

Fertility requirements and types of fertilizers for bermudagrass and zoysiagrass

TURFGRASS ESTABLISHMENT

Turfgrasses may be established by seeding or vegetative propagation. Regardless of the method used, proper establishment practices, including site preparation, should be performed. Correct establishment practices will assure adequate turfgrass quality and persistency.

Turfgrass Selection

Selection of the correct turfgrass species is one of the most important decisions in the establishment process. Selecting a turfgrass seed blend or seed mixture should be based on the intended use and performance data of the turfgrasses. A *seed blend* is a combination of different cultivars of the same species. A *seed mixture* is a combination of two or more species. State universities evaluate different turfgrass species and new cultivars and provide information on their performance (Figure 24-17).

Site Preparation

Proper site preparation may include any or all of the following activities:

1. Debris removal
2. Nonselective weed control
3. Installation of a subsurface drainage and/or irrigation system

CULTIVAR	April 1	May 4	June 8	July 15	September 2	October 17
Adelphi	4.0*	5.6	5.4	4.5	5.9	5.7
Baron	3.8	5.1	4.8	4.4	5.4	5.5
Bensun (A-34)	3.6	5.3	6.0	5.9	6.1	5.4
Cheri	3.2	5.2	4.7	4.9	5.8	6.0
Enmundi	4.0	4.9	4.6	4.5	5.4	5.7
Glade	4.2	5.1	5.4	5.5	6.3	6.3
Majestic	3.9	6.1	6.0	5.8	6.0	5.8
Newport	3.7	3.6	3.8	4.0	4.2	4.9
Parade	4.5	6.8	5.9	4.4	5.7	5.8
Ram 1	4.3	6.9	5.9	6.0	6.2	6.4
Sydsport	3.8	6.2	6.0	5.9	6.4	6.1
Touchdown	4.1	6.3	5.9	5.7	6.2	6.5

*It is important to note that quality ratings can vary significantly from one region or location to another. A rating of 1 = no live turf, 9 = ideal turf, >5 = acceptable quality.

FIGURE 24-17

Quality ratings for Kentucky bluegrass cultivars

4. Soil modification, tillage, and grading

The removal of woody vegetation, leftover construction debris, and any large stones will reduce future maintenance problems. If woody debris is buried on the site rather than removed, the soil will settle and diseases such as fairy ring may result. Excessive amounts of stone and rock present in the seedbed will cause localized dry spots and interfere with future cultivation practices.

Nonselective weed control may be necessary if perennial grassy weeds or difficult-to-control weeds are present. A nonselective herbicide or a soil fumigant should be applied prior to seeding. The herbicide glyphosate (Round Up) is often used to provide nonselective weed control. A soil fumigant, such as metham, may also be used.

If the site is to be used as a sports turf, the installation of a drainage and an irrigation system may be required. This should be done prior to final grading.

Soil modifications, followed by tillage and grading, will be the last steps in site preparation. Soil amendments, such as organic matter and sand, may be incorporated to improve nutrient retention and drainage. Topsoil may also be used to provide a favorable growth medium for turfgrasses. The soil should be worked to a depth of 6″ and then lightly tilled to provide a uniform seedbed and adequate surface drainage (Figure 24-18). Incorporation of lime and fertilizer should be done during this time. Excessive tillage will destroy soil tilth. A seedbed with soil aggregates of ¼–1″ in diameter is ideal.

Planting Turfgrasses can be established by seeding, sodding, stolonization, sprigging, and plugging. The last four methods are vegetative means of planting. Vegetative establishment may be done if the grasses produce infertile seed or have low seed yields. Also, if quick establishment is required, sod may be installed.

Seeding Seeding is the principal method for turfgrass establishment, since it is the least expensive. Figure 24-19 shows the seeding rates for the different grasses. These rates will vary depending on seed size and percentage of seed germination.

The seed label informs the buyer or consumer of the quality and ingredients of a seed blend or mixture. Information on purity percentage, germination, weed seed content, inert matter, and test date is present on the label (Figure 24-20). Certification programs are available for turfgrass seed. *Seed certification programs* are administered by state governments, and they ensure that seed is true to type. They also require that the seed meet other minimum quality standards concerning percentage of germination and weed content.

The best time to seed turfgrass will vary, depending on the type of turfgrass. The optimum time

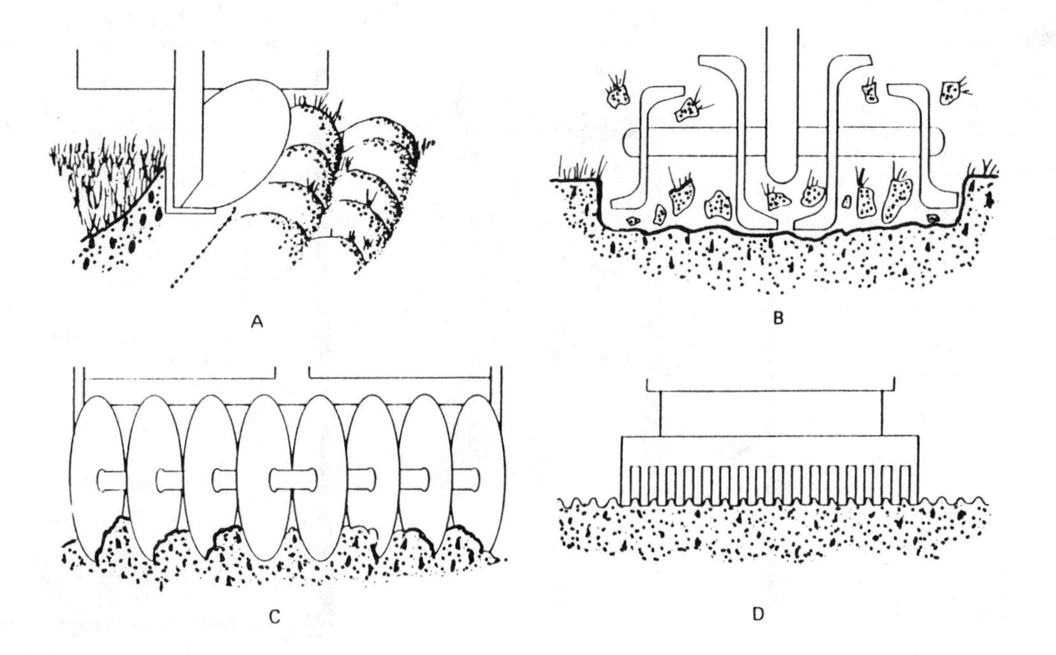

FIGURE 24-18
Tillage equipment for turfgrass establishment: (a) plow, (b) rototiller, (c) disk cultivator and (d) harrow.

Species	Pounds of Seed per 1,000 ft² (93 m²)
Bahiagrass	3–8
Bentgrass	
Colonial	0.5–1.5
Creeping	0.5–1.5
Bermudagrass (hulled)	1–2
Bluegrass	
Kentucky	1–2
Rough	1–2
Buffalograss	3–7*
Carpetgrass	1.5–5
Centipedegrass	0.25–2*
Fescue	
Fine	3–5
Tall	5–9
Grama, blue	1.5–2.5
Ryegrass	
Annual	5–9
Perennial	5–9
Wheatgrass, crested	3–6
Zoysiagrass	1–3

*The higher rates are best but lower rates are commonly used because the seed is expensive.

FIGURE 24-19
Seeding rates for the major turfgrass species

for establishing cool-season grasses is in the fall. Ideal weather conditions and less annual weed competition are the principal reasons for fall establishment. The plant will also have sufficient time to develop an adequate root system prior to summer stress conditions. Spring seeding is less desirable because there is insufficient time to develop a mature stand capable of competing with undesirable grasses.

Warm-season turfgrasses are established during the spring and summer months. The best time is in late spring. Then the new grass has the longest time period for optimum growth and development after establishment.

Different types of equipment may be used to apply the seed. Fertilizer drop spreaders, overseeders, hydroseeders, and cultipacker seeders are some types of equipment used for seeding. Shallow seed placement (¼″) is important for proper germination. This can be done by hand raking for small areas or the use of a drag mat for larger areas. Specialized seeding equipment, such as a hydroseeder or cultipacker, will place the seed at the right soil depth (Figure 24-21).

After seeding, the use of a mulch will provide a more favorable environment for seed germination. Mulch can conserve soil moisture and prevent soil erosion. This is extremely important, since

Recommended Seed Mixture

Suburban Brand Grass Seed Mixture

Lot 410-11
FINE TEXTURED GRASSES
29.96% South Dakota Certified Kentucky Bluegrass
85% germination
30.06% Certified Merion Kentucky Bluegrass
85% germination
28.92% Certified Adelphi Kentucky Bluegrass
85% germination
8.96% Certified Pennlawn Creeping Red Fescue
85% germination
COARSE KINDS
None Claimed
OTHER INGREDIENTS
0.05% Other Crop Seed
1.96% Inert Matter Tested: July 1976
(Must be within 90 days of purchase date)
0.09% Weed Seed
Suburban Seed Company
99 Westland Avenue
Seedville, Maryland

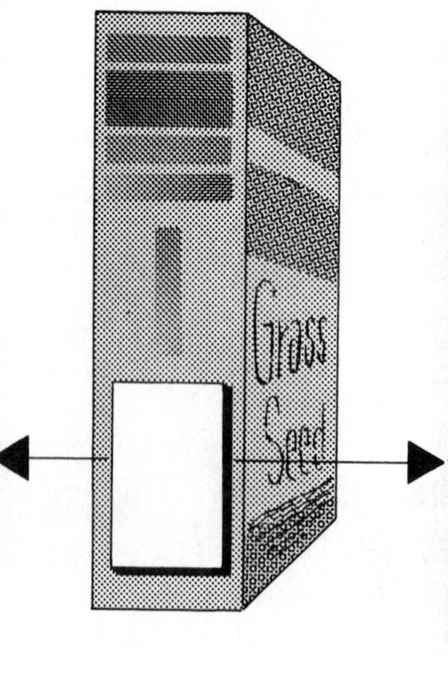

Poor Seed Mixture

Suburban Brand Special Grass Seed

Lot SP-24
FINE TEXTURED GRASSES
None or None Claimed
COARSE KINDS
50.62% Kentucky 31 Tall Fescue
85% germination
19.00% Perennial Ryegrass
90% germination
25.00% Annual Ryegrass
90% germination
OTHER INGREDIENTS
3.10% Other Crop Seed
1.78% Inert Matter Tested: Jan. 1971
0.50% Weed Seed
Noxious Weeds
3 Poa annua per ounce

Suburban Seed Company
99 Westland Avenue
Seedville, Maryland

FIGURE 24-20
Sample labels for lawn and turf seed

adequate surface moisture must be present for seed germination. A straw mulch is preferred and is usually applied at the rate of two bales per 1,000 ft.^{2} Other mulches are wood and paper byproducts, net or fabric, and peat moss.

Sodding *Sodding* refers to removing a rectangular piece of grass and a slice of the soil beneath it and moving it to another location. The grass and the soil immediately beneath are referred to as *turf*.

FIGURE 24-21
A cultipacker seeder used for turfgrass seeding

Sodding offers the ability to establish turf at any time of the year. It also provides an instant cover. However, the cost is extremely high and the sod must be installed shortly after harvesting. Specialized equipment for sod harvesting has been developed (Figure 24-22). Sod is cut into 12–24″ widths and lengths of up to 3′. The depth of cut will vary from 0.3 to 0.5″. The cut sod is then placed on a carefully prepared seed bed, rolled to ensure good contact with the soil, and watered thoroughly.

Sprigging *Sprigging* is the planting of a section of a rhizome or stolon, referred to as a sprig. A sprig may be up to 6–8″ long. For successful establishment, a section of the sprig with several nodes must be inserted into the soil (Figure 24-23). After sprigging, irrigation must be applied to prevent desiccation, or drying out. Equipment for sprigging has been developed and will plant sprigs in rows spaced 6–18″ apart.

Stolonizing *Stolonizing* is similar to sprigging, in that 6–8″ sections of rhizomes or stolons are used. However, in stolonizing, sprigs are broadcast onto the soil surface. These sprigs may be lightly top-dressed with soil and rolled prior to irrigation. Survival rates are lower than with sprigging. Therefore, stolonizing requires greater quantities of sprigs.

FIGURE 24-22
A sod harvester for cutting sod *(From Emmons/Turfgrass Science & Management, copyright 1984 by Delmar Publishers Inc.)*

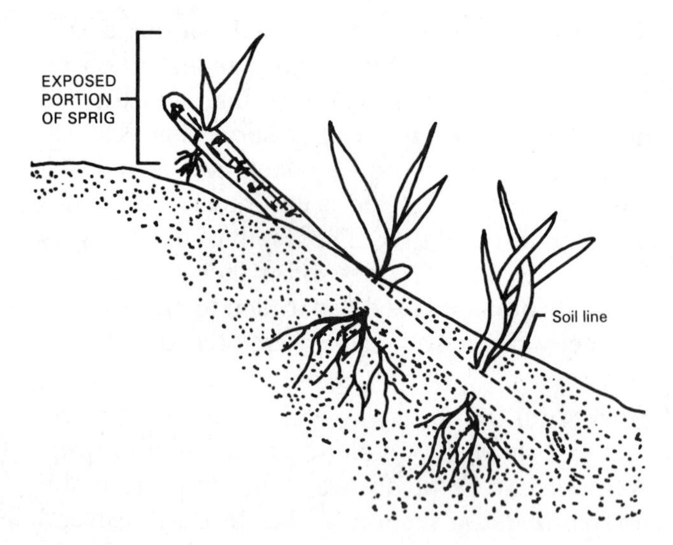

FIGURE 24-23
Planting of a sprig into the soil

Plugging *Plugging* is the establishment of a turfgrass stand by using plugs or small pieces of existing turf. Plug sizes will vary, as noted in Figure 24-24. Spacing of plugs can be on 6–18″ centers, depending on how quickly you need the turfgrass established. Plugging can be done by hand or with specialized equipment. It will require a longer time to cover the soil than the other establishment techniques.

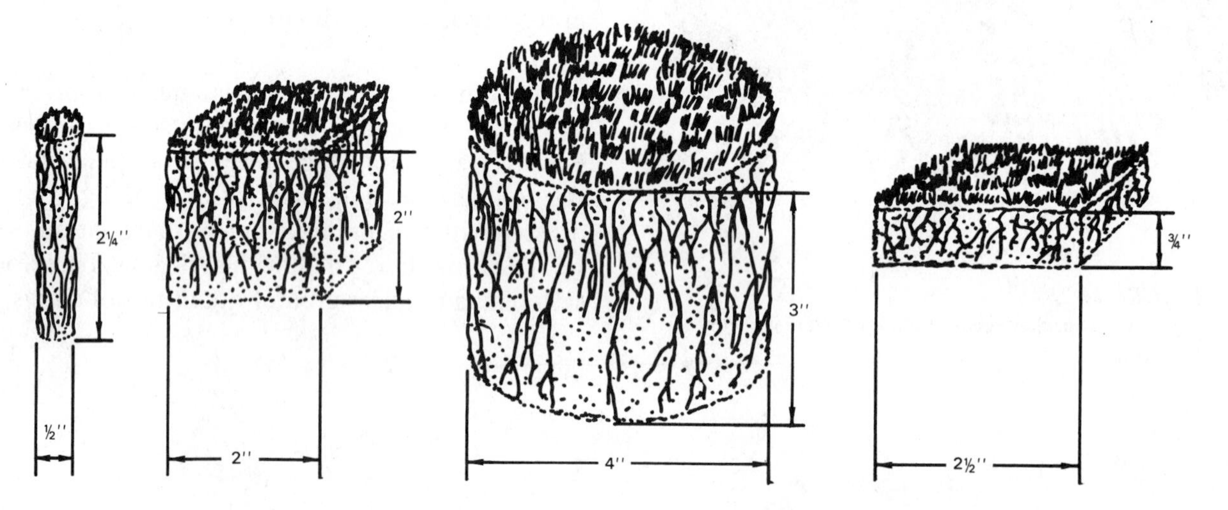

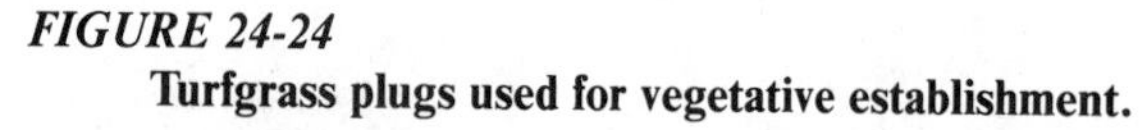
FIGURE 24-24
Turfgrass plugs used for vegetative establishment.

Student Activities

1. Define the "Terms To Know."
2. Determine if your community is located in a warm-season, transitional, or cool-season zone.
3. Visit or call your county cooperative extension office and obtain copies of the various publications available on turfgrass or lawn production and maintenance.
4. Determine the recommended cultural practices for establishing and maintaining a major turfgrass species or mixture for your locality. Report your findings to the class.
5. Arrange to visit a golf course and discuss with the superintendent the duties of turfgrass workers on the golf course.

6. Obtain a map of your county and attach it to the bulletin board. Place a pin at the location of all businesses, schools, government buildings, and institutions that have extensive lawn or turfgrass areas around them. Ask your classmates to help identify them. Note: Use different colored pins to identify different types of institutions.

7. Obtain a grass identification key from your teacher or the cooperative extension service office. Collect 10 specimens of different turfgrasses and identify them by using the identification key.

8. Work with your teacher or cooperative extension service to set up turfgrass demonstration plots. Show the affect of various fertilizer applications and mowing heights on turfgrass health and vigor.

Self-Evaluation

A. MULTIPLE CHOICE

1. The major meristematic tissue of the turfgrass plant is the

   a. infloresence.
   b. leaf blade.
   c. rhizome.
   d. crown.

2. The root system of a cool-season turfgrass is actively growing during the

   a. summer.
   b. fall and winter.
   c. fall and spring.
   d. late spring.

3. The estimated value of the turfgrass industry in the United States is

   a. $25 billion.
   b. $50 billion.
   c. $100 billion.
   d. $200 million.

4. Which irrigation practice will develop a deeper, more extensive root system?

   a. frequent and light applications
   b. infrequent and heavy applications
   c. frequent and heavy applications
   d. infrequent and light applications

5. Turfgrasses are in the ______ family.

   a. *Compositae*
   b. *Cyperacese*
   c. *Cruciferue*
   d. *Poacea*

6. ______ is a warm-season turfgrass.

   a. Kentucky bluegrass
   b. Zoysia grass
   c. Red fescue
   d. Perennial ryegrass

7. Which cultural practice will have an adverse effect on rooting depth?

   a. low mowing
   b. heavy, infrequent irrigation
   c. moderate fertilizer applications during the time of root growth
   d. high mowing

8. A tiller is an example of

   a. rhizome growth.
   b. extravaginal growth.
   c. stolon growth.
   d. intravaginal growth.

9. Which of the following is not a turfgrass stem?

   a. rhizome
   b. stolon
   c. leaf sheath
   d. crown

10. Which cool-season turfgrass is adapted to shady, dry locations?

    a. red fescue
    b. Kentucky bluegrass
    c. tall fescue
    d. creeping bentgrass

B. MATCHING

______	1. Rhizome	a.	An induction process for inflorescence development
______	2. Stolon	b.	A turfgrass stem that grows horizontally aboveground
______	3. Creepingbentgrass	c.	A cool-season turfgrass that is very drought tolerant
______	4. Sprigging	d.	A wide-leaf blade
______	5. Photoperiod	e.	A cool-season turfgrass used on putting greens
______	6. Tall fescue	f.	A turfgrass stem that grows horizontally below ground
______	7. Coarse texture	g.	A buildup of organic matter on the soil around turfgrass plants
______	8. Syringing	h.	A complete fertilizer
______	9. 10-10-10	i.	Light application of water to a turfgrass
______	10. Thatch	j.	Vegetative establishment

C. COMPLETION

1. ______ is a warm-season turfgrass having good winter hardiness.
2. ______ roots are present at the time of seed germination.
3. ______ is an example of intravaginal growth.
4. A seed ______ consists of different cultivars of the same species.
5. ______ is a turfgrass species with rapid seed germination.
6. As mowing heights decrease, mowing frequency will ______ .
7. ______ is considered the most important element in turfgrass fertilization.
8. A planting depth of ______ is recommended for turfgrasses.
9. ______ is an establishment practice of applying the sprig to the soil surface.
10. A ______ mulch is preferred over other types of mulch for seeding turfgrasses.

UNIT 25 Trees and Shrubs

OBJECTIVE To use trees and shrubs for beautification, improved air quality, and pollution control.

Competencies to Be Developed

After studying this unit, you will be able to

- ☐ identify ornamental trees and shrubs.
- ☐ select trees and shrubs for appropriate landscape use.
- ☐ classify trees and shrubs according to growth habit, growth habitat needs, and other requirements.
- ☐ identify trees and shrubs by using proper nomenclature.
- ☐ purchase plant material for installation in a landscape.
- ☐ plant and maintain plant material.

TERMS TO KNOW

Trees
Shrubs
Specimen plant
Border plant
Group planting
Hardy
Plant hardiness zone map
Nomenclature
Genus
Species
Variety
Cultivar
Bare-rooted plants
Healing in
Balled and burlapped (B & B)
Root pruning
Tree spade
Container grown
Staking
Guying
Drip line
Pruning
Canopy

MATERIALS LIST

nursery catalog
pictures of plants commonly grown in an area
list of insects and diseases of trees and shrubs that are normally problems in the area

Trees and shrubs are major components of the environment. They are important in providing natural beauty and in providing oxygen to humans and animals. A well-planned landscape will improve the economic value of an area. The need for trees and shrubs has increased in recent years. This need has increased the requirement for horticulture technicians that understand the production and care of plants after they are established in a landscape.

VALUE OF TREES AND SHRUBS A well-designed landscape increases the value of a property (Figure 25-1). Residential properties are much more attractive when trees and shrubs are used to enhance their beauty and balance.

Trees and shrubs provide more than natural beauty. An acre of good, healthy trees and/or shrubs will produce enough oxygen to keep be-

FIGURE 25-1
A well-designed landscape adds economic value to a property.

tween 16 and 20 people alive each year. These plants are also valuable in helping to keep our air clean by using the carbon dioxide that is produced by people, autos, and factories.

Trees and shrubs will cut noise pollution by acting as a barrier to sound. They can deflect sound as well as absorb it. When used properly in the landscape, trees and shrubs can act as an insulator to keep the house cooler in the summer and warmer in the winter.

Many urban areas depend on trees and shrubs to soften the concrete and steel environment. In fact, many of these areas now employ urban foresters to help design, install, and maintain trees and shrubs in the large cities.

PLANT SELECTION

Trees and shrubs are separate groups of plant material. *Trees* are woody plants that produce a main trunk and a more or less distinct and elevated head (a height of 15′ or more). *Shrubs* are woody plants that normally grow low, produce many stems or shoots from the base, and do not reach more than 15′ high.

USE

Ornamental trees and shrubs in the landscape are both beautiful and functional. They are used as specimen plants, border planting, or as a grouping. A *specimen plant* is used as a single plant to highlight or provide some other special feature to the landscape. A *border planting* is used to separate some part of the landscape from another. Or, it might be used as a fence or a windbreak. A *group planting* is when a number of trees or shrubs are planted together so they point out some special feature, provide privacy, or create a small garden area.

The location of plant material in the landscape will play an important part in the selection process. Color of the leaves, texture of the plant, and color of the flowers are just some of the considerations that must be made before purchasing or planting trees or shrubs.

Geographic Location

The location in a country is an important factor in the selection of a tree or shrub. Some plants are not *hardy*, that is, they may not be able to survive or even grow properly in a given environment. The hardiness of a plant is affected by the intensity and duration of sunlight, the length of the growing season, minimum winter temperatures, the amount of rainfall, summer droughts, and humidity.

The U.S. Department of Agriculture has issued a *plant hardiness zone map* (Figure 25-2). There are 10 zones in the United States. Each represents an area of winter hardiness that is based on the average minimum winter temperatures. It is possible that local climates may vary from the general zone map. They may be colder or warmer than is indicated on the map. Local nursery personnel are generally willing to help in the selection of the best plant material. Most nursery catalogs list the plant for the coldest zone in which it will grow normally. Two examples of such catalog listings are:

Example 1: *Cornus florida* (Flowering Dogwood)—**Zone 5**
A low branched, flat topped tree. It has a horizontal branching habit. Will grow 20 to 30 feet high and 25 to 35 feet wide. Growth rate is slow to medium and the texture is fine to medium. White flowers appear in April to May. Red Berries appear in the fall.

Example 2: *Pyrus calleryana* 'Bradford' (Bradford Pear)—**Zone 4**
A dense, pyramidal tree that becomes brittle with age. Will grow 30 to 50 feet high and 30 to 35 feet wide. The growth rate is medium and the texture is medium-fine. Make an excellent all purpose tree useful as a shade or street tree. Profuse white flowers 1/3 inch in diameter appear in late April or early May. Are in clusters of 3 inches in diameter.

A plant can also be expected to live in a warmer zone than indicated if rainfall, soil, and summer conditions are comparable. In some cases it may be necessary to adjust these conditions by irrigation, correction of soil conditions, wind protection, and adjustment of the shade or sun exposure conditions. It is also possible to grow some plants in areas north of the indicated zone. Such plants may need special attention to protect them from wind or cold. Without such protection, they may not perform normally and are likely to suffer winter injury.

Site Location

When planning the location for planting trees and shrubs, many factors must be given consideration. Some are:

- □ Flower color. Is it compatible with the house, fences, patios, and any other plants in the area?

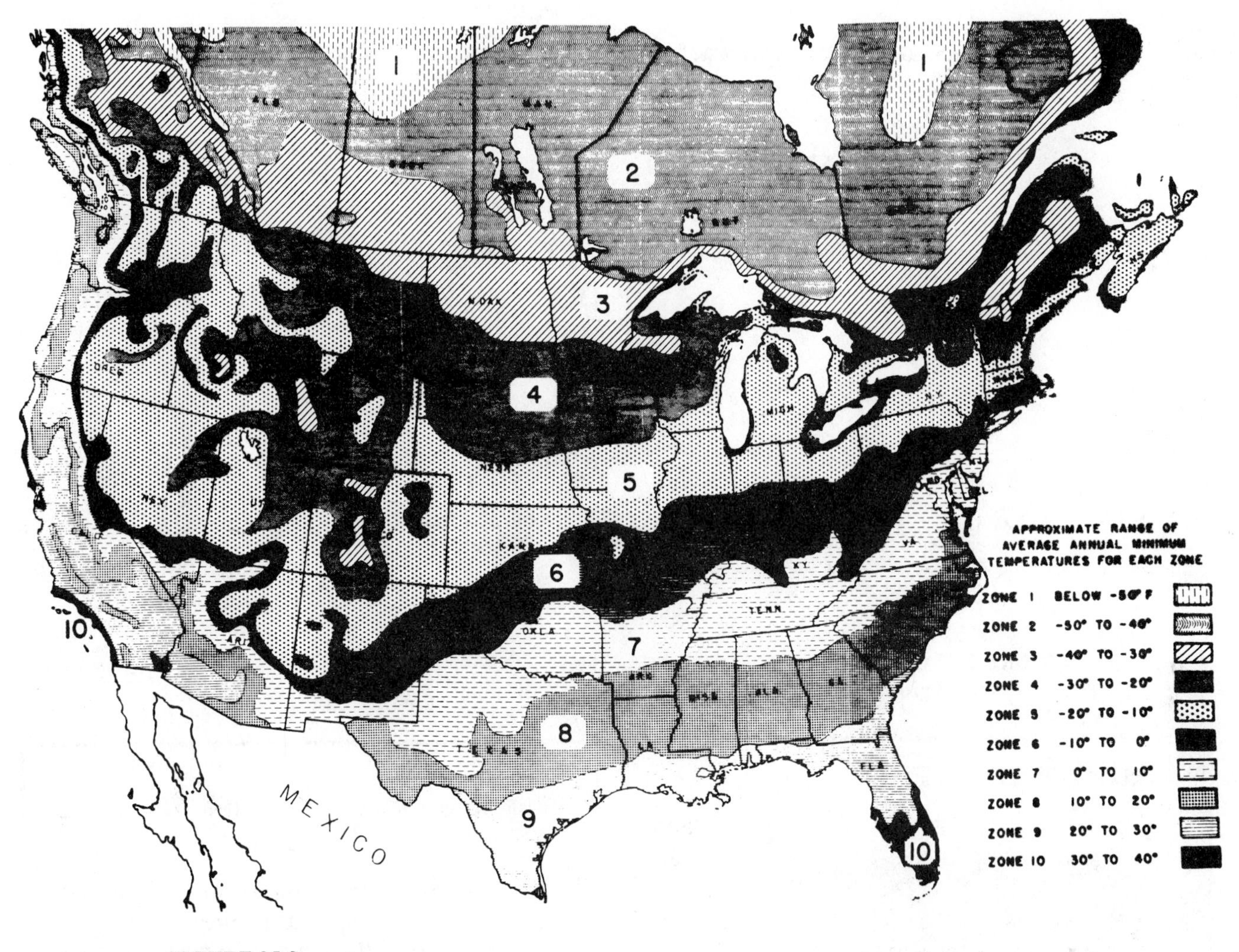

FIGURE 25-2
The plant hardiness zone map

- □ Fruit size and type. Many trees, such as some crabapples or cherries, drop messy fruit. Therefore, they should not be placed near an area that will be walked on or used by people. Such areas include driveways, walks, patios, and swimming pools.
- □ Other structures. Do not plant trees directly in front of doors, near wells, cesspools, or field drains; or under utility lines where interference is likely when the plant matures.
- □ The plant's ornamental characteristics. Flowering time, shape, foliage texture, fall color, pest resistance, landscape suitability, and mature size are all part of any consideration when selecting a plant.

Type and Growth Habit Plants have many different types of growth habits. The type or growth habit is an important consideration when selecting plants (Figure 25-3). A good landscape planner or designer will be aware of the types of growth habits for that plant used in their area.

Plant size in relation to the structure that is being landscaped is very important. Maintenance and pruning of plants is often expensive. Pruning of shrubs is frequently done without climbing, while most trees need to be climbed when pruned. The use of a tree that is not in proportion to nearby structures can result in an expensive, frequent pruning program. Figure 25-4 illustrates some common trees and their mature sizes. A tree that is too tall will require extra work to keep it in proper relationship to adjacent structures. Such a design error would cancel out the goal of property enhancement.

Shape of the Plant The shape of a plant is an important factor in the selection of plant material. Some important undesirable effects can be avoided by asking, Does the plant grow straight up

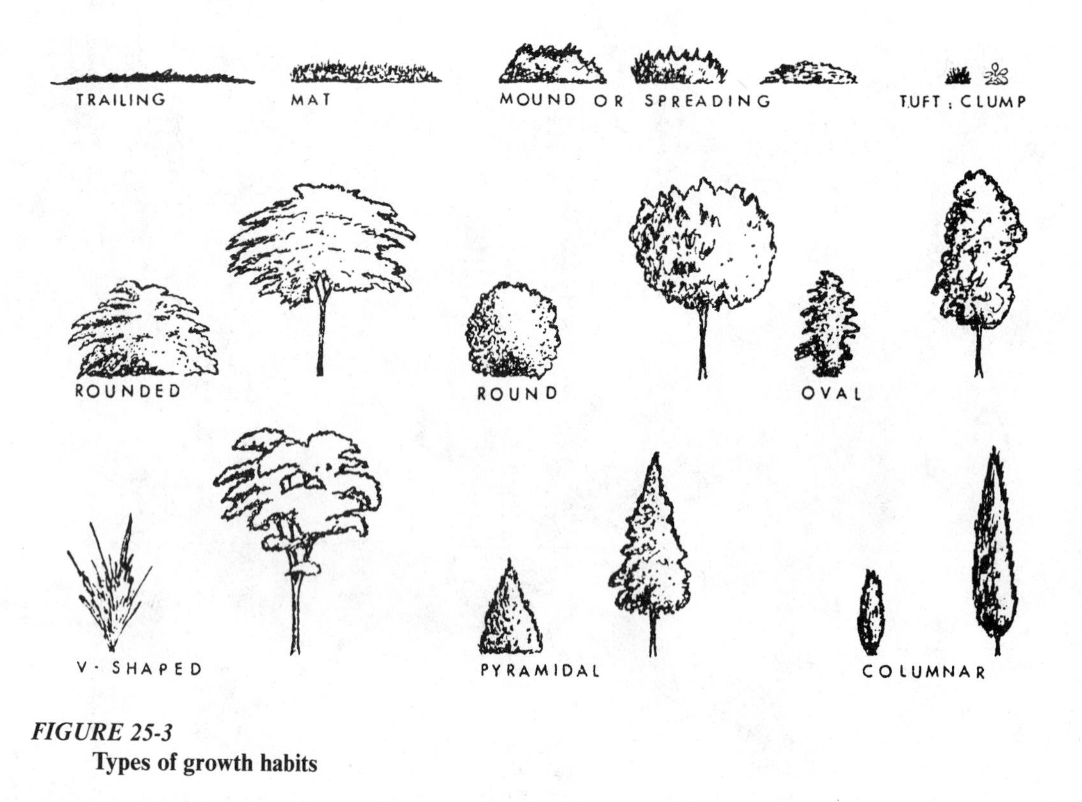

FIGURE 25-3
Types of growth habits

AGRI-PROFILE

CAREER AREAS: LANDSCAPE ARCHITECT/ LANDSCAPER/LANDSCAPER TECHNICIAN

Years of plant culture may be wiped out by one severe cold snap. Researcher George Yelenosky attaches a temperature sensor to a blossom to study the effect of surrounding temperatures on tissue freezing. ***(Courtesy USDA Agricultural Research Service)***

Ornamental trees and shrubs vary greatly in size, shape, temperature preference, light preference, fertility needs, and pest tolerance. This variety stimulates diversity in jobs and specialties in the area of ornamental horticulture. Career opportunities span both the arts and the sciences.

Landscape architects practice the art of design as they plan landscapes that are pleasing to their clients. They must know plant species and plant materials to develop plans that are attractive yet functional in their environment. Landscape architects generally have bachelor's degrees in landscape design, and considerable experience in ornamental horticulture.

Horticulture specialists and technicians have careers centering on nursery or greenhouse management, landscape contracting, groundskeeping, wholesaling, retailing, public information, writing, and consulting. They may specialize on just trees or just shrubs or both. Some specialize in pruning, tree planting, pest control, or landscape maintenance. Ornamental horticulture seems to provide an endless array of career opportunities in many localities.

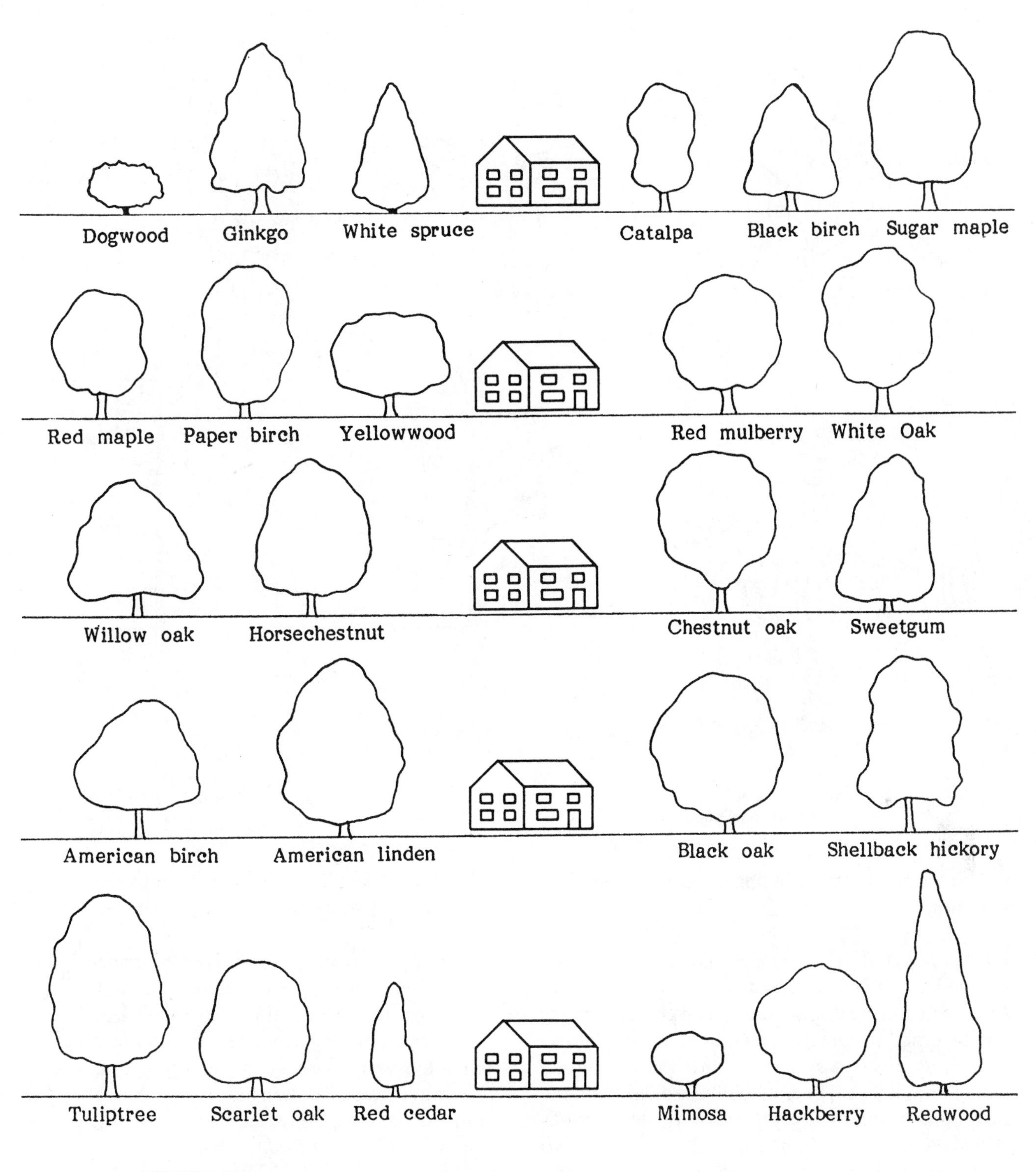

FIGURE 25-4
Tree size in relation to a two-story house

and give little shade? Does the plant have a trunk that divides and spreads to cast unwanted shade? and is the plant a low-growing tree that gives very little shade? Consider which shapes of trees and shrubs are needed in the landscape. Then select the appropriate plant to achieve the objective (Figure 25-5).

Before plants are selected, research is necessary to better understand the types of plants used in a given area. To do this, study the types of plants sold and used in the locality and note the various characteristics of each plant. Catalogs are available from nurseries that sell plants in the locality. Also, a visit to local garden centers and nurseries

FIGURE 25-5
Shapes of some plant material

will pay dividends in gathering information on plant selection.

PLANT NAMES Trees and shrubs must be ordered by their proper names. Common names such as flowering dogwood and upright juniper are not governed by any formal code of nomenclature. *Nomenclature* is a systematic method of naming plants or animals. The botanical or scientific name is recognized internationally. Scientific designations are always written in Latin and consist of two names. The first name is the genus and the second is the species. The genus name always begins with a capital letter and is a noun. The species name is usually written in all small letters and is an adjective.

The *genus* (plural is genera) is defined as a group of closely related and definable plants comprising one or more species. The common, definable characteristics are fruit, flower, and leaf type and arrangement. The *species* (plural is also species) is the basic unit in the classification system whose members have a similar structure and common ancestors, and maintain their characteristics.

Variety (var.) is a subdivision of species that has various inheritable characteristics of form and structure that are continued through both sexual and asexual propagation. The varietal term is written in lowercase letters and underlined or *italicized.* Often in catalogs and on plant labels the abbreviation **var**. is used. An example of how this is used might be *Cornus florida rubra* or *Cornus florida var. rubra*.

A *cultivar* (cv.) is a group of plants within a particular species that has been cultivated and is distinguished by one or more characteristics that, through sexual or asexual propagation, will keep these characteristics. The term is written inside single quotation marks. An example is *Pyrus calleryana* 'Bradford' or *Pyrus calleryana* cv. 'Bradford'.

It is important to become familiar with this type of naming plants because common names will vary from area to area. There are no standards for creating common names. Professional horticulturists order plants by the method described above. Landscape architects place scientific names on landscape drawings. Finally, nurseries use the scientific name in their catalogs to avoid confusion by anyone who orders the plant material.

OBTAINING TREES AND SHRUBS

After determining the specific plants needed in a landscape by studying their growth habits and how they'll fit in, the plants must be purchased. Plants are normally dug and shipped as bare rooted, balled and burlapped (B & B), and container grown (Figure 25-6).

Bare-Rooted Plants

Bare-rooted plants are dug and the soil is shaken or washed from the roots. Normally only deciduous trees, shrubs, and trees with taproots are shipped this way. They are dug while the trees are dormant. Plants that are ordered from nursery catalogs and can survive bare rooted are shipped this way because of lower transportation costs and easier handling. It is not economical to ship soil with the roots because it is so heavy.

Bare-rooted plants are often planted while they are dormant; however, the roots must be protected to keep them from drying out. If the plant cannot be planted when received, the roots must be protected by putting the plants in a container of water, wrapping them in plastic after a good watering, or placing wet newspaper around them. If they cannot be planted for many days, the plant may be healed in. *Healing in* is accomplished by simply digging a trench in the soil deep enough to hold the roots of the plants, placing the plants in the trench, and covering them with soil. It is important to wet the soil well after healing in. Frequently, plants that are bare rooted will need additional pruning at planting time.

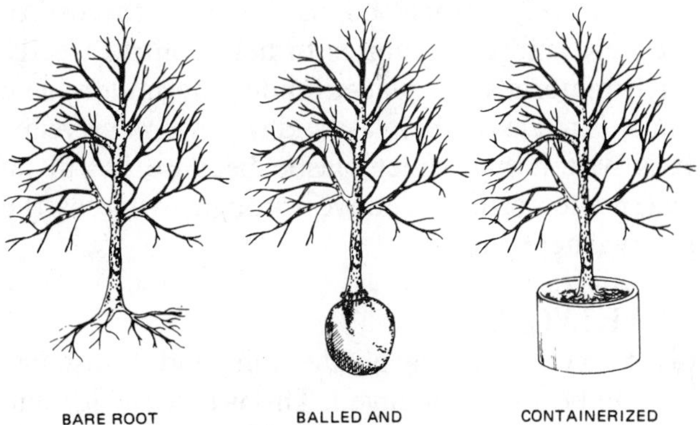

FIGURE 25-6
Ways of obtaining a plant *(From Reiley & Shry/Introductory Horticulture, 3rd edition, copyright 1988 by Delmar Publishers Inc.)*

Balled and Burlapped Plants

Balled and burlapped (B & B) are plants that are dug with a ball of soil remaining with the roots. This is wrapped with burlap and laced with twine. Normally these plants have been root pruned. *Root pruning* is a process where roots are cut close to the trunk so that a good root system develops close to the trunk before the plant is dug. The result is that when the ball is dug, the tree or shrub will have a compact root system in the ball. This gives the plant a good chance to reestablish itself in a new environment.

It is important that transplanted trees and shrubs have a good chance to grow after transplanting. Plants typically balled and burlapped for transplanting are deciduous trees with branching root systems, conifers, azaleas, rhododendrons, and other plants that have fibrous root systems.

As is true with bare-root plants, B & B plants cannot have their root systems dry out. But, since there is soil around the roots, the drying-out process is more gradual. When purchasing B & B plants, avoid those that do not have a sound ball. Handling tends to break the ball up and strip the hair roots from the plants. Therefore, a good root ball will help substantially to avoid damage when handling. Plant material that is shipped as B & B may be planted anytime, as long as the soil can be worked.

The mechanical tree spade is becoming more popular in the industry today. A *tree spade* is an expensive piece of equipment that will dig a tree in a matter of a few minutes with a very specific-sized ball. Additionally, it is used to dig holes for trees and shrubs quickly and efficiently. The tree spade saves many hours in digging and planting trees. Frequently, B & B trees and shrubs are shipped in burlap with a wire cage, specially prepared basket or other container. The tree spade has helped make this an efficient way to ball and burlap.

These types of plants require very little pruning at planting time. This makes B & B a popular way of handling plants by the mechanized professional horticulturist.

Container-Grown Plants

The use of container-grown plants is increasing in the nursery industry. *Container grown* means plants are grown and shipped in a pot or can. Normally, the smaller types of plants are handled in this manner. They are grown for a reasonable period of time in the container in which they are shipped and purchased. They are easy to handle in the field, require little or no pruning when they are planted, and can be planted any time the soil can be worked.

The disadvantage in using container-grown plants is that they need to be planted carefully. The roots of the plant have developed in a very limited space and are generally rootbound. The roots may also have grown back around the trunk. To prevent the plant from strangling itself and also to encourage the root to leave the confined area, these plants must have the "container ball" broken. This is done by placing a sharp shovel through the root mass or by breaking and spreading the root mass as it is planted in the new hole.

PLANTING TREES AND SHRUBS

This may be done in spring, summer, or fall as long as soil can be worked. A good practice for preparing the hole is to dig it about one-third wider than the ball or container. If the soil is hard, compacted, or of poor quality, it is advisable to dig the hole 4–5″ deeper than the ball. This will allow peat moss or another soil conditioner to be added to the soil and placed under and around the root ball. Fill the hole with enough good soil to allow the ball to be the proper depth when it is put in the hole. The ball or container should be about 2–5″ above the top of the hole. After cutting the twine that holds the ball together, peel back and remove the burlap. Backfill the hole around the plant, tamping the soil lightly to ensure removal of all air pockets. Continue backfilling the hole until it is level with the surrounding soil. Place a ring of soil about 4″ high around the backfill to retain water.

In the case of a container plant, remove the plant from the container and slice the root zone with a knife or shovel. Then place the plant in the hole and backfill carefully.

After backfilling, fill the ring with several inches of water to saturate the soil to the bottom of the backfill. This will settle the soil and provide moisture for the plant (Figure 25-7).

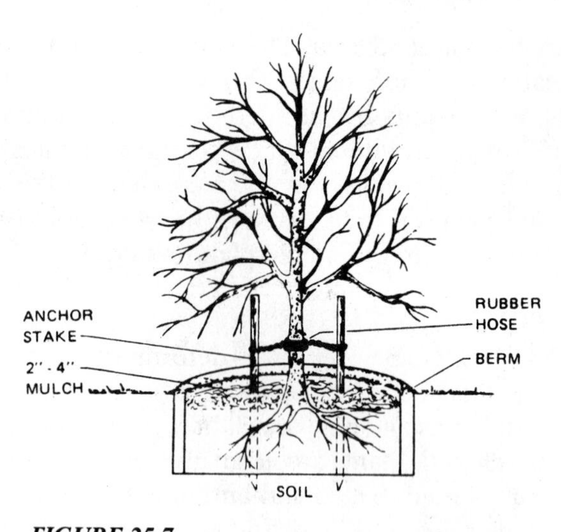

FIGURE 25-7
Proper planting of a tree or shrub ***(Courtesy USDA)***

MULCHING

Trees and shrubs will benefit from the addition of mulch around the planted area. Mulch in the form of shredded hardwood or pine straw will help hold in the moisture. A newly planted tree will take about 15–20 gallons of water about twice a week in a hot, dry environment. The conservation of water is important. Other advantages of mulch is that it will help provide a more even soil temperature, help control weeds, prevent erosion, and help prevent soil compaction. Some species of plants have shallow root systems and will compete with any ground cover for nutrients. The addition of mulch will reduce this competition.

The use of mulch will prevent damage from lawn mowers or string weeders by keeping grass and ground covers away from the trunk. Mulch should be applied 3–4″ deep. For estimating the mulch needs for a bed of plants, use the rule of thumb that 1 yd^3 of mulch can be spread over 100 ft^2 of area. Mulch will last for 1 to 3 years, depending on the type applied. Pine bark will need replacing annually, while shredded hardwood mulch will not need additional mulch for about 3 years.

As an additional bonus, consider the use of the new landscape fabrics to help control weeds. Such materials are placed under the mulch. The fabric is made of fiberglass and will last many years. It is better than sheet plastic because it will let water and fertilizers move through, whereas the plastic sheet will not.

STAKING AND GUYING

Newly planted trees and shrubs may have to be staked or guyed. This is to avoid loosening and disfiguration of the plant by the wind. *Staking* is to drive a wooden or metal pole into the ground near the plant and tie the upper part of the plant to the stake. *Guying* is to tie a tree to stakes with wire or rope (Figure 25-8). Stakes and guys

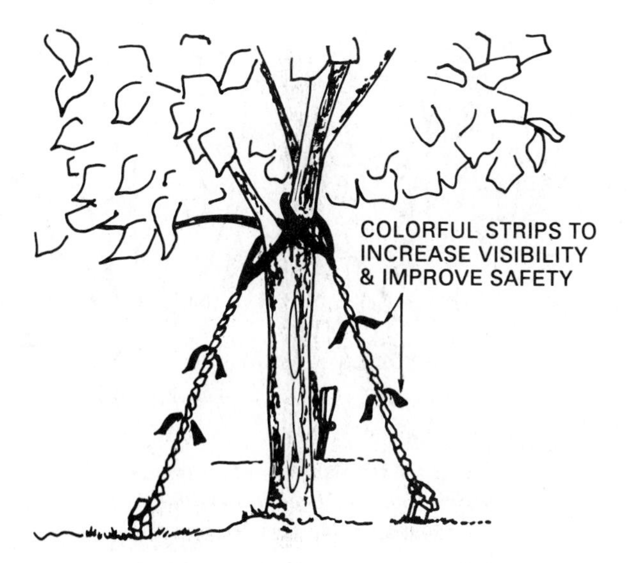

FIGURE 25-8
Staking of trees ***(From Ingels/Ornamental Horticulture, copyright 1985 by Delmar Publishers Inc.)***

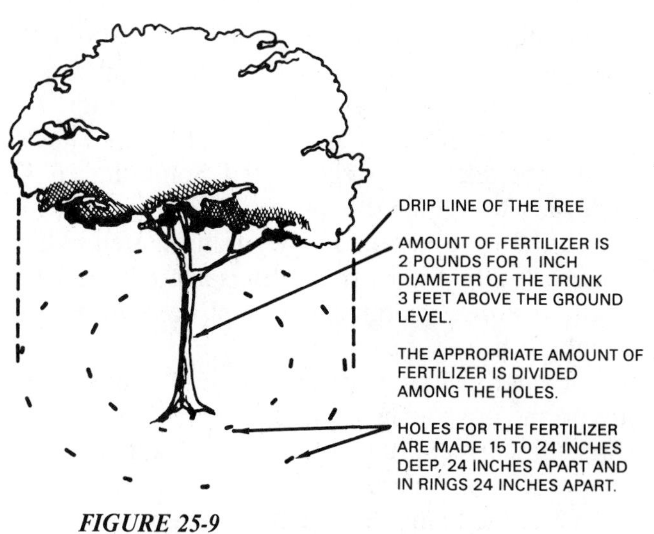

FIGURE 25-9
Fertilizing a tree ***(From Reiley & Shry/Introductory Horticulture, 3rd edition, copyright 1988 by Delmar Publishers Inc.)***

should be left in place for at least one growing season. When using guy wires, make sure the wires do not come in contact with the trunk. Use of old water hose with the wire running through is a good practice to protect the trunk. Another good practice is to tighten the wires at the stake and not at the trunk. The best method is to use a double strand of wire twisted together for tightening. Guy wires should be checked for tightness several times during the growing season.

FERTILIZING Plants need nutrients to maintain their vigor and to make healthy new growth. If the soil is fertile, it is not necessary to add fertilizer at planting time. Trees and shrubs should not be fertilized during the first year of growth. This practice is followed to prevent the plant from developing too much top growth in relation to the root growth. Plants should be fertilized every 3-5 years, starting with the growing season after the first year. Generally speaking, the best time is in the early spring. It is not a good practice to fertilize a plant after middle to late July. This practice would force new growth that will not mature enough to escape damage by the winter cold.

Unless the plants are in a bed, it is not a good practice to place fertilizer on the surface of the soil. It will wash away in the rains or will not penetrate into the root-zone area. To fertilize a tree properly, it is necessary to place the fertilizer into a hole 1″ in diameter and 18″ deep. Care should be taken to avoid underground utility wires and pipes such as electric lines, telephone wires, gas pipes, and water pipes.

Holes for fertilizer are made with a 1″ steel stake placed in concentric circles starting, about halfway out from the trunk and to the drip line. The *drip line* is the outer edge of the tree where the branches stop. Holes should be about 8″ apart and the circles should be about 18″ apart (Figure 25-9). Each hole receives the appropriate amount of fertilizer and is sealed with soil or by closing the hole with a heel of the foot.

The amount of fertilizer needed by a tree is determined by measuring the trunk about 4½′ above the ground. If the trunk of that plant is under 8″, multiply the number of inches of diameter by 3. If the trunk is over 8″, multiply by 6. This will give the number of pounds of fertilizer the tree needs.

For example, if the tree measures 6″ at 4½′, multiply 6×3. The answer is 18 lbs. of fertilizer. If the tree measures 12″ at 4½′, then $12 \times 6 = 72$ lbs. of fertilizer. If the tree has more than one trunk, combine the diameters of all the trunks and multiply by the appropriate factor.

A fertilizer with an analysis of 10–6–4 or 10–10–10 is generally acceptable. Never apply more than 100 lbs. of fertilizer to any given tree in a year. If the rate to be applied is over 100 lbs. make the application over a 2-year period. Trees should be fertilized every 3–5 years.

PRUNING *Pruning* is the removal of dead or undesirable limbs from a tree or shrub. Removing dead, broken, diseased, and insect-infested wood helps to protect the plant from additional damage. Trees may have waterspouts, bad-angle crotches, branches that cross over each other, or branches that form an asymmetrical habit. These need to be taken out or corrected. Sometimes it is necessary to reduce the top growth of the plant to match the root ball on a newly transplanted plant (Figure 25-10). After a tree is transplanted, the size of the *canopy* or branches should be reduced to equal the size of the root ball.

Proper pruning techniques and timing are important in growing good ornamental trees and shrubs. Incorrect pruning can leave a plant in worse condition than before. A plant in weak condition is susceptible to insects and diseases. Poor pruning can cause the loss of a season of flowers or fruit. In general, flowering plants should be pruned just after they *bloom* to produce flowers. This is done so that the next season's flowering wood is not cut off.

It is easier to prune deciduous trees and shrubs in the late fall or winter after the leaves drop. The framework is bare and easier to see.

Specific plants have specific pruning requirements. It is best to determine how each plant should be pruned before the pruning operation is started. Local plant specialists are usually willing to help or give suggestions for proper pruning. Good reference books are available with accurate information on pruning a particular plant.

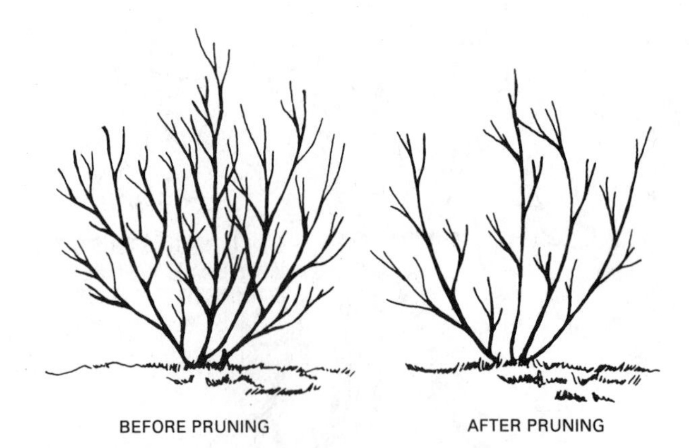

FIGURE 25-10
Pruning of trees and/or shrubs ***(From Reiley & Shry/Introductory Horticulture, 3rd edition, copyright 1988 by Delmar Publishers Inc.)***

INSECTS AND DISEASES

Most plants are subject to damage by insects and diseases. It is easier to prevent the damage than to control it. Pest-resistant varieties should be used, and they should be kept healthy and vigorous. Weakened trees and shrubs are easily attacked and damaged by insects and diseases.

It is important that plants be selected that are suited to the environment in which they are to be planted. Plants cannot tolerate stressful conditions continuously. Plants will adapt to various conditions, but some plants adapt better than others. Some plants can withstand air pollution or drought, or heat, or even wet soils. Other types of plants can tolerate infertile conditions, but not hot, dry weather. Still others may not survive in heavy shade, and/or in full sun.

In its lifetime, any plant can be expected to experience some insects and diseases. To be an informed technician, it is best to become aware of the plants in your area and to consult the cooperative extension service for more specifics on the types of insects and diseases that can be expected in a given locality.

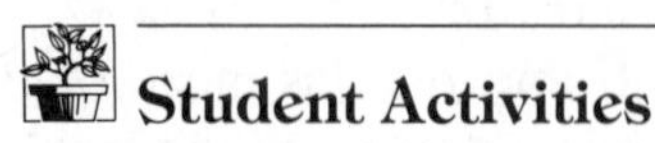

Student Activities

1. Contact a local nursery and obtain a nursery catalog.
2. Visit a garden center or nursery and compile a list of trees and shrubs that are available and recommended for your community.
3. Prepare a chart of popular ornamental trees and shrubs for your community. List the scientific name, common name, mature size of plant, time of flowering, color of flowers, spring leaf color, and full leaf color.
4. Survey your home or other assigned area and make a list of trees and shrubs present. Use books, nursery catalogs, and other resources to help identify the plants.

5. For a given area, determine the diameter of each tree 4½′ above the ground. Work up a recommended fertilizer program for the trees, including amount of fertilizer per tree, local fertilizer prices, cost of fertilizer for each tree, time of year to fertilize each, and the total cost of fertilizer for the area.
6. Prepare a pruning schedule for plants around your home or other area based on local recommendations. Include in the schedule the name of each scientific plant, time of pruning, and special requirements for pruning each plant.
7. Prepare a sketch or a scale drawing of your home property or other assigned area. Locate the existing trees and shrubs on the sketch, using a circle with the initials of the scientific name to represent each plant. Add new plants that you believe will enhance the property you have surveyed.

Self-Evaluation

A. MULTIPLE CHOICE

1. An acre of plant material can produce enough oxygen to keep

   a. 5–10 people alive each year.
   b. 10–16 people alive each year.
   c. 16–20 people alive each year.
   d. nobody alive. It does not produce enough oxygen to be of any value.

2. A border planting

   a. is used as a single plant to highlight a fence or some other special feature of the landscape.
   b. is used to separate some part of the landscape from another.
   c. is a number of trees or shrubs planted together as a point of interest.
   d. is a collection of plants that are placed in the landscape as needed.

3. When planning the location for planting trees and shrubs, which is not a major consideration?

   a. fruit size and type
   b. flower color
   c. other structures
   d. bare-rooted plants

4. The term *Cornus florida rubra* is a

   a. common name.
   b. scientific name.
   c. name that was developed in Florida.
   d. type of annual deciduous plant.

5. Planting of trees and shrubs may be done

   a. in spring, summer, or fall.
   b. in spring only.
   c. in fall only.
   d. in spring and summer.

6. Mulch should be applied

   a. 1–2″ deep.
   b. 2–2½″ deep.
   c. 3–4″ deep.
   d. 6″ deep to keep out the weeds.

B. MATCHING

_____ 1. Urban foresters
_____ 2. Shrubs
_____ 3. Specimen plant
_____ 4. Nomenclature
_____ 5. Pruning
_____ 6. Canopy

a. the removal of dead, broken, unwanted, diseased, and insect-infested wood
b. Used as a single plant to highlight it or some other special feature of the landscape
c. A systematic method of naming plants
d. Woody plants that normally grow low and produce many stems or shoots from the base
e. The top of the plant that has the framework and leaves
f. Help install and maintain trees in large cities.

C. COMPLETION

1. Plant material can cut noise pollution by _____.
2. A rule of thumb is that 1 yd^3 of mulch can be spread over a _____ ft^2 area.
3. Newly planted trees may have to be _____ to prevent wind damage.
4. Plants should be fertilized every _____ to _____ years, starting with the growing season after the first year.
5. Proper pruning techniques and timing are important in growing good ornamental trees and _____.
6. _____ is the systematic cutting of the roots by hand or machine to encourage the roots to develop close to the trunk.

D. TRUE OR FALSE

_____ 1. Trees and plants are valuable only for beauty.
_____ 2. Trees can act as an insulator and keep a house cool in the summer and warm in the winter.
_____ 3. A well-designed landscape does not increase the value of a property.
_____ 4. All plants have the same types of growth habits.
_____ 5. It is not a good practice to fertilize a plant after middle to late July.
_____ 6. Most plants are subject to damage from insects and/or diseases.

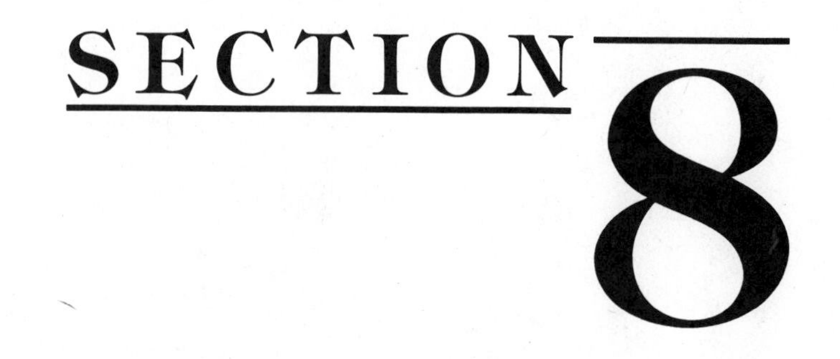

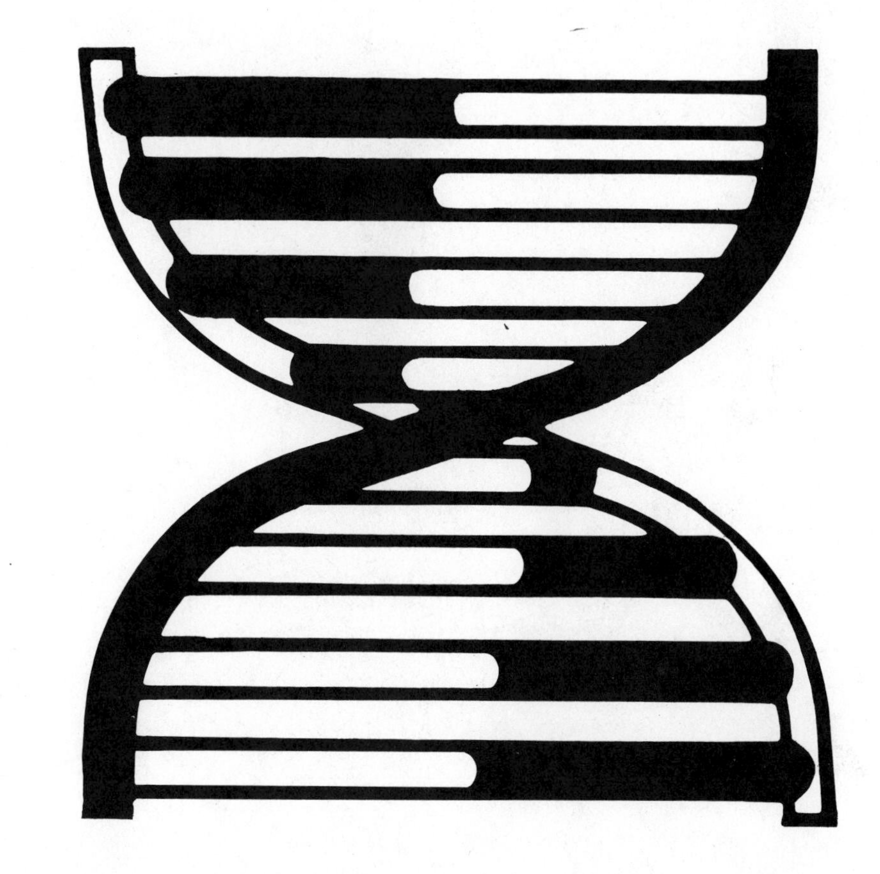

Animal Sciences

UNIT 26 Animal Nutrition

OBJECTIVE To determine the nutritional requirements of animals and how to satisfy those requirements.

Competencies to Be Developed

After studying this unit, you will be able to

- ☐ compare animal digestive systems.
- ☐ understand the basics of animal physiology.
- ☐ understand how nutrients are used by animals.
- ☐ identify classes and sources of nutrients.
- ☐ identify symptoms of nutrient deficiencies.
- ☐ explain the role of feed additives in livestock nutrition.
- ☐ compare the composition of various feedstuffs.

TERMS TO KNOW

Feed
Nutrition
Scurvy
Anorexia
Obesity
Ration
Balanced ration
Deficiency diseases
Vitamins
Minerals
Physiology
Skeletal system
Bone
Bone marrow
Muscular system
Voluntary muscles
Involuntary muscles
Proteins
Circulatory system
Lymph glands
Carbohydrates
Respiratory system
Central nervous system
Peripheral nervous system
Urinary system
Endocrine or hormone system
Hormones
Digestive system
Ruminants
Rumens
Roughage
Monogastric
Concentrates
Lactation
Glucose
Fructose
Galactose
Sucrose
Maltose
Lactose
Starch
Cellulose
Fat
Supplement
Synthetic nutrients
Oil meal
Urea
Rickets
Osteomalacia
Stiff lamb disease
White muscle disease
Dermatitis
Anemia
Goiter
Parakeratosis
Feed additive
Antibiotics
Dry matter
TDN
By-product
Hay
Legume
Green roughages
Silage

MATERIALS LIST

commercial feed tags from various feeds
charts of various animal digestive systems

Feed is animal food. It represents the largest single cost item in the production of livestock. Therefore, it is important to understand the complex nature of animal nutrition and how animals use the feed that they eat. *Nutrition* is the process by which animals eat food and use it to live, grow, and reproduce.

"You are what you eat" is an axiom that is true to a considerable extent, especially as it relates to good health of both humans and animals. This unit explores the relationship between good nutrition and good health.

NUTRITION IN HUMAN AND ANIMAL HEALTH The relationship between proper nutrition and health has long been recognized. Early sailors stocked their sailing vessels with limes when going to sea for long periods. This was to prevent the dreaded disease, scurvy. *Scurvy* is a disease of the gums and skin caused by a deficiency of vitamin C in the diet. Even today, the effects of poor nutrition are seen in the human problems of anorexia and obesity. In simple terms, *anorexia* is a result of too little nutrition, and *obesity* the result of too much or improper types of food being eaten.

In animals, proper nutrition is just as important as it is in humans. Feed efficiency, rate of gain, and days to market weight are all uppermost in the minds of those people who raise livestock for meat. Proper nutrition is just as important for animals being grown for milk and wool or fur production. Slow growth, poor reproduction, lowered production, and poor health are generally the result of less-than-adequate animal rations. The amount and content of food eaten by an animal in one day is referred to as the animal's *ration*. When the amount of feed consumed by an animal in 24 hours contains all of the needed nutrients in the proper proportions and amounts, the ration is referred to as a *balanced ration*.

Numerous diseases may result from improper amounts or balances of vitamins and minerals. Such diseases are called *deficiency diseases*. *Vitamins* are complex chemicals and *minerals* are elements essential for normal body functions of human and animal alike. Not all types of animals require the same vitamins and minerals to maintain good health.

ANIMAL PHYSIOLOGY The internal functions and vital processes of animals and their organs is referred to as animal *physiology*. The various body systems, such as the skeletal, muscular, circulatory, respiratory, nervous, urinary, endocrine, digestive, and reproductive systems, must all be properly fed and working together in order for the animal to be healthy and productive. To this end, proper nutrition is a must.

Skeletal System The *skeletal system* (Figure 26-1) is made up of bones joined together by cartilage and ligaments. The purpose of the skeletal system is to provide support for the body and protection for the brain and other soft organs of the body.

Bone is the main component of the skeletal system. It is composed of about 26% minerals. This mineral material is mostly calcium phosphate and calcium carbonate. Another 50% of bone is water, 20% protein, and 4% fat.

The material inside bones is called *bone marrow*, and it produces blood cells. The growth and strength of bones are greatly affected by minerals and vitamins in animal rations.

Muscular System The *muscular system* (Figure 26-2) is the lean meat of the animal and the part of the body that is used for human consumption (food). The purposes of muscles are to provide for movement in cooperation with the skeletal system, and to support life.

Muscles may be voluntary or involuntary, depending on whether or not they can be physically controlled by the animal. *Voluntary muscles* can be controlled by animals to do such things as walk and eat food. *Involuntary muscles* operate in the body without control by the will of the animal and function even while the animal sleeps.

Muscles are composed largely of protein. Large amounts are required for the maintenance of the animal and for growth and reproduction. *Proteins* are nutrients made up of amino acids and are building blocks of muscles.

Circulatory System The heart, veins, arteries, and lymph glands comprise the *circulatory system* (Figure 26-3). This system provides food and oxygen to the cells of the body and filters waste materials from the body. *Lymph glands* secrete disease-fighting materials from the body.

Vitamins, minerals, proteins, and carbohydrates are all essential for the smooth running of the circulatory system. *Carbohydrates* are sugars and starches that supply energy to the animal.

Respiratory System The *respiratory system* provides oxygen to the blood of the animal. It is composed of the nostrils, nasal cavity, pharynx, larynx, trachea, and lungs (Figure 26-4). This sys-

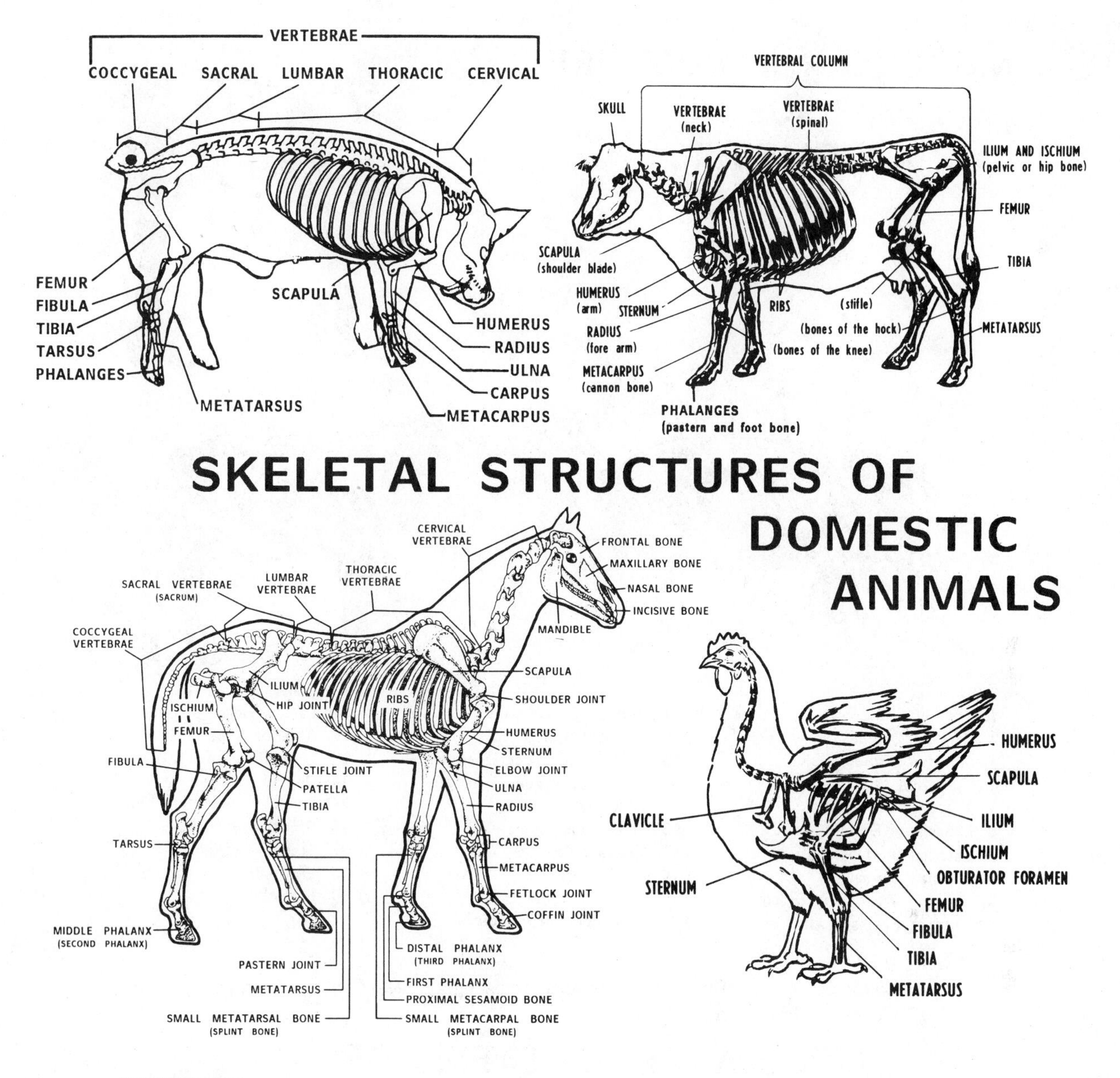

FIGURE 26-1
The skeletal system provides support for the body and protection for the soft organs. ***(Courtesy Instructional Materials Service, Texas A&M University)***

tem allows for breathing and makes use of the muscular and skeletal systems to draw air in and out of the lungs. Oxygen is then taken from the lungs and distributed to the cells of the body by the circulatory system.

Nervous System The nervous system of animals is composed of the central nervous system and the peripheral nervous system (Figure 26-5).

The *central nervous system* includes the brain and the spinal cord. It is responsible for coordinating the movements of animals and also responds to all of the senses. The senses are hearing, sight, smell, touch, and taste.

The *peripheral nervous system* controls the functions of the body tissues, including the organs. The nerves transmit messages to the brain from the outer parts of the body.

Because the nervous system is composed primarily of soft tissues, proteins are particularly important in maintaining its health.

Urinary System The function of the *urinary system* is to remove waste materials from the blood. The primary parts are the kidneys, bladder, ureters, and urethra (Figure 26-6). The kidneys also help regulate the makeup of blood and maintain other internal systems.

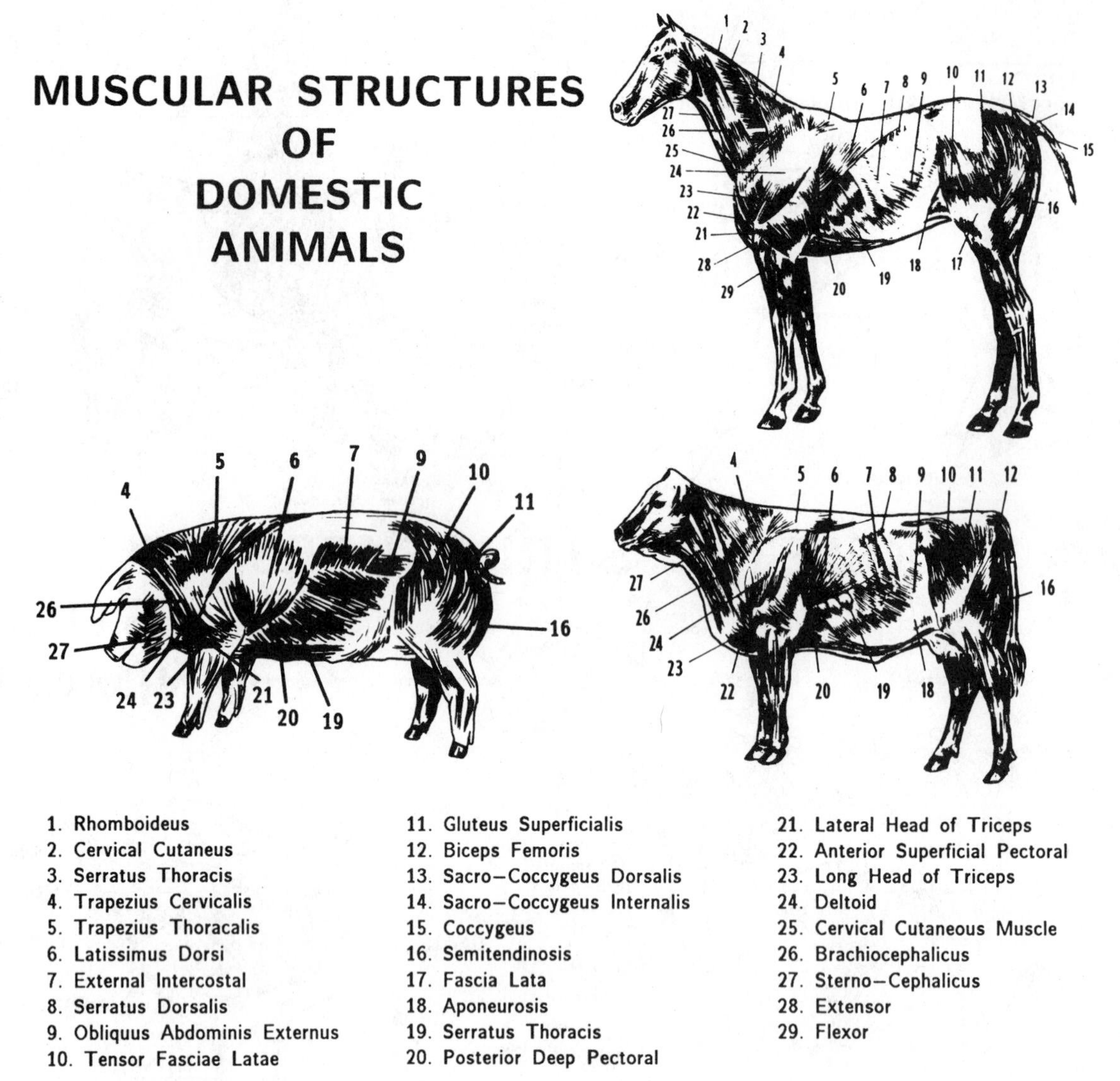

FIGURE 26-2

The muscular system provides for movement and supports life. ***(Courtesy Instructional Materials Service, Texas A&M University)***

CIRCULATORY SYSTEMS OF DOMESTIC ANIMALS

HEART
LUNGS
KIDNEYS
ARTERIES
VEINS
LIVER
VESSELS OF THE LARGE INTESTINES

HEART
LUNGS
KIDNEYS
ARTERIES
VEINS
LIVER
VESSELS OF THE LARGE INTESTINES

FIGURE 26-3

The circulatory system carries food and oxygen to the body. ***(Courtesy Instructional Materials Service, Texas A&M University)***

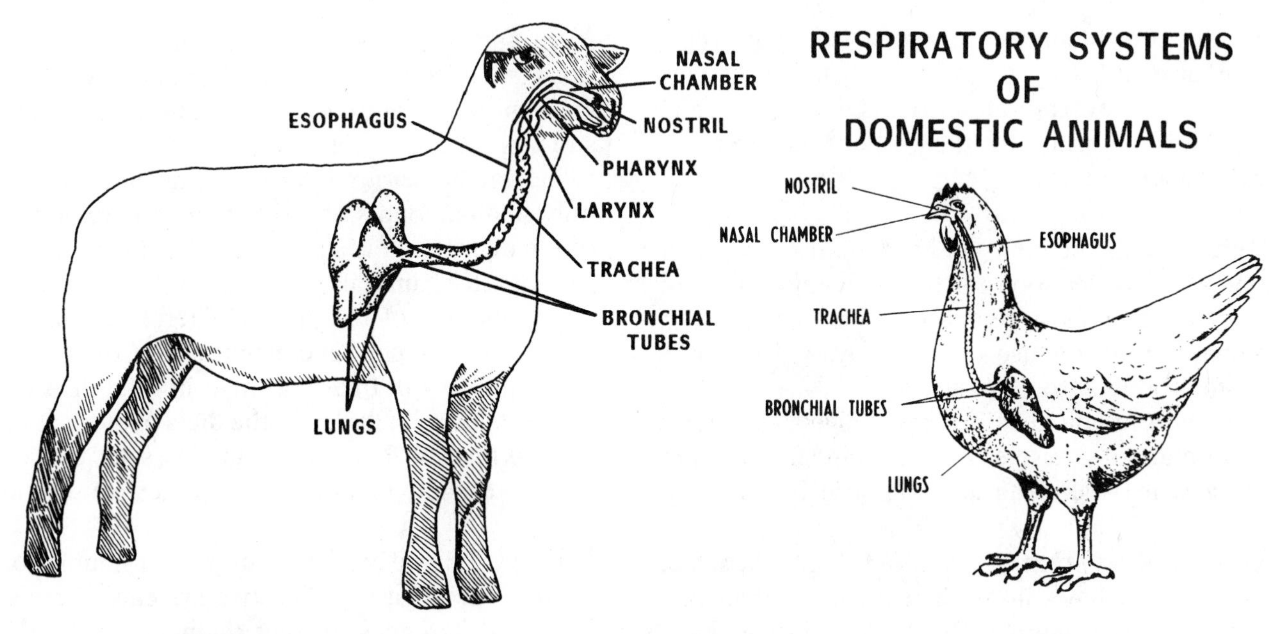

FIGURE 26-4

The respiratory system provides oxygen to the body. ***(Courtesy Instructional Materials Service, Texas A&M University)***

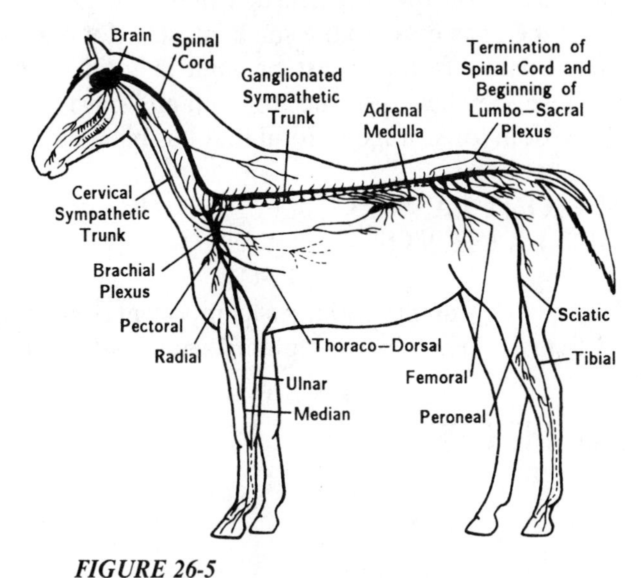

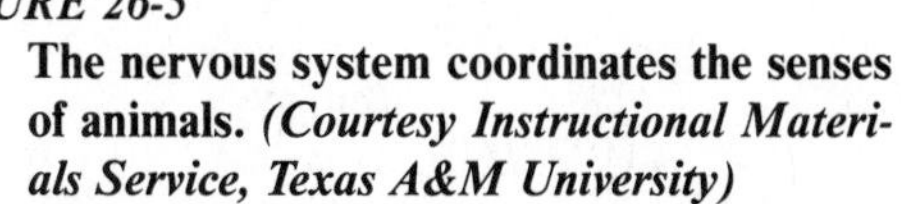

FIGURE 26-5

The nervous system coordinates the senses of animals. ***(Courtesy Instructional Materials Service, Texas A&M University)***

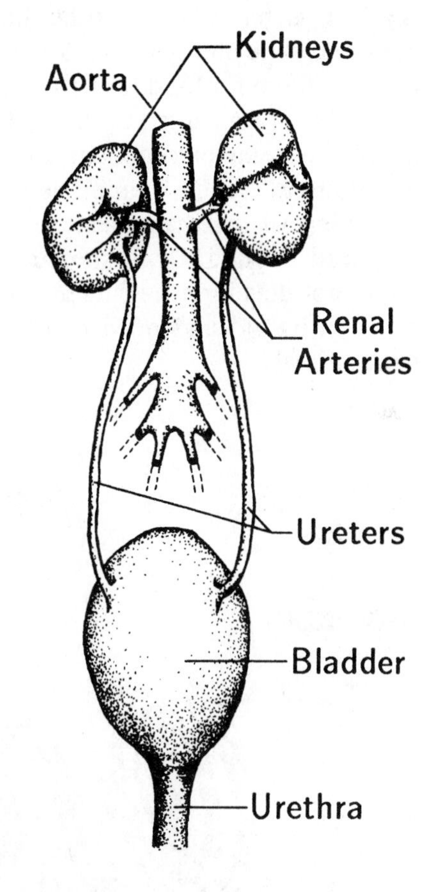

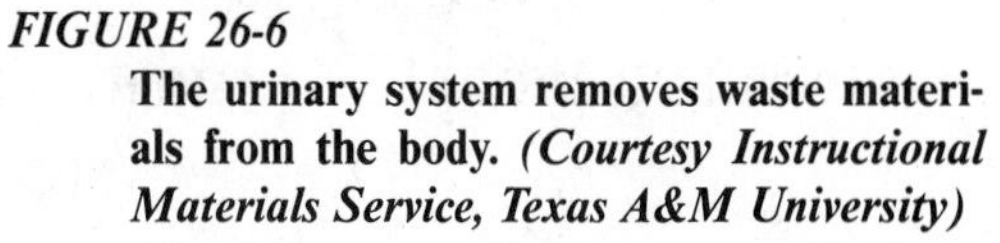

FIGURE 26-6

The urinary system removes waste materials from the body. ***(Courtesy Instructional Materials Service, Texas A&M University)***

Abnormal levels of proteins fed to animals have been known to cause stress to the urinary system, which rids the body of excess protein. Similarly, higher-than-recommended levels of minerals may also cause kidney problems.

Endocrine System The *endocrine or hormone system* is a group of ductless glands that release hormones into the body. *Hormones* are chemicals that regulate many of the activities of the body. Some of these are growth, reproduction, milk production, and breathing rate. Hormones are needed in only very minute amounts. For example, only 1/100,000,000 g of oxytocin hormone will stimulate the almost immediate letdown of milk in females. Oxytocin is a hormone from the hypothalmus gland.

Proper levels of all nutrients, especially minerals, are important for the proper functioning of the endocrine system.

Digestive System The *digestive system* of animals provides food for the body and for all of its systems. This system stores food temporarily, prepares food for use by the body, and removes waste products from the body.

There are three basic types of digestive systems in animals of agriscience importance. They are polygastric or ruminant, monogastric, and poultry.

Ruminant System *Ruminants* are classes of animals that have stomachs with more than one compartment (Figure 26-7). Cattle and sheep have multicompartment stomachs called *rumens*. These rumens can store very large amounts of roughages. A *roughage* is grass, hay, silage, or other high-fiber feed. Rumens have the ability to break down plant fibers and to use them for food far better than can animals that are not ruminants. Also manufactured in the digestive systems of ruminant animals are B-complex vitamins. Such vitamins need not be added to the diets of these animals, even though they are required by the body. It should be noted that calves do not develop true rumens until they are several months old and need to be fed like nonruminant animals.

Monogastric System The digestive system of swine, horses, and many other animals is called monogastric (Figure 26-8). *Monogastric* means a stomach with one compartment. The stomachs of swine and horses are relatively small and can store only small amounts of food at any one time. Most of the digestion takes place in the small intestines. Monogastric animals are unable to break down large amounts of roughages. Therefore, their rations must be higher in concentrates. *Concentrates* are grains low in fiber and high in total digestible nutrients. Also included in the diets of monogastric animals must be B-complex vitamins, since they cannot make such vitamins in their digestive systems.

Poultry Digestive System Although poultry have monogastric digestive systems, there are enough differences to treat them separately (Figure 26-9). Chickens have no teeth and must swallow their food whole. The food is stored in a crop and passed on to the gizzard, where it is ground up. It then passes on to the small intestine for digestion. Poultry rations must be high in food value because they have no true stomach and have very little room for storage of food that has been eaten.

MAJOR CLASSES OF NUTRIENTS

Water Water is the largest component of nearly all living things. Growing plants are usually 70 to

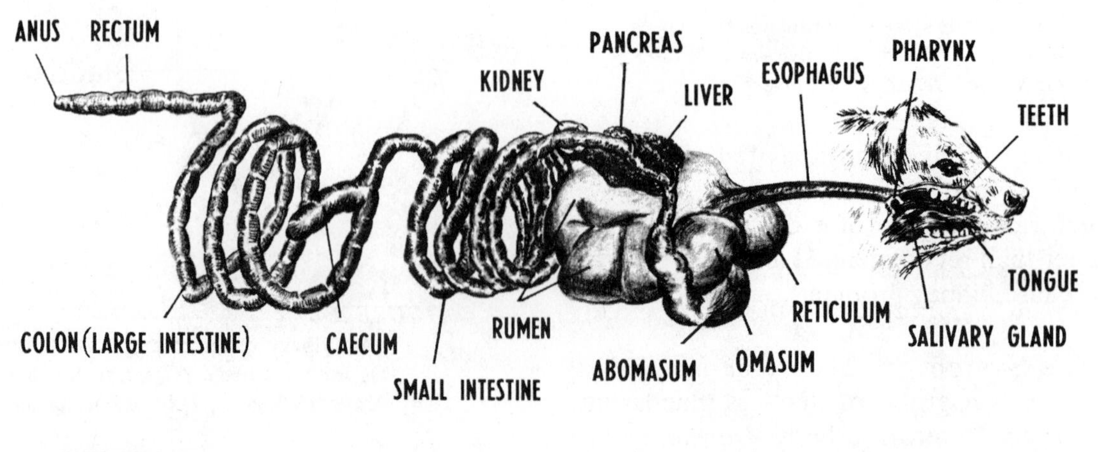

FIGURE 26-7
The ruminant digestive system can utilize large amounts of roughages. ***(Courtesy Instructional Materials Service, Texas A&M University)***

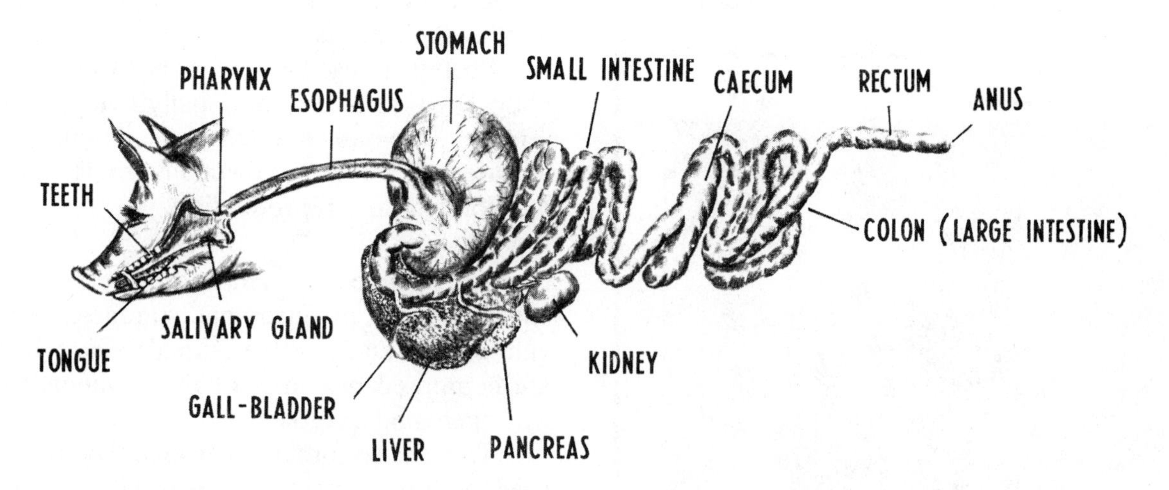

FIGURE 26-8
The monogastric digestive system has a simple stomach. ***(Courtesy Instructional Materials Service, Texas A&M University)***

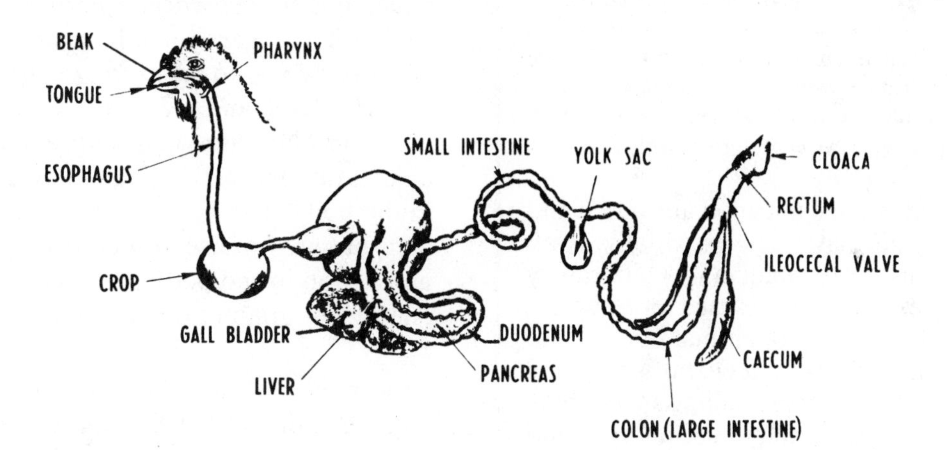

FIGURE 26-9
The poultry digestive system has no true stomach. ***(Courtesy Instructional Materials Service, Texas A&M University)***

80% water. Similarly, the muscles and internal organs of animals contain 75% or more of water.

Water is the solution in which all nutrients for animals are dissolved or suspended for transport throughout the body. Water reacts with many chemical compounds in the body to help break down food into products usable by the body.

Water provides rigidity to the body, allowing it to maintain its shape. The liquid solution in each cell is responsible for this rigidity.

Water is also important in regulating the body temperature of animals through perspiration and evaporation. Because of the ability of water to absorb and carry heat, body temperatures of animals rise and fall slower than would be possible otherwise.

Water is the least expensive nutrient for animals. Yet, most animals can live only a matter of days if they do not have access to it.

Protein Protein is the major component of muscles and tissues. Proteins are very complex materials and are made of various nitrogen compounds called amino acids. Some amino acids are essential for animals and some are not. Therefore, the quality of proteins fed to animals must be considered.

Monogastric animals need very specific amino acids. So it is important that they receive high-quality proteins containing the appropriate amino acids.

In ruminant animals, quantity of protein is more important than quality. They can convert amino acids in their rumens to other amino acids to meet their needs.

Protein is used by animals to maintain the body. Body cells are continually dying and being replaced. In young animals, large amounts of protein are used for body growth. Protein is also important for healthful reproduction.

AGRI-PROFILE

CAREER AREAS: ANIMAL NUTRITIONIST/ FEED FORMULATOR/ANIMAL MANAGER

Nutrition is one of the most critical factors in animal production. Animal nutritionists are frequently called on to evaluate feeds and make feeding recommendations to animal managers. ***(Courtesy United States Department of Agriculture)***

Careers in animal nutrition are varied and interesting. One may work essentially as an organic chemist or technician in a laboratory where complex equipment is used for research, analysis, and discovery of animals' needs and the nutritional values of feedstuffs. Or, one may be a business person selling feeds, may produce feedstuffs, or may raise animals and manage nutrition, along with other management practices.

Certain elements of nutrition are basic, such as the composition of feed grains, animal by-products, and the basic nutrients needed by animals. However, the nutritive content of forages and other feedstuffs vary considerably according to stage of growth, condition, and quality; and, the digestive capabilities of animals vary considerably from species to species. Additionally, nutritional needs of animals vary with age, stage of development, production, pregnancy, and the like.

Careers in nutrition may be predominantly in the basic sciences or may be applied. They may focus on fish, small animals, pets, equines, poultry, livestock, dairy, or wild animals.

Carbohydrates These are a class of nutrients composed of sugars and starches. They provide energy and heat to animals. Carbohydrates are composed primarily of the elements carbon, hydrogen, and oxygen.

The energy obtained from carbohydrates is used for growth, maintenance, work, reproduction, and *lactation* (milk production). Carbohydrates come in several forms, with the sugars being the simplest. Examples of simple sugars used in animal feeds are *glucose*, *fructose*, and *galactose*. Compound sugars include *sucrose*, *maltose*, and *lactose. More complex forms of carbohydrates include starch* and *cellulose*.

Carbohydrates make up about 75% of most animal rations. Yet there is very little carbohydrate in the body at any one time. Carbohydrates in the diet that are not used quickly are converted to fat and stored in the body. *Fat* has 2¼ times as much energy per gram as do carbohydrates.

Minerals The functions of minerals in the animal are many. The skeleton is composed mostly of minerals. They are important parts of soft tissues and fluids of the body. The endocrine system is heavily dependent on various minerals, as are the circulatory, urinary, and nervous systems.

There are 15 minerals that have been identified as being essential to the health of animals. These are calcium, phosphorus, sodium, chlorine, potassium, sulfur, iron, iodine, cobalt, copper, fluorine, manganese, molybdenum, selenium, and zinc. In the past, most of these minerals were provided naturally by feeds grown on fertile soils and by contact with the soil itself. Today it is increasingly important to provide additional mineral matter to the diet of animals. Mineral supplements are especially important for animals that spend their lives in confinement. Additional minerals in feed is called a *supplement*.

Vitamins Vitamins are acquired by animals in several different ways. Some are available in roughages and concentrates. Some are available in feeds made from animal by-products. Finally, some are made by the body itself.

Vitamins are required in only minute quantities in animals. They act mostly as a catalyst in other body processes. There are large variations in the necessity for vitamins in various species of animals important to agriscience.

Some of the specific ways that vitamins are used in animals include clotting of blood, forming bones, reproducing, keeping membranes healthy, producing milk, and preventing certain nervous-system disorders.

Fat Only small amounts of fat are required in most animal diets. The addition of fat to the diets of animals improves the palatability, flavor, texture, and energy levels of feed. The addition of small amounts of fat to the diet has also been shown to increase milk production and to aid in the fattening of meat animals. Fats are also necessary in the body as a carrier for fat-soluble vitamins.

SOURCES OF NUTRIENTS

The sources of nutrients for animals are many and extremely varied. Important animal feed components include roughages, concentrates, animal by-products, minerals from mineral deposits, and nutrients made chemically. These are called *synthetic nutrients*.

Proteins The major sources of protein for animals include oil seeds such as soybeans, peanuts, cottonseed, and linseed. These seeds are processed by cooking and other procedures to remove the bulk of the oil from them. The remainder of the seed content is then dried and ground up for feed. The feed consisting of ground oil seeds with the oil removed is called *oil meal*.

Cereal grains provide lesser amounts of protein, but are also important protein sources. Good quality legume hay, such as alfalfa and clover, is also a good plant source of protein for ruminant animals.

Animal protein is generally of higher quality than plant protein. More specifically, animal protein usually contains more of the essential amino acids than does protein from plants. Sources of animal protein include tankage, fish meal, blood meal, skim milk, feather meal, and meat scrap.

Nonprotein nitrogen in the form of urea can be used as a substitute for protein for ruminant animals. *Urea* is a synthetic source of nitrogen made from air, water, and carbon. The feeding of urea should be limited to not more than 1% of the total dry matter fed. Young ruminants and all nonruminants are unable to digest urea.

Carbohydrates Carbohydrates are found in all plant materials. The major sources of carbohydrates for animal feed are the cereal grains. Corn is the most important of these in the United States, followed by wheat, barley, oats, and rye.

Other sources of carbohydrates include nonlegume hay such as orchard grass, timothy, other grasses, and molasses.

Normal animal rations generally contain adequate levels of carbohydrates.

Fats Because fats are needed in fairly small amounts in the diets of animals, it is seldom necessary to identify specific sources of dietary fat. Most sources of protein are also sources of fat. This is especially true for the oil seeds and animal by-products.

Vitamins and Minerals Vitamins and minerals are part of all the normal feeds of animals. Ruminants manufacture B-complex vitamins in their rumens. Exposure to sunlight allows the body to manufacture vitamin D. Contact with the soil, coupled with other feeds grown on fertile land, provides most of the mineral requirements of animals that have access to pasture and high-quality feeds.

However, it is sometimes necessary to supplement natural sources of vitamins and minerals. Commercial vitamin and mineral supplements are formulated for specific classes of animals and their special needs. Such supplements are available wherever animals exist in the developed countries of the world.

SYMPTOMS OF NUTRIENT DEFICIENCIES

Animals must be fed appropriate types and amounts of feed regularly to remain healthy and produce milk, meat, wool, eggs, fur, work, or healthy young. Shortages or deficiencies of various nutrients will generally produce observable effects on the animal. Some common symptoms of nutrient deficiencies follow.

Vitamin Deficiency

Vitamin A

1. Night blindness
2. Loss of young
3. Poor growth
4. Nasal discharge
5. Diarrhea

Vitamin C

1. Scurvy
2. Gum inflammation
3. Hemorrhages
4. Slow healing of wounds

Vitamin D

1. Bone weakness and deformities
2. *Rickets* and *osteomalacia* are bone disease in young and old animals, respectively
3. Thin egg shells
4. Weak, deformed young
5. Lowered milk production

Vitamin E

1. Reproductive failures
2. Degeneration of certain muscles
3. *Stiff lamb disease*, or muscle degeneration in lambs
4. *White muscle disease*, or muscle degeneration in young calves
5. Poor egg hatchability

Vitamin K

1. Poor blood clotting
2. Internal hemorrhages

Thiamine

1. Poor appetite
2. Slow growth
3. Weakness
4. Nervousness

Riboflavin

1. Slow growth
2. *Dermatitis*, or skin disorder
3. Eye abnormalities
4. Diarrhea
5. Weak legs in pigs

Niacin

1. Dermatitis
2. Retarded growth
3. Digestive troubles

Pyridoxine

1. *Anemia*, or low red-blood-cell count
2. Poor growth
3. Convulsions in pigs

Pantothenic acid

1. "Goose-stepping" in pigs
2. Unhealthy appearance
3. Digestive problems

Biotin

1. Dermatitis
2. Loss of hair
3. Retarded growth

Choline

1. Poor coordination
2. Poor health
3. Fatty liver
4. Poor reproduction in swine

Folic acid

1. Blood disorders
2. Poor growth

Vitamin B_{12}

1. Slow growth
2. Poor coordination
3. Poor reproduction

Mineral Deficiency

Calcium

1. Rickets
2. Poor growth
3. Deformed bones
4. Milk fever

Phosphorus

1. Lameness
2. Stiff joints
3. Rickets
4. Poor milk production

Sodium chloride

1. Lack of appetite
2. Unhealthy appearance
3. Slow growth

Potassium

1. Slow growth
2. Joint stiffness
3. Poor feed efficiency

Sulfur

1. General unthriftiness (Lack of strong growth)
2. Poor growth

Iron

1. Anemia
2. Labored breathing
3. Edema (swelling) of the head and shoulders
4. Flabby, wrinkled skin

Iodine

1. *Goiter*, or enlarged thyroid gland in the neck
2. Weak or dead offspring at birth
3. Hairlessness
4. Infected navels, especially in foals

Cobalt
1. Delayed sexual development
2. Poor appetite
3. Slow growth
4. Decreased milk and wool production

Copper
1. Abnormal wool growth
2. Poor muscular coordination
3. Anemia
4. Weakness at birth

Fluorine—Poor teeth

Manganese
1. Poor fertility
2. Deformed young
3. Poor growth

Molybdenum—Poor growth rate

Selenium
1. Muscular degeneration
2. Heart failure
3. Paralysis
4. Poor growth

Zinc
1. Poor growth
2. Unhealthy wool or hair
3. Slow healing of wounds
4. *Parakeratosis*, or a skin disease similar to mange

FEED ADDITIVES A *feed additive* is a nonnutritive substance that is added to feed to promote more rapid growth, to increase feed efficiency, or to maintain or improve health. Feed additives fall into two major groups—growth regulators (mostly hormones) and antibiotics. *Antibiotics* are substances used to help prevent or control diseases.

Some common growth regulators include stilbestrol, progesterone, and testosterone. They increase growth rates and feed efficiency by as much as 5%.

There are a wide range of antibiotics that are added in very low levels to the diets of animals such as swine and poultry. Antibiotics keep certain low-grade infections at bay. Antibiotics added to feed allow growing animals to gain weight at their highest potential.

In recent years there has been a good deal of controversy over including antibiotics and growth hormones in the feed of animals. Of major concern is the possibility of these substances remaining in the meat of animals slaughtered for human consumption. To reduce this possibility, the substances generally must be removed from the feed well before the animal is marketed.

COMPOSITION OF FEEDS All feeds are composed of water and dry matter. The material left after all water has been removed from feed is *dry matter*. Water makes up 70 to 80% of most living things. However, dry feeds generally contain only 10 to 20% water.

Dry matter is made up of organic matter and ash or mineral. The organic-matter portion of animal feeds consists of protein, carbohydrates such as starch and sugar, fat, and some vitamins. The proportion of these materials vary widely with different feeds.

CLASSIFICATION OF FEED MATERIALS In general, feed for animals is classified into two types—concentrates and roughages. Concentrates are low in fiber and high in total digestible nutrients, abbreviated as *TDN*. Total digestible nutrients include all of the digestible protein, digestible nitrogen-free extract, digestible crude fiber, and 2¼ times the digestible fat contained in the ration. On the other hand, roughages are high in fiber and low in TDN.

Concentrates Included under the classification of concentrates are the feed or cereal grains. These include corn, wheat, oats, barley, rye, and milo, as well as many others. These grains make up the bulk of most concentrates.

Also, grain by-products such as wheat bran, wheat middlings, brewers grain, and distillers grain are concentrates. They are materials left over from the production processes used in making flour and alcohol. A *by-product* is a secondary product left from the production of a primary commodity.

The oil meals are by-products left from making vegetable oil. Both oil meals and sugar in the form of cane molasses and beet molasses are considered concentrates used in animal feeds.

Finally, animal by-products are important concentrates. They include tankage, fish meal, meat scraps, blood meal, feather meal, and dried dairy products. These products are some of the by-products of the food processing industry.

Roughages Roughages can be divided into three categories—dry, green, and silage. The most important of the dry roughages is *hay*. Some types of legume hay are alfalfa, clover, lespedeza, soybean, and peanut. A *legume* is a plant in which certain bacteria can transform nitrogen in the air to nitrogen that plants can use. Grass hays include timothy, orchard grass, bromegrass, Bermuda grass, and others. The hulls of cottonseed, peanuts, and rice also fall into the category of dry roughages.

Green roughages are plant materials with a high moisture content, such as grasses in pastures and root plants, including sugar beets, turnips, and rutabagas. Tubers such as potatoes are also a green roughage.

Silage is the feed that results from the storage and fermentation of green crops. The fermentation takes place in the absence of air. Corn silage is the most important member of this group. Other examples are grass, legume, and small grain silages.

The successful production of animals for fun, profit, or sport requires proper animal nutrition. A knowledge of animal physiology, feed materials, and nutrition helps keep animals healthy and productive. Further, a knowledge of nutrition deficiency symptoms will permit the animal manager to take corrective steps when the animal suffers from improper nutrition.

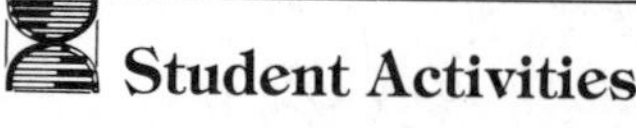

Student Activities

1. Define the "Terms To Know."
2. Obtain several different commercial feed tags and compare the percentage of protein, fat, TDN, and other nutrients listed on the tag. Also make note of any vitamins and mineral supplements that are part of the ingredients. Types of feed additives should also be noted. If the price is known, try to determine what causes variations in the price of the various feeds.
3. Compare the digestive systems of ruminants, nonruminants, and poultry. Note the parts that are alike and those that are different.

Self-Evaluation

A. MULTIPLE CHOICE

1. ______ are most important for the formation of bone.

   a. Vitamins
   b. Proteins
   c. Minerals
   d. Carbohydrates

2. Sugar, starch, and cellulose are all components of

   a. carbohydrates.
   b. minerals.
   c. fats.
   d. proteins.

3. The purpose of additional fat in the diet is to

   a. improve palatability.
   b. increase energy levels.
   c. improve feed texture.
   d. all of the above.

4. Night blindness may be caused by a deficiency of

   a. zinc.
   b. vitamin A.
   c. protein.
   d. biotin.

5. Improved feed efficiency can be accomplished by adding ______ to animal feeds.

   a. fats
   b. minerals
   c. additives
   d. roughages

6. Grain, oil meal, molasses, and meat by-products are all forms of

   a. roughage.
   b. concentrate.
   c. carbohydrate.
   d. dry matter.

7. Dry matter is made up of organic matter and

   a. mineral.
   b. water.
   c. fat.
   d. concentrate.

B. MATCHING (Group I)

______	1. Goiter	a. Iron
______	2. Parakeratosis	b. Vitamin D
______	3. Rickets	c. Vitamin E
______	4. White muscle disease	d. Zinc
______	5. Anemia	e. Iodine
______	6. Dermatitis	f. Niacin

MATCHING (Group II)

______	1. Ration	a. Feed high in TDN, low in fiber
______	2. Ruminant	b. Disease-control substance
______	3. Carbohydrate	c. Essential element
______	4. Mineral	d. Amount of feed fed in one day
______	5. Antibiotic	e. Fermented green roughage
______	6. Silage	f. Sugars and starches
______	7. Concentrate	g. Multicompartment stomach

UNIT 27 Animal Health

OBJECTIVE To determine how to best maintain animal health.

Competencies to Be Developed

After studying this unit, you will be able to

- ☐ identify signs of good and poor health.
- ☐ identify symptoms of animal diseases and parasites.
- ☐ understand how to prevent health problems.
- ☐ explain various methods of treating animal health problems.

TERMS TO KNOW

Diseases
Parasites
Syringe
Disinfectant
Feedlots
Host animal
Contagious
Noncontagious
Abortion
Roundworms
Flukes
Protozoa
Secondary host
Mange
Vaccination
Balling gun
Drenching
Injection
Intravenous
Intramuscular
Subcutaneous
Intradermal
Intraruminal
Intraperitoneal
Infusion
Udder
Teats
Cannula
Dipping
Rectum
Immune
Veterinarian

MATERIALS LIST

various gauges of needles and syringes
various containers or labels from containers of animal drugs
discarded ears from a meat processing plant

Maintaining animal health is the key to a profitable and satisfying animal enterprise. There are several considerations that need to be made in dealing with the health of animals. These include being able to recognize signs of good and poor health, maintaining a healthy environment, being able to identify animal diseases and parasites, and knowing how to treat health problems when they occur. These items will be explored in this unit. *Diseases* are infective agents that result in lowered health in living things. *Parasites* are animals that live on other animals and derive their food from their host.

SIGNS OF GOOD AND POOR HEALTH

Having the ability to recognize the signs of good health or the symptoms of health problems is the single most important key to being efficient in maintaining good animal health. A keen sense of observation is important, as well as the innate ability to know when something is not right with an animal.

Signs of Good Health

One of the best signs of good health is simply a contented animal. Of course, a good deal of experience in dealing with animals is necessary to recognize contentment. Alertness and the chewing of the cud in ruminant animals is a good sign. A shiny hair coat, bright eyes, and pink membranes are other signs that an animal is healthy. Normal body discharges of urine and feces are further evidences that animals are not suffering from serious health problems. On the technical side, healthy animals should have a normal body temperature, pulse rate, and respiration or breathing rate (Figure 27-1).

Signs of Poor Health

Often it is easier to tell when an animal is sick than to tell when it is healthy. A rough hair coat and dull, glassy eyes are

Class of Livestock or Poultry	Degree F Average	Degree F Range
Cattle	101.5	100.4-102.8
Sheep	102.3	100.9-103.8
Goats	103.8	101.7-105.3
Swine	102.6	102.0-103.6
Horses	100.5	99.9-100.8
Poultry	106.0	105.0-107.0

FIGURE 27-1
Normal body temperatures for animals

often the first signs that an animal is not well. Sick animals usually stay alone, with their heads down. They may be drawn up and walk slowly when forced to walk. Abnormal feces, either too hard or too soft, as well as discolored urine may also indicate that an animal is suffering from some health problem. Lowered production, especially in dairy cattle, is often the first sign that the animal is not well. High temperatures, labored breathing, and rapid pulse rates are other indications of poor health in animals (Figure 27-2).

HEALTHFUL ENVIRONMENTS FOR ANIMALS

Maintaining a healthy environment for animals is a key factor in a complete animal-health program. It is often much less expensive to maintain a healthy environment for animals than it is to treat animals that are unhealthy due to poor conditions that occur.

Sanitation Good sanitation is important to good health. Factors related to good sanitation include keeping facilities for animals clean. Sanitation also requires the use of clean equipment when dealing with animals. This includes milking equipment, artificial breeding equipment, needles and syringes, and surgical equipment. A *syringe* is an instrument used to give injections of medicine or to draw body fluids from animals (Figure 27-3). Simple on-farm surgical procedures should always be performed with the strictest sanitation possible. The liberal use of disinfectants in dealing with animals is also important. A *disinfectant* is a material that kills disease-causing organisms.

Housing Maintenance of proper housing is also an important consideration in maintaining good animal health. Housing should be clean and free from cold drafts. Good air circulation throughout the housing is important to help lower high temperatures in the summer and reduce humidity in the cold of winter. Extremely dry and dusty conditions are also to be avoided when possible. Proper maintenance of animal housing is also important. Loose boards, roofing materials, and nails often pose problems in poorly maintained facilities.

Handling Manure Piles of manure, dirty pens, and dirty feedlots are often sources of serious health problems in animals. It is important that manure not be allowed to accumulate in areas frequented by animals (Figure 27-4). Manure piles

FIGURE 27-2
An unhealthy animal is an unprofitable animal. ***(From Gillespie/Modern Livestock & Poultry Production, copyright 1989 by Delmar Publishers Inc.)***

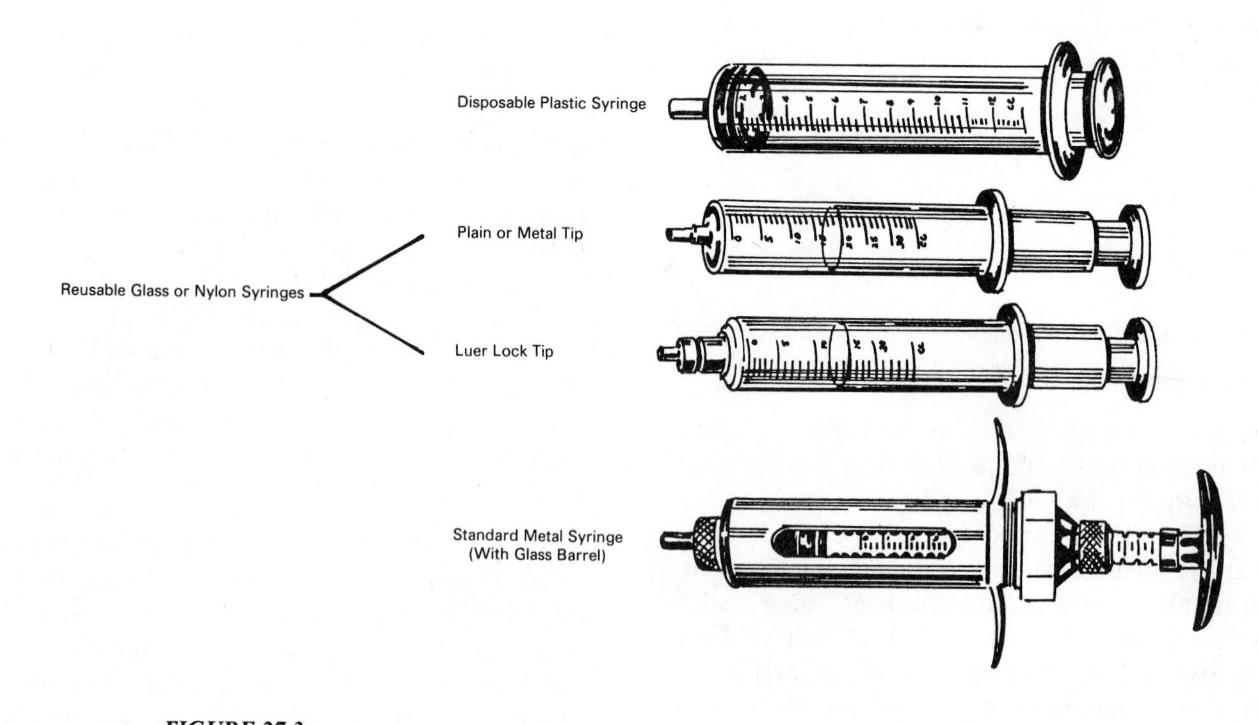

FIGURE 27-3
Parts of a syringe ***(Courtesy Instructional Materials Service, Texas A&M University)***

often harbor diseases and parasites. They also attract flies, which may spread diseases. Cages and pens soiled continually with animal waste products may also lower the quality of the air that animals breathe. Wet, poorly drained, manure-soiled feedlots usually reduce the rate of gain of beef cattle and swine. *Feedlots* are areas in which large numbers of animals are grown for food. Feet and leg problems can often be traced to poorly maintained feedlots.

Controlling Pests The control of pests and parasites is also an important consideration in the maintenance of animal health and welfare. Regular use of disinfectants to control parasites such as lice and flies is necessary in a good disease-prevention program. Regular, close observation of animals may be necessary to determine when outbreaks of parasites occur. Prevention of such parasites is preferable to controlling outbreaks that occur. To that end, the development of a good prevention program is a wise decision.

The control of other pests, such as birds and wild animals, is also part of a good animal health program. Many birds carry parasites on their bodies and in their droppings. When they move from infected animals to healthy ones, they often carry diseases and parasites with them. Wild animals and pets may also cause serious health problems when allowed to roam freely around farm animals. Dogs and coyotes will often chase animals and cause them to injure themselves. Bites from these animals may also cause infection and other health problems. Just the presence of pets around farm animals may cause them to be nervous and affect how rapidly they grow and produce.

Isolation The isolation of animals new to the herd is an important part of any good health-prevention program. Such animals may be harboring disease or parasites that are not readily apparent. It is wise to keep them isolated from other animals for a period of time, usually a minimum of 30 days. This gives the new owner time to observe the animals closely for health problems.

Similarly, isolation of diseased animals is important. Animals with contagious diseases that can be spread by contact should never be allowed to come into contact with healthy animals. It is difficult to treat unhealthy animals when they are living with large groups of animals. Healthy animals tend to pick on unhealthy ones, making it especially difficult for such animals to regain health.

Pasture Rotation The rotation of pastures is a consideration in maintaining a healthy environment for animals and in preventing health problems. Many diseases of animals are harbored in the soil and are killed only by not being able to come into contact with host animals for extended

FIGURE 27-4
An efficient manure handling system helps control disease and prevent parasites. *(Courtesy of Gilbert High School, Agriculture Education Department, photo by Joe Granio)*

periods of time. A *host animal* is a species of animal in or on which diseases or parasites can live. Moving animals to different pastures on a regular basis also allows for the breakdown of animal wastes and for pasture regrowth (Figure 27-5).

ANIMAL DISEASES AND PARASITES

Diseases The diseases of animals can be divided into two major classes, contagious and noncontagious. *Contagious* diseases are those that can be passed on to other animals. *Noncontagious* diseases cannot be spread to other animals.

The handling of these two classes of diseases varies somewhat. It is important that animals suffering from contagious diseases be isolated from other animals in the herd as soon as the disease is identified. Since some contagious diseases of animals can be transmitted to humans, care must be taken when handling animals so infected.

Noncontagious diseases pose no threat to humans or other animals, except to the animals with the disease. Therefore, there is more leeway in

FIGURE 27-5
Rotation of pastures is necessary to break internal parasites' life cycles. ***(Courtesy of Gilbert High School, Agriculture Education Department, photo by Joe Granio)***

dealing with animals suffering with noncontagious diseases. It is still a good idea to isolate these animals from the herd for their own good.

Causes Contagious diseases are caused mostly by bacteria and viruses. They can be spread by direct contact with infected animals, from shared housing, or from contaminated feed or water. In some cases, the spread of infectious diseases takes place through intermediary hosts, such as birds, rodents, or insects.

Noncontagious diseases may be caused by nutrient deficiencies or nutrient excesses. Poisonous plants and animals, injection of foreign material, and open wounds may cause or lead to noncontagious disease.

Symptoms General symptoms of disease are extremely varied and may include:

1. poor growth and/or reduced production.
2. reduced intake of feed.
3. rough, dry hair coat.
4. discharge from the nose or eyes.
5. coughing or gasping for breath.
6. trembling, shaking, or shivering.
7. unusual discharges such as diarrhea, bloody feces, or urine.
8. open sores or wounds.
9. unusual swelling of the body, including lumps and knots.
10. *abortion*, or the loss of a fetus before it is born.
11. peculiar gait or walking pattern, or other odd movements.

Some diseases may have little or no external symptoms and may even progress so rapidly that death of the animal occurs before noticeable symptoms occur.

Parasites Parasites may also be grouped into two general classifications. They are internal—inside the animal—and external—living on the outside of the animal.

The most important group of internal parasites infesting animals are the *roundworms* (slender worms that are tapered on both ends). Other types of internal parasites include flukes and protozoa. *Flukes* are very small, flat worms and *protozoa* are microscopic, one-celled animals. Most internal parasites spend at least some of their life cycle outside of the host animal (Figure 27-6). It is during this period that the parasite may most easily be spread to other animals. Contact with discharges from infested animals, contaminated feed, water, housing, or by contact with secondary hosts may result in the spread of internal parasites. A *secondary host* is a plant or animal that carries a disease or parasite during part of the life cycle. Some internal parasites are also spread by insects such as flies and mosquitoes.

External parasites include flies, ticks, mites, and fleas. They are spread in the same ways as are internal parasites.

Symptoms of parasite infestation may include:

1. poor growth.
2. weight loss.
3. constant coughing and gagging.
4. anemia.
5. lowered production and reproduction.
6. diarrhea or bloody feces.
7. worms in the feces.
8. swelling under the neck.
9. poor stamina.
10. loss of hair and *mange*, the presence of a crusty skin condition caused by mites.
11. visibility of the parasite itself.

PREVENTING AND TREATING HEALTH PROBLEMS

There are a number of activities and procedures that are used to prevent and treat health problems found in animals. Some of these include administering drugs, dipping, and restraining animals. The role of feed additives and vaccination will also be explored. *Vaccination* is the injection of an agent into an animal to prevent disease.

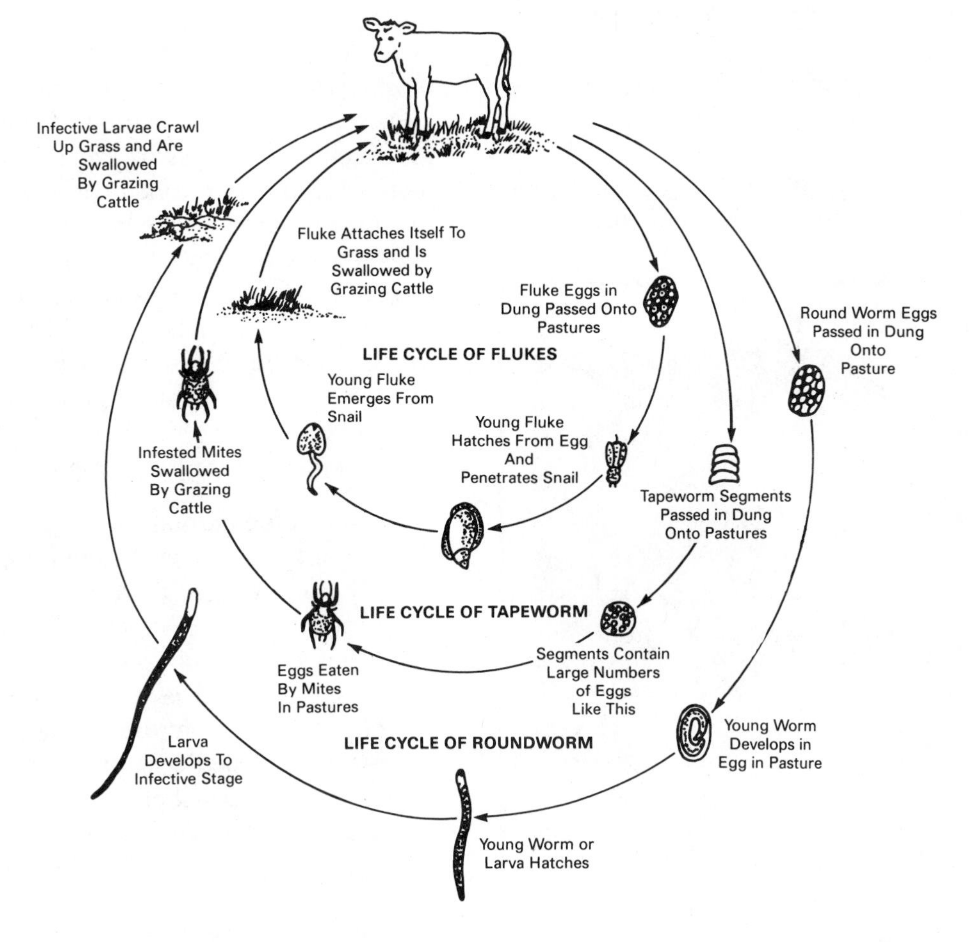

FIGURE 27-6
Life cycles of some common internal parasites

Administering Drugs There are several factors to be considered before administering drugs to an animal. They include determination of the amount to be administered, type of drug to use, purpose of the drug, site of administration of the drug, and type of animal to be treated. Most of this information can be found on the drug container. It is important that the drug manufacturer's recommendations be followed closely. Another factor that needs to be considered is the amount of time that the drug remains in the animal. This is important when determining how long milk from the animal will be contaminated by the drug. Contaminated milk must be discarded. Also, it must be determined how long to wait before a treated animal can be slaughtered for meat.

Drugs may be manufactured and sold as pills, powder, paste, and liquid.

Pills The procedure for giving a pill to an animal is to restrain the animal and lift its head so that the mouth opens. Force the pill as far down the side of the mouth as possible, using either your hand or a *balling gun* (a device used to place a pill in an animal's throat). Then massage the animal's throat until it swallows the pill.

Powders These drugs are normally mixed in the feed or water of the animal. Often it is necessary to withhold feed or water for a period of time before administering the drug. Otherwise, the animal may refuse to eat or drink the drugged food.

Paste Paste is normally used for treating horses for worms. The preparation is placed on the back of the horse's tongue with a caulking gun, and the horse is forced to swallow. Pastes are used for horses because it is often nearly impossible to treat them for worms by any other method.

Liquids Liquid drugs administered orally (by mouth) are often placed directly in the animal's stomach by drenching. *Drenching* is the process of administering fairly large amounts of liquid to an animal. A syringe or drenching gun is used. In the process, the animal is restrained, with the head held level. The upper lip of the animal is lifted and the tube is inserted along the side of the tongue. The drug is released and the animal is allowed to swallow. Care must be taken not to get the drug into the animal's lungs.

The injection of drugs into animals takes many forms, based on the location of the injection. *Injection* is the process of administering drugs by needle and syringe. Some of the injection sites include *intravenous* (in a vein), *intramuscular* (in a muscle), *subcutaneous* (under the skin), *intradermal* (between layers of skin), *intraruminal* (in the rumen), and *intraperitoneal* (in the abdominal cavity). One determining factor as to where injections are made is how fast the drug needs to work. A drug injected into the blood is available faster than one injected under the skin. Often it is desirable for drugs to be released slowly over a long period of time. Growth hormones are generally administered in this way.

The procedure for giving an injection is to:

1. restrain the animal.
2. select the location for the injection.
3. fill the syringe, making sure that all air is removed.
4. disinfect the area to be injected.
5. if the injection is to be made intradermally, clip the hair from the area to be injected.
6. insert the needle in the desired area without the syringe attached. This prevents the loss of the drug if the animal jumps.
7. attach the syringe to the needle and inject the liquid.

Infusion *Infusion* is another method of getting drugs to the site of the infection. It is used most often to treat dairy animals with udder and teat problems. The *udder* is the milk-secreting glands of the animal. *Teats* are the appendages of an udder. A sterile *cannula* (blunt needle) is inserted into the opening of the teat, and the drug is forced into the teat canal (Figure 27-7).

Dipping *Dipping* is a process for treating animals, mostly cattle and sheep, for external parasites. It involves filling a vat with medicated water and forcing the animal to walk or swim through it. This process is also used to treat dogs for ticks and fleas. Dipping is popular where large numbers of animals must be completely covered with the medication.

Taking Temperatures Taking an animal's temperature is basically the same as taking a human's temperature. It is usually taken in the rectum. The *rectum* is the last organ in the digestive tract. Animal thermometers are normally longer and heavier than ones used in human medicine. They also have an eye at one end and should have a string attached to it to prevent loss of the thermometer into the body cavity.

AGRI-PROFILE

CAREER AREAS: VETERINARIAN/ANIMAL HEALTH TECHNICIAN

Veterinarians and their associates are working constantly for better ways to maintain animal health and treat diseases. ***(Courtesy USDA Agricultural Research Service)***

Animal pathologists, animal behaviorists, physiologists, biologists, zoologists, microbiologists, geneticists, nutritionists and others must work together to understand the complexities of animals. Animals exist as pets, production animals, work animals, pleasure animals, fish, fowl, birds, wild animals, and specimen animals in zoos. The need for health services for animals varies with the species and use of the animal.

The desire to become veterinarians has been high among youth in recent years. This interest has permitted the supply of veterinarians to increase despite the rigor of the college curriculum and competition to get into veterinarian schools. Colleges of animal sciences offer curriculums in other career areas in animal sciences, such as nutrition, breeding, education, and production.

In urban and suburban areas, animal shelters, hospitals, kennels, and pet stores provide many career opportunities for those interested in animal health at the technician level. In rural areas, large animal veterinarian, vet assistant, laboratory veterinarian, and laboratory technician are typical positions in the animal health industry.

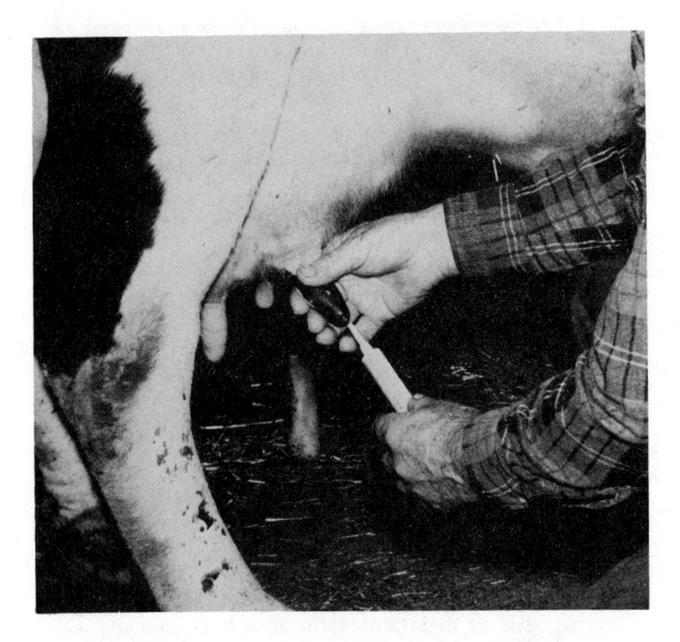

FIGURE 27-7
Infusion is a method of treating udder problems. ***(From Quinn/Dairy Farm Management, copyright 1980 by Delmar Publishers Inc.)***

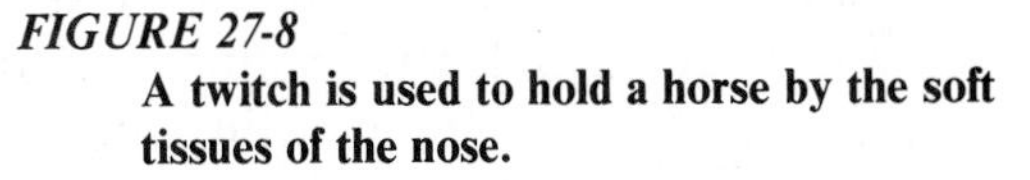

FIGURE 27-8
A twitch is used to hold a horse by the soft tissues of the nose.

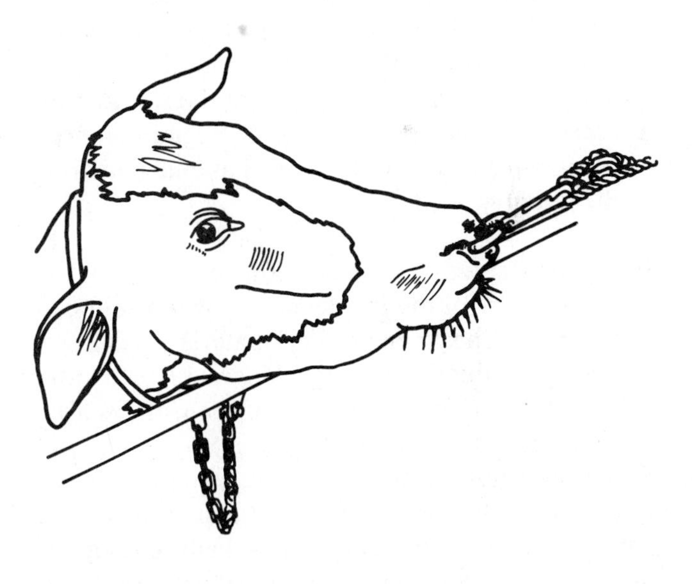

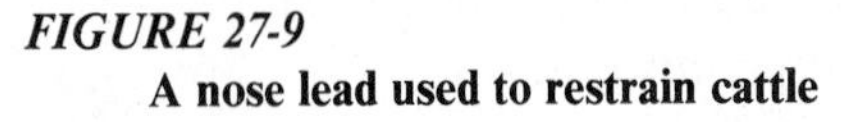

FIGURE 27-9
A nose lead used to restrain cattle

To use an animal thermometer, first shake down the column of mercury. Coat the thermometer with sterile jelly to make insertion easier. Do not force the thermometer into the rectum. If there is resistance, injury may result. Correct the conditions that are causing the resistance and then reinsert the thermometer. After several minutes, remove the thermometer and read the temperature on the scale. Thermometers with digital readouts are also available, but are not in general use by livestock managers.

Determining Pulse and Respiration Rates The pulse rate for large animals can be taken by holding your ear against the animal's chest and listening to the heartbeat. The number of heartbeats in 1 minute is the pulse rate.

The respiration rate of an animal can be determined by watching its rib cage move. Counting the number of breaths that the animal takes in 1 minute will indicate the rate of respiration.

Restraining Animals There are a number of ways to restrain animals for observation and treatment of diseases and parasites. These include head gates, squeeze chutes, halters, twitches, nose leads, and casting harnesses, to mention a few. Head gates trap the head of large animals, whereas squeeze chutes hold the whole animal. When halters are used, they are usually tied to a post or something else substantial to hold the animal. Twitches hold the tender lip of a horse (Figure 27-8). Nose leads hold cattle by the nose (Figure 27-9). Sometimes large animals need to be examined lying down. An easy way to do this is with a casting harness (Figure 27-10). Properly applied, a

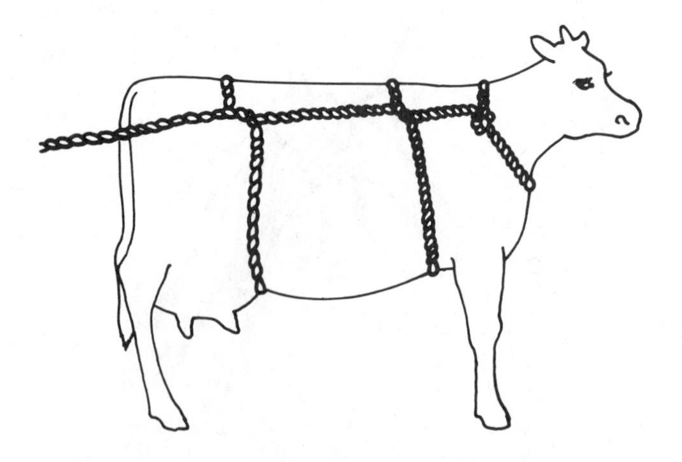

FIGURE 27-10
A casting harness properly used makes throwing an animal a simple task.

casting harness will cause an animal to fall down with just a gentle tug of the rope.

Vaccination The prevention of diseases is nearly always less expensive than treating animals once they have the disease. A good disease-prevention program should include vaccination of animals. Vaccination is the injection of an agent into an animal to prevent disease. The agent causes the animal's body to become immune to the disease. *Immune* means not affected by something. Vaccination programs are usually part of the services of a *veterinarian* (an animal doctor). Vaccination programs vary widely with the type of animal and area of the country.

Feed Additives These additives are used primarily to control the incidence of low-level infections in growing animals. The materials used are primarily antibiotics that help increase feed efficiency and rate of gain, as well as control disease. Feed additives are sometimes used to control internal parasites. Caution must be taken to always follow the manufacturer's recommendations concerning the use of these materials. Failure to do so may lead to contamination of animal products used for human consumption.

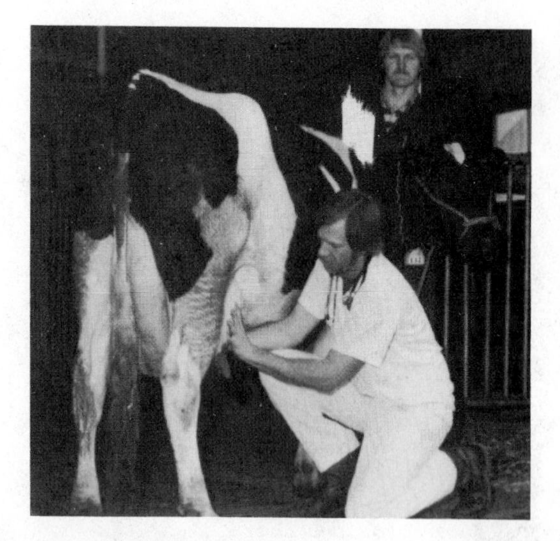

FIGURE 27-11
The services of a good veterinarian are essential in maintaining the health of animals. ***(Courtesy of the American Veterinary Medical Association)***

VETERINARY SERVICES

The veterinarian is an essential part of any good health program for animals (Figure 27-11). It is important to know when to call the veterinarian for help and when to deal with the problem yourself. There are no hard and fast rules in this regard, and it will vary greatly depending on the experience of the individual. In general, it pays to call the veterinarian any time that you are not absolutely sure of the problem and how to handle it.

A veterinarian should normally be consulted when you are planning and executing a disease-prevention program. Any time that an animal is having reproductive problems, a veterinarian should be consulted. Such problems include failure to conceive, abortion, or great difficulty in giving birth. When an animal dies suddenly and there is no apparent reason, a veterinarian should also be consulted to determine the cause of the death. A veterinarian should also be contacted when animals have symptoms of a contagious disease. This will help minimize spread of the disease.

Student Activities

1. Define the "Terms To Know."
2. Compare the methods of administration, dosage rates, and time of withdrawal of several drugs for animals.

3. Practice giving injections of water using discarded animal ears from a meat processing plant.
4. Develop a complete disease-prevention program for a livestock operation.
5. Check the housing of production animals at home, on someone else's farm, or at someone's business for animal safety and proper sanitation.
6. Visit a farm supply store or a pet center and record the names of animal medicines, animal pest controls, and disinfectants that are for sale there. Also list the use(s) of each item.
7. Interview the manager of a small animal or livestock operation concerning the disease prevention and health maintenance practices used. Report your findings to the class.

Self-Evaluation

A. MULTIPLE CHOICE

1. Subcutaneous injections are made
   - a. in a vein.
   - b. under the skin.
   - c. between layers of skin.
   - d. in the body cavity.
2. An animal that lives and feeds on other animals is a
   - a. parasite.
   - b. disease.
   - c. vaccination.
   - d. host.
3. When taking the temperature of an animal, use a/an
   - a. catheter.
   - b. oral thermometer.
   - c. rectal thermometer.
   - d. syringe.
4. Which of the following is *not* a means of administering drugs orally?
   - a. drench
   - b. balling gun
   - c. pills
   - d. infusion
5. The purpose of vaccination is to
   - a. prevent parasites.
   - b. prevent diseases.
   - c. control parasites.
   - d. treat diseases.
6. Which of the following is *not* a sign of good health in animals?
   - a. smooth hair coat
   - b. elevated pulse
   - c. alertness
   - d. contentment
7. The purpose of isolation of sick animals is to
   - a. prevent spread of contagious diseases.
   - b. keep other animals from hurting the sick animal.
   - c. allow for easier treatment of the problem.
   - d. all of the above.
8. Which of the following is *not* an internal parasite?
   - a. fluke
   - b. mite
   - c. stomach worm
   - d. protozoa

9. Call a veterinarian when the animal

a. has bright eyes.
b. chews its cud.
c. aborts.
d. starts to lose hair.

10. A twitch is used to

a. restrain horses.
b. treat internal parasites.
c. give injections.
d. isolate infected animals.

B. COMPLETION

1. ______ diseases cannot be passed from one animal to another.
2. An ______ injection is one that is made in the vein of an animal.
3. Parasites that are very small, flat worms are called ______ .
4. ______ ______ refers to the number of heartbeats in 1 minute.
5. An agent that prevents the growth of a disease organism is called an ______ .
6. Protozoa are a form of ______ parasites.
7. Oral administration of medicine refers to administering the drug through the animal's ______ .
8. Vaccination is a form of disease ______ .
9. An animal doctor is called a ______ .
10. A nose lead is used to ______ animals.

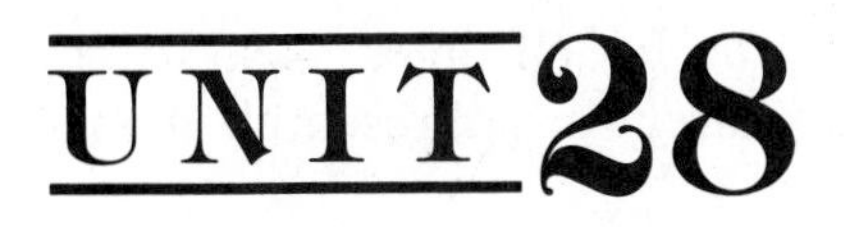

UNIT 28 Genetics, Breeding, and Reproduction

OBJECTIVE To determine the role of genetics, breeding, and reproduction in animal agriscience today.

Competencies to Be Developed

After studying this unit, you will be able to

- ☐ define terms associated with genetics and reproduction.
- ☐ understand the principles of genetics.
- ☐ identify systems of breeding.
- ☐ identify parts of reproductive systems.
- ☐ understand new technologies in animal reproduction.
- ☐ evaluate animals for type and production.
- ☐ understand types of testing programs.

TERMS TO KNOW

Artificial insemination
Eggs
Sperm cells
Embryo transfer
Zygotes
Geneticist
Genetics
Heredity
Breed
Gene
Chromosome
Gamete
Cell
Mitosis
Meiosis
DNA
Homozygous
Heterozygous
Dominant
Recessive
Incomplete dominance
Genotype
Phenotype
Sex-linked
Mutations
Polled
Lethal
Inheritability
Genetic engineering
Testes
Testosterone
Estrogen
Estrus
Progesterone
Mammary system
Ovaries
Ovulation
Embryo
Parturition
Fetus
Gestation
Freemartin
Sterility
Purebreed
Registration papers
Crossbreeding
Hybrids
Hybrid vigor
Grade
Grading up
Progeny
Inbreeding
Sire
Dam
Closebreeding
Linebreeding
Outcrossing
Natural service
Pasture mating
Hand mating
Insemination
Semen
Ejaculate
In vitro
Pedigree

MATERIALS LIST

copies of male and female reproductive systems

animal magazines
bulletin board materials

Probably the fastest growing area of technology in agriscience is genetics. Artificial insemination has allowed for more improvement in milk production in the last 20 years than had occurred in the previous 200 years. *Artificial insemination* is the placing of sperm cells in contact with female reproductive cells called *eggs* by a method other than natural mating. *Sperm cells* are male reproductive units. Embryo transfers have made it possible for superior females to produce far larger numbers of offspring than would be possible otherwise. Artificial insemination allows for use of a superior male to father many times as many offspring than would be possible naturally. *Embryo transfer* is a process that removes fertilized eggs or *zygotes* from a female and places them in another female, who carries them until birth. Embryo transfers are even being used to reproduce endangered species of animals faster than would normally be possible (Figure 28-1).

Genetic engineering has made it possible to increase resistance to diseases, improve production, and improve efficiency of animals. Gene splicing, recombinant DNA, and biotechnology are "buzz words" of the geneticist of today. A *geneticist* studies *genetics* or heredity. *Heredity* is the passing on of traits or characteristics from parents to offspring.

This unit explores the new technologies that exist today and reviews the basics of animal breeding that have been true for all times. Also to be noted will be current directions of research in animal breeding.

FIGURE 28-1
Successful embryo transfer requires the services of a highly trained veterinarian.

THE ROLE OF BREEDING AND SELECTION ON ANIMAL IMPROVEMENT

Robert Bakewell, an Englishman, is generally credited with being the father of animal husbandry. His work in selection of Merino sheep for fine wool production and quality encouraged other farmers of his era to try to improve their livestock. Bakewell and others took great pains to always cross the most desirable females with the best males, with the expectation that the offspring would be as good as or superior to their parents. These practices have continued through the years and have resulted in advances in animal agriscience that were not imagined even 25 years ago.

By continually selecting animals for a specific type or characteristic, the resulting generations of animals tend to conform to the characteristics for which they were selected. For example, 200 years ago cattle were not separated into dairy and beef types. Through careful selection of those animals with superior milk production and those types with excellent meat production, two distinct types of animals emerged from the same ancestors. The many breeds of animals have also been developed in the same way. A *breed* is a group of animals having similar physical characteristics that are passed on to their offspring. It should also be noted that selection is an extremely important part of animal agriscience today. This is especially true as consumer demands for products of animal agriscience change and the margin of profit continues to decline in this area.

PRINCIPLES OF GENETICS

Gregor Mendel, an Austrian monk, is generally given credit for having discovered the basic principles of genetics. He did this through keen observation as he raised peas in his garden. These principles have become the foundation of modern genetics. They are summarized as follows:

1. In every living thing there is a pair of genes in every cell that controls the appearance of every trait in that individual. A *gene* is a unit of hereditary material located on a chromosome. A *chromosome* is the rodlike carrier for genes.

2. Individuals receive one gene for each trait from each parent.
3. Genes are transmitted from parent to offspring as an unchanging unit.
4. In the production of reproductive cells, gene pairs separate and only one gene for each trait is contained in each gamete. A *gamete* is a reproductive cell.
5. When an individual has different genes for a trait, one usually shows while the other does not.

Cells and Cell Division Cells are the basis of all genetic activity. A *cell* is a unit of protoplasmic material with a nucleus and walls (Figure 28-2). It is the basic structure of all living things. Cells are microscopic in size. All plant and animal life begins as a single cell. The nucleus of the cell contains pairs of chromosomes on which rests genes at specific locations. The gene for a specific trait is always located in the same place on the same pair of chromosomes in a species of animals.

Animal growth and reproduction takes place by cell division. In simple cell division for growth called *mitosis*, each chromosome first divides in two. The wall of the cell nucleus disappears and the chromosomes move to opposite sides of the cell. A new nucleus wall forms around each of the groups of chromosomes. Finally, the cell wall divides, resulting in two new cells complete with nuclei and pairs of chromosomes (Figure 28-3).

The cell division that results in the formation of gametes is called *meiosis* (Figure 28-4). It differs from mitosis primarily in that instead of the chromosomes dividing and moving in pairs to the opposite sides of the cell, they separate and move individually to the cell walls. When the new cells are formed, each cell contains only one of each chromosome rather than pairs. Meiosis occurs only in the reproductive organs of animals.

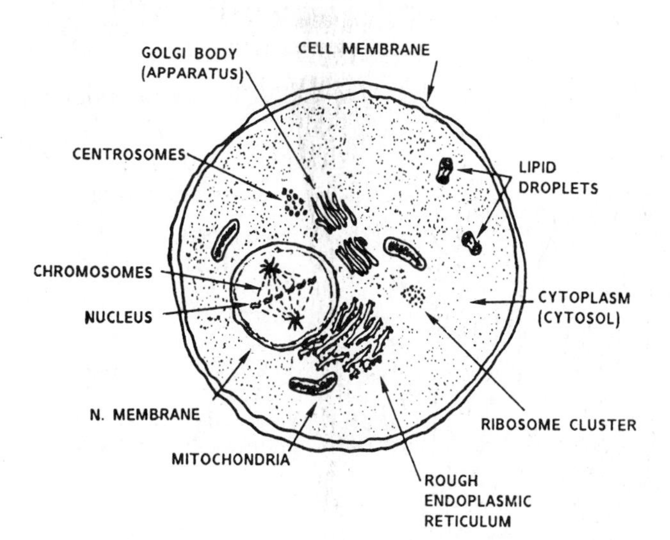

FIGURE 28-2
A cell of an animal *(Courtesy Instructional Materials Service, Texas A&M University)*

Genes Genes are the units of genetic material that are responsible for all of the traits or characteristics of animals. Genes occur at specific locations on chromosomes. Chromosomes control certain enzyme and protein production that controls some traits in animals. The chromosomes themselves are composed of a protein covering surrounding two chains of *DNA*, deoxyribonucleic acid. This substance serves as the coding mechanism for heredity.

The two genes, one each of a pair of chromosomes, may be either alike or different. Pairs of genes that are alike are said to be *homozygous*, whereas those pairs that are different are called *heterozygous*. When the two genes in a pair are different, one gene usually expresses itself and the other remains hidden. The gene that expresses itself is referred to as *dominant*. The gene that remains hidden and expresses itself in the absence of a dominant gene is called *recessive*. Sometimes neither gene of a pair expresses itself to the exclusion of the other. When this happens, the gene pair is referred to as expressing partial or *incomplete dominance*. The actual configuration of genes in an animal is called the *genotype*. On the other hand, *phenotype* is the term that describes the physical appearance of the animal. All of this is important when exploring the basics of genetics and the use of genetics in animal breeding.

Some traits are controlled by genes that are located on the chromosomes that control the sex of the animal. These are called *sex-linked* traits. The chromosomes that control sex in most animals are not perfectly matched. The result is that not all of the genes on these chromosomes occur in pairs. When this happens, some traits show only in males and some only in females.

Genes normally duplicate themselves accurately. However, sometimes accidents or changes occur. These genetic accidents or changes in genes are called *mutations*. Sometimes these mutations

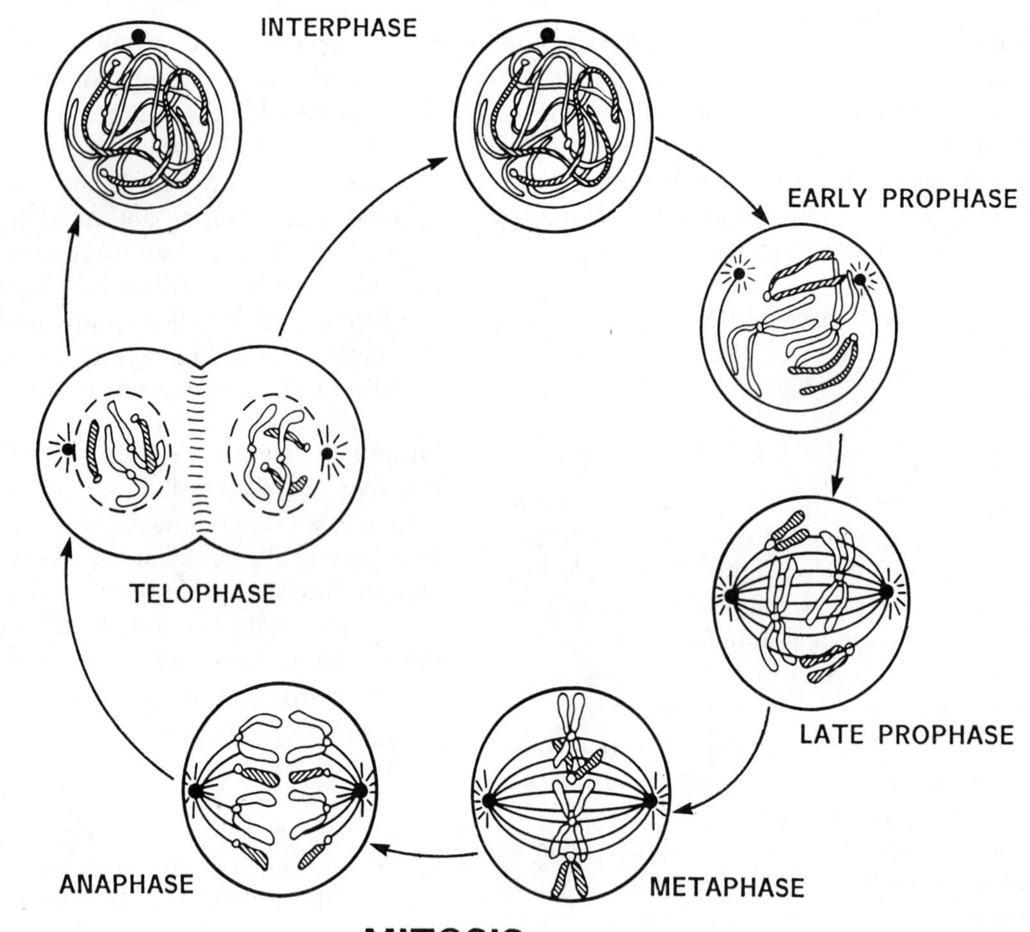

FIGURE 28-3
Mitosis or normal cell division ***(Courtesy Instructional Materials Service, Texas A&M University)***

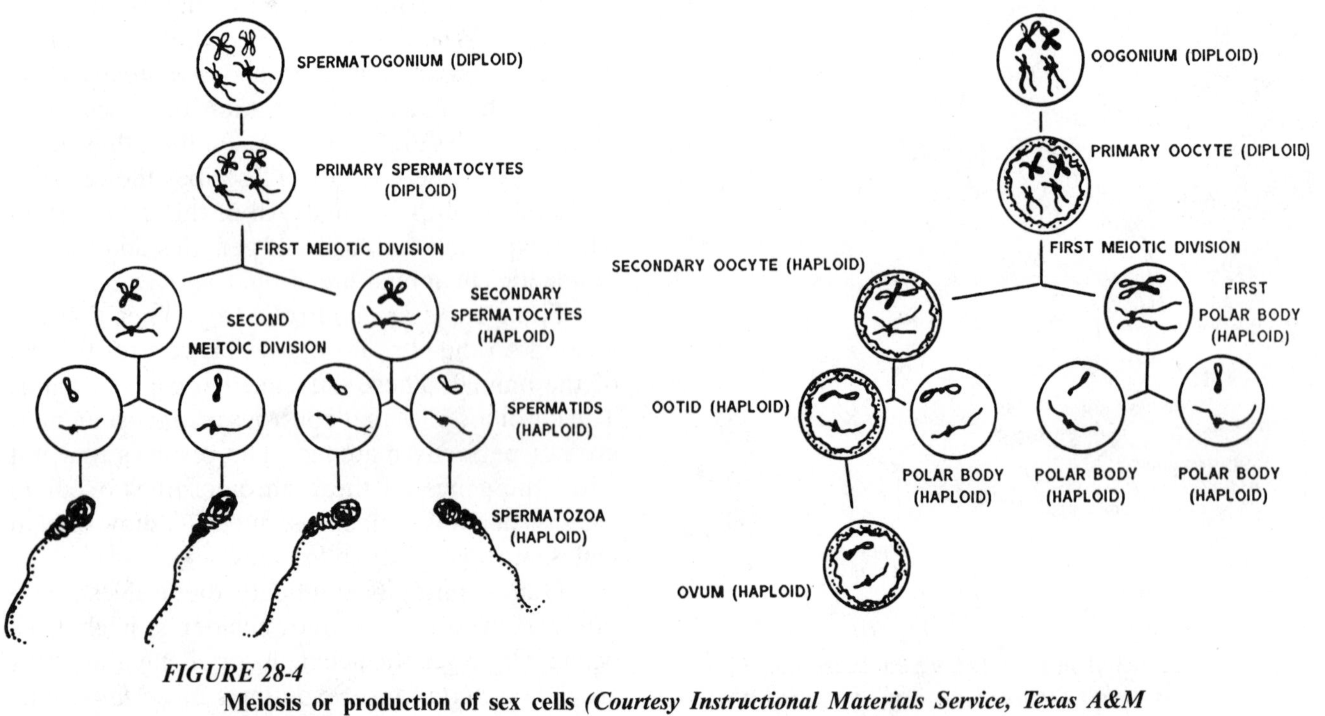

FIGURE 28-4
Meiosis or production of sex cells ***(Courtesy Instructional Materials Service, Texas A&M University)***

result in changes in animals that are desirable. One such example is the polled characteristic in breeds of cattle that are normally horned. *Polled* is naturally or genetically hornless. In other cases, the mutation results in a *lethal* characteristic, which causes an animal to be born dead or to die shortly after birth.

Genetics in the Improvement of Animals The improvement of animals through genetics can be either natural or planned. In natural selection, the "survival of the fittest" occurs. In other words, as changes in genes occur naturally in animals, only the animals with changes that make them better adapted to their environment will survive. Popular examples include protective colorations, ability to digest certain feeds, and ability to survive in extreme heat or cold.

In planned or artificial selection, people decide which traits they want in animals. They then use the animals with the desirable traits in the breeding program. Over a period of time, the animals that result from such selection show more and more of the desired traits.

Unfortunately, most of the traits for which people are selecting animals are the result of a combination of many pairs of genes. Because of this, few traits are 100% inheritable from parents. For example, the extent of inheritability for loin-eye size of pigs is 50%. *Inheritability* means the capacity to be passed down from parent to offspring. A boar with a 6″ loin-eye is crossed with a sow that has a 5″ loin-eye. The expected average loin-eye size for the resulting offspring would be 5½″ if loin-eye size was 100% inheritable. However, because loin-eye size is only 50% inheritable, the offspring can only be expected to have 5¼″ loin-eyes.

Other percent inheritability rates can be found in (Figure 28-5). These rates should be used as a guide only when attempting to improve animals through genetics.

Environmental factors often play a part in the expression of genetic traits, masking to some extent the true potential of the animal. For example, an animal that is improperly fed or cared for may never reach the size or weight that its genetic potential would indicate.

Genetic Engineering This is a new field in agriscience, with much potential for improving ani-

	PERCENT HERITABILITY					
Trait	**Cattle**	**Sheep**	**Swine**	**Poultry**	**Rabbits**	**Horses**
Fertility	0–15	0–15	0–15	0–15	—	Low
Number of young weaned	10–15	10–15	10–15	—	3	—
Weight of young at weaning	15-25	15-20	15-20	—	35	—
Postweaning rate of gain	50-55	50-60	25-30	—	60	—
Postweaning gain efficiency	40-50	20-30	30-35	—	—	—
Fat thickness over loin	45-50	—	40-50	—	—	—
Loin-eye area	50-60	—	45-50	—	60	—
Percent lean cuts	40-50	—	30-40	—	60	—
Milk production, lb	25-30	—	—	—	—	—
Milk fat, lb	25-30	—	—	—	—	—
Milk solids, nonfat, lb	30-35	—	—	—	—	—
Total milk solids, lb	30-35	—	—	—	—	—
Body weight	—	—	—	35-45	40	—
Feed efficiency	—	—	—	20-25	—	—
Total egg production	—	—	—	20-30	—	—
Age at sexual maturity	—	—	—	30-40	—	—
Viability	—	—	—	5-10	—	—
Speed	—	—	—	—	—	25-50
Wither height	—	—	—	—	—	25-60
Body length	—	—	—	—	—	25
Heart girth circumference	—	—	—	—	—	34
Cannon bone circumference	—	—	—	—	—	19
Points for movement	—	—	—	—	—	40
Temperament	—	—	—	—	—	25

FIGURE 28-5

The rates of inheritability for certain traits of domestic animals ***(Courtesy Instructional Materials Service, Texas A&M University)***

mals for the use of humans. *Genetic engineering* is the process of transferring genes from one individual to another individual or organism without mating male and female cells. Geneticists have been able to link specific genes to specific traits. They have also developed procedures for removing the genes from the cells of one animal and inserting them into the cells of another animal.

The potential for change in animals is tremendous using genetic engineering. For example, if a species of animal is genetically resistant to a certain disease, genes that make that animal resistant could be inserted into cells of an animal species that is not resistant. Because genes are passed on to offspring from parents, resulting generations of animals would be resistant to that disease. Unfortunately, genetic engineering is still in its infancy and many problems still need to be solved.

Some of the areas being explored by geneticists working on genetic engineering include disease resistance, cancer research, vaccines, increased growth and production, and immunology.

REPRODUCTIVE SYSTEMS OF ANIMALS

Male The male reproductive system functions to produce, store, and deposit sperm cells. Secondary functions include production of male sex hormones and elimination of urine from the body (Figure 28-6).

The actual structural makeup of the male reproductive system varies widely with different species of animals. The *testes* are the organs that produce sperm cells. They also produce *testosterone*, the male sex hormone. The testes are attached to the body by the spermatic cord and are protected by the scrotum. A coiled tube called the epididymis stores and transports the sperm cells. The def-

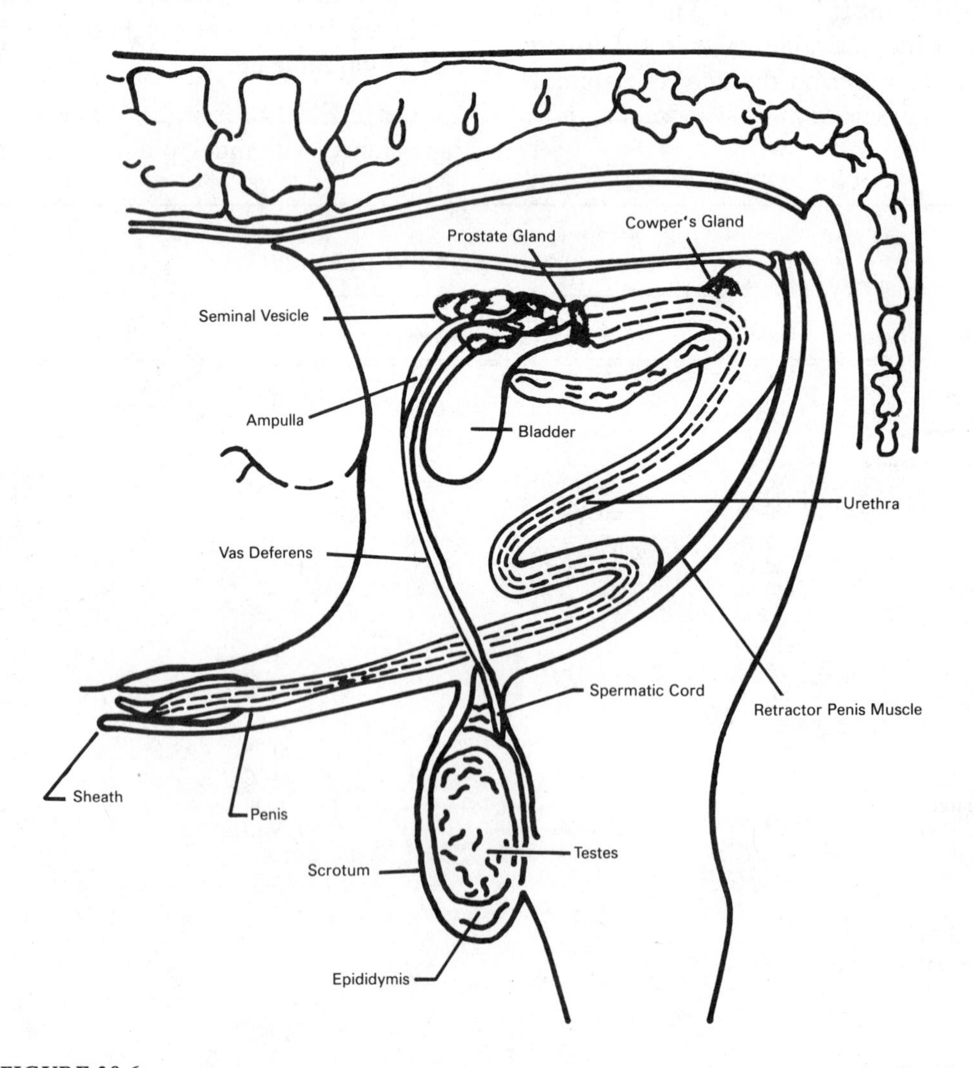

FIGURE 28-6
Male reproductive system *(Courtesy of Maryland State Instructional Guide, University of Maryland)*

erent duct moves the sperm cells to the urethra, which extends through the penis. Other glands of the male reproductive system include the seminal vesicles and the prostate and Cowper's glands. The seminal vesicles secrete seminal fluid. The prostate gland provides nutrition to the sperm cells, and the Cowper's gland prepares the urethra for the passage of the sperm cells. The penis serves to deposit the sperm cell into the female reproductive tract.

Female The female reproductive system produces the egg and the female sex hormones estrogen and progesterone (Figure 28-7). *Estrogen* regulates the heat period, *estrus*, whereas *progesterone* prevents estrus during pregnancy and causes development of the mammary system. The *mammary system* produces milk.

The parts of the female reproductive system include two *ovaries*, which produce eggs. Funnel-shaped devices called the infundibulums catch the eggs during *ovulation* (the process of releasing a mature egg from the ovaries). The egg then passes to the fallopian tubes, also called oviducts. The fallopian tubes are the site where fertilization of the egg takes place. The fertilized egg or *embryo* travels to the uterine horn, where it attaches to the wall of the uterus and remains until birth, *parturition*. When the embryo attaches itself to the uterine wall, it becomes known as a *fetus*. The period of time between fertilization of the egg by a sperm cell and birth is called the *gestation* period. The vagina, which is separated from the uterine horn by the cervix, serves as the passageway for the sperm cells. The external opening or vulva protects the rest of the female reproductive system from infection from the outside.

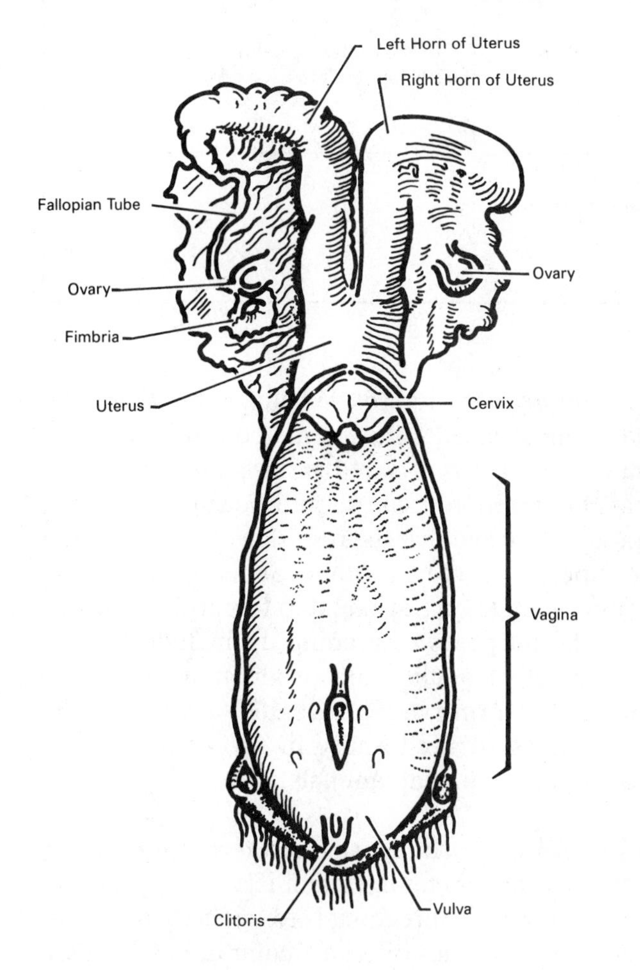

FIGURE 28-7
Female reproductive system ***(Courtesy of Maryland State Instructional Guide, University of Maryland)***

Reproductive Problems There are a number of conditions that may result in reproductive problems or failures. Some of these problems are either physical or genetic. Examples include: (1) *Freemartin*, a sterile female born as a twin of a male in cattle, (2) scrotal hernia (a muscle tear), (3) undeveloped or missing ovaries, and (4) malformed penis.

Infections and diseases are also important causes of sterility in animals. *Sterility* is the inability of an animal to reproduce. Physical damage to the reproductive system and nutritional deficiencies may also contribute to reproductive failures.

SYSTEMS OF BREEDING

There are a number of breeding systems that are important to animal agriscience. Which system or systems to use depends on many factors. Some of the considerations include type of operation, markets available, resources available, climatic conditions, size of operation, goals of the breeder, and personal preference.

Commonly recognized systems of breeding include purebreeding, crossbreeding, grading up, inbreeding, and outcrossing (Figure 28-8).

Purebreeding Purebreeding occurs when a purebred animal is bred to another purebred animal. A *purebred* animal is an animal of a recognized breed with registration papers. *Registration papers* are records of ancestry.

Although there are no guarantees, purebred animals are usually considered to be superior to non-purebreds. They are used as show animals and are important parts of the crossbreeding and grading-up systems of breeding.

Persons who elect to use the purebreeding system of breeding should have ample resources and

	Relationship of Mates	Advantages	Disadvantages
Purebreeding	Unrelated	1. Concentration of selected traits 2. Breed assn to create demand	1. May result in less desirable traits 2. Loss of hybrid vigor
Crossbreeding	Unrelated	1. Increased growth 2. Increased prod. 3. Increased hybrid vigor 4. Higher fertility 5. Disease resistance	1. Less uniformity of offspring 2. Not eligible for registry
Grading up	Unrelated	1. Herd improvement w/o purchasing purebreds 2. Develop uniformity	1. Slow process of improvement
Closebreeding	Sire-mother Son-dam Brother-sister	1. Concentrates desirable traits	1. Concentrates undesirable traits 2. Expression of abnormal traits
Linebreeding	Not closer than ½ brother to ½ sister	1. Concentrates desirable traits of one individual	1. Concentrates undesirable traits 2. May result in expression of abnormal traits
Outcrossing	Unrelated	1. Produces hybrid vigor within a breed	

FIGURE 28-8
Systems of breeding for animal improvement

a good knowledge of genetics, and be a good salesperson.

Crossbreeding *Crossbreeding* is the breeding of one recognized breed of animals to another recognized breed. The resulting offspring are called *hybrids*.

There are a number of advantages of hybrid animals. They tend to be faster growing, stronger, and higher producing as a result of the combination of desirable traits from the two breeds. This is called *hybrid vigor*. They also tend to be more fertile and more disease resistant.

Crossbreeding is generally used by commercial producers who are more interested in offspring that are efficient producers than in maintaining a specific breed of animals.

Grading Up Livestock producers who are raising animals that are not purebred often use grading up to improve their herds. When a nonpurebred female animal called a *grade* is mated with a purebred male, the process is called *grading up*. The idea is that the purebred male should be superior to the grade female and that the resulting offspring should be superior to their mother. Succeeding generations of females are also mated to purebred males.

The purposes of grading up include the improvement of quality and production in the offspring. Offspring are also called *progeny*. The development of uniformity in the herd is also a reason for grading up animals.

Inbreeding In the simplest terms, *inbreeding* is the crossing or mating of animals that are related. The purpose of inbreeding is to intensify the desirable characteristics of a particular animal or family of of animals. Unfortunately, inbreeding also intensifies the undesirable and abnormal characteristics as well.

There are various degrees of inbreeding based on how closely related the individuals being mated are. When a father or *sire* is mated with his daughter, a son is mated with his mother or *dam*, or a brother is mated to his sister, the term *closebreeding* is used.

Linebreeding is the mating of less closely related individuals that can be traced back to one common ancestor. Normally, the most closely related cross made in linebreeding is half brother to half sister.

Inbreeding must be used carefully, because inbred animals tend to exhibit more undesirable characteristics than animals produced by other systems. Unless a breeder is willing to carefully select the outstanding individuals resulting from inbreeding, this system should be avoided.

Outcrossing *Outcrossing* is the mating of unrelated animal families in the same breed. This is probably the most popular system of breeding used in purebred herds of animals. It also has many of the advantages of crossbreeding, including increased production and improved type.

METHODS OF BREEDING

There are three general methods of breeding animals—natural service, artificial insemination, and in vitro. *Natural service* occurs when the male is allowed to mate directly with the female. There are several ways in which this may be accomplished, depending on the type of operation, amount of labor available, number of animals in the herd, dictates of the breed associations, and personal preference.

Pasture mating is a system of natural service where the male animal is allowed to roam freely with the females in the herd. The male is responsible for detecting the heat period of each female in the herd and for mating with her at the appropriate time. One disadvantage of this system is that one male can mate with only a limited number of females. If more than one female is in heat at the same time, they may not all get bred. Breeding records are also more difficult to keep. Finally, the male may become sterile and the females not get bred at all.

Hand mating is the bringing of the male to the female for mating. More labor and better management is required, because someone has to determine when the female is in heat and get the male to the female so that mating can take place. Advantages of hand mating include being able to keep more accurate breeding records and being able to mate more females to a single male.

There are numerous advantages of mating animals by artificial insemination. There are a few disadvantages as well. Artificial *insemination* involves collecting semen from a male animal and placing it in the reproductive tract of the female. *Semen* is the sperm cells and accompanying fluids. The technique of artificial insemination has been responsible for tremendous increases in animal production in recent years.

AGRI-PROFILE

CAREER AREAS: GENETICIST/GENETIC ENGINEER/VETERINARIAN

Robert Bellows flushes seven-day-old embryos from a cow treated with a hormone that causes her to release numerous eggs at one time. The embryos are implanted into other cows serving as surrogate mothers or are frozen for implanting at a future date. ***(Courtesy USDA Agricultural Research Service)***

The greatest improvements in animal production and performance have come about as a result of selection and breeding. Recent advances in the knowledge of genetics and the ability to stimulate multiple ovulation, transplant embryos to surrogate mothers, and freeze embryos for future incubation have greatly increased the genetic output of superior individuals.

However, genetic engineering, where the genetic coding of cells can be manipulated, holds forth the greatest opportunities for changing animal characteristics and capabilities in the future. These are products of biotechnology.

The greatest number of new positions in agriscience is expected to be for scientists, engineers, and related specialists in the foreseeable future. The animal industry will provide many of these positions.

Advantages include the fact that semen that is collected from the male can be used fresh or stored frozen in liquid nitrogen for later use. One *ejaculate*, the amount of semen produced at one time by a male, can be diluted and used to breed thousands of females. Because semen can be frozen and stored for long periods of time, the use of an outstanding male can be greatly extended. Artificial insemination also greatly reduces the need to keep male animals and reduces the danger of having males around. Other advantages include less chance of injury to breeders and their animals, reduced spread of reproductive diseases, and improved record keeping.

There are some disadvantages of artificial insemination as a method of mating animals. A person trained to inseminate females must be available when the animal is in heat. Semen collection from the male can be a dangerous activity, requiring the services of a trained technician, special equipment and facilities, and excellent management. Finally, genetic defects of the male can be spread faster. *In vitro* mating occurs outside of the animal's body. Mature eggs are flushed from the female and fertilized by sperm cells collected from the male. The fertilized eggs are then placed in a host female for development into offspring. Although this process is very exacting in its requirements for cleanliness and requires special equipment and facilities, there are times when this is the only way that a viable fetus can be obtained. This means of mating is often used as part of the new technology of genetic engineering.

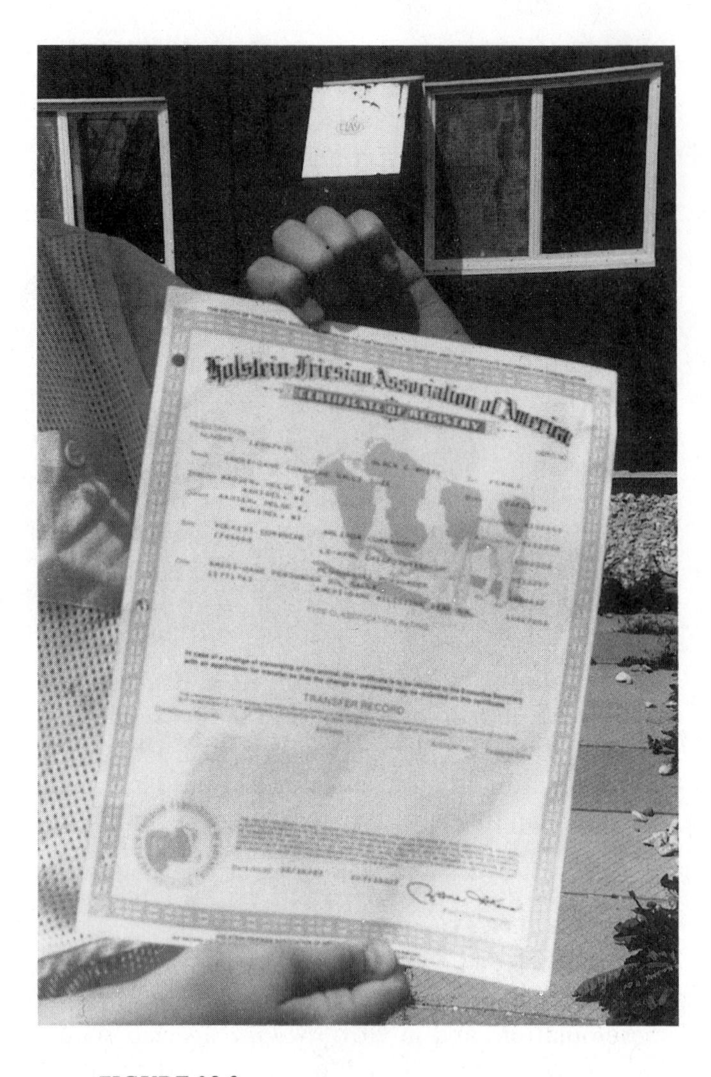

FIGURE 28-9
Pedigree and registration papers are valuable documents when buying purebred animals. ***(Courtesy of Denmark High School, Agriculture Education Department, photo by Jim Jones)***

SELECTION OF ANIMALS

There are a number of methods by which animals may be selected. They fall into two major types—selection based on physical appearance and selection based on performance or production of either the individual or its progeny.

Selection based on physical appearance is generally used when choosing purebred animals. Often the sole criterion for selection of an animal is how well he or she performs in the show ring. This is a very acceptable means of selection when animal breeders are raising animals for show and to sell to others for the same purpose. However, this method often leaves much to be desired when the animal so chosen is expected to produce a product or perform a desired activity. Animals fitted to perform or look their best in the show ring often fall sadly short of expectations at home.

Selection based on production or performance is usually a more reliable means of choosing animals. If you are selecting dairy cows to produce milk, it makes more sense to select cows with high-production records and high-production relatives. In meat animals, progeny testing, or the testing of performance of the offspring, is often the only way to predict the value of the parents. Other measures of production on which various animals may be selected include rate of gain, feed efficiency, butterfat production, back-fat thickness, loin-eye area, yearly egg production, and pounds of wool produced.

Sometimes selection of animals is based on pedigree. A *pedigree* is a record of an animal's ancestry and is included on registration papers for purebred animals (Figure 28-9). Although the consideration of pedigrees can be important in the selection process, it should always be used in combination with other methods of selecting animals.

In summary, tremendous gains have occurred in the productivity of animals in the past 100 years. Fewer and fewer animals are providing the needs of an ever-growing human population. Genetics and gains in animal breeding have been responsible for much of the increase in productivity. This will continue as technology in the field of animal agriscience continues to advance.

Student Activities

1. Define the "Terms To Know."
2. Label the parts of the male and female reproductive systems.
3. Sketch and label the various stages in mitosis and meiosis.
4. Study the genetic principles that determine eye color in humans. Compare the factors that control how eye color is passed from parent to child with those that control the tendency toward baldness.
5. Suppose you mate a Hereford bull with horns to six females that were born without horns (polled). If the females were homozygous for the polled characteristic, how many would probably bear a calf without horns? Why?
6. Select a species of animal and determine the origin of the popular breeds in that species.
7. Clip pictures of animals from magazines and newspapers and make a collage of popular breeds for the bulletin board.
8. Develop a bulletin board with the popular breeds of sheep, cattle, horses, swine, dogs, rabbits, and cats in your community.
9. Suppose that you mated a black male rabbit with a white female rabbit and that the female gave birth to eight bunnies. If white is dominant to black and the female is heterozygous for hair color, how many of the bunnies are likely to be black?
10. If a red pig with floppy ears were crossed with a white pig with erect ears and all possible characteristics were homozygous, what is the probability that the offspring will be red with erect ears? In pigs, white hair color and erect ears are dominant.

Self-Evaluation

A. MULTIPLE CHOICE

1. The male sex hormone is called

   a. estrogen.
   b. testosterone.
   c. progesterone.
   d. oxytocin.

2. The gestation period is the

   a. length of pregnancy.
   b. time during which an animal is in heat.
   c. period when an animal is fertile.
   d. time it takes the egg to mature.

3. Simple cell division is called

a. meiosis.
b. parturition.
c. mitosis.
d. estrus.

4. A chromosome

a. is composed of DNA.
b. is a carrier of genes.
c. occurs in pairs.
d. is all of the above.

5. A pair of genes for characteristics that are alike are said to be

a. heterozygous.
b. homozygous.
c. genotype.
d. recessive.

6. Mating of an animal of one breed to an animal of another breed is

a. purebreeding.
b. grading up.
c. crossbreeding.
d. outcrossing.

7. The phenotype of an individual is

a. what the genes look like.
b. the physical appearance.
c. the type of animal.
d. the expected production.

8. When genes are transferred from one individual to another other than through mating, it is referred to as

a. artificial insemination.
b. in vitro fertilization.
c. genetic engineering.
d. hybrid vigor.

9. The inability to reproduce is

a. ovulation.
b. meiosis.
c. biotechnology.
d. sterility.

10. Mating a sire to his daughter is

a. inbreeding.
b. crossbreeding.
c. purebreeding.
d. grading up.

B. MATCHING

______ **1.** Gamete
______ **2.** Genotype
______ **3.** Fetus
______ **4.** Progeny
______ **5.** Dam

a. Developing embryo
b. Mother
c. Sex cell
d. Offspring
e. Configuration of genes

C. COMPLETION

1. The ______ manufacture male sex cells.

2. A ______ studies genetics.

3. The passing of traits from parents to offspring is ______ .

4. The crossing of half brother to half sister is ______ .

5. ______ is the coding mechanism for heredity.

UNIT 29 Small Animal Care and Management

OBJECTIVE To determine the types, uses, care, and management of small animals.

Competencies to Be Developed

After studying this unit, you will be able to

- ☐ describe the domestication and history of small animals.
- ☐ determine the economic importance of the various classes of small animals.
- ☐ list the types and uses of the various classes of small animals.
- ☐ describe the approved practices in feeding and caring for small animals.

TERMS TO KNOW

Domesticate
Jungle fowl
Waterfowl
Poultry
Broilers
Layers
Chicks
Roasters
Capons
Castration
Cockerols
Cocks
Roosters
Pullets
Hens
Bantams
Toms
Poults
Drakes
Ducklings
Ganders
Goose
Goslings
Hatchery
Flock
Pelt
Angora
Buck
Does
Hutch
Litter
Tattoo
Honey
Hive
Comb
Swarm
Queen
Super
Worker
Drones
Apiary

MATERIALS LIST

an observation beehive
bulletin board materials
magazines and other materials with pictures of poultry and rabbits
labels from poultry and rabbit feed bags

As our world population continues to expand and there is less and less room for humans and large animals to coexist, the importance of small animals continues to increase. They are more efficient than larger animals are in converting food eaten into usable feed and other products for humans. They are less intrusive on the lives of people and are therefore more readily accepted. This unit explores poultry, rabbits, pets, and honeybees, and their contributions to our daily lives.

POULTRY

History and Domestication The domestication of chickens occurred about 4000 B.C. in Southeast Asia. To *domesticate* means to tame for the use of people. The association of *jungle fowl*, ancestors of our present-day chickens, with humans benefitted both. Humans made small clearings in the jungle that attracted insects and other food for the jungle fowl.

The jungle fowl provided some eggs and meat for humans. This association over centuries gradually led to the domesticated chicken of today. Chickens came to the New World with the earliest settlers, and the people of Jamestown settlement had their pens of chickens.

Turkeys are the only domesticated animal of agriscience importance to have originated and been domesticated in the New World. When early explorers arrived in the New World, they found that the Indians of Central America had highly domesticated turkeys that were being grown for food for animals and humans. While present-day turkeys are direct descendants of the wild turkey of the United States, they have been domesticated to the point where they are totally dependent on humans and cannot survive in the wild.

The various types and breeds of ducks and geese have originated from places all over the world. Ducks and geese are also known as *waterfowl.*

Economic Importance The consumption of red meat has declined somewhat in recent years. This is believed to be due to negative publicity regarding fat and cholesterol. Second, the cost of red meat such as beef and pork may have caused a decrease in its demand. As a result, consumption of poultry and poultry products, with the exception of eggs, has been increasing. *Poultry* is a group name given to all domesticated birds. Present consumption of poultry meat is more than 60 lbs. per person each year. Americans also eat about 225 eggs per person each year. As a result, the poultry industry is and will continue to be an important part of the American agriscience industry. Currently, poultry production ranks third behind beef and swine production in dollar sales of meat. Three of the four largest farms in the United States are poultry operations.

Centers of production in the United States for the more than 4 billion *broilers* (young chickens grown for meat) produced each year are Arkansas, Georgia, Alabama, North Carolina, and Mississippi. California is the leading producer of eggs by a wide margin.

The production of turkeys is spread over a wide area, with North Carolina, Minnesota, California, Arkansas, and Missouri as the five leading states. Nearly 60% of the more than 10 million ducks produced in the United States each year come from Long Island, New York. New York, Missouri, Iowa, South Dakota, and Minnesota are major producers of geese.

Types and Uses of Poultry The types of domesticated poultry can be divided into the following general groups: chickens, turkeys, ducks, geese, and captive game birds.

Chickens are usually classified as either layers or broilers, depending on their intended use. *Layers* are chickens developed to produce large numbers of eggs (Figure 29-1). They may produce either white or brown eggs, depending on the breed. Laying chickens are also maintained to produce eggs to be hatched for the production of broiler chicks. *Chicks* are newborn chickens.

Chickens produced for meat are usually classified according to age. Broilers are young meat chickens usually not more than eight weeks old (Figure 29-2). *Roasters* are mature chickens used for meat. *Capons* are castrated male chickens that are 14 to 17 weeks old when they are marketed. *Castration* is the removal of the male sex organs. This can be accomplished either surgically or chemically. Game or Cornish chickens are also breeds of chickens raised for meat.

Young male chickens are called *cockerols*, whereas adult males are called *cocks* or *roosters*. These terms also apply to male pheasants. Young female chickens are called *pullets*, and adult female chickens are called *hens*. Adult female turkeys, ducks, and pheasants are also called hens.

Other classes of chickens include *bantams*, or miniature chickens, and ornamental chickens, which are of value strictly for show.

FIGURE 29-1
Most laying chickens can be traced back to the white leghorn.

FIGURE 29-2
Broiler type chicken

There are more than 200 recognized breeds of chickens in the United States. However, nearly all of the layer and broiler types used are the result of crossbreeding to maximize production. The foundation breed of most laying-type chickens is the White Leghorn. Most broilers can trace their ancestors back to Cornish or game chickens.

There is really only one type of turkey used commercially in the United States, the Broad Breasted White. This breed accounts for more than 90% of the more than 170 million turkeys produced in the United States each year. Other turkey breeds include Broad Breasted Bronze, Bourbon Red, Holland White, and Beltsville Small White. Male turkeys are called *toms* and young turkeys are *poults*.

Ducks can be classified as meat producers or egg producers. The primary meat breed is the Pekin (Figure 29-3). They reach a market weight of about 7 lbs. in 8 weeks. This makes them faster growing than broilers, which reach 3½ to 4 lbs. in the same period of time. Other breeds of ducks used for meat production are Aylesbury, Muscovy, rouen, and Call.

Egg-laying ducks are generally either Khaki Campbell or Indian Runner. The Khaki Campbell is the champion egg layer of the bird world, often averaging more than 350 eggs per year (Figure 29-4). This compares to an average of about 200 eggs laid per year for laying chickens. Male ducks are called *drakes* and young ducks are *ducklings*.

Geese are primarily raised for meat. There is a limited market for geese used for weeding cer-

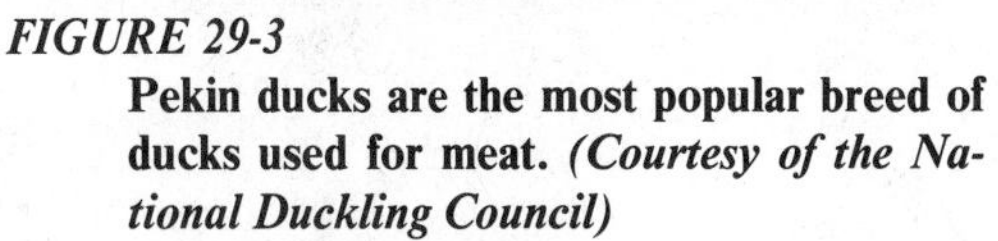

FIGURE 29-3
Pekin ducks are the most popular breed of ducks used for meat. ***(Courtesy of the National Duckling Council)***

tain crops. The Chinese breed is popular for this use. Other breeds of geese are Toulouse, Emden, Pilgrim, and African (Figure 29-5). Male geese are called *ganders*, a female is a *goose*, and young geese are *goslings*.

Captive game birds include pheasants, quail, chukor partridge, and pigeons (Figure 29-6). The uses of game birds include meat and eggs. Some game birds are also raised to release to the wild or on game preserves for hunting. Pigeons may be raised for sport or for meat. The young pigeons, called squabs, are used for meat before they learn to fly.

Approved Practices for Poultry Production

In the most general terms, approved practices for the production of poultry include the following:

1. Purchase young poultry with a specific use in mind.
2. Purchase young poultry or eggs for hatching from reputable hatcheries or breeders only. A *hatchery* is a business that hatches young poultry from eggs.

3. Purchase young poultry at the proper time. Broilers should be 7 to 8 weeks old before marketing them. Layer chicks should be purchased 20 to 22 weeks before you expect them to produce eggs. Ducks should be 7 to 8 weeks old before marketing them; turkeys, 12 to 14 weeks old; and geese, 12 to 14 weeks old.
4. Ensure that proper housing is available for the type and number of poultry you are planning to raise. Housing considerations include size, ventilation, ease of cleaning, lighting, heating and cooling, feed storage, and maintenance

FIGURE 29-4
Khaki Campbell ducks are the champion egg layers of the bird world. *(Courtesy of the National Duckling Council)*

FIGURE 29-5
Pilgrim geese *(Courtesy of the National Goose Council)*

required.

5. Secure and maintain the proper equipment for the type of poultry operation planned. Consider feeder and waterer space, and brooder size.
6. Feed a commercial, balanced ration designed especially for the type of poultry being grown.
7. Plan a flock health program. A *flock* is a group of birds.
8. Plan for marketing at the optimum time.
9. Properly clean and disinfect facilities between flocks of poultry.

Bob-White, *Colinus v. virgianus*

Common Quail, *Coturnix c. coturnix*

Southern Caucasian Pheasant, *Phasianus c. colchicus*

Chukor Partridge, *Alectoris graeca chukar*

Chinese Ring necked Pheasant, *Phasianus colchicus torquatas*

FIGURE 29-6

Some types of captive game birds *(Courtesy Instructional Materials Service, Texas A&M University)*

RABBITS

History and Domestication

Much of the early history of the rabbit is obscure. It is believed that the Phoenicians brought rabbits to Spain about 1100 B.C. They are also given credit for having introduced rabbits to most of the then-known world.

Romans kept rabbits in special enclosures. Roman women were known to have eaten large quantities of rabbit meat. They felt that it enhanced their beauty.

Early monasteries produced large amounts of rabbit meat and furs. These religious institutions are given credit for having domesticated the rabbit. It is known that great pride was taken in producing good-quality rabbits and that much rabbit trading existed between monasteries.

Rabbit meat has long been an important component of the diets of people in densely populated countries of Europe. Rabbits are efficient converters of feed to meat. They take up relatively little space and reproduce rapidly.

Some rabbits have been raised in the United States since the time of early settlers. But serious rabbit production did not begin until the turn of this century. An intense advertising campaign was conducted for "Belgian Hare" at that time to promote commercial production of rabbits.

Rabbit production also got a boost in the United States during the two world wars. At a time when shortages and rationing of food products occurred, rabbits became an inexpensive source of lean, red meat.

Economic Importance

Rabbit production is an important agriscience enterprise in the United States. Each year, 7 to 10 million rabbits are raised. About 300,000 growers produce more than 50 million lbs. of rabbit meat each year. Laboratories also use another 600,000 to 800,000 rabbits in research yearly.

Rabbit production is an ideal enterprise for a young person, because it can be started with limited capital. With only a small investment in housing and equipment, a person with one pair of rabbits can produce 60 to 80 rabbits each year to eat or to sell. Because they are small and generally accepted by people, rabbits are better adapted for production in more urbanized areas than most other types of animals.

The outlook for rabbit production in the future looks promising, as demand for rabbits and their products continues to increase each year. As the

AGRI-PROFILE

CAREER AREAS: ANIMAL TECHNICIAN/ GROWER/MANAGER

Small animal enterprises require relatively small space and can provide good career opportunities. ***(Courtesy United States Department of Agriculture)***

The career opportunities for small animal care and management are good in many areas. The extensive use of laboratory animals for research, small animals for pets, fish for home aquariums and garden pools, small animals for fur, and animals for zoological parks assures attractive jobs in the future.

The operation of animal hospitals, kennels, grooming services, pet stores, training programs, boarding facilities, public aquariums, animal shelters, and humane societies provide opportunities in a number of fields. These include animal nutrition, facilities construction and maintenance, feeds, health services, care and management, production, breeding, and marketing.

Curriculums in small animal care and management have been added to many high school agriscience programs. Similarly, programs are available in many technical schools, community colleges, and universities.

competition between humans and animals for grain products increases, the rabbit will play an important role in meeting the protein needs of humans in the future.

Types and Uses Rabbits are divided into two families and three genera, with very distinct differences. These types are rabbits, cottontails and hares. Rabbits bear their young in underground burrows in the wild. The young are born blind, hairless, and completely helpless. In contrast, cottontails and hares usually give birth in nests above ground. The young are born with their eyes open and with hair. They are able to fend for themselves shortly after birth. Hares also have larger hind legs and longer ears.

Hares and cottontails include the jackrabbit, arctic hare, and snowshoe rabbit. Because they belong to a different genera, cottontails, hares and rabbits cannot interbreed.

Domestic rabbits can be divided into a number of groups based on use. These groups include meat, fur, pets, show, and laboratory use. Many of the breeds will fall into several of these use groups.

The primary use of rabbits in the United States is for the production of meat, with pelts being a by-product. A *pelt* is the skin of an animal with the hair attached. Almost 100 million rabbit pelts are used in the United States each year. Most of these are imported from other countries because, in the United States, rabbits are slaughtered at too young an age to have desirable pelts.

Although all of the breeds of rabbits will produce meat, some breeds are far more efficient in producing desirable-quality meat. The New Zealand breed of rabbit is the most popular in the United States for meat production. This rabbit occurs in three colors: white, red, and black. It is of medium size. It may be grown into a 4 lb. rabbit at 8 weeks of age, using about 4 lbs. of feed for each pound of rabbit produced. Californian and Champage D'Argent are also popular breeds used for meat production.

Some breeds of rabbits are grown for their lustrous fur, used in the manufacture of fur coats and many other rabbit-fur products. The Satin, Rex, and Havana are examples of breeds of rabbits grown for their fur.

Rabbits have long been important for use in laboratory work. They are used for research in the development of drugs for treating a wide range of diseases. They are also important participants in nutritional studies and in genetic research. Commonly used breeds of rabbits for laboratory work include New Zealand white, Dutch, and Florida white. Producers who breed rabbits for laboratory work should be aware that many labs will use only white rabbits of a medium size.

The Angora rabbit is used strictly for the production of wool called *angora*. The wool from Angora rabbits is sheared or pulled from the rabbit about every 10 to 12 weeks. Mature Angora bucks may produce 1 to 1½ lbs. of wool each year. A *buck* is a male rabbit. Female rabbits are called *does*. There are two breeds of Angora rabbits—French and English.

All of the 40 breeds of rabbits recognized by the American Rabbit Breeders Association can be used for pets and/or show. They range in size from the Flemish Giant, which can weigh nearly 20 lbs., to the Netherland Dwarf, which seldom weighs more than 2 lbs. (Figure 29-7). Personal preferences and the availability of breeding stock usually determine what breed or breeds of rabbits to raise for pets or show.

Approved Practices for Rabbit Production Approved practices for the production of rabbits include the following:

1. Select the correct breed for the intended use.
2. Use purebred stock if you plan to sell breeding stock and to maintain uniformity in your herd.
3. Purchase breeding stock only from reputable breeders with accurate records.

FIGURE 29-7
The Netherland dwarf rabbit is popular as a pet. ***(Courtesy of American Rabbit Breeders Association)***

4. Build or choose a hutch of the proper size for the breed of rabbit that you are growing. A *hutch* is a cage or house for a rabbit. For small- and medium-sized breeds, provide hutches 30″ wide × 36″ long × 18″ high. For large breeds, provide hutches 30″ wide × 48″ long × 18″ high.
5. Place the hutch where the rabbit will have adequate ventilation and be protected from heat, wind, rain, sleet, and snow.
6. Provide adequate feeder and waterer space. Rabbits should have access to fresh, clean water at all times.
7. Provide a separate hutch or cage for each mature rabbit.
8. Breed rabbits when does are 5 to 6 months old and bucks are 6 to 7 months old. It may be wise to delay breeding of large breeds until they are 9 to 12 months old, because they are slower in reaching sexual maturity.
9. Take the doe to the buck's cage for breeding and return her to her own cage immediately after breeding.
10. Maintain one mature buck for every 8 to 10 does.
11. Place a nesting box in the doe's cage 25 days after mating occurs.
12. Keep the handling of rabbits to a minimum to avoid injury. When handling rabbits, hold them by the skin on the back of the neck, with the other hand supporting the weight of the rabbit.
13. Feed a commercial pelleted-feed free choice to does and *litters* (a group of young born at one time to the same parents). Feed single bucks and does 3 to 6 ounces of feed each day. Rabbits need to be fed only once a day, and they should be fed in the evening if possible.
14. Maintain accurate breeding and health records on all rabbits.
15. Tattoo all breeding rabbits for identification. A *tattoo* is a means of marking rabbits and other animals for identification. Rabbits are tattooed in the ear.
16. Plan for and maintain a strict herd-health program.
17. Dispose of sick and dead rabbits promptly.
18. Market rabbits as soon as they reach market size or weight.

HONEYBEES

History and Domestication Honeybees have been a part of history for at least 15,000 years. Cavemen drew pictures on cave walls of bees and of collecting honey. *Honey* is a thick, sweet substance made by bees from the nectar of flowers. In Egypt, mummies were embalmed and stored in a liquid based on honey. Jars of honey have been found in many of the Egyptian tombs.

The Bible makes many references to honey and the use of honey for food. During biblical times, honey was not produced in nice, neat combs as it is today. Rather, in most cases the hive was destroyed in the process of removing the honey and comb. A *hive* is a home for honeybees. *Comb* is the wax foundation in which bees store honey.

Greeks and Romans were very familiar with honeybees and honey. Pompey used poisoned honey to defeat his enemies in at least one engagement. Aristotle wrote about bees and their production of honey in great detail.

Most early civilizations considered honey to be the food of gods. Athletes competing in Olympic games often ate honey before their events in order to gain extra strength and endurance.

Early beekeepers kept their bees in hollow logs, straw hives, or even in crude clay cylinders. All of these containers had to be destroyed in order to remove the honey.

With the invention in the 1850s of movable combs with wax foundations to encourage the bees to make neat, straight honeycomb, the whole beekeeping industry changed and honey was finally a commodity to be enjoyed by nearly everyone. Soon after the development of movable comb, the discovery that honey could be whirled out of the comb led to the invention of the honey extractor. It was no longer necessary to destroy the comb in order to get to the honey. The comb, after having been emptied of honey, could now be placed back in the hive to be refilled by the bees.

Today, the production of honey in the United States is a large and profitable business. Far more important than the production of honey is the work that honeybees do in the pollination of crops that are important to agriscience.

Economic Importance It is very difficult to accurately

gauge the true economic importance of the honeybee. It is responsible for about 80% of insect pollination of plants. Without honeybees, many of the crops important to agriscience would simply disappear from Earth.

Pollination of orchard crops such as citrus, peaches, and apples by honeybees is so important that many beekeepers rent their bees to orchardists when their trees are in bloom. With $10 to $30 in rent per hive, and flatbed trailers to move hives, commercial beekeepers make more money from bee rental than from honey. Such beekeepers operate from Florida to Maine and from Texas to Washington State.

There are about 300,000 beekeepers in the United States, of which about 99% are hobby or part-time beekeepers. These 300,000 people care for about 6 million hives of bees. In a normal year, a hive of bees will produce 100 to 150 lbs. of honey in excess of the approximately 150 lbs. needed for the bees to live on. At $1.50 to $2.00 per pound of honey retail, it is easy to see that the production of honey is big business.

Approved Practices for Beekeeping

The following is a list of approved practices to be used in the keeping of bees.

1. Check local regulations before starting a beekeeping operation.
2. Locate bees out of direct contact with people and neighbors' yards and gardens. It is desirable to locate hives so the bees must fly straight up upon leaving the hive.
3. Place hives facing away from prevailing winds. They should also be protected from hot summer sun.
4. Thoroughly clean and disinfect hives before allowing new groups of bees to use them.
5. Purchase bees from reputable sources. It is usually far more profitable to purchase 3-lb. packages of bees with a purebred queen than to rely on swarms to populate new beehives. A *swarm* is a group of bees with a queen that leaves an overcrowded hive to find a new home.
6. Replace queens every 2 years. A *queen* bee is the only fertile, egg-laying female in each hive.
7. Have your bees inspected by a federal bee inspector each year for contagious diseases.
8. Make sure that each hive of bees has a store of at least 75 lbs. of honey for the winter. Hives that do not have enough surplus honey stored for winter should be fed a sugar-water mixture to supplement their own honey stores.
9. Always be sure that bees have ample room to store the honey being produced.
10. Remove surplus honey as soon as the bees have capped it over with wax.
11. Remove honey in the evening or at night when nearly all of the bees are in the hive. Supers containing surplus honey can be freed from bees by blowing cool smoke over the bees and brushing them off the comb with a bee brush. You can also use a bee excluder between the honey to be removed and the hive body. A *super* is a box filled with a movable foundation that is used by the bees to store honey.
12. After moving the hive, put a deflector in the entrance of the hive so the bees will notice that they have been moved. Hives must be moved at least 5 miles to prevent bees from returning to the former site of their hive.
13. Inspect beehives at least monthly to determine the strength of the hive and the queen. Be sure to observe the number of eggs being layed by the queen. Also note whether the worker bees are building drone or queen cells in the hive. Drone and queen cells look like peanuts. Such cells should be destroyed. *Worker* bees are undeveloped females and constitute all of the working force of the hive. *Drones* are males whose only purpose in life is to fertilize the queen once in her life.
14. Reduce or prevent swarming of bees by providing ample hive space for the bees and eliminating queen cells as they are found. Overcrowding often causes bees to develop a second queen. New queens will attract a group of worker bees and leave the hive to start a new colony. This process is called swarming. Bees will not swarm without a queen because she is their only hope of survival.
15. Be aware of pesticides being used in the area that could kill bees or be stored in the honey being produced.
16. Secure the proper equipment before starting an apiary. An *apiary* is an area for the keeping of beehives.
17. Keep honey that has been removed from bees in an area that bees cannot get to. Otherwise, they will steal all of the honey in a short time.

18. Extract honey from the comb as soon as possible after harvesting it. Honey stored for long periods of time in the comb may granulate, which makes it impossible to extract.
19. Develop a market for your honey.

Raising small animals provides the opportunity for persons with limited capital and facilities to get a start in animal agriscience. Most small animals are better adapted to production in urban and suburban areas than are larger animals. The same experiences in planning for, caring for, managing, and marketing can be learned with small-animal enterprises without the large outlay of cash needed for the production of large animals.

Student Activities

1. Define the "Terms To Know."
2. Make a bulletin board display of breeds and types of poultry and rabbits.
3. Attend a fair or show and record the names of the breeds of poultry and rabbits shown there.
4. Interview a local beekeeper about beekeeping practices in your area.
5. Compare the label from a bag of poultry feed with one from a bag of rabbit feed. Determine the differences in ingredients, percentage of protein, additives, and fiber.
6. Set up an observation beehive in the school.
7. Make a list of local crops of importance to agriscience that bees pollinate.
8. Develop a crossword puzzle or word search using the Terms to Know.

Self-Evaluation

A. MULTIPLE CHOICE

1. A hive of honeybees needs about ______ lbs. of honey stored in order to live during the winter.

   a. 25 c. 75
   b. 50 d. 100

2. A gosling is a baby

   a. chicken. c. pigeon.
   b. duck. d. goose.

3. A pair of rabbits can produce ______ lbs. of meat per year.

   a. 10–30 c. 50–80
   b. 30–50 d. 200 or more

4. A castrated male chicken is a

   a. capon. c. cockerol.
   b. cock. d. rooster.

5. The only bee in a hive capable of laying eggs is the

   a. king. c. worker.
   b. queen. d. drone.

6. Honeybees account for about ______ % of all insect pollination.

   a. 20 c. 60
   b. 40 d. 80

7. The ______ breed of chickens is the foundation of nearly all types of laying hens.

   a. Cornish c. leghorn
   b. Pekin d. game

8. Two breeds of rabbits produce wool called

   a. pelt. c. angora.
   b. fur. d. wool.

9. The champion egg-laying breed of birds is the

   a. leghorn. c. Khaki Campbell.
   b. quail. d. Toulouse.

10. More than ______ lbs. of rabbit meat are produced in the United States each year.

    a. 1 million c. 25 million
    b. 10 million d. 50 million

B. MATCHING

______	1. Drone	a. Male goose
______	2. Cockerol	b. Young chicken
______	3. Gander	c. Male bee
______	4. Doe	d. Female bee
______	5. Drake	e. Female pheasant
______	6. Chick	f. Young turkey
______	7. Buck	g. Male rabbit
______	8. Worker	h. Female rabbit
______	9. Hen	i. Male chicken
______	10. Poult	j. Male duck

C. COMPLETION

1. More than 90% of the turkeys produced in the United States are ______ .
2. Ducks raised for meat reach a weight of 7 lbs. in about ______ weeks.
3. The most popular breed of rabbits for meat production is the ______ .
4. A ______ is a home for honeybees.
5. The average laying hen produces about ______ eggs each year.

UNIT 30 Dairy and Livestock Management

OBJECTIVE To determine the history, types, uses, care, and management of dairy and livestock.

Competencies to Be Developed

After studying this unit, you will be able to

- ☐ determine the history and economic importance of dairy and livestock.
- ☐ recognize major types and classes of livestock.
- ☐ list major uses of livestock.
- ☐ understand basic approved practices in the care and management of dairy and livestock.

TERMS TO KNOW

Mammals
Milk
Veal
Beef
Butterfat
Calf
Colostrum
Milk replacers
Free choice
Wean
Heifer
Calve
Cow
Dry period
Cull
Draft
Bull
Implant
Steer
Pork
Lard
Render
Sows
Piglet
Boar
Gilt
Farrow
Hand-mating system
Farrowing crates
Dock
Ear notching
Creep feed
Shepherd
Wool
Lamb
Mutton
Flock
Ewe
Ram
Lambing
Shearing
Mohair
Chevon
Lactation period
Kids
Doe
Buck

MATERIALS LIST

copies of various livestock breed magazines
bulletin board materials
paper glue
scissors
paper for notebook covers
examples of various animal products or things made from animal products

Large animals, including dairy, beef, sheep, goats, and swine, are the backbone of agriscience. Keeping wild animals captured for food made it necessary for early humans to provide food for them. It also changed their lives, in that it forced them to stop being nomadic and follow the herds. The capturing and domestication of animals changed humans from hunters to farmers. Today, much of agriscience is centered around the production of animals and animal products and the production of food for those animals.

DAIRY CATTLE

Origin and History In the wild, mammals normally produce only enough milk to feed their offspring. *Mammals* are animals that produce milk. *Milk* is a white or yellowish liquid secreted by the mammary glands of animals for the purpose of feeding young. When early humans realized that milk was good to drink and that some types of animals produced more milk than others, the eventual domestication of animals began. Although the cow, buffalo, goat, ewe, mare, and sow have been and are currently being used for the production of milk in various parts of the world, this unit will concentrate on cattle in discussing dairy production.

There are records of the use of cattle to produce milk as early as 9000 B.C. There are a number of references in the Bible to milk and the production of milk. Even Hippocrates, the father of modern medicine, recommended milk as a medicine in his writings around 400 B.C.

There were no dairy cows in the New World until Columbus brought them with him on his second voyage in 1493. Cattle also came to the New World in 1611 with the Jamestown colonists. The production of milk was limited to a few cows per family during the colonial period. It was not until the late 1800s that dairying became an important agricultural industry in the United States.

Economic Importance The production of milk is the second most important animal enterprise, if sales dollars is the criterion for importance. The consumption of milk and other dairy products has remained nearly steady during the past several years, after several decades of steady decline. The average American uses slightly more than 300 lbs. of milk and dairy products each year. There were more than 10 million cows in the United States producing about 145 million lbs. of milk in 1988. With an investment of over $4,000 per cow required, dairying is truly big business.

The production of milk is not the only income-generating part of dairy production. Calves not needed as replacements for the dairy herd are sold as *veal* (the meat of young calves). Similarly, cows that are no longer profitable producers of milk are also sold for *beef* (meat from cattle).

Dairy cattle operations are generally divided into two types—Class A and Class B. The class refers to the intended use for the milk produced. Class A milk is produced under very strict standards and is intended for consumption as fluid milk. Fluid milk includes whole milk, reduced-fat milk, and cream. Class B milk, which can be produced under less strict standards, is intended to be used to make butter, cheese, ice cream, nonfat dry milk, and other manufactured dairy products.

Types and Breeds More that 90% of all dairy cattle in the United States are of the holstein breed (Figure 30-1). The familiar black and white cattle are the highest average producers of milk of any breed in the country. Because of the large numbers of cattle involved, the breed has also been able to make the most genetic improvement in recent years.

The second most popular breed of dairy cattle is the jersey. Even though they are the smallest of the dairy breeds, they rank number one in butterfat production. *Butterfat* is the fat in milk. Another popular breed of dairy cattle is the Guernsey, which is known for the yellow color of its milk. Ayrshire and Brown Swiss round out the top five breeds of dairy cattle in the United States.

Approved Practices

Raising Calves General approved practices for raising dairy calves include the following:

1. Make sure the newborn *calf* receives colostrum as its first food and for at least the first 36 hours of its life (Figure 30-2). *Colostrum* is

FIGURE 30-1
The Holstein is the most popular breed of dairy cattle in the United States. ***(Courtesy of Denmark High School, Agriculture Education Department, photo by Jim Jones)***

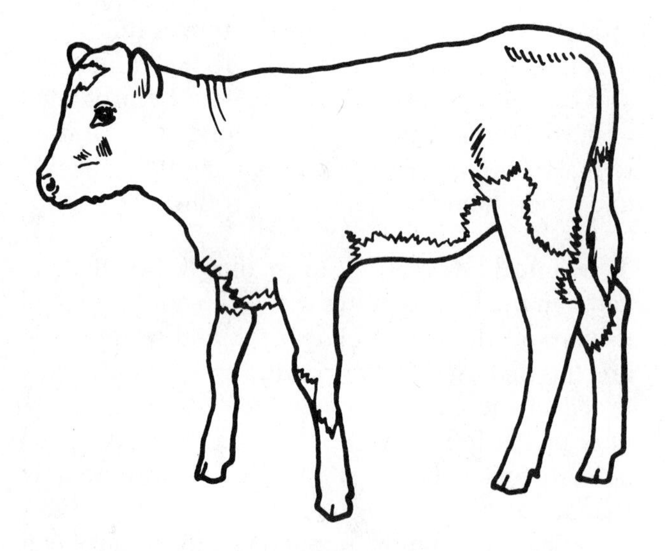

FIGURE 30-2
A newborn dairy calf should receive colostrum from its mother to protect it from disease.

the milk that a cow produces for a short time after calving. It contains antibodies that protect the newborn animal from disease until it can build up its own natural defenses.

2. Feed milk or milk replacer at 8 to 10% of the calf's weight daily until the calf is 4 weeks old. *Milk replacers* are dry dairy or vegetable products that are mixed with warm water and fed to young calves in place of milk.
3. Start feeding calf starter, a grain mixture, free choice at about 10 days. *Free choice* means all that the animal wants to eat.
4. Wean calves from milk when they are eating 1½ lbs. of calf starter per day. To *wean* means to remove and keep away from.
5. Feed calves hay and water free choice from 3 to 9 months. Up to 4 lbs. of grain and some silage can be fed daily.
6. Make up the bulk of the ration with forages fed free choice after 9 months.
7. Remove horns at an early age, preferably as soon as they begin to develop.
8. Remove extra teats at an early age.
9. Identify calves permanently as soon as possible after birth.
10. Prevent calves from sucking each other.
11. Keep hooves properly trimmed.
12. Vaccinate for calfhood diseases at the recommended times.
13. Maintain the calf in clean and sanitary conditions.
14. Plan for and maintain a disease and parasite prevention and control program.
15. Breed heifers to calve at 2 to 2½ years of age. A *heifer* is a female that has not given birth to a calf. With cattle, to *calve* means to give birth.
16. Maintain heifers and calves in uniform groups according to size and weight.

Dairy Cows   Some approved practices for dairy cows include:

1. Rebreed cows 60 to 90 days after calving. A *cow* is a female of the cattle family that has calved. Cows should be bred to calve once every 12 months.
2. Observe cows for evidence of heat period twice daily (Figure 30-3).
3. Check cows to determine if they are pregnant. This should be done 45 to 60 days after breeding them.
4. Provide a dry period of about 60 days before calving to allow the cow to rebuild her body. The *dry period* refers to the time when a cow is not producing milk.
5. Feed dairy cows according to their level of production and stage of pregnancy.
6. Maintain complete health, breeding, and production records on every cow in the herd (Figure 30-4).
7. Establish and maintain a disease and parasite prevention and control program.
8. Contact a veterinarian anytime that you are unsure of how to treat a dairy-herd problem.
9. Milk dairy cows at a regular interval each day. Two milkings each day, approximately 12 hours apart, is a normal routine. However, milking three times per day is common in many herds.
10. Maintain a regular routine in handling dairy cattle in order to maintain maximum production.
11. Cull unprofitable dairy cows. To *cull* means to remove from the herd.
12. Properly maintain dairy housing and milking equipment.

BEEF CATTLE

Origin and History   It is likely that cattle were domesticated in Europe and Asia at some time during the New Stone Age. Domesticated cattle of today are probably all descended from one of two wild species, *Bos taurus* or *Bos indicus* (Fig-

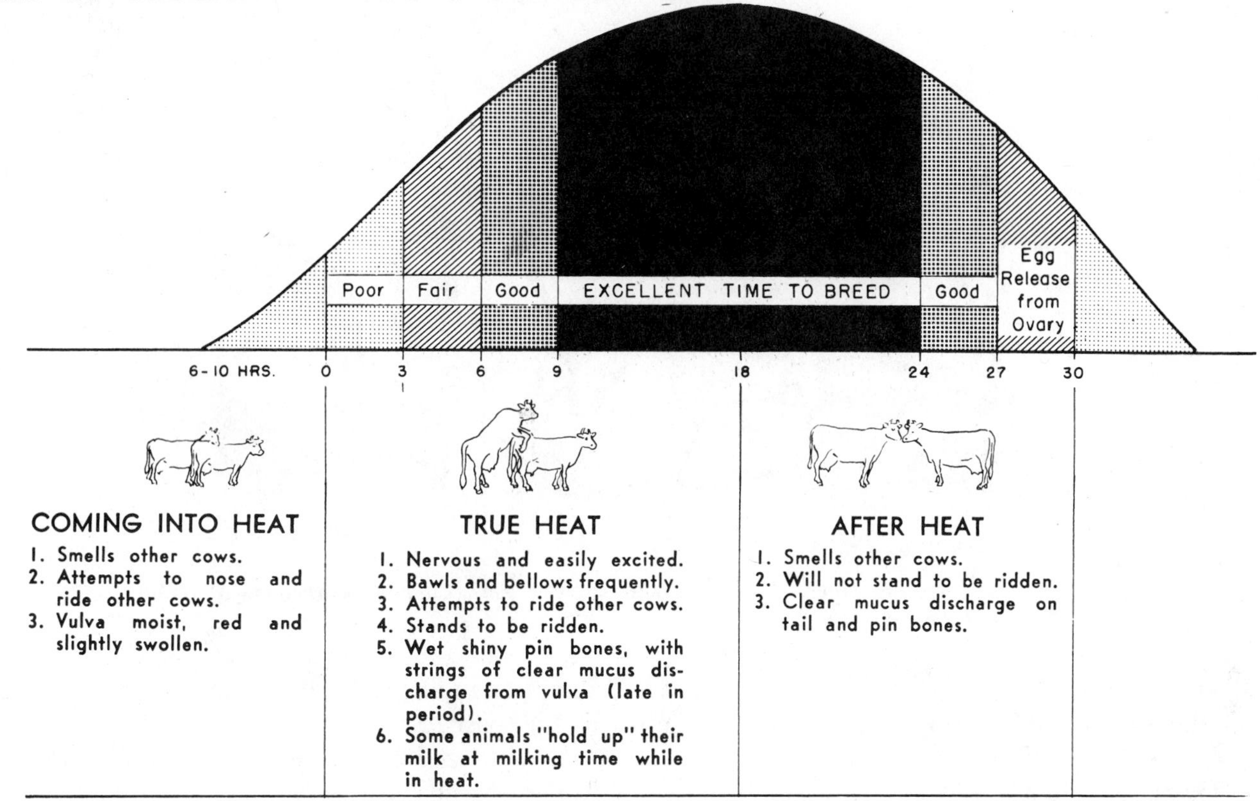

FIGURE 30-3

Breeding at the proper stage of the heat period is essential to maximize pregnancy rates.

FACTORS DETERMINING EFFICIENCY

Factor	Results	Goals	State Average
_______ 1. No. of cows in herd			
_______ 2. Total lbs. of milk produced			
_______ 3. Total lbs. of butterfat produced			
_______ 4. Average annual milk production/cow (lbs.)			
_______ 5. Average annual butterfat production/cow (lbs.)			
_______ 6. Average annual % butterfat for student's herd			
_______ 7. Feed cost/lb. of butterfat produced			
_______ 8. Average feed cost/cow			
_______ 9. % of calves sold or kept until 3 months of age			
_______ 10. Profit or loss			
_______ 11. Pounds of milk sold/hour of total labor			
_______ 12. Dollar returns/hour of self labor			
_______ 13. Production cost/pound of milk			

FIGURE 30-4

Maintenance of a complete record system is important in a profitable dairy enterprise.

FIGURE 30-5
Bos tauras **was one of the wild ancesters of the cattle of today.** ***(Courtesy Instructional Materials Service, Texas A&M University)***

ure 30-5). Some of these wild cattle stood as tall as 7′ at the shoulders. They were domesticated for meat, milk, and draft. *Draft* means used for work.

Owning cattle was a symbol of wealth in early times. They were worshipped, used, and often abused. Several early civilizations had "gods" fashioned to look like cattle. The use of cattle in sports such as bullfighting also developed during early times.

Cattle came to the New World with the earliest settlers, who were more interested in animals that could do heavy work than in those that could produce meat. As the European explorers came to the New World, they also brought cattle with them. Spanish settlers brought longhorn-type cattle to serve as food for Christian missions in the Southwest. As demand increased for beef, the cattle industry developed on the frontier, where grass and the large open spaces required for cattle were abundant. Great cattle drives originated in these areas as cattlemen drove their cattle to human population centers in order to market them.

Today, the beef cattle industry is concentrated in the Midwest and South, where plenty of feed is available and where production of other crops may not be profitable.

Economic Importance The beef cattle industry is the number one red-meat production industry in the United States. In 1985, Americans ate about 79 lbs. of beef per person. This is part of a total consumption of about 215 lbs. of meat and poultry on a retail basis.

With sales of nearly $30 billion in cattle and calves, the beef cattle business is one that will be likely to remain important for many years to come. There are more than 150 million head of cattle and calves on U.S. farms today. Leading states in the sales of beef are Iowa, Texas, California, Wisconsin, and Nebraska.

The production of meat is not the only use for beef cattle. Cattle convert inedible grasses into food for people. Cattle manure provides fertilizer for crops. They also contribute meat by-products that are made into many nonfood products for use by people (Figure 30-6).

Types and Breeds The general types of beef cattle operations include purebred breeders, cow-calf operations, and slaughter cattle or feedlot operations. In a purebred operation, only cattle of a single, pure breed are raised. Operations are geared to produce purebred bulls for cow-calf operations and to produce animals to be sold to other purebred operations. A *bull* is a male of the cattle family. Breeders of purebred cattle have been responsible for much of the genetic improvement in beef cattle in recent years.

Cow-calf operations serve to produce feeder calves for slaughter cattle producers. They are located mostly in the upper Great Plains states and in the western range states, where grass is in abundance and much of the land is unsuited to

1. Bone for bone china.
2. Horn and bone handles for carving sets.
3. Hides and skins for leather goods.
4. Rennet for cheese making.
5. Gelatin for marshmallows, photographic film, printers' rollers.
6. Stearin for making chewing gum and candies.
7. Glycerin for explosives used in mining and blasting.
8. Lanolin for cosmetics.
9. Chemicals for tires that run cooler.
10. Binders for asphalt paving.
11. Medicines such as various hormones and glandular extracts, insulin, pepsin, epinephrine, ACTH, cortisone, and surgical sutures.
12. Drumheads and violin strings.
13. Animal fats for soap and feed.
14. Wool for clothing.
15. Camel's hair (actually from cattle ears) for artists' brushes.
16. Cutting oils and other special industrial lubricants.
17. Bone charcoal for high-grade steel, such as ball bearings.
18. Special glues for marine plywoods, paper, matches, window shades.
19. Curled hair for upholstery. Leather for covering fine furniture.
20. High-protein livestock feeds.

FIGURE 30-6

Byproducts of the production of meat are important contributions to the quality of life for people.

produce many other crops. Usually calves are born in the spring, stay with their mothers during the summer, and are weaned in the fall. They are then sold to slaughter cattle producers. Often these operations use purebred bulls on the grade cows in the herd.

The slaughter cattle or feedlot operator buys calves from cow-calf operators and feeds them until they reach slaughter weight. These operations are generally located in the Midwest, where there is an abundance of corn and other grain available for feed.

Until about 30 years ago, there were basically three breeds of beef cattle in the United States—Hereford, Angus, and shorthorn. Although these three breeds are still important today, they have been joined by more than 50 other breeds from all over the world. These breeds can be divided into many classifications. However, for this study, they are designated as English, exotic, and American.

English breeds originated in the British Isles and include Hereford, Angus, shorthorn, Galloway, devon, and red poll, to name a few (Figure 30-7). In general, English breeds of cattle are of medium size and are noted for the excellent quality of meat that they produce.

Exotic breeds of cattle were first imported into the United States from all over the world when consumers began demanding leaner beef. Beef producers also became aware that calves from these breeds of cattle grew faster and much more efficiently than most of the English breeds. Exotic-breed bulls are often used in grade cow-calf operations to increase the weight of calves being produced. Examples of exotic breeds of cattle include Charolais, Limousin, Simmental, Blond D'Aquitaine, and Maine Anjou (Figure 30-8).

American breeds of beef cattle were developed from necessity. In order to grow beef cattle in the South and Southwest, heat tolerance and

FIGURE 30-7
The most popular breed of beef cattle is the Hereford. Both horned and polled varieties are popular. ***(From Gillespie/Modern Livestock & Poultry Production, copyright 1989 by Delmar Publishers Inc.)***

FIGURE 30-8
The importation of Charolais cattle to the United States changed the beef industry of today. ***(From Gillespie/Modern Livestock & Poultry Production, copyright 1989 by Delmar Publishers Inc.)***

disease and parasite resistance was a must. To develop cattle that met these requirements and still had desirable quality meat, breeders crossed Brahman cattle from India with the English breeds (Figure 30-9). Examples of American breeds of beef cattle or Brangus, Beefmaster, Santa Gertrudis, and Barzona.

Approved Practices for Beef Production

The following are some of the considerations in the production of beef cattle.

1. Select beef cattle according to intended use, area of the country, and personal preferences.
2. Buy cattle only from reputable breeders.
3. Select purebred cattle according to physical appearance, pedigree, and available records.
4. Isolate new animals from the herd for at least 30 days to observe for possible diseases and parasites.
5. Provide enough human contact so that beef cattle can be handled when necessary.
6. Break calves to lead as soon as possible if they are going to be exhibited at fairs and shows.
7. Plan a complete herd health program and follow through with it.
8. Use the services of a veterinarian for serious health problems.
9. Vaccinate to prevent diseases of local concern.
10. Castrate, dehorn, and permanently identify calves at an early age.
11. Breed heifers to calve at 2 years of age.
12. Implant steers and heifers grown for slaughter with growth hormones. To *implant* is to place a substance under the skin. The substance is released slowly over a long period of time. A *steer* is a castrated male of the cattle family.
13. Wean calves at 205 days of age and at 450 to 500 lbs.

FIGURE 30-9
The Braman was an integral part of the formation of most American breeds of beef cattle. ***(From Gillespie/Modern Livestock & Poultry Production, copyright 1989 by Delmar Publishers Inc.)***

14. Provide supplemental nutrition to cattle when natural forages are in short supply.
15. In a pasture-breeding system, allow 1 mature bull for every 25 to 30 cows. With pen mating, 1 bull can be used for every 30 to 50 cows.
16. Time breeding so that all calves are born in a 30- to 60-day period.
17. Allow cows to calve in clean stalls or pastures.
18. Group cattle being fed for slaughter according to size and sex.
19. Provide shelter from inclement weather.
20. Provide access to clean, fresh water at all times.
21. Feed slaughter cattle to reach market weight at 15 to 24 months of age.
22. Have proper facilities and equipment available for the type of operation planned.
23. Market animals at the optimum time to maximize profits.
24. Maintain complete and accurate records.

SWINE

Origin and History Swine were apparently domesticated in China during the Neolithic Age. The first written record of keeping swine appears in 4900 B.C. Mention of swine occurs in the Bible as early as 1500 B.C.

Domestic swine originate from two wild stocks, the European wild boar, *Sus scrofa*, and the East Indian pig, *Sus vittatus*. There are several other wild types of swine that exist today. Even domesticated swine can revert quickly to the wild when the opportunity or need arises.

AGRI-PROFILE

CAREER AREAS: FARM MANAGER/HERDSMAN/RANCHER/BREED ASSOCIATION FIELD REPRESENTATIVE
Farming, ranching, and feedlot management have become scientific and big business. Louis Gasbarre conducts research on disease resistance at Beltsville, Maryland, using twin cows resulting from split-embryo transplants. ***(Courtesy USDA Agricultural Research Service)***

Science and production can no longer be separated. Individuals and teams of scientists visit farms, ranches, and feedlots frequently to gather data, analyze problems, and seek solutions. Farmers, ranchers, and feedlot managers are frequently graduates of agricultural colleges and may have advanced degrees in business, management, or animal sciences.

In addition to scientists and managers, technicians find meaningful careers in the animal production businesses of the nation. Livestock enterprises include beef, dairy, sheep, and swine. There are many off-farm career opportunities as field representatives for breed associations, feed companies, marketing cooperatives, supply companies, animal health products, herd improvement associations, and financial management firms.

Many individuals utilize their farming and ranching backgrounds in agriscience writing, publishing, and telecasting careers. Further, farm and ranch experience provides excellent background for careers in agriscience teaching and extension work.

Swine came to the New World in 1493 on the second voyage of Columbus. As European explorers came to what is now the United States, they brought swine with them as a food supply. The original 13 head of swine brought to the New World by the explorer deSoto multiplied to over 700 head in just 3 years and provided food for the explorers.

European settlers also brought swine with them as they settled on the East Coast of the United States. Because swine could find food on their own and reproduce rapidly, they were soon known as "mortgage lifters." Swine in excess of local needs were exported as pork and lard. *Pork* is meat from swine. *Lard* is rendered pork fat. *Render* means to cook and press the oil from.

As westward expansion occurred, swine production followed. In time, the center of swine production settled in the Corn Belt. This area provided large amounts of corn and grain necessary for large-scale pork production.

Economic Importance Pork production is the number two red-meat industry in the United States. It generates about $10 billion in sales each year. With a population of about 60 million hogs, the United States ranks fourth in the world in swine production behind China, the European Community, and Russia. There are about 700 million head of swine in the world today, with more than 40% in China alone.

On a retail weight basis, Americans consumed about 60 lbs. of pork per person in 1985. The consumption of red meat in general, and pork in particular, has been slowly decreasing in recent years. Fears about health and the rising price of red meat has contributed to this decline.

The modern hog of today is vastly different from its ancestors. It is very lean and trim compared to the "two-cans-of-lard" type of hog in demand 30 years ago. Hogs are the most efficient converters of feed into meat among the large red-meat animals. It takes about 3½ lbs. of feed to 1 lb. of gain.

Leading states in the United States in the production of swine include Iowa, Illinois, Minnesota, Indiana, and Nebraska. More than 20% of the swine production in the country takes place in Iowa alone.

Types and Breeds The basic types of swine operations in the United States are feeder-pig producers and market-hog producers. Operations that produce feeder pigs usually maintain large herds of sows that produce 2 to 2½ litters of piglets each year. *Sows* are females of the swine family that have given birth. A *piglet* is a young member of the swine family. Piglets are usually sold to other producers, who feed them until they reach market weight. In most operations, crossbred sows are bred to purebred boars to produce offspring with more hybrid vigor than would otherwise result. A *boar* is a male member of the swine family.

Market-hog operations normally purchase pigs at 5 to 8 weeks of age from feeder-pig producers. They feed the pigs until they reach a market weight of 220-260 lbs. They are then marketed at local slaughter plants for processing.

Another type of swine operation is the purebred producers. They are responsible for producing boars for feeder-pig operations. They also produce purebred stock for other purebred operations and

FIGURE 30-10
The Duroc breed of swine is one of the most popular breeds in the United States. ***(From Gillespie/Modern Livestock & Poultry Production, copyright 1989 by Delmar Publishers Inc.)***

contribute much to the genetic improvement of swine in general.

Until recently, two types of purebred swine existed in the United States—the lard type and the meat type. With the decreased demand for lard and the demand for lean pork nearly devoid of fat, the lard-type hog has been bred out of existence.

Popular breeds of swine that were developed in the United States include Duroc, Hampshire, Chester White, Poland China, Spotted Hog, and Landrace (Figure 30-10). The Berkshire, Yorkshire, and Tamworth breeds were developed in England. Durocs and Hampshires are the two most popular breeds of swine in the United States today.

Approved Practices in Swine Production Some of the approved practices in the production of swine include:

1. Buy pigs only from reputable producers or at certified feeder-pig sales.
2. Observe newly purchased animals for signs of diseases and parasites.
3. Group pigs according to size in groups of not more than 20 to 25 animals.
4. Feed a complete, balanced ration based on the age and weight of the animals being fed.
5. Ensure that access to an unlimited supply of fresh water is available at all times.
6. Keep facilities and equipment clean and sanitary.
7. Clean and disinfect all facilities and equipment as each group of animals leaves and before the next group arrives.
8. Select replacement gilts for the breeding herd at an early age and raise them separately from market hogs. A *gilt* is a female of the swine family that has not given birth.
9. Breed gilts at 8 months of age and 250 to 300 lbs. so they farrow at 1 year of age. To *farrow* means to give birth to pigs.
10. Use a hand-mating system to breed gilts and sows. In a *hand-mating system*, the boar and sow are kept separate except during mating. Use a boar to check for animals in heat.
11. Put bred gilts or sows in farrowing facilities or farrowing crates 3 days before they are due to farrow (Figure 30-11). Farrowing crates are specially made cages or pens in which swine give birth. This 3-day period gives the female time to adjust to new surroundings before the piglets are born.
12. Perform the following to the piglets at birth:
    a. Clip needle or wolf teeth.
    b. Clip or tie navel cord and dip the end in iodine.
    c. Provide supplemental iron.
    d. Dock tails of pigs to be marketed for meat. To *dock* refers to removing all but about 1″ of the tail.
    e. Weigh all pigs in the litter.
    f. Ear notch all pigs for identification. *Ear*

FIGURE 30-11

Farrowing crates are in common use in the production of pigs. ***(From Gillespie/Modern Livestock & Poultry Production, copyright 1989 by Delmar Publishers Inc.)***

notching is a system of permanently marking animals for identification by cutting notches in their ears at specific locations.

13. Provide creep feed for the baby pigs by the time they are 1 week old. *Creep feed* is feed provided especially for young animals to supplement milk from their mothers.
14. Castrate males at an early age.
15. Wean pigs at 5 to 8 weeks of age. Weaning at about 6 weeks is normal in most herds of swine.
16. Rebreed sows on the first heat period after weaning the pigs. This usually occurs about 3 days after the pigs are weaned.
17. Limit the feed for sows to prevent them from getting too fat.
18. Provide protection from heat and cold, especially heat. Swine have no sweat glands and care must be taken to keep them cool in hot weather.
19. Maintain complete health and production records on each animal in the breeding herd.
20. Set realistic production goals and cull animals not meeting the goals.

SHEEP

Origin and History The domestication of sheep apparently occurred before the time of recorded history. Fibers of wool have been found in ruins of villages in Switzerland more than 20,000 years old. Egyptian sculptures showing the importance of sheep date back 7,000 years. Even the Bible is filled with mentions of sheep and *shepherds* (one who cares for sheep).

Sheep were probably domesticated from wild types in Europe and Asia by early humans to use for meat, wool, pelts, and milk. *Wool* is the hair from sheep. Sheep have become so dependent on humans that they can no longer survive on their own in the wild.

As civilization advanced, the production of wool became a priority in sheep production. As a result, specific wool-producing breeds were developed in Europe. Most of today's breeds can be traced back to these breeds developed 500 to 1,000 years ago.

Columbus had sheep with him on his second voyage to the West Indies in 1493. Cortez brought sheep when he explored Mexico in 1519. Spanish missionaries also kept sheep and taught Indians of the Southwest how to weave wool into cloth.

English settlers on the East Coast also raised sheep for the production of wool. Lamb and mutton were of secondary concern. *Lamb* refers to young sheep as well as the meat of young sheep. *Mutton* is meat from mature sheep.

Centers of sheep population gradually moved from the Northeast to the West as populations expanded. Areas of open spaces and abundant grasses proved to be ideal for the production of sheep.

Economic Importance Production of sheep in the United States is relatively unimportant compared to dairy, beef, and swine. Americans eat only about 1.5 lbs. of lamb and mutton per person each year. This amount is increasing slowly, but is not expected to seriously challenge beef and pork producers. Each person in the United States also uses about 0.6 lbs. of wool each year.

Sales of sheep and lambs total about $400 million each year. The sale of wool adds about another $100 million to the income of sheep producers.

Because sheep have the ability to survive in areas of limited feed and harsh climates, they are of more economic importance than would be expected. In many such areas, only sheep can survive and constitute a major animal enterprise.

Types and Breeds Sheep operations can be divided into two basic types—farm flocks and range operations. A *flock* is a group of sheep. Farm-flock operations are generally small and are often a part of diversified agriscience operations. They may raise either purebred or grade sheep. These flocks usually average less than 150 animals. They are responsible for about one-third of the sheep and wool produced in the United States.

The other two-thirds of the approximately 13 million sheep in the United States are produced on range operations. Many flocks contain 1,000 to 1,500 head. They are nearly 100% grade sheep. Range production is concentrated in the 12 western states. Texas, California, Wyoming, South Dakota, and Colorado are leading states in sheep and wool production.

There are five basic classifications of sheep according to wool type in the United States. They are fine wool, medium wool, long wool, crossbred wool, and fur sheep.

Fine-wool breeds of sheep produce wool that is very fine in texture with a long staple length. It has a wavy texture, is very dense, and is used to make fine quality garments. Fine-wool sheep

FIGURE 30-12
The Suffolk is one of the most popular medium wool breeds of sheep in the United States. ***(Courtesy of the Colombia Sheep Breeders Association of America)***

often produce as much as 20 lbs. of wool per sheep per year. Breeds of fine-wool sheep all originated from the Spanish Merino breed. Fine-wool breeds in the United States include the American Merino, Delaine Merino, Debouillet, and Rambouillet.

Medium-wool breeds of sheep were developed for meat, and little emphasis was placed on the production of wool. Popular medium-wool breeds include Suffolk, Shropshire, Dorset, Hampshire, and Southdown (Figure 30-12).

The long-wool breeds of sheep were developed in England. They tend to be larger than most of the other breeds. The wool produced tends to be long and coarse in texture. Long-wool breeds in the United States include Leicester, Lincoln, Romney, and Cotswold.

Crossbred-wool breeds are the result of crossing fine-wool breeds of sheep with long-wool breeds. They were developed to combine good quality wool with good quality meat. Because they tend to stay together as a group better than other breeds of sheep, they are popular in the western range states. Crossbred-wool breeds include Corriedale, Columbia, Panama, and Targhee (Figure 30-13).

There is only one breed of fur sheep in the United States. The karakul is grown for the pelts of its lambs. The pelts are taken from very young lambs and made into expensive Persian lamb coats.

FIGURE 30-13
The Columbia is a crossbred wool breed of sheep popular in the western range states. ***(Photos courtesy of the Colombia Sheep Breeders Association of America)***

The production of wool and meat is of little importance in this breed.

Approved Practices in Sheep Production Some of the approved practices in the production of sheep and wool include the following:

1. Select lambs that are large and growthy for their age.
2. Select purebred stock based on physical appearance and pedigree.
3. Select breeding stock with a history of multiple births.
4. Provide shelter from severe weather conditions.
5. Provide good quality forages and unlimited fresh, clean water.
6. Vaccinate for disease problems of local concern.

7. Treat for internal and external parasites.
8. Breed ewes to lamb at no more than 2 years of age. A *ewe* is a female of the sheep family.
9. Use a marking harness on rams to tell when ewes have been bred. A *ram* is a male of the sheep family.
10. Use a system of identification to distinguish ewes.
11. Do not disturb ewes during lambing. With sheep, *lambing* means to give birth.
12. Shear ewes at least 1 month before lambing. *Shearing* is the process of removing wool from sheep.
13. Provide clean, warm, dry stalls for lambing.
14. Make sure that ewes accept their new lambs.
15. Dock lambs' tails at 7 to 10 days of age.
16. Castrate ram lambs to be marketed.
17. Keep hooves properly trimmed.
18. Maintain a complete flock health-prevention and control program.
19. Cull ewes that do not lamb or those that have health problems.
20. Maintain a complete and accurate record-keeping system.

GOATS

Origin and History The domestication of goats probably took place in western Asia during the Neolithic Age, between 7000 and 3000 B.C. Remains of goats have been found in Swiss lakes villages of that time period. Mention of the use of mohair from goats is made in the Bible. *Mohair* is hair from goats.

Goats were imported into the United States from Switzerland for milk production early in the colonial period. Angora goats to be used for mohair production were also imported from Turkey.

Milk or dairy-type goats can be found all over the United States, with concentrations on the East and West coasts. Nearly 95% of the mohair-producing goats are located in Texas.

Economic Importance Goats are of little economic importance to agriscience in the United States. Most dairy goats are raised in very small numbers by suburbanites and small farmers to produce milk and meat for their own families. There are few large herds of milk goats. Finding a processor to bottle goat's milk is often difficult as the number

FIGURE 30-14
The Angora goat produces mohair for use in upholstery, rugs, and other products. ***(Courtesy of the Angora Goat Breeders Association)***

of health standards for food production and handling continue to increase.

The United States produces nearly 60% of the mohair in the world. Texas provides more than 95% of this mohair.

Goats provide little competition for food for cattle and sheep. They prefer to eat twigs and leaves from woody plants rather than grass.

Types and Breeds There are two types of goats in the United States. They are hair producing and milk producing.

The only hair-producing goat is the Angora (Figure 30-14). This produces 6–7 lbs. of mohair per year. Mohair ranges in length from 6 to 12″. Angora goats are best adapted to a dry climate with moderate temperatures. In addition to producing mohair, Angora goats are used for meat and to help control weeds and brush. Goat meat is called *chevon*.

Milk-producing or dairy goats are found in every state in the United States. Normal production (per goat) averages 3 to 4 quarts of milk per day during a 10-month lactation period. The *lactation period* is the time during which an animal produces milk. The common breeds of dairy goats are Nubian, Saanen, French Alpine, LaMancha, and Toggenburg (Figure 30-15).

Approved Practices in Goat Production Some approved practices in the production of goats and goat products are:

1. Select goats according to intended use.
2. Use physical appearance, pedigrees, and records as a basis of selection.
3. Purchase replacement animals from reputable breeders.
4. Provide additional feed for hair goats on the range in winter.
5. Feed dairy goats supplemental grains based on amount of milk production.
6. Breed goats in the fall to have *kids* (young goats) in the spring.
7. Breed *does* (female goats) for the first time at 10 to 18 months of age.
8. Use 1 *buck* (male goat) for every 25 to 50 does.
9. Shear or clip hair goats twice each year.
10. Maintain clean and sanitary conditions for the production of milk.

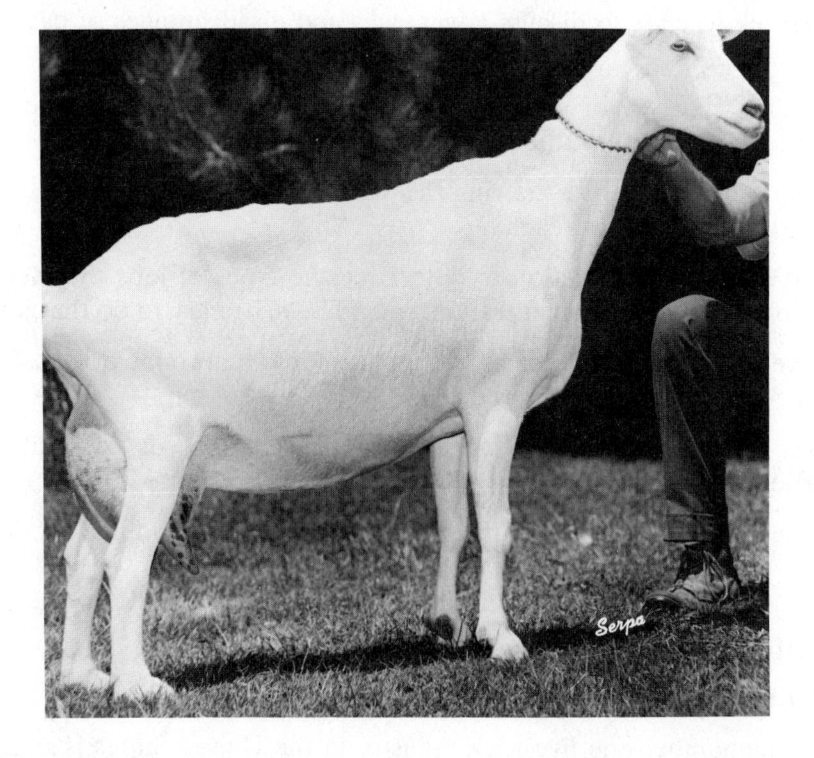

FIGURE 30-15
The Saanen is one of the highest milk producing breeds of dairy goats. ***(Courtesy of the American Goat Society)***

11. Castrate bucks at an early age that are not to be used for breeding.
12. Maintain a herd health program.
13. Milk dairy goats twice a day (the normal schedule).
14. Dehorn dairy-type goats at an early age.
15. Maintain complete and accurate records of reproduction, production, and health.

The production of dairy and livestock in the United States is big business, generating more than $68 billion in revenues in 1981. There are types of livestock production that are adapted to almost every locality and situation. Nearly every facet of life is affected in one way or another by animals and animal products. Although food from animals is more expensive than food from crops, animal products add variety and quality to human diets. Similarly, animals are the source of high-quality fabrics, leather, and many other products in high demand. Therefore, animals and the production of animals will be important enterprises in the future.

Student Activities

1. Define the "Terms To Know."
2. Develop a word search or other puzzle using the "Terms to Know." Trade your puzzle with someone else in class and solve his or hers.
3. Make a chart showing when and where the various types of animals were domesticated.
4. Participate in a class discussion on how and why certain animals were domesticated and others were not.
5. Take a class survey of all of the breeds of animals owned by students and their families.
6. Invite a breeder of purebred livestock and a breeder of commercial or grade livestock to class to discuss advantages and disadvantages of each type of operation.
7. Conduct a survey of your local school district to determine the types of livestock being raised, the breeds, and the numbers of each.
8. Make a bulletin board showing the various types of livestock in your area and the importance of each.
9. Visit a livestock operation to determine the types of jobs that need to be performed there and what training would be necessary to do that job.
10. Survey various purebred livestock magazines to determine prices of animals being sold at various purebred sales.
11. Make a notebook-cover collage showing as many breeds as possible of the particular type of livestock in which you are interested.
12. Make a bulletin board showing some items made from animal products.

Self-Evaluation

A. MULTIPLE CHOICE

1. The number one livestock industry in the United States is ______ production.

   a. dairy
   b. beef
   c. swine
   d. sheep

2. Milk for the purpose of making cheese is produced in ______ operations.

a. Grade A
b. Grade B
c. Class A
d. Class B

3. Merino sheep belong to the ______ wool type.

a. fine-
b. medium-
c. long-
d. crossbred-

4. The average Angora goat produces about ______ lbs. of mohair each year.

a. 2–3
b. 3–5
c. 6–7
d. 10–15

5. The most popular breed of dairy cattle in the United States is

a. jersey.
b. Ayrshire.
c. Brown Swiss.
d. holstein.

6. The leading state in swine production is

a. California.
b. Texas.
c. Iowa.
d. Pennsylvania.

7. Docking refers to

a. fees charged for selling animals.
b. removal of tails.
c. backing a cattle trailer to a loading ramp.
d. none of the above.

8. Sows normally have ______ litters of piglets per year.

a. one
b. two
c. three
d. four

9. The average American eats about ______ lbs. of beef each year.

a. 12.1
b. 55.5
c. 62.5
d. 79

10. Dairy heifers should be bred to calve when they are ______ year(s) old.

a. 1
b. 2
c. 3
d. 4

B. MATCHING

______ 1. Buck
______ 2. Mutton
______ 3. Milk
______ 4. Calve
______ 5. Veal
______ 6. Chevon
______ 7. Farrow
______ 8. Shepherd
______ 9. Shear
______ 10. Butterfat

a. Give birth to cattle
b. Remove wool from sheep
c. Calf meat
d. Male goat
e. One who cares for sheep
f. Meat from mature sheep
g. Fat in milk
h. Give birth to piglets
i. Liquid from mammary glands
j. Goat meat

C. COMPLETION

1. ______ were domesticated in China.

2. Goats are of two types—hair and ______ .
3. ______ is the most common system of identification of swine.
4. ______ is rendered pork fat.
5. Milk-producing animals are called ______ .
6. The ______ period is the time when cows are not producing milk.
7. Three classifications of breeds of beef cattle are English, exotic, and ______ .
8. Consumption of red meat in the United States has been ______ in recent years.
9. ______ feed is feed provided for young animals to supplement milk from their mothers.
10. Breeds of sheep grown especially for meat belong to the ______ wool class.

UNIT 31 Horse Management

OBJECTIVE To determine the role of horses in our society and how to care for and manage them.

Competencies to Be Developed

After studying this unit, you will be able to

- ☐ understand the origin and history of the horse.
- ☐ determine the economic importance of the horse in the United States.
- ☐ recognize the various types and breeds of horses.
- ☐ understand approved practices for the care and management of horses.
- ☐ understand the basics of English and western riding.
- ☐ list rules of safety for handling horses.
- ☐ understand the vocabulary generally associated with horses.

TERMS TO KNOW

Light horses
Draft horses
Donkey
Coach horses
Horses
Ponies
Hand
Color breeds
Mule
Jack
Mares
Hinny
Stallion
Jennet
Tack
Gait
Equitation
Saddle
Stirrups
Walk
Trot
Canter
Gallop
Rack
Pace
Reins
Bit
Horsemanship
Jog
Lope
Filly
Foal
Colt
Gelding
Shod
Farrier
Blemish
Unsoundness
Vice

MATERIALS LIST

bulletin board materials
horse breed magazines
examples of horse tack

Horses are an important part of many industries in the United States. They are often nearly indispensable on working cattle ranches. In sports, they generate millions of dollars in the horse-racing industry. Horses are also important components of rodeos and hunting competitions. They are often status symbols. Horses even serve as food for pets and humans in some cases. But probably most important, horses serve as friends, companions, pets, and a form of relaxation to millions of people.

ORIGIN AND HISTORY There are fossil remains of the horse family that have been found in the United States dating back nearly 58 million years. However, by the time of the discovery of America, no horses remained in the New World. They apparently had all died out only a few thousand years before. It should be noted that these early horses were only about the size of a collie dog.

Even though the horse became extinct in the New World a few thousand years ago, it had previously migrated to Europe and Asia when Alaska and Siberia had been connected. It is from these horses that humans domesticated the horses of today.

The horse was probably one of the last animals to be domesticated. This occurred first in Central Asia or Persia about 5,000 years ago. Horses came to Egypt in about 1680 B.C. from Asia. From there, the Egyptians introduced horses over the known world.

The Arab horse is the ancestor of most modern light breeds of horses. *Light horses* are breeds used for riding. However, horses were not used to any great extent until A.D. 500 to 600.

Draft horses probably originated from the heavy Flanders horse of Europe. *Draft horses* are used for work. They were often used in religious rites for sacrifice and for food at the time of domestication.

Donkeys were domesticated in Egypt some time before 3400 B.C. when they appeared on slates of the First Dynasty. A *donkey* is a member of the horse family with long ears and a short, erect mane. Mention of the use of donkeys appears in many places in the Bible. They were generally used as saddle animals or beasts of burden.

Horses were imported to the New World by Columbus in 1493. When Cortez came to Mexico in 1519, he brought nearly 1,000 horses with him. The death of the explorer deSoto in the upper Mississippi region led to the abandonment of many of his horses. These horses, coupled with some of those lost or stolen from Spanish missionaries, probably formed the nucleus of the horses used by the Plains Indians.

The introduction of the horse to the Plains Indian culture completely changed it. Horses permitted easier hunting of buffalo in wider areas. This led to competition between various tribes of Plains Indians and a nearly constant state of war in an area where peace had previously prevailed.

The European settlers on the East Coast were far less dependent on the horse than were the explorers. It was of little advantage to have a riding horse with no place to ride. As draft animals, oxen or cattle were more popular because they were stronger and produced meat and milk as well as work.

It was only as Americans prospered that horses became an important part of their life-styles. Horses have had an illustrious past, and a bright future looms ahead.

ECONOMIC IMPORTANCE

There are approximately 10 million horses in the United States today. Approximately 75% of these horses are owned by nonfarmers and nonranchers. They are used primarily for pleasure, racing, breeding, and companionship.

Horse owners spend more than $16 billion on horses and their use each year. Horse racing is the most popular spectator sport in the United States. Individual racing events may offer as much as $1 million in purses.

Horse shows are also popular events, especially for young people. One breed association alone sponsored more than 4,000 shows in a recent year.

There are numerous noneconomic benefits of horses. They help develop a sense of responsibility in young people. They provide physical activity for the young and old alike. They provide the opportunity for families to participate in outdoor activities together, and they provide a companion when one is needed.

TYPES AND BREEDS

For the purpose of this discussion, members of the horse family are divided into horses, ponies, and donkeys and mules. Horses are further divided into light horses, coach horses, and draft horses. *Coach horses* are developed to pull coaches.

Members of the horse family that are 14.2 hands or more tall are called *horses*. Animals of the same family that are less than 14.2 hands tall are *ponies*. A *hand* is a unit of measurement for members of the horse family, and it is equal to 4″.

The light breeds of horses are by far the most popular type of horses in the United States. They are further divided into true breeds, or purebreds, and color breeds. True breeds have registration papers and parents of the same breed. *Color breeds* need only to be a specific color or color pattern, although other characteristics may be considered for registration. They need not have purebred parents.

Examples of the nearly 50 true breeds of horses in the United States include Arabian, Morgan, thoroughbred, quarter horse, standardbred, Tennessee walking horse, and American saddle horse. The most popular of the light breeds of horses in the United States is the quarter horse (Figure 31-1). The quarter horse is used for riding, hunting, and sports, and on cattle ranches to work with cattle.

Some of the color breeds of horses include the palomino, American white, American creme, and pinto (Figure 31-2). They may or may not breed true to color. The uses of the color breeds are the same as those for the true breeds of light horses.

When coaches disappeared from the American scene coach horses became very rare in the United

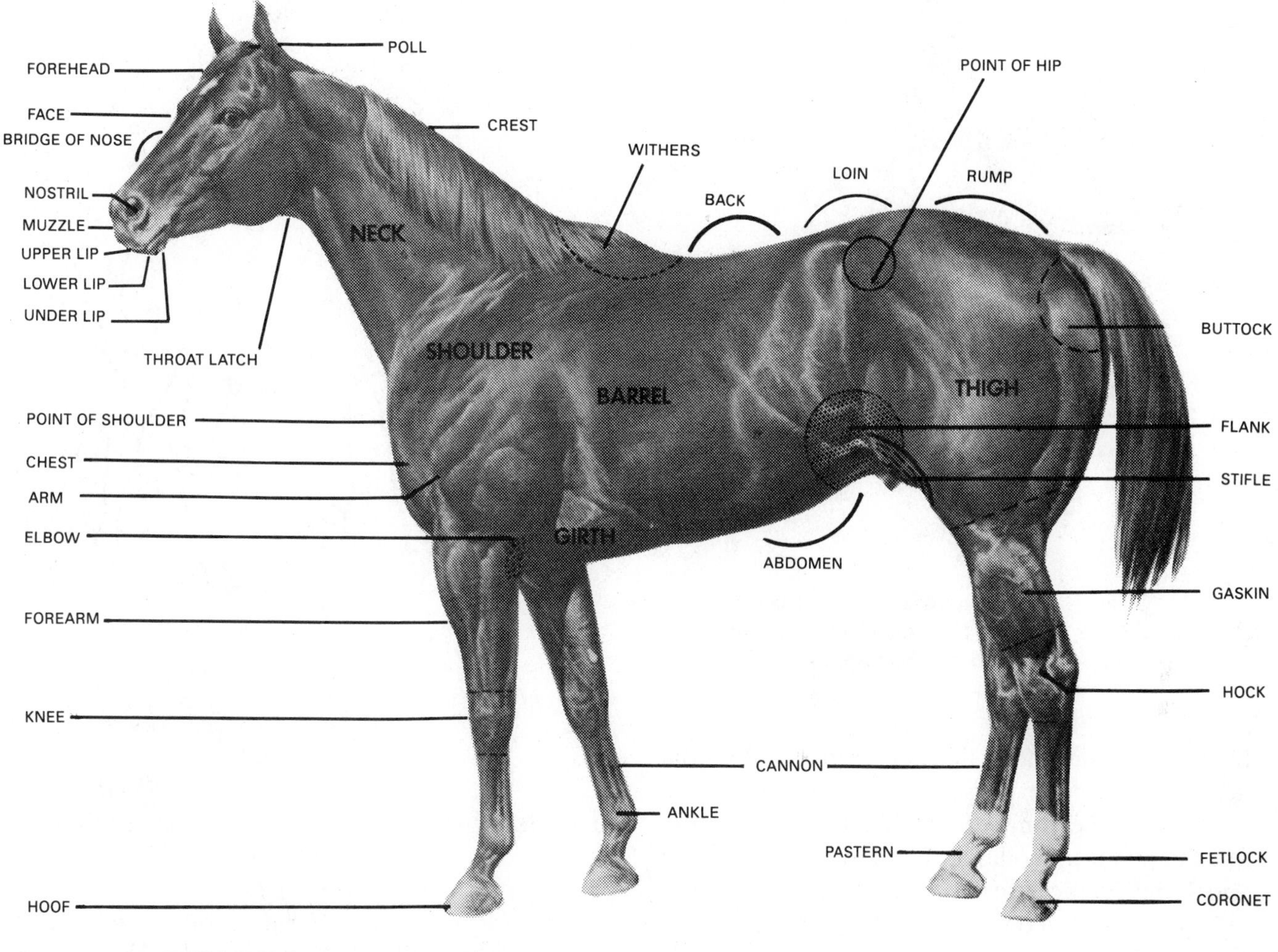

FIGURE 31-1
The Quarter Horse is the most popular breed of horse in the United States. ***(Courtesy of the American Quarter Horse Association)***

FIGURE 31-2
The palomino is probably the most recognized breed of horse in the United States due to its unique coloring.

States, since there was little need for them. While they were used, though, they combined the qualities of both the light horse and draft breeds. They were fast (like the light horse) and strong enough (like a draft horse) to pull the heavy stage coaches of their time. The only example of the old coach-style horse in the United States today is the Cleveland Bay.

Draft or work horses were once the backbone of American agriculture. Today they are used primarily for show and recreation. Their doom was the invention of the internal combustion engine, which only had to be fed when it was actually working and could do the work of many horses. Breeds of draft horses include Belgian, Clydesdale, Percheron, shire, and Suffolk.

Ponies are smaller versions of the horse. They are used for riding and driving, and as pets. Some

FIGURE 31-3
The Shetland pony is a very popular children's mount in the United States. ***(Courtesy of The American Shetland Pony Club, Peoria, IL)***

FIGURE 31-4
The mule is a cross between a jack and a mare.

breeds were originally developed to work in coal mines and in other pursuits where small sizes are important. The common breeds of ponies include Shetland, Welsh, Gotland, Appaloosa, Connemara, and Pony of the Americas. The Shetland is the most popular breed of ponies in the United States (Figure 31-3).

Donkeys are used in the United States for work and as pack animals. Miniature donkeys also make excellent pets. Donkeys can be used as the male parent of mules. A *mule* is a cross between a jack and a mare (Figure 31-4). A *jack* is a male donkey or mule. *Mares* are mature female horses or ponies. The opposite cross, a stallion with a jennet, results in a *hinny*. A *stallion* is an adult male horse or pony. A *jennet* is a female donkey or mule.

Mules are more intelligent than horses and can do more work than comparable-sized horses. There are several types and sizes of mules, depending on the breed or type of horse that serves as the female parent. Mules may be used for work, sports, or pleasure.

RIDING HORSES The riding style for horses can be divided into two general classifications, English and Western. The style of riding and the *tack* (horse equipment) required vary greatly between the two classifications. Even the *gaits* (the way an animal moves) of the horses involved have different names.

English Equitation In English equitation, the saddle is smaller and lighter, and has shorter stirrups than the Western saddle (Figure 31-5). *Equitation* is the art of riding on horseback. A *saddle* is the padded leather seat for the rider of a horse. *Stirrups* are footrests used when riding horses. The English riding clothing includes close-fitting breeches, jodhpurs, jacket, and hard hat or derby (Figure 31-6).

The gaits of the English equitation horse are walk, trot, canter, gallop, rack, and pace (Figure 31-7). The *walk* is a slow, four-beat gait. The *trot* is a fast, two-beat diagonal gait. The *canter* is a slow, three-beat gait, whereas the *gallop* is a fast, three-beat gait. The *rack* is a fast, four-beat gait and the *pace* is a side-to-side, two-beat gait.

The English equitation horse is controlled by reins that may be held in either one or both hands. *Reins* are leather or rope lines attached to the bit to control the horse. The *bit* is a metal mouthpiece used for that purpose.

Western Horsemanship Western horsemanship differs from English equitation in many ways. *Horsemanship* is the art of riding and knowing the needs of the horse. The Western saddle is heavy and has a horn that is used when roping livestock (Figure

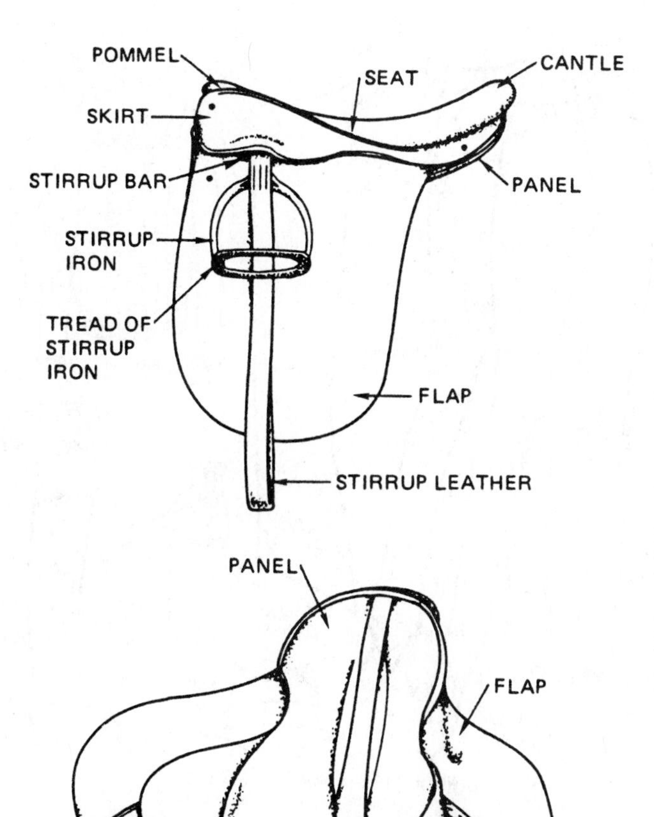

FIGURE 31-5
The English saddle is lighter and has shorter stirrups than the Western saddle. ***(From Gillespie/Modern Livestock & Poultry Production, copyright 1989 by Delmar Publishers Inc.)***

31-8). This saddle has longer stirrups than the English saddle and may be plain or have very fancy ornamentation. The Western saddle was designed for the comfort of cowboys, who spend much of their time in a saddle.

Western riding attire was also designed for comfort. It consists of jeans and a western shirt and hat. Cowboy boots and chaps complete the typical Western outfit.

The gaits of the Western horse are walk, jog, lope, and gallop. The *jog* is a slow, smooth, two-beat diagonal gait. The *lope* is a very slow canter.

The Western rider controls the horse with the reins held in one hand. The hand holding the reins cannot be changed during competition riding.

HORSE SAFETY RULES

Some of the safety rules to observe

AGRI-PROFILE

CAREER AREAS: HORSE BREEDER/TRAINER/RANCHER/JOCKEY/FARRIER

The number of horses for racing, working, and pleasure is increasing in the United States. A prominent horse breeder of Ocala, Florida, displays an excellent thoroughbred race horse from among the many on the farm.

Careers in horse management are available in a variety of settings. Urban police forces and park services have rediscovered the advantages of patrolling from horseback, and horse-drawn carriages have captured the fancy of sightseers in many areas. Additionally, many families have adopted the horse for teenage animal production projects in semi-rural as well as rural settings. These frequently involve the entire family in fairs, shows, or rodeos. Horses are used extensively for herding and handling cattle on ranches and in cattle feedlot operations. Pack mules and riding horses are used extensively for hunting and other recreational pursuits. Finally, the horse racing industry is huge in the United States, involving many people and extensive career possibilities.

While many raise horses as an avocation, there are rich and varied career opportunities to raise, manage, or care for horses. Veterinarians, trainers, managers, attendants, jockeys, and farriers are common around race tracks and breeding farms. Tack, feed, and equipment supply centers provide business opportunities for those wishing to interface with the horse industry but preferring not to handle horses directly.

Horse health, soundness, breeding, behavior, and performance command the attention of those in research, veterinarian services, and education. Horse-related jobs may be indoor or outdoor, and run the spectrum from laborer to scientist.

FIGURE 31-6
Proper riding attire for English and Western riders.

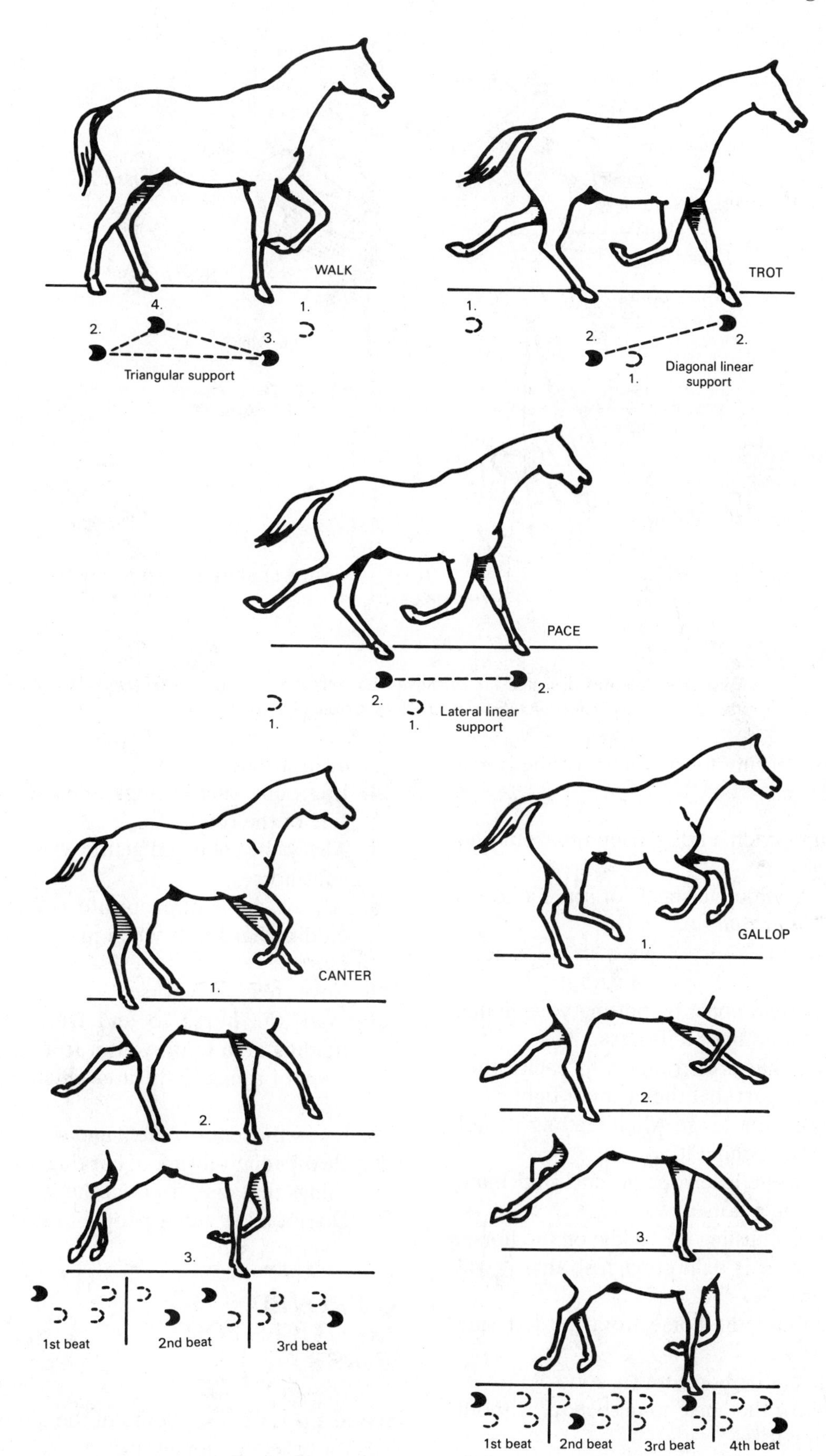

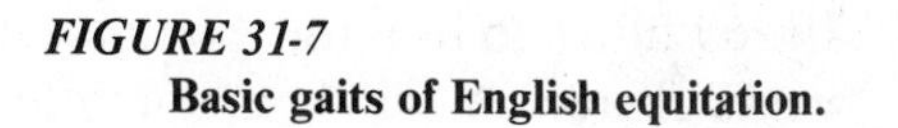

FIGURE 31-7
Basic gaits of English equitation.

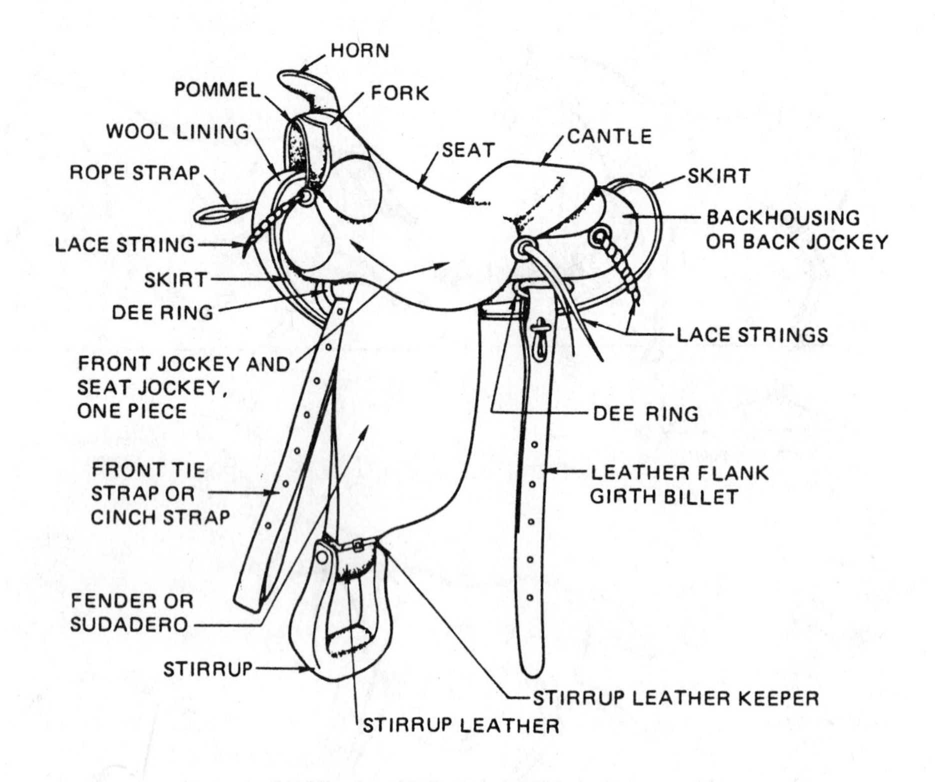

FIGURE 31-8
The Western saddle was designed for cowboys who worked cattle. ***(From Gillespie/Modern Livestock & Poultry Production, copyright 1989 by Delmar Publishers Inc.)***

when riding and caring for members of the horse family include:

1. Always approach a horse from the front and the left side.
2. Never do anything to startle or scare a horse. It may kick or rear up.
3. Be sure that the horse always knows exactly what you intend to do to or with it.
4. Always pet the horse by putting your hands on its shoulder, not on its nose.
5. Tie horses that are strangers to each other far enough apart that they cannot fight.
6. Walk beside the horse when leading it, not in front of or behind it.
7. Never wrap the lead rope around your hand when leading a horse.
8. Use care in adjusting the saddle on the horse. Be sure that it is tight enough so that it will not slip or slide.
9. Always mount the horse from the left side (Figure 31-9).
10. Always keep the horse under control.
11. Do not allow the horse to misbehave without disciplining it.
12. Walk the horse up and down steep slopes, on rough ground, and across paved roads.
13. Reduce speed when riding on rough terrain or in wooded areas.
14. In groups, ride in single file and on the right side of the road.
15. Always be calm and gentle when dealing with your horse.
16. Always wear appropriate riding attire, including hard hats when jumping over obstacles.
17. Never tease the horse.
18. Walk the horse to and from the barn or holding area to prevent it from developing a habit of riding to the area when it comes into sight.
19. Know the temperament and vices of the horse.
20. Be especially aware of barking dogs and other things that may frighten the horse.
21. Do not allow other people to ride your horse unsupervised.

APPROVED PRACTICES FOR HORSES

Breeding Horses Some of the approved practices for breeding horses are:

1. Breed fillies to foal for the first time at 3 to 4 years of age. A *filly* is a young female horse

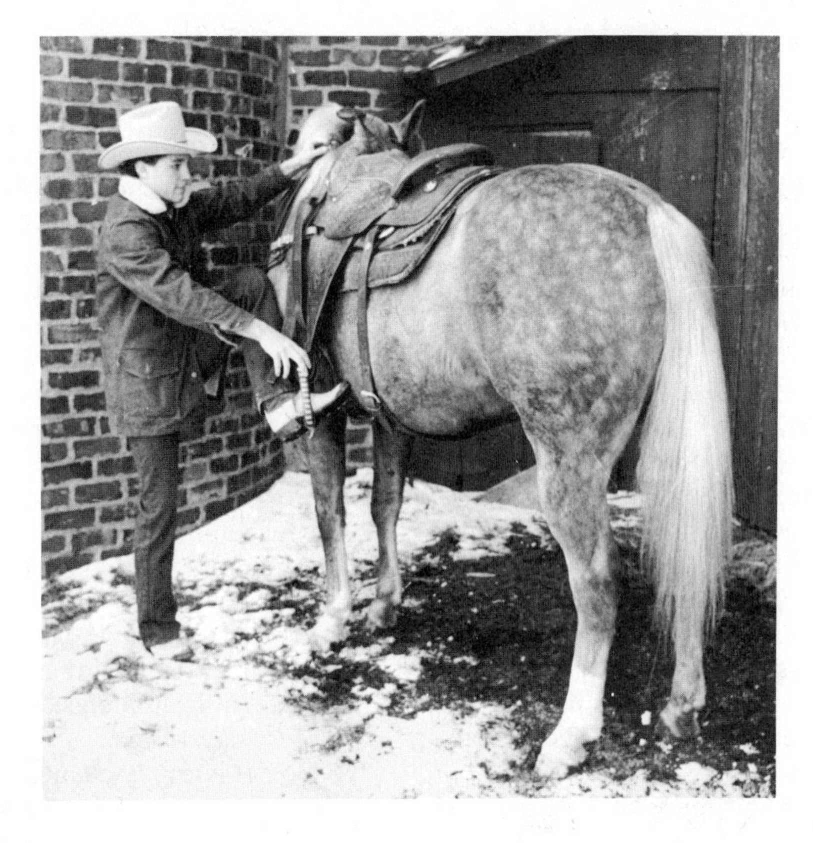

FIGURE 31-9
Always mount the horse from the left side. *(From Gillespie/Modern Livestock & Poultry Production, copyright 1989 by Delmar Publishers Inc.)*

or pony. A *foal* is a newborn horse or pony. To foal is also to give birth in the horse family.

2. Breed mares in April, May, or June when their fertility is normally highest. Breeding at this time also allows the mare to foal in the spring when pastures are actively growing.
3. Rebreed mares 25 to 30 days after foaling, if they are in good condition with no reproductive-tract problems.
4. Make sure that the pregnant mares and fillies get plenty of exercise.
5. Allow the mare to foal in a clean pasture free from internal parasites or in a large, clean pen or stall (Figure 31-10).
6. Put the mare in the facility in which she will foal several days before she is expected to foal. This gives her some time to get used to the facility.
7. Remain out of the mare's sight when she is foaling. Mares prefer to be alone with no interruptions during the foaling process.
8. Contract the services of a veterinarian if the mare shows signs of difficulty in foaling or if the position of the foal is abnormal for delivery.
9. At birth:
   a. Make sure that the foal is breathing. Tickling the foal's nose with a piece of straw will often stimulate it to start breathing after it is born. More drastic measures, such as artificial respiration, may be needed in more serious situations.

FIGURE 31-10

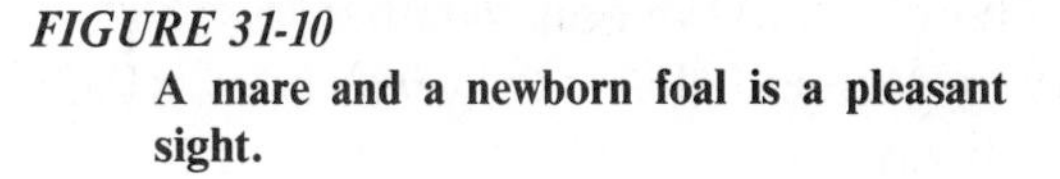

A mare and a newborn foal is a pleasant sight.

   b. Be sure to remove the mucus from the nose of the foal and dry it off.
   c. Dip the end of the navel cord in iodine after tying it off. This prevents infection from entering the foal through the open navel cord.
   d. Make sure that the foal nurses as soon as it can stand. It should be allowed to stand on its own. Normally, nursing first takes place within one-half hour after birth.
   e. Check to see that the foal has a bowel movement within the first 12 hours of birth. The foal may need an enema if the bowel movement does not occur naturally.
   f. If the foal shows signs of diarrhea, reduce the amount of milk that it is allowed to drink.
10. Provide creep feed for the nursing foal when it reaches 10 days to 2 weeks of age.
11. Begin training the foal as soon as possible. A foal that is trained from an early age seldom needs to be broken.
12. Do not mistreat the mare or foal at any time.
13. Wean the foal at 4 to 6 months of age. The foal and the mare should not be allowed to see each other for several weeks after weaning to break the bond between mother and offspring.
14. Castrate colts that are not to be used for breeding purposes. A *colt* is a young male horse or pony. A castrated male horse is called a *gelding*. Geldings are usually much more docile and easy to handle than stallions.

All Horses Approved practices for the care and management of all types of horses and ponies include the following:

1. Groom the horse daily or weekly at a very minimum.
2. Always use soft brushes when grooming a horse. Its skin is very sensitive and easily damaged by rough treatment.
3. Shorten the mane and tail when needed by pulling the hairs from the underside.
4. Be especially careful not to get water in the horse's ears when washing it. Also, be sure to dry the horse off quickly so that it does not catch a cold in cool weather.
5. Inspect and clean the horse's hooves daily and always before and after riding it.
6. Trim the horse's hooves every 4 to 6 weeks if the horse is not shod. *Shod* means wearing shoes.
7. Replace the shoes of a shod horse every 4 to 6 weeks. A *farrier* (a person who shoes horses and cares for their feet) should be engaged to perform this task.
8. Shoe a horse for the first time when it is 2 years of age or when it starts being worked.
9. Cool the horse out thoroughly after every exercise period and after riding. This should be done before the horse is allowed to drink large quantities of water.
10. Do not overfeed a horse. The total daily consumption of concentrates and roughages should not total more than 2 to 2.5% of the horse's weight.
11. Feed a horse on a regular routine. The number of times that the horse is fed per day makes little difference, as long as it is the same every day.
12. Do not abruptly change the ration being fed. The stomach of the horse is very temperamental and adjusts to changes in feeding practices very slowly.
13. Never feed moldy feed to a horse.
14. Provide unlimited access to fresh, clean water at all times.
15. Maintain a strict health care program for all horses.
16. Maintain clean and sanitary facilities at all times.
17. When buying a horse, be sure to:
   a. Note blemishes and unsoundnesses of the horse. A *blemish* is an abnormality that does not affect the use of the horse. An *unsoundness* is an abnormality that does affect the use of the horse.
   b. Check the age of the horse by looking at its teeth (Figure 31-11). Examining the teeth can also indicate how long a horse might be usable.
   c. Determine that the horse is not blind or have other problems that may affect its use value and use.
   d. Check for evidence of good health.
   e. Try to determine if the horse has undesirable vices. A *vice* is a bad habit.
   f. Try to determine the personality and spirit of the horse.
   g. Be sure to consider the price of the horse and whether or not you can afford the cost of owning a horse.

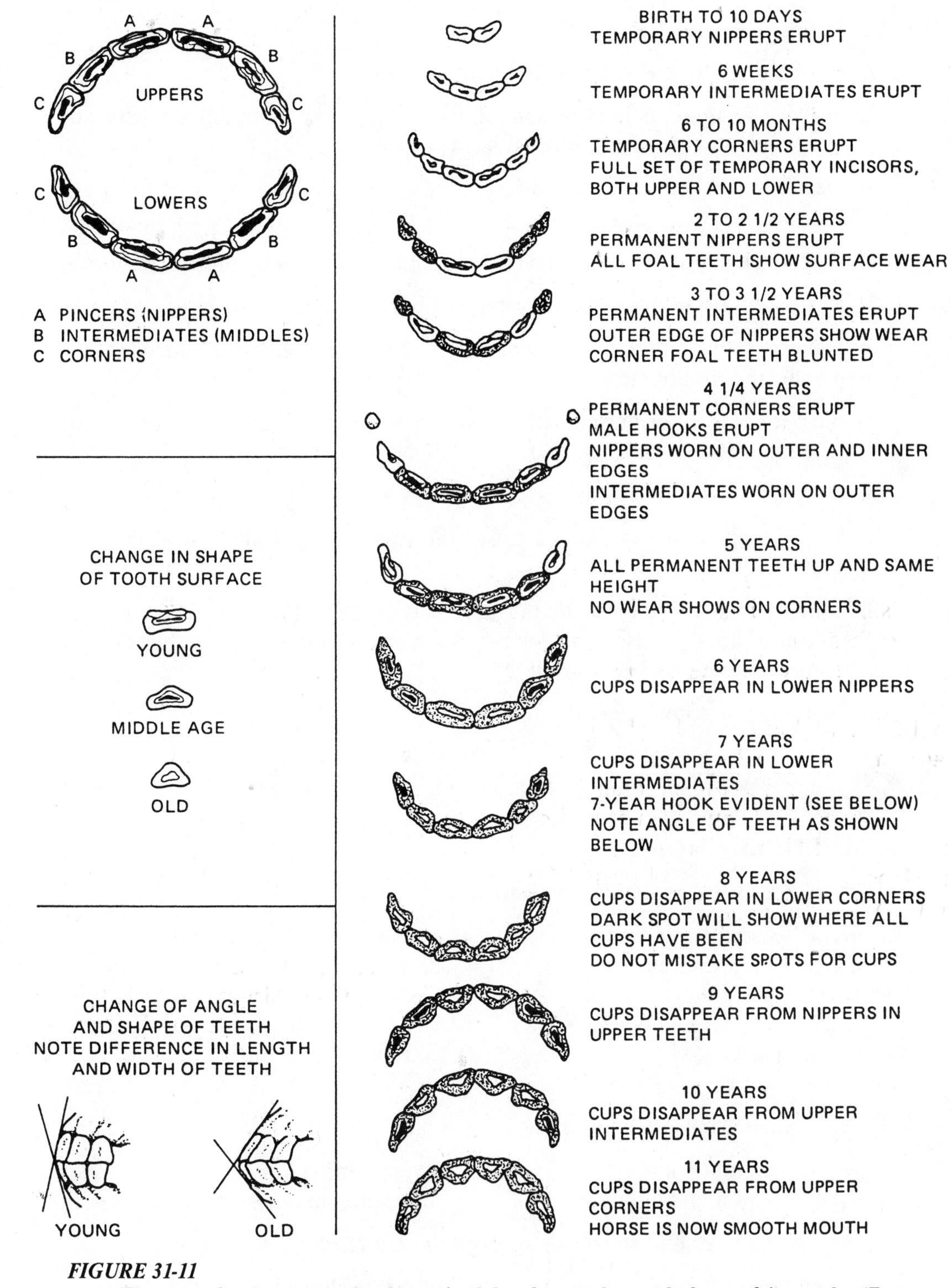

FIGURE 31-11
The age of a horse can be determined by the number and shape of its teeth. ***(From Gillespie/Modern Livestock & Poultry Production, copyright 1989 by Delmar Publishers Inc.)***

h. Check the pedigree of the horse if it is a purebred.

Owning and caring for horses and ponies is an excellent way to get outdoor exercise. The companionship of a horse has given many young people a friend to confide in during the trials of youth. Competitors find horses and racing excellent outlets for their excess energy. Finally, cowboys find horses to be an essential part of a team when working with cattle. The horse was, is, and will continue to be an important part of the world of many people in the United States.

Student Activities

1. Define the "Terms To Know."
2. Survey the students in your school concerning how many have horses and what type and breed of horses that they own.
3. Develop a bulletin board showing as many breeds and types of horses as you can.
4. Attend a horse show and interview participants about why they show horses.
5. Write a report about a breed or type of horse that interests you.
6. Have a horse breeder or owner talk to the class about the care and management of horses.
7. Have a horse veterinarian talk to the class about health care and first aid procedures for horses.
8. Develop a word search using the Terms to Know.
9. Visit a tack shop and make a list of the various types of horse equipment sold there.
10. Look at horse breed and care magazines to determine the uses of the various breeds of horses. Also look for the types of job opportunities that may be available in the horse industry.
11. Demonstrate to the class the techniques of grooming a horse.

Self-Evaluation

A. MULTIPLE CHOICE

1. Horses were probably domesticated in (the)

   a. Russia. c. Persia.
   b. United States. d. China.

2. Light horses are used for

   a. riding. c. racing.
   b. rodeos. d. all of the above.

3. A mule is a cross between a

   a. jack and a mare. c. jack and a jennet.
   b. stallion and a jennet. d. stallion and a mare.

4. The number one spectator sport in the United States is

   a. rodeos. c. horse shows.
   b. horse racing. d. riding.

5. Fillies should be bred to foal at ______ years of age.

   a. 1–2 c. 3–4
   b. 2–3 d. 4–5

6. Always approach horses from (the)

   a. rear. c. left side.
   b. right side. d. none of the above.

7. Horses should be reshod every ______ weeks.

a. 2–4
b. 4–6
c. 6–8
d. 8–10

8. The most popular breed of light horses in the United States is the

a. thoroughbred.
b. Appaloosa.
c. Morgan.
d. quarter horse.

9. A fast, three-beat gait is the

a. gallop.
b. canter.
c. pace.
d. trot.

10. There are approximately ______ horses in the United States.

a. 100,000
b. 1 million
c. 10 million
d. 100 million

B. MATCHING

______	**1.** Stallion	**a.**	Way of moving
______	**2.** Jennet	**b.**	Young female horse
______	**3.** Gelding	**c.**	Female donkey
______	**4.** Mare	**d.**	Newborn horse
______	**5.** Gait	**e.**	Adult male horse
______	**6.** Filly	**f.**	Adult female pony
______	**7.** Tack	**g.**	Male donkey
______	**8.** Jack	**h.**	Horse equipment
______	**9.** Foal	**i.**	Young male horse
______	**10.** Colt	**j.**	Castrated male horse

C. COMPLETION

1. A ______ is a person who shoes horses.
2. Horses originated in what is now the ______ .
3. ______ horses were developed for work.
4. ______ breeds of horses need only to be a certain color or color pattern in order to be registered.
5. The art of riding is called ______ .
6. An abnormality that affects the use of a horse is called a(an) ______ .
7. Bad habits are called ______ in horses.
8. The result of the cross between a stallion and a jennet is a ______ .
9. The most popular breed of ponies is the ______ .
10. The unit of measurement for horses is the ______ .

Food Science & Technology

UNIT 32 Marketing Animal Products

OBJECTIVE To determine the strategies and procedures for marketing animals and animal products to maximize profits from animal enterprises.

Competencies to Be Developed

After studying this unit, you will be able to

- ☐ describe the marketing strategies that maximize profits.
- ☐ distinguish between wholesale and retail marketing.
- ☐ describe the methods of marketing at the farm, roadside stands, and farmers markets.
- ☐ discuss advantages and disadvantages of terminal markets, auctions, and direct marketing.
- ☐ recognize fees, commissions, and other costs of marketing.
- ☐ list the procedures for handling livestock to minimize losses during marketing.
- ☐ understand the grades of popular livestock and animal products.
- ☐ recognize marketing trends and cycles.

TERMS TO KNOW

Supply
Demand
Retail marketing
Consumers
Middlemen
Wholesale marketing
Terminal market
Yardage fee
Commission
Auction markets
Auctioneer
Direct sales
Cooperatives
Vertical integration
Pencil shrink
Calves
Cattle
Veal
Slaughter calves
Feeder calves
Steer
Heifer
Bull
Yearlings
Two-year-olds
Cow
Stag
Slaughter cattle
Quality grades
Finish
Yield grades
Pigs
Hogs
Gilts
Barrows
Boars
Sows
Hothouse lambs
Spring lambs
Lambs
Rams
Ewes
Wethers

MATERIALS LIST

various newspapers and marketing reports

The first question to ask when deciding on a production enterprise should be, Can this enterprise be profitable? In order for a profit to be made, successful marketing of a product is a must. This unit explores the basics of marketing animals and animal products to maximize profits.

There are a number of factors that need to be considered when deciding on how to market animals and animal products. Some of these factors include the following:

1. Demand for the product to be produced
2. Supply of the product already available
3. Types and availability of markets in the area
4. Competition from similar products
5. Buying power of intended consumers
6. Seasonal variations in demand
7. Government price supports available

Supply and demand have long been the factors that determine whether or not the production of a product can be profitable. *Supply* is the amount of a product available at a specific time and price. Some of the things that may determine supply are how many people there are in the business of producing that product in the market area, how much of the product is coming into the market area from other areas, and what the past history of profitability is for that product in your area.

Demand for a product is also determined by numerous factors. *Demand* is the amount of a product wanted at a specific time and price. It is often determined largely by price. The less expensive a product is, the more of it is wanted. However, there are other factors that influence demand for a product. The amount of money available to consumers to buy that product is a factor often overlooked. Competition from similar products may reduce demand. Seasonal variations in demand also need to be considered. It is easier to sell ice cream in the summer than in the winter. Advertising to create demand is also a factor to be taken into consideration.

MARKETING STRATEGIES FOR MAXIMIZING PROFITS

In order to make the most money from a production enterprise, successful marketing is a must. Strategies that can be used to market animals and animal products most profitably include:

1. Determine what types of markets are available to you.
2. Determine the costs of various types of marketing.
3. Determine transportation costs to market at each of the markets available to you, and sell where transportation costs are low.
4. Determine the most profitable form in which to market your product (age, size, weight, degree of preparation).
5. Advertise to create markets where none existed before.
6. In seasonal markets, market your product at the peak of demand.

RETAIL MARKETING

Retail marketing is selling a product directly to *consumers* (people who use a product). This may take place on the farm, at roadside markets, or at farmers markets that have developed around most centers of population. Retail marketing is generally used for processed products. For example, ham and sausage, rather than live pigs, are usually sold at a retail outlet. This creates many other problems that need to be dealt with and that have an effect on the profitability of a production enterprise. When an animal product is processed for retail sales, various standards and state and federal regulations must be met. Also, changes in facilities are often necessary. Sanitary storage of processed products and the increased labor costs associated with retail customers must also be considered.

On-the-farm retail sales present special problems. If retail marketing is to take place on the farm where the product is produced, the farm must be well maintained and attractive to the customers buying the product. Facilities for parking are necessary. Animal and sales facilities must be clean, and the animals present must be well cared for and contented. An education program to make consumers aware of good management techniques is often necessary so that they understand what is happening on the farm. Salespersons or the farm family itself must be available to wait on customers. Privacy is often difficult to achieve, because retail customers often feel that since products are sold at the farm, it is all right to drop by at any time.

Retailing at the Farm

On the positive side of farm sales, profits are often higher because retail prices are higher. Also, middlemen are eliminated in the marketing procedure. *Middlemen* are people who handle an agricultural product between the farm and the consumer. Examples of middlemen include buyers, processors, and salespeople. Higher prices can often be charged when a superior product is produced. For farm families who like to meet new people, on-the-farm sales is an excellent opportunity to do so. Also, these types of sales provide the opportunity for urban and suburban people to see the agriscience way of life firsthand.

Roadside Markets

Roadside marketing retains nearly all of the advantages of on-the-farm retail sales of animal products while eliminating some of the disadvantages. The sales unit is usually somewhat removed from the actual farm operation when roadside marketing is employed. By removing the customers from the production area, more efficient use of labor can take place. Less perfect care and maintenance of facilities can be tolerated.

On the less-positive side, separate facilities for the retail sales unit must be maintained. More labor may be required because it may be inconvenient to use the farmer to staff the roadside stand. The positive effects of the consumer seeing the actual farm in operation are diminished.

Farmers Markets Farmers markets have appeared in many large metropolitan areas to cater to the demands of urban consumers (Figure 32-1). They are normally operated one or two days per week and give urban and suburban consumers access to fresh agriscience products directly from the producer. Farmers markets give the producer an access to markets that would seldom be available otherwise. The producer also has the opportunity to educate the consumer concerning the values of good agriscience products.

Of course, there are several costs and inconveniences associated with marketing agriscience products at farmers markets. There are often fees to be paid for the privilege of participating in farmers markets. Competition may be higher, especially if several producers are selling the same type of product. Vehicles with heating and/or cooling may be necessary to get the product to the market as fresh as possible. Products may be subject to certain food regulations and packaging requirements to meet state and federal standards. Facilities to hold and display the product must also be available at the market.

In summary, the decision whether or not to market products of animal agriscience by retail means requires much consideration. Although the returns are generally higher when products are marketed retail, the costs are also higher. Consideration of the product being produced, availability of markets, labor availability, and personal desires need to be made before a decision concerning marketing method can be made.

WHOLESALE MARKETING

Wholesale marketing is the marketing of a product through a middleman, who eventually forwards the product to the consumer. Most agriscience products produced in the United States are marketed in this way. Wholesale markets allow for marketing large volumes of products with comparatively little labor. Types of wholesale markets include terminal markets, auction markets, and direct sales.

Terminal Markets A *terminal market* is usually a stockyard that acts as a place to hold animals until they are sold to another party (Figure 32-2). The terminal market never actually owns the animals. The animals that are delivered to the terminal market are consigned to a selling agent, who does the actual selling of the animals. The terminal market charges the seller a fee for caring for the animals until they are sold. This is called a *yardage fee*. The selling agent also receives a fee called a *commission* for work in selling the animals.

There are several advantages of selling animals through a terminal market. Because large numbers of animals are massed in one location for sale, volume buyers are more likely to be available to purchase the animals. Professional selling agents

FIGURE 32-1
Farmers' markets allow for the producer to meet the ultimate consumer firsthand.

FIGURE 32-2
Terminal markets allow for the massing of large numbers of animals for sale by selling agents. ***(Courtesy of Omaha Livestock Market, Inc.)***

negotiate the sale of the animals and are more likely to get the highest prices possible. Animals can also be graded into more uniform lots to better meet the demands and requirements of the purchaser.

The use of terminal markets to market animals has always been mostly confined to the midwestern and western states. In recent years, the use of terminal markets to market animals has greatly decreased, and most livestock are currently being marketed by other methods.

Auction Markets At *auction markets*, animals are sold by public bidding on individual lots of animals. An *auctioneer* generally conducts the sales at auction markets (Figure 32-3). These markets are very widespread and are convenient to most local communities. They have grown in popularity and now represent the most common means of marketing animals of importance to agriscience. Auction markets are usually most practical for the smaller livestock producer.

As is true when selling in terminal markets, commissions are charged for selling animals at auction markets. The amount of commission charged varies with the type and size of the animal. Because many auction markets are small, there may not be the competition that is present at terminal markets to buy smaller numbers of animals.

Direct Sales *Direct sales* refers to the producer selling animals directly to processors. This method of marketing has several distinct advantages. There are no commission fees to be paid to selling agents, and there are no yardage fees. Transportation costs are kept to a minimum because the buyer generally comes to the farm to make purchases. The animals look their best for the buyer because they have not been exposed to the stresses of hauling and contact with strange facilities and animals.

FIGURE 32-3
Auction markets are a convenient way for most small producers to market their animals. ***(From Gillespie/Modern Livestock & Poultry Production, copyright 1989 by Delmar Publishers Inc.)***

There are a few negative aspects to direct marketing of animals. A producer must have fairly large numbers of animals to be marketed at one time before the purchaser comes to the farm or ranch to buy. The producer is generally at the mercy of the buyer concerning the price received for the animals sold. Often the producer is not paid for the animals until after they have been processed.

Direct sales of animals, especially beef cattle, have increased significantly in recent years.

Other Wholesale Marketing Techniques Approximately 75% of the milk marketed in the United States is sold through farmer milk-marketing cooperatives. *Cooperatives* are groups of producers who join together to market a commodity. The cooperatives then either process the milk and sell it directly to consumers or sell it to other large processing plants. Marketing cooperatives have the ability to maintain product quality, arranging for transportation of products from farm to market, balancing supply and demand of agriscience products, and planning advertising to increase sales of agriscience products.

In the poultry market, almost 99% of the chickens produced for meat are grown under a system called *vertical integration* (several steps in the production, marketing, and processing of animals are joined together). The use of vertical integration in the production and marketing of agriscience products allows for extremely large systems of production that can be very efficient. Only the anticipated numbers of animals that will be demanded by consumers are produced. There is less competition from other producers, and all phases of production can be controlled.

There are a number of methods of marketing animals of importance to agriscience. The one or more methods chosen by an individual producer are often a matter of what is available and what the producer prefers. Care should be taken to carefully choose the means of marketing the animals produced. Intelligent marketing practices may well make the difference between profit and loss in a very competitive business.

MARKETING FEES AND COMMISSIONS

Because there are numerous methods of marketing animals, the fees and commissions vary fairly widely. Livestock are usually sold by the head, and a fee is charged for each animal sold. This fee varies from $1.50 to $3.00 per animal. If animals are kept at a terminal market, yardage fees are also charged for the feed and care of the animals. Charges for insurance may also be deducted from the seller's check.

For animals that are purchased on the farm for slaughter, a fee called pencil shrink is sometimes deducted from the selling price of the animals. *Pencil shrink* is the estimated amount of weight that an animal will lose when it is transported to market. It usually varies from 1 to 5%.

Fees may also be charged for the sale of animals based on a percentage of the gross amount received for the animals.

MARKETING PROCEDURES

How animals are prepared for and transported to markets can often make the difference between profit and loss. Some of the procedures to follow are:

1. Always handle animals quietly and carefully. Animals that are calm and quiet lose less weight during transportation.
2. Move animals when temperatures are moderate, if possible. Usually this means at night during the summer.
3. Do not overcrowd animals on trailers or in lots while they are waiting to go to market.
4. Do not overfeed animals just before hauling them to market. Animals do not travel well on full stomachs. Do provide ample fresh, clean water.
5. Avoid injuring or bruising animals when loading and unloading them. Dead or injured animals are worth very little at the market.
6. Sort animals according to sex and size before shipping them to market.
7. Precondition animals for several days before marketing them.

AGRI-PROFILE

CAREER AREAS: BREEDER/AUCTIONEER/DEALER/PACKER/MEAT GRADER/MEAT CUTTER

Before being marketed, animals and animal products must meet health standards set by the United States Department of Agriculture and other government agencies. Vaccinations, medications, shots, and inspections help assure healthful animal products for the consumer. ***(Courtesy of the United States Department of Agriculture)***

The marketing of animal products occurs in all segments of our society. The producers and breeders of pets, laboratory animals, livestock, horses, dairy animals, poultry, fish, wildlife, milk, eggs, wool, and fur all market the products of their businesses. But these are only the starters in the process of animal products marketing. From the farm, ranch, feedlot, or other production unit, there is a whole system of professional and scientific workers.

Marketing cooperatives, dealers, auction markets, auctioneers, livestock handlers, jobbers, brokers, clerks, truckers, and railroad personnel all get in the marketing act. Government as well as company inspectors monitor the animals and products as they make the numerous transitions from production site to finishing site, pet shop, by-products or dinner table.

Animal and animal product marketing careers may be accessed through technical school, college, or university studies in the animal sciences, food science, agricultural economics, marketing, or communications. Additionally, studies in the biological sciences would be appropriate for laboratory work associated with animal product marketing.

GRADES AND MARKET CLASSES OF ANIMALS

Cattle Beef animals are usually classified as either calves or cattle. *Calves* are less than 1 year old, whereas *cattle* are more than 1 year of age. Calves are further divided into veal calves, feeder calves, and slaughter calves. *Veal* calves are less than 3 months old and are slaughtered for meat that is also called veal. They usually weigh less than 200 lbs.

Calves that are between 3 months and 1 year old and are marketed for meat are called *slaughter calves*. They have usually been fed at least some grain. *Feeder calves* are 6 months to 1 year of age and are sold to people who feed them to market weight as slaughter cattle. The sex classes for feeder and slaughter calves are steers, heifers, and bulls. *Steers* are castrated males, *heifers* are young females, and *bulls* are unaltered males.

Cattle are divided into feeder cattle and slaughter cattle. Feeder cattle are further categorized into age groups of yearlings and two-year-olds and over. *Yearlings* are between 1 and 2 years of age and *two-year-olds* are two or more years old. These two classes of cattle can also be divided into five sex classes—steer, heifer, bull, cow, and stag. A *cow* is a female that has had a calf, and a *stag* is a male that was castrated after reaching sexual maturity.

Slaughter cattle are marketed for the purpose of being slaughtered for meat. They are divided into the same age and sex classes as feeder cattle. They are also divided into quality and yield grades. *Quality grades* refer to the amount and distribution of *finish* (fat) on the animal (Figure 32-4). The quality grades for cattle are prime, choice, good, standard, commercial, utility, canner, and cutter. *Yield grades* are based on the amount of lean meat an animal will yield in relation to fat and bone (Figure 32-5). The yield grades are 1 through 5, with yield grade 1 producing the most lean meat.

Swine The use classes of swine are feeder pigs, stocker hogs, slaughter pigs, and slaughter hogs. In general, *pigs* are swine less than 4 months of age, whereas *hogs* are more than 4 months old. Feeder pigs and stocker hogs are sold to be fed to higher weights before being slaughtered. Slaughter pigs and slaughter hogs are sold for immediate slaughter.

FIGURE 32-4
Quality grades of slaughter cattle are based on the amount and distribution of finish. *(From Gillespie/Modern Livestock & Poultry Production, copyright 1989 by Delmar Publishers Inc.)*

The sex classes of swine are gilt, barrow, boar, sow, and stag. *Gilts* are young females. *Barrows* are castrated males. *Boars* are males, and *sows* are mature females.

Swine are also graded according to quality, with the official USDA grades being U.S. No. 1 through U.S. No. 4. Animals with lean meat of an unacceptable quality are graded U.S. Utility. The highest grade of swine is U.S. No. 1 (Figures 32-6 and 32-7).

Sheep Sheep that are marketed are classified according to age, use, sex, and weight. The age

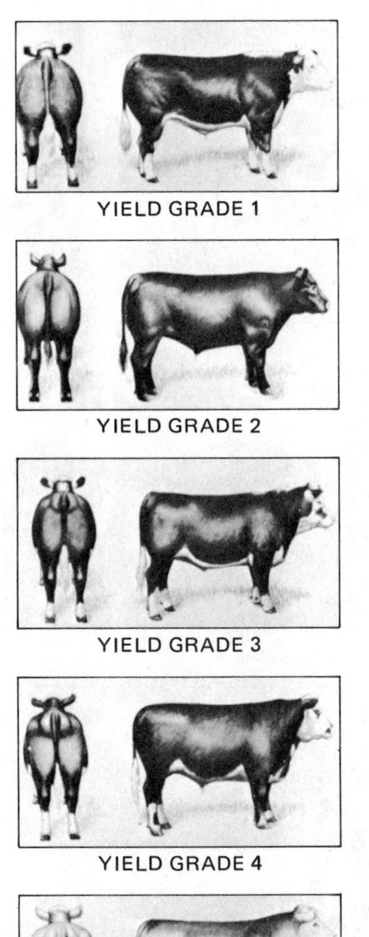

FIGURE 32-5
Yield grades of slaughter cattle are based on the yield of lean meat in relation to the amount of fat and bone. ***(From Gillespie/Modern Livestock & Poultry Production, copyright 1989 by Delmar Publishers Inc.)***

classes are lambs, yearlings, and sheep. Lambs can be divided into hothouse lambs, spring lambs, and lambs. *Hothouse lambs* are under 3 months of age. *Spring lambs* are 3 to 7 months of age, and *lambs* are 7 to 12 months old. Lambs can also be classified as feeder lambs or slaughter lambs, depending on whether they are to be fed to heavier weights or slaughtered immediately.

Yearlings are between 1 and 2 years of age, and sheep are more than 2 years of age. Sheep are divided into three sex classes—ram, ewe, and wether. *Rams* are males, *ewes* are females, and *wethers* are castrated males.

Sheep are also graded according to yield and quality. The yield grades are 1 through 5, with 1 being the most desirable. Quality grades are prime, choice, good, utility, and cull (Figure 32-8).

MARKETING TRENDS AND CYCLES

Livestock prices previously tended to rise and fall on a fairly well-defined cycle. The cause of general price cycling can be blamed on supply and demand. When prices are high, producers increase production of animals and animal products. At first, this action pushes the prices even higher, because females that normally would have been marketed are retained in the breeding herd. When the results of the increased production reach the market and supply exceeds demand, the prices begin to fall. When prices fall to the point where production is not profitable, livestock producers reduce production by selling some of the breeding herd. Of course, this pushes prices even lower. With decreased supplies, the demand increases and the cycle begins again.

Today, there are many other factors that influence prices. As a result, the traditional livestock price cycles have less variation than in the past. Some factors that currently influence livestock prices are:

1. Importation and exportation of livestock products.
2. Development of new uses for livestock products.
3. Increased advertising of livestock products.
4. Government support prices.
5. World weather conditions.
6. General economic conditions in the world.
7. Changes in consumer demands. In recent years, demand for beef and pork has decreased while demand for poultry has been increasing. Egg consumption is down and milk consumption is increasing slightly.

The marketing of animals and animal products is a complex operation if maximum profits are to be realized. There are many different markets for animals and many ways of marketing them and their products. Planning for marketing should take place before entering into the production of animals, and it should be kept in mind during the entire production process.

U. S. NO. 1

U. S. NO. 2

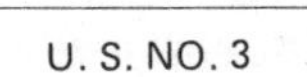

U. S. NO. 3

U. S. NO. 4

U. S. UTILITY

FIGURE 32-6

Quality grades of feeder pigs ***(From Gillespie/Modern Livestock & Poultry Production, copyright 1989 by Delmar Publishers Inc.)***

U. S. NO. 1

U. S. NO 2

U. S. NO. 3

U. S. NO. 4

U. S. UTILITY

FIGURE 32-7

Quality grades of slaughter hogs *(From Gillespie/Modern Livestock & Poultry Production, copyright 1989 by Delmar Publishers Inc.)*

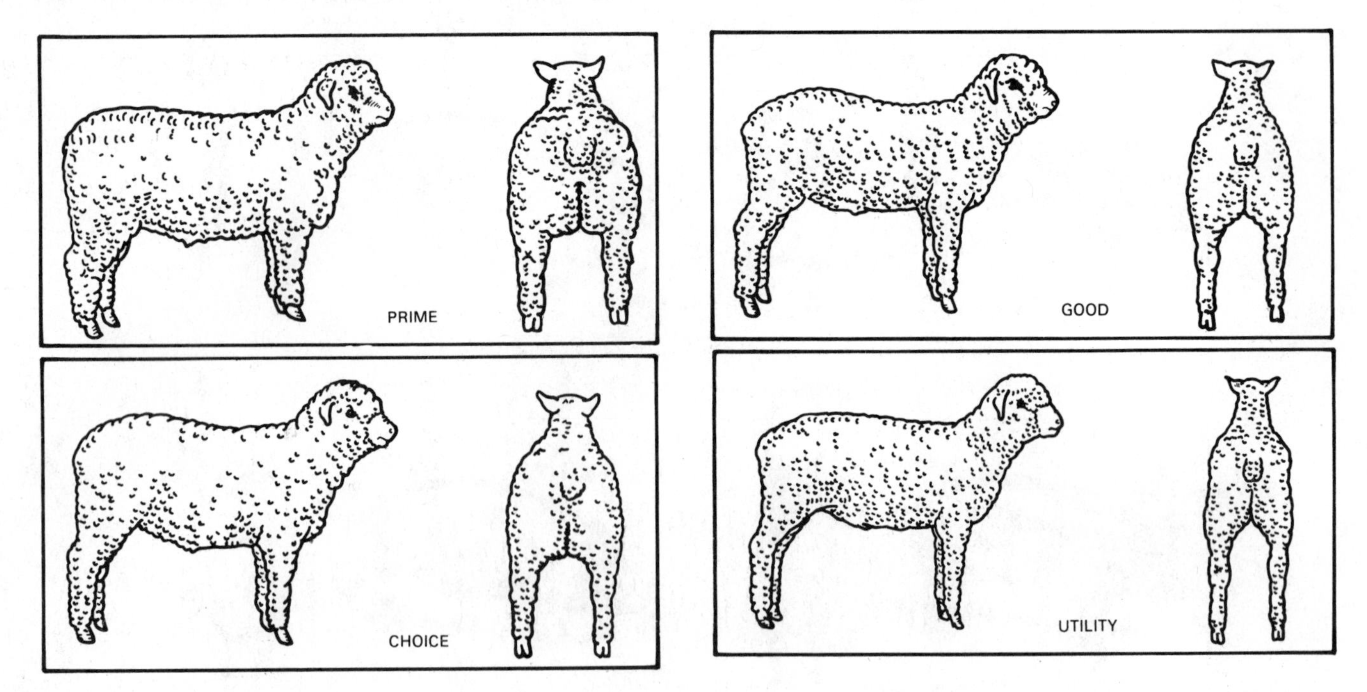

FIGURE 32-8
Quality grades of slaughter lambs *(From Gillespie/Modern Livestock & Poultry Production, copyright 1989 by Delmar Publishers Inc.)*

Student Activities

1. Define the "Terms To Know."
2. Obtain several copies of marketing reports and newspapers that have marketing reports. Compare the prices received for the various classes and grades of animals listed. Also compare the prices received from various markets.
3. Interview a livestock buyer. Ask how prices are determined and where animals are bought and sold. Also ask how to go about getting a job as a livestock buyer. Report your findings to the class.
4. Go to a wholesale market and talk to a livestock grader about how to determine the various grades and classes of livestock. Try to grade some animals yourself.
5. Invite the operator of a retail market to class to talk about the advantages and disadvantages of marketing animals and animal products retail.
6. Make a list of the types of markets for animals and animal products in your area.
7. Develop an advertisement for an animal or animal product.

Self-Evaluation

A. MULTIPLE CHOICE

1. The grading system based on the amount and distribution of finish on an animal is called

   a. prime.
   b. quality.
   c. yield.
   d. commercial.

2. The amount of a product that is available at a specific time and price is the

   a. supply. c. commission.
   b. demand. d. production.

3. The estimated amount of weight that an animal loses during transportation to market is

   a. yardage. c. loss.
   b. stress. d. pencil shrink.

4. Yardage is a fee paid to terminal markets for

   a. feed. c. selling.
   b. insurance. d. none of the above.

5. Lambs that are slaughtered when they are less than 3 months old are called

   a. mutton. c. feeder lambs.
   b. chevon. d. hothouse lambs.

6. The least desirable yield grade for cattle is

   a. 1. c. U.S. No. 4.
   b. Utility. d. 5.

7. The sex classes for sheep are

   a. stag, boar, and sow. c. ram, ewe, and wether.
   b. lamb, sheep, and mutton. d. none of the above.

8. Roadside stands are

   a. wholesale markets. c. retail markets.
   b. terminal markets. d. direct markets.

9. A male castrated after reaching sexual maturity is a

   a. wether. c. stag.
   b. steer. d. barrow.

10. An example of a wholesale market is a/an

   a. terminal market. c. on-the-farm market.
   b. roadside stand. d. farmers market.

B. MATCHING

______ **1.** Prime
______ **2.** Ewe
______ **3.** U.S. No. 1
______ **4.** Barrow
______ **5.** Veal

a. Castrated male pig
b. Female sheep
c. Quality grade of cattle
d. Quality grade of swine
e. Calf meat

C. COMPLETION

1. The lowest quality grade of slaughter hogs is ______ .
2. ______ is the amount of a product desired at a specific place and time.
3. A professional livestock seller receives a ______ for selling animals for producers.
4. ______ markets are those where animals are sold by competitive bidding.
5. Demand for beef has ______ in recent years.

UNIT 33 The Food Industry

OBJECTIVE To explore elements, trends, and career opportunities in the food industry.

Competencies to Be Developed

After studying this unit, you will be able to

- ☐ explain what is meant by the term food industry.
- ☐ determine the importance of the food industry to the consumer.
- ☐ describe the economic scope of the food industry.
- ☐ identify government requirements and other assurances of food quality and sanitation.
- ☐ compare the major crop and animal commodity production areas in the nation and the world.
- ☐ discuss the major food commodity groups and their predominant origins.
- ☐ explain the major operations that occur in the food industry.
- ☐ describe career opportunities in food science.
- ☐ discuss future developments predicted for the food industry.

TERMS TO KNOW

Gourmet
Food industry
Retailer
Wholesaler
Distributor
Processor
Grader
Packer
Trucker
Harvester
Producer
Grades
Climatic conditions
Technology
Aquaculture
Harvesting
Maturity
Underripe
Overripe
Spoiled
Microorganisms
Migratory labor
Processing
Bran
Endosperm
Germ
Edible

MATERIALS LIST

bulletin board materials
state department of agriculture reports on commodities grown and foods processed in particular state

Food is all around us—in the school cafeteria, on the dinner table at home, at the fast-food chains that dot our nation's highways, and in our nation's supermarkets. Less visible are the gourmet and specialty food stores and restaurants that satisfy special dietary needs and tastes. *Gourmet* means sensitive and discriminating taste in food preferences. Learning about other cultures is frequently accomplished by tasting their foods and learning how those foods are prepared.

The *food industry* is that industry involved in the production, processing, storage, preparation, and distribution of food for consumption by living things (Figure 33-1). Pet and animal food, as well as human food, requires a chain of people, places, equipment, regulations, and resources to change farm products into edible foods.

This unit explores the many operations in the food processing industry. Careers are plentiful in this industry. As the different operations are explored, remember that new people are needed to maintain and expand the vital functions that keep the abundant food before us.

THE ECONOMIC SCOPE OF THE FOOD INDUSTRY

When you purchase groceries in the food store or a hamburger at the local fast-food restaurant, does most of your food dollar go directly to the farmer who grew the beef? How about the lettuce, tomato, pickle, bun, and sesame seed that adorns your hamburger? There are many

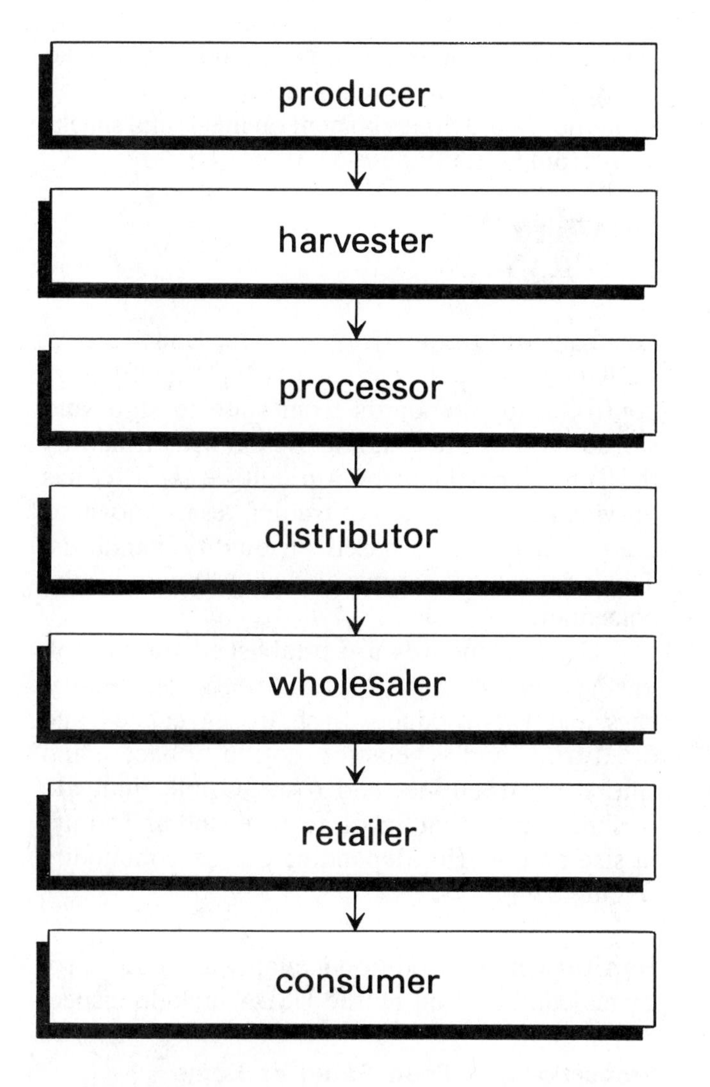

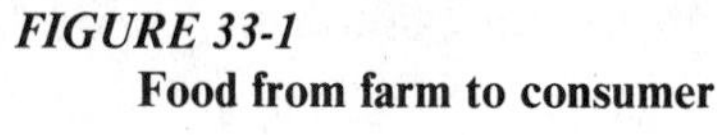

FIGURE 33-1
Food from farm to consumer

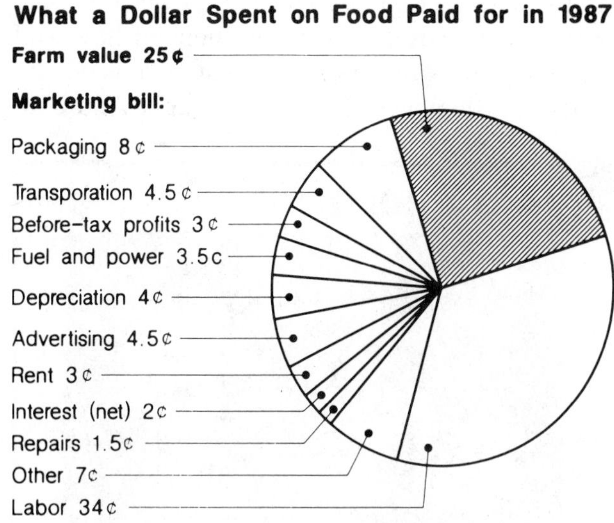

1987 preliminary. Other costs include property taxes and insurance, accounting and professional services, promotion, bad debts, and miscellaneous items.

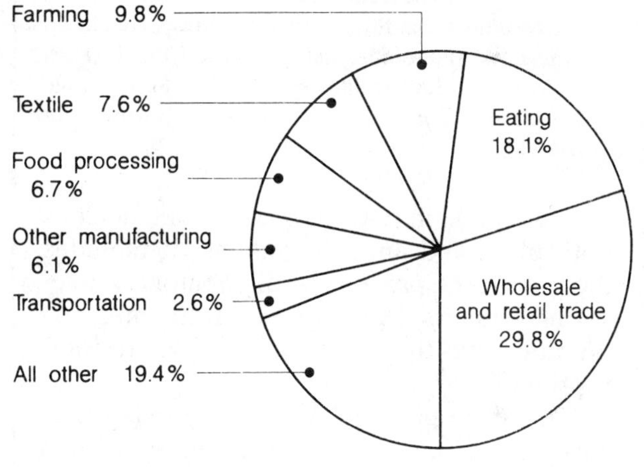

1986 data. Total does not add due to rounding.

FIGURE 33-2
Your food dollar provides careers. ***(Courtesy USDA)***

businesses and individuals that join the farmer in dividing your food dollar. The economic chain reaction that begins with your food purchase sends signals to the retailer, wholesaler, distributor, processor, grader, packer, trucker, harvester, producer, and others to replace that food for your next purchase (Figure 33-1).

A *retailer* is the person or store that sells directly to the consumer. The retailer is the end of the marketing chain, whereas a *wholesaler* is a person who sells to the retailer, having purchased fresh or processed food in large quantities. A *distributor* stores the food until a request is received to transport the food to a regional market. A *processor* is anyone involved in cleaning, separating, handling, and preparing a food product before it is ready to be sold to the distributor. A *grader* is the person who inspects the food for freshness, size, and quality, and determines under what criteria it will be sold and consumed. A *packer* is the person or firm that is responsible for putting the food into containers such as boxes, crates, bags, or bins for shipment to the processing plant, and a *trucker* is the person responsible for transportation of the product anywhere along the way from farm to consumer. A *harvester* is the person who removes the edible portions from plants in the field. Finally, the *producer* grows the crop and determines its readiness for harvest. It seems as if everyone gets a piece of your food dollar (Figure 33-2)!

Where you spend your food dollar also has an influence on who gets how much of your dollar. A meal purchased in a restaurant costs considerably more than a meal prepared from raw food products at home. How has our change in life-styles and the shift to families with two or more people employed outside the home influenced how and what we eat? More meals are eaten outside the home than they were a generation ago. And convenience foods for use at home are more in demand today. Approximately one-fourth of the food dollar in the United States is spent on meals and snacks away from home (Figure 33-3).

AGRI-PROFILE

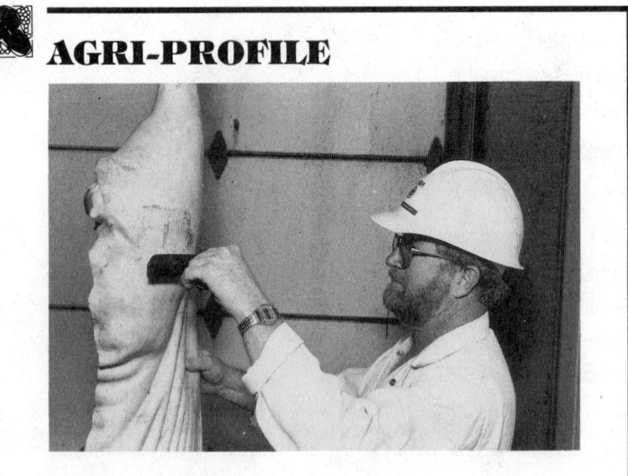

CAREER AREAS: SCIENTIST/QUALITY CONTROLLER/RETAILER/BUTCHER

The stamp of approval or grade assures the consumer that this meat has been inspected by an official, is wholesome, and meets the United States Department of Agriculture specifications for the grade indicated. ***(Courtesy United States Department of Agriculture)***

The food industry is massive and includes both plant and animal products. It includes the producers, processors, distributors, wholesalers, retailers, fast food establishments, restaurants, and the home kitchens where food is prepared.

Career opportunities include many that have been discussed previously and some new ones too. The meat processing and packaging industry is massive and employs a large number of individuals in the United States. Field supervisors and coordinators direct the work of crews to harvest crops at the peak of their quality and transport them to processing or packing plants. Similarly, crews take fish, oysters, clams, lobsters, and other seafood from production habitat to processing centers. In many cases, huge packing and/or processing machines are used in the field, orchard, or boat. Quality control personnel collect food specimens, label them, test, and maintain records to assure quality control on each batch of food coming out of the plant.

Careers in food science, store management, produce management, meat cutting, laboratory testing, field supervision, research, diet and nutrition, health and fitness, and promotion are all possibilities in the food industry.

QUALITY ASSURANCE

Grading and Inspecting In the United States, we have become accustomed to high-quality food controlled by standards from state to state and store to store. The grading system established by the U.S. Department of Agriculture (USDA) has provided a uniform set of trading terms known as grades. *Grades* are based on quality standards. These improve acceptability of products by the consumer.

Grade standards are established for the following commodities: meat, cattle, wool, poultry, eggs, and dairy products; fresh, frozen, canned, and dried fruits and vegetables; cotton, tobacco, and spirits of turpentine; and rosin. Grades indicate freshness, potential flavor, texture, and uniformity in size and weight, depending on the commodity (Figure 33-4).

Sanitation Additional quality assurance programs administered by the USDA include inspec-

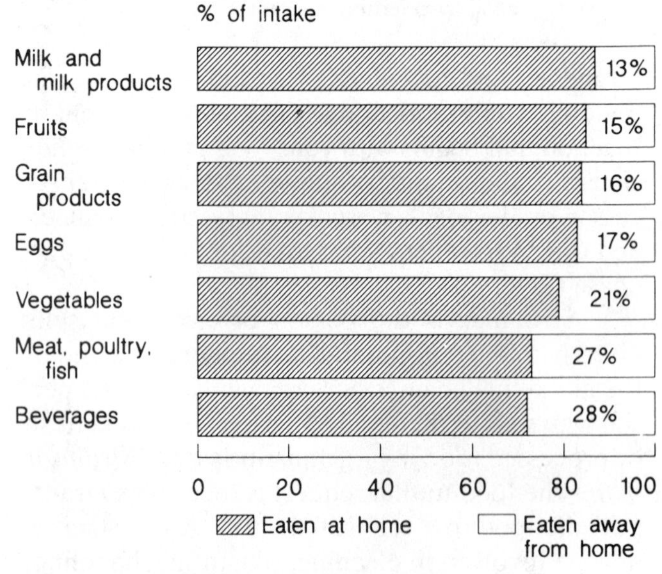

FIGURE 33-3
Where we spend our dollars ***(Courtesy USDA)***

Food	Current Grade Names	What the Grades Mean
Meat		
Beef	*USDA PRIME	Very tender, juicy, flavorful; has abundant marbling (flecks of fat within the lean).
	*USDA CHOICE	Quite tender and juicy, good flavor; slightly less marbling than Prime.
	USDA SELECT	Fairly tender; not as juicy and flavorful as Prime and Choice; has least marbling of the three.
Lamb	*USDA PRIME	Very tender, juicy, flavorful; has generous marbling.
	*USDA CHOICE	Tender, juicy, flavorful; has less marbling than Prime.
Veal	*USDA PRIME	Juicy and flavorful; little marbling.
	*USDA CHOICE	Quite juicy and flavorful; less marbling than Prime.
Poultry		
Chickens Turkeys Ducks Geese	*U.S. Grade A	Fully fleshed and meaty; uniform fat covering; well formed; good, clean appearance.
	U.S. Grade B	Not quite as meaty as A; may have occasional cut or tear in skin; not as attractive as A.
	U.S. Grade C	May have cuts, tears, or bruises; wings may be removed and moderate amounts of trimming of the breast and legs are permitted.

*Indicates grades most often seen at retail

Food	Current Grade Names	What the Grades Mean
Eggs	*U.S. Grade AA	Clean, sound shells; clear and firm whites; yolks practically free of defects; egg covers small area when broken out—yolk is firm and high and white is thick and stands high.
	*U.S. Grade A	The same as AA except egg may cover slightly larger area when broken out and white is not quite as thick.
	U.S. Grade B	Sound shells, may have some stains or shape may be abnormal; white may be weak and yolk enlarged and flattened; egg spreads when broken out.
Dairy Products		
Instant nonfat dry milk	*U.S. Extra Grade	Sweet, pleasing flavor; natural color; dissolves readily in water.
Butter	*U.S. Grade AA	Delicate sweet flavor and smooth texture; made from high quality fresh sweet cream.
	*U.S. Grade A	Pleasing flavor; fairly smooth texture; made from fresh cream.
	U.S. Grade B	May have slightly acid flavor or other flavor or body defects.
Cheddar cheese	*U.S. Grade AA	Fine, pleasing Cheddar flavor; smooth, compact texture; uniform color.
	U.S. Grade A	Pleasing flavor; more variation in flavor and texture than AA.

*Indicates grades most often seen at retail

Food	Current Grade Names	What the Grades Mean
Fish	*U.S. Grade A	Uniform in size, practically free of blemishes and defects, in excellent condition, and having good flavor for the species.
	U.S. Grade B	May not be as uniform in size or as free of blemishes or defects as Grade A products; general commercial grade.
	U.S. Grade C	Just as wholesome and nutritious as higher grades; a definite value as thrifty buy for use where appearance is not an important factor.
Fresh Fruits and Vegetables		
The grade is more likely to be found without the shield.	*U.S. Fancy	Premium quality; only a few fruits and vegetables are packed in this grade.
	*U.S. No. 1	Good quality; chief grade for most fruits and vegetables.
	U.S. No. 2	Intermediate quality between No. 1 and No. 3.
	U.S. No. 3	Lowest marketable quality.
Processed Fruits and Vegetables and Related Products		
Canned and frozen fruits and vegetables.	*U.S. Grade A	Tender vegetables and well-ripened fruits with excellent flavor, uniform color and size, and few defects.
	U.S. Grade B	Slightly mature vegetables; both fruits and vegetables have good flavor but are slightly less uniform in color and size and may have more defects than A.
	U.S. Grade C	Mature vegetables; both fruits and vegetables vary more in flavor, color, and size and have more defects than B.
Dried or dehydrated fruits. Fruit and vegetable juices, canned and frozen. Jams, jellies, preserves. Peanut butter. Honey. Catsup, tomato paste.	*U.S. Grade A	Very good flavor and color and few defects.
	U.S. Grade B	Good flavor and color but not as uniform as A.
	U.S. Grade C	Less flavor than B, color not as bright, and more defects.

*Indicates grades most often seen at retail

FIGURE 33-4
Grades of food commodities *(Courtesy USDA)*

tion of slaughtering houses and processing plants, and oversight of processing operations (Figure 33-5). The USDA oversees food labeling and enforces regulations regarding representation on such labels (Figure 33-6). The National Shellfish Sanitation Program, the U.S. Public Health Service, and the U.S. Food and Drug Administration work with USDA to assure the safety of food and food products. States, counties, and municipalities also have inspectors. They regulate local conditions to assure sanitation and safe food handling, especially in restaurants and food preparation areas.

COMMODITY GROUPS AND THEIR ORIGINS

What Foods Are Grown Where Food is grown all over the world. Climatic conditions and available technology dictate that some foods grow better and in greater abundance in certain areas of the world. *Climatic conditions* refer to average temperature, number of days with a certain temperature range, length of growing season, and amount of precipi-

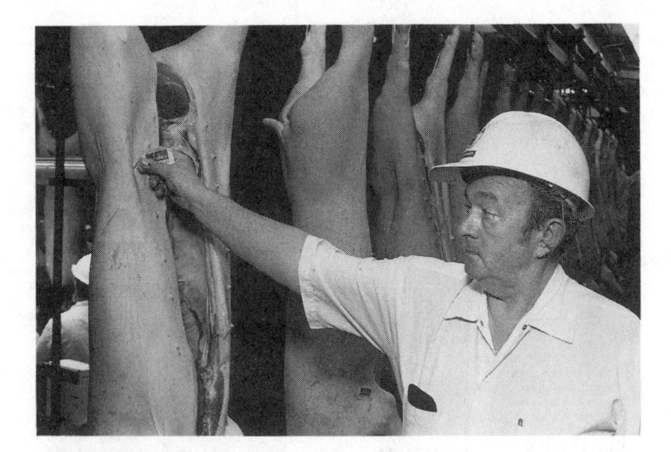

FIGURE 33-5
The professional food inspector

NUTRITION INFORMATION PER SERVING

SERVING SIZE	2 CUPS	CARBOHYDRATES	26 GRAMS
SERVINGS PER PACKAGE	6	SUGARS	24 GRAMS
CALORIES	280	OTHER CARBOHYDRATES	2 GRAMS
PROTEIN	6 GRAMS	FAT	17 GRAMS
		SODIUM	180 MG

PERCENTAGE OF U.S. RECOMMENDED DAILY ALLOWANCES (U.S. RDA)

PROTEIN	10	VITAMIN C	*	RIBOFLAVIN	6	CALCIUM	2
VITAMIN A	*	THIAMINE	*	NIACIN	8	IRON	2

*CONTAINS LESS THAN 2 PERCENT OF THE U.S. RDA OF THESE NUTRIENTS

INGREDIENTS: MILK CHOCOLATE (MILK CHOCOLATE CONTAINS SUGAR; COCOA BUTTER; MILK; CHOCOLATE; AND SOYA LECITHIN, AN EMULSIFIER), PEANUTS, SUGAR, DEXTROSE, SALT, TBHQ AND CITRIC ACID (TO PRESERVE FRESHNESS).

FIGURE 33-6
The commodity label is the consumer's best assurance of quality food.

tation for a given geographic area. *Technology* refers to the equipment and scientific expertise available to cultivate, store, process, and transport the crop for consumption in a variety of forms after harvest.

Since early times when humans traveled and traded, foods have been introduced outside the areas where they are grown naturally. The origins of the soybean, for example, can be traced over 3,000 years to China, where it is still produced and consumed today. However, major growing areas for the soybean today include the United States, Brazil, and western Europe.

Modern technology has allowed producers to raise crops somewhat artificially with irrigation, and in greenhouses where the conditions of temperature and moisture are controlled. Different varieties of foods have been developed to grow under different climatic conditions, such as extreme heat or cold. Seafood and fish are also produced under controlled conditions. *Aquaculture* is the science of producing fish in large numbers where fish are not present naturally.

In the United States, we are accustomed to having almost every food available at any time of the year. All foods are not grown in all parts of the country. Citrus fruit, including oranges and grapefruit, require warm climates, such as those found in Texas, California, Florida, and South America. In fact, approximately 60% of the fresh fruits and vegetables that are consumed in the United States are grown in Texas, California, and Florida. However, people in North Dakota and Maine enjoy the nutrients and good taste of fresh or processed citrus and vegetable products such as orange juice almost daily.

The many operations of the food industry that are discussed later in this unit will explain how we can enjoy foods that are not grown naturally in our particular region of the country and world. Realizing where some of our food products originate also makes you appreciate the size and scope of the food industry.

Crop Commodities

Grains

Various grains have different growing requirements (Figure 33-7). Therefore, they are produced in different parts of the United States and world. Wheat, which originated in Asia, is grown in the cooler climates of the United States. Different varieties have been developed to accommodate different growing seasons and climatic conditions around the world. Corn is a warm-weather crop, but the many varieties and types permit its growth in every state in the United States. However, rice has special moisture requirements, so its production is limited to specific areas of the country.

Oil Crops

These crops are sometimes thought of as the invisible food product. Soy and peanut oil are the predominant oils from oil crops in the United States. However, neither crop is native to this country. Today, the United States is a leader in the production of soy and peanut oil. Soybean products are referred to in ancient Chinese literature, and the origin of the peanut may be traced to Brazil and Paraguay. Sunflowers are native to the United States, but sunflower oil has been introduced in Spain, China, Russia, and the Mediterranean. Safflower oil is a relatively minor oil in terms of proportion to the total oil crop worldwide. It originated in northern India, North Africa, and the Middle East. Its origin is indicative of its drought tolerance.

FIGURE 33-7
Food sources are determined to a large extent by climate, topography, cost of labor and distance to market.

Sugar Crops Sugar beets and sugar cane are the principal sugar crops in the United States. Corn is a secondary source of sugar. Sugar beets are grown in temperate areas, whereas sugar cane is grown in tropical and subtropical locations around the world.

Citrus Oranges, limes, lemons, and grapefruit all require warm temperatures to survive. Consequently, the southern states with warmer climates, such as Florida, Texas, Arizona, and California, are the major producers of citrus.

Tree Fruits The many varieties of fruits that grow on trees require different weather conditions. Therefore, various fruits are adapted to different parts of the country. Apples and pears require cooler temperatures and do well in mountainous areas. Bananas require very warm conditions and survive best in tropical areas.

Vegetables Vegetables are consumed immediately or are processed by canning, drying, or freezing for future consumption. Some vegetables require warm climates and some require cool climates. Vegetables that require cooler climates include cabbage, broccoli, potatoes, and cauliflower. Vegetables requiring a warmer environment include beans, tomatoes, and sweet corn.

Meat Commodities Animals, like crops, are typically raised in locations with the same regard to climatic conditions as crops are. Artificial cooling or heating of livestock areas is costly. Further, large amounts of water must be available for livestock. Where fewer artificial conditions are introduced, the cost of production is minimized. The type and availability of livestock feed is a factor that influences where animals are raised in the United States and around the world.

Beef Most beef is raised near corn, their main feed source. Approximately 58% of the corn in the United States is grown in Nebraska, Iowa, Illinois, and Minnesota. Therefore, beef is raised extensively in the Midwest. The open range in the western parts of the United States provides another important area where beef is raised.

Pork The primary food for hogs is also corn. Much like beef, the primary area where hogs are raised is the midwestern parts of the United States. Grazing on open range is not required for hogs, so the mid-Atlantic and southern states are also important hog-producing areas.

Lamb Sheep are animals that require large amounts of grazing pasture area. Therefore, they are raised extensively in the range states of the Far West. However, as is true with beef and pork, lamb products are produced in other states.

Dairy Products Wisconsin has long been called the "Dairy State." Indeed, many of our dairy products are produced in Wisconsin and other northern-tier states. However, California has a very large dairy industry with many cooperatives and processing plants. These provide the dairy products to consumers nationwide. Dairy animals prefer cooler environments, so the industry is extensive in the northern parts of the United States.

Game Each state has native game. Whether or not it is harvested as an agricultural product depends on the demand for the products. Venison is the tasty and popular meat of deer. In some states deer are being raised in captivity to help meet the market demand for venison.

Seafood States that border the Atlantic and Pacific oceans and the Gulf of Mexico are considered the primary suppliers of seafood in the United States. However, the science of aquaculture permits the production of fish in interior states.

Poultry Poultry can be raised in a variety of settings. Typically chickens and turkeys are raised indoors, where ventilation and temperature are carefully controlled. In the East, most poultry is raised in the mid-Atlantic and southern states. However, important poultry-producing areas are also found in California and other states.

OPERATIONS WITHIN THE FOOD INDUSTRY

The food industry begins with the process of photosynthesis in plants. It progresses through plant and animal growth and on to commodity processing and distribution. The discussion in this unit will focus primarily on the food industry after the crops and animals have been grown.

Harvesting *Harvesting* means taking a product from the plant where it was grown or produced. This may involve taking potatoes out of the ground, picking oranges off a tree, removing bean pods from bean plants, or removing and threshing grain from stalks (Figure 33-8). It is most important that the crop be harvested in a timely and careful fashion. The plant should be harvested at the correct stage of maturity. *Maturity* means the state or quality of being fully grown. Harvesting a crop when it is mature means that it is not underripe, overripe, or spoiled. *Underripe* means that it has not reached maturity; *overripe* means that the plant is beyond maturity, and that stalks or limbs can possibly break or shatter or fruit can drop. *Spoiled* means that chemical changes have taken place in the food or food product that either reduce its nutritional value or render it unfit to eat.

Spoilage is usually caused by microorganisms. *Microorganisms* that contribute to food spoilage are bacteria, fungi, and nematodes. Bacteria are a group of one-celled plants. Fungi are plants that lack chlorophyll and obtain their nourishment from other plants, thus causing rot, mold, and plant diseases. Nematodes are small worms that feed on or in plants or animals. They live in moist soil, water, or decaying matter and pierce cells of plants and such the juices. The moisture content of a product determines many changes that the product can undergo. It also dictates the types of handling pro-

FIGURE 33-8
Machine harvesting has lowered the cost of food in the United States and the world. ***(Courtesy of the American Soybean Association)***

cedures and storage facilities that are required for certain foods. There are health considerations that are influenced by the various levels of crop maturity. Some fruits, such as bananas and tomatoes, continue to ripen after they have been picked from the tree and vine. Other foods, such as beans and oranges, do not continue to ripen once picked and must be handled accordingly. Knowledge of the complete growth process of a commodity is essential to the producer and harvester.

Harvesting involves the use of equipment and labor. Since timing is critical, migratory labor is often used at harvest time. *Migratory labor* refers to workers who move to the places where harvesting is occurring throughout the year. As crops ripen, laborers migrate to new geographic locations. Workers are not always available. They may be poorly trained and may not harvest crops as efficiently and economically as modern machines do.

Therefore, engineers and business people have developed and now sell machines to perform many harvesting operations. Such machines make harvesting easier than it was in the past. However, machines may not be as gentle as human hands. Consequently, plant breeders have developed new crop varieties that lend themselves better to mechanical harvesting. Such varieties may not be as good to the human touch or taste. For instance, tomato varieties that have skins tough enough for mechanical harvesting are harder to slice and may not be as juicy as varieties suitable for the home garden.

Processing and Handling The steps involved in turning the raw agricultural product into an attractive and consumable food is known as *processing*. Processing factories or plants clean, dry, weigh, refrigerate, preserve, store, and turn the commodity into a variety of other products.

Wheat is cleaned, dried, weighed, and graded for quality. It is then ground into flour. However, before this occurs it may be separated into bran and germ. The skin or covering of a wheat kernel is known as *bran*. Inside the bran is the *endosperm*, which will become flour, and the *germ*, which is a new wheat plant inside the kernel. Wheat flour is used for breads, cereals, cakes, and pasta. Other grains have similar parts and are used to make similar products.

The processing of tomatoes results in a variety of products. Some people claim that the fresh tomato defines summer. The peak of the North American growing season occurs when the tomato is ripe. To the gardener, this means picking tomatoes directly from the vine in a backyard garden and consuming them immediately. However, tomatoes are harvested year round somewhere in the world and the food industry can get them to us in edible form. *Edible* means fit to eat or consume by mouth. Processing permits tomatoes to be kept for future use. This begins after the tomato is harvested in the field or greenhouse and arrives at the processing plant.

Once tomatoes are cleaned and separated for size and quality, they may be canned whole; chopped, cooked, and strained for juice; or made into other products. Such products include spaghetti and hamburger sauces, relishes, and many other foods.

Transporting Trucks, planes, boats, cars, trains, carts, and bicycles are all vehicles used in the food industry of the world (Figure 33-9). The transporting of fresh and processed food products comprises 5.5% of the marketing cost within the food industry in the United States. Timing, and the distance foods must travel, contribute to the

FIGURE 33-9

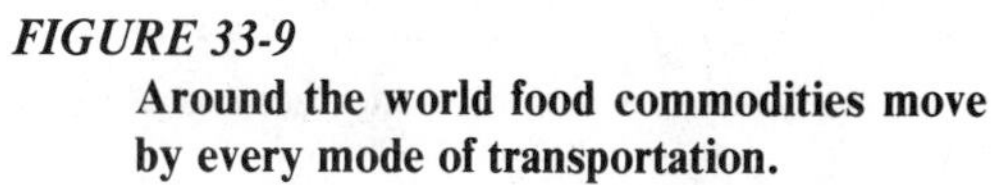

Around the world food commodities move by every mode of transportation.

ultimate cost of the foods. The efficiency of transportation can influence food quality in terms of freshness and spoilage.

Approximately 90% of our perishable food is shipped by truck. Much of the less-perishable foods, such as wheat, potatoes, and beets, are shipped by rail. Air transportation allows us to enjoy perishable foods from distant regions and different countries. For example, pineapples and papayas from Hawaii are enjoyed all over the world because of air transportation. Because of modern transportation, fresh seafood is enjoyed in many areas where fish do not live.

The consumer who drives to and from the grocery store is a member of the food industry and provides the final link from farm to table. How far the food is shipped, how the food was wrapped for transportation, how long the food was in transit, and how warm the food became during transport all influence ultimate food quality. From the milk truck driver taking milk from farm to processing center, to the trucker delivering products to the local store, transportation is a key component in the food industry. Knowledgeable and competent employees are a great asset when the cash value of the load on their truck or rail car is considered. A delay could be costly for both business and the consumer.

Even if food is in perfect condition when you buy it, the quality can decrease substantially before you get it home if precautions are not taken. Food should be packaged correctly in the store, kept cool in the car or truck, and refrigerated or frozen upon arrival home.

RETAIL FOOD PURCHASE CENTERS

Type of Market	Size	Number of Items
Superstores	30,000 sq ft.	15,000 or more
Conventional Supermarkets	17,000 sq ft.	8,000–12,000
Small Supermarkets	4–8,000 sq ft.	5,000–7,000
Corner Stores	500–3,000 sq ft.	2,000
Convenience Stores	2,500 sq ft.	3,000
Warehouse Supermarkets		1,500–7,500
Limited Assortment and Box Stores	9,000 sq ft.	<1,500

FIGURE 33-10
Stores are the chief source of food commodities.

Marketing Wholesalers purchase food products from packing houses, processors, fish markets, and produce terminals. They, in turn, sell to institutions such as hospitals, schools, restaurants, and retail stores. Grocery stores and fast-food chains are important links in the food chain before the food is purchased by the consumer.

There are many types of retail stores from which consumers can purchase their food items. Such marketing sources meet the needs of consumers in different locations and situations. Superstores, conventional supermarkets, limited assortment and box stores, convenience stores, nonconventional

Business
- accounting
- marketing
- finance
- banking and credit
- distribution
- statisticians
- sales

Processing
- cleaning
- packing
- storage

Communications
- advertising
- public information through media such as tv, radio, newspapers and magazines

Research and Development
- engineers for new equipment and techniques
- new product development
- packaging
- processing techniques
- distribution
- new uses of products
- technicians

Quality Assurance
- inspectors
- graders
- plant and animal quarantine

Transportation
- truckers
- rail operators
- merchant marines
- dispatchers

Education
- teachers
- extension workers
- industry educators

FIGURE 33-11
There are many career opportunities in the food industry.

food stores, small stores, corner stores, food cooperatives, farmers markets, roadside stands, pick-your-own businesses, and other farm outlets are the most common places that consumers purchase their food items (Figure 33-10). The major differences between the various types of stores are in the number of items stocked and the physical size of the facilities.

CAREER OPPORTUNITIES IN THE FOOD INDUSTRY Each of the areas discussed in this unit and the following unit require people to manage, operate, and carry out the many and varied elements of the food industry. As with all careers in agriscience, those in the food industry present many challenges and rewards. When you consider the chain-reaction nature of the food industry, career opportunities await individuals at the local, county, state, national, and international levels. Careers in the food industry can be divided into seven, often overlapping, categories. Examples of career opportunities in each area are numerous (Figure 33-11).

THE FOOD INDUSTRY OF THE FUTURE The food industry is ever changing, with new developments each day. Some areas that will attract the food researcher may include new food products, new processing and preserving techniques, and new equipment to harvest labor-intensive crops.

Aquaculture is meeting the increasing demand for fish and will continue to supplement the catch of commercial fishermen. The use of extreme heat and cold in processing has provided many food items that meet the demand for convenience foods. The convenience food store will continue to play a larger role in the food chain. Economic efficiency in convenience foods and convenience stores is under constant review. Additionally, the USDA and other agencies continue their vigilance regarding safety and nutritional standards at all steps of the food chain.

Improved harvesting equipment for products such as grapes are being tested to lower the labor cost of such crops. Fuel alternatives for cost-effective transportation and refrigeration with carbon dioxide snow, instead of conventional diesel-powered mechanical refrigeration, are some of the many developments under constant review in the food industry. This industry will meet the demands of the United States and the world through dedicated research and qualified employees.

Student Activities

1. Define the "Terms To Know."
2. Keep a food-dollar diary to document where your food dollars are spent. Record the cost of meals and snacks eaten in and outside of the home.
3. Do a cost comparison of meals prepared at home and similar meals consumed at fast-food places and restaurants.
4. Trace the activities that occur in transforming wheat in the field to a hamburger roll consumed in your home.
5. Draw a diagram tracing the individual food components of a deluxe hamburger back to the places where the components were produced. Label each component, process, and commodity name along the way.
6. Ask your instructor to arrange a field trip to a butcher shop or supermarket to observe demonstrations on meat cutting and packaging.
7. Invite the manager of a food processing plant in your community to your class to speak on the food processing industry.
8. Make a collage illustrating the various activities of the food industry.

9. Using resources from your state, identify on a state map the major food commodities produced in different regions of your state.

Self-Evaluation

A. MULTIPLE CHOICE

1. Approximately what percentage of the American food dollar is spent on meals away from home?

   a. 15% c. 35%
   b. 25% d. 45%

2. Which of the following products is native to North America?

   a. soybeans c. sunflowers
   b. wheat d. peanuts

3. In the United States, more than one-half of the fresh fruits and vegetables are grown in which states?

   a. Montana, Oregon, and Washington
   b. California, Florida, and Texas
   c. New Jersey, North Carolina, and Georgia
   d. Arizona, Nebraska, and Ohio

4. Migratory workers would harvest wheat last in which state?

   a. Arizona c. Montana
   b. Nebraska d. Ohio

5. When you spend one dollar for food, approximately how much goes into the labor required to harvest and process that food once it leaves the farm?

   a. $0.94 c. $0.34
   b. $0.64 d. $0.04

6. Approximately what percentage of all the jobs in the food and fiber system are related to wholesale and retail sales?

   a. 60% c. 40%
   b. 50% d. 30%

7. Which product is consumed away from home the most?

   a. fruits c. vegetables
   b. beverages other than milk d. meat

8. Superstores are likely to carry how many items?

   a. 15,000 c. 150
   b. 1,500 d. 15

B. MATCHING

______	**1.** Harvester	**a.**	Purchases food in large quantities
______	**2.** Grader	**b.**	Stores food until requested
______	**3.** Retailer	**c.**	Follows crop harvesting geographically
______	**4.** Wholesaler	**d.**	Involved in the transportation of food
______	**5.** Migrant worker	**e.**	Takes the crop from the field
______	**6.** Trucker	**f.**	Those involved in cleaning, sorting, and preparing a product
______	**7.** Processor	**g.**	Inspects food and determines how it will be sold
______	**8.** Distributor	**h.**	Sells directly to the public

C. COMPLETION

1. Three careers that you could pursue in the quality assurance area of the food industry might include ______ , ______ , and ______ .

2. Differences in how grocery stores are categorized are primarily determined by ______ and ______ .

3. When buying food at a pick-your-own farm, the producer is also the ______ .

4. A beef grade of ______ would indicate very tender, juicy, flavorful, and abundant marbling.

5. Grocers purchase their supplies through ______ .

UNIT 34 Food Science

OBJECTIVE To explore the nutrient requirements for human health and the processes used in food science to ensure an adequate and wholesome food supply.

Competencies to Be Developed

After studying this unit, you will be able to

- ☐ discuss nutritional needs of humans and the food groups that meet these needs.
- ☐ discuss food customs around the world.
- ☐ relate methods used in processing and preserving food.
- ☐ list the major steps used in slaughtering meat animals.
- ☐ list the major cuts of red-meat animals.
- ☐ identify methods of processing fish.
- ☐ describe techniques used to enhance retail sales of food commodities.

TERMS TO KNOW

Food
Nutrients
Nutrition
Fermentation
Controlled atmosphere
Refrigeration
Hydrocooling
Blanching
Canning
Dehydration
Freeze drying
Oxidative deterioration
Dehydrofrozen
Humidity
Retortable pouches
Food additive
Shelf life
Dry-heat cooking
Moist-heat cooking
Conventional ovens
Microwave ovens
Convection ovens
Dehydrators
Smokers
Milk
Butterfat
Cream
Cheese
Cottage cheese
Condensed milk
Evaporated milk
Vacuum pan
Slaughter
Rendering insensible
Shackles
Hoist
Stuck
Bleeding out
Hide
Viscera
Carcass
Offal
Split carcass
Shrouded
Age or ripen
Block beef
Disassembly process
Fabrication and boxing
Processed meats
Pelt
Evisceration
Singe
Leaf fat
Lard
Jugular vein
Giblets
Kosher
Shehitah
Shohet or shochet
Chalaf
Dressing percentage
Sweetbreads
Tripe
Rumen
Tankage
Collagen

MATERIALS LIST

bulletin board materials

You are what you eat. Have you ever stopped to consider what that statement means? *Food* is defined by Webster as a material containing or consisting of carbohydrates, fats, proteins, and supplementary substances such as minerals used in the body of an organism to sustain growth, repair, and vital processes and to furnish energy, especially parts of the bodies of animals and plants consumed by humans and animals. This unit explores the foods that humans need to sustain growth (Figure 34-1). Additionally, it explores how those foods reach our tables from their beginning as raw products.

NUTRITIONAL NEEDS

The body is a complex system that has many nutritional demands. *Nutrients* are substances necessary for the functioning of an organism. There are more than 50 specific nutrients required for bodily functions. *Nutrition* involves the combination of processes by which all body parts receive and utilize materials necessary for function, growth, and renewal. Nutrition includes the release of energy, the building up of body tissues (both hard and soft), and the regulating of body processes.

Once food is digested and in the blood system, nutrients are able to do their work. Nutrients are classified into six major groups. The different groups represent different purposes in the body (Figure 34-2).

FIGURE 34-1
Food is needed to sustain growth. ***(Courtesy of Denmark High School, Agriculture Education Department, photo by Jim Jones)***

Carbohydrates

Carbohydrates serve as the main source of energy for the body. There are three different types of carbohydrates: sugars, starches, and fiber. Sugars are simple carbohydrates and are found naturally in many foods, such as fruit, milk, and peas. Refined sugar, or the substance used in many households, comes from sugar beets and sugar cane. Starch is a complex carbohydrate that is found in foods such as bread, potatoes, rice, and vegetables. Starches and sugars are converted to glucose in the body and serve as the major body fuel. Some of the fuel that is generated is stored by the body for later use. However, when the glucose is not used by the body, it is changed to fat. Fiber is also a complex carbohydrate and is found on the walls of plant cells. Humans are unable to digest fiber, yet it plays an important role in moving food through the body and expelling waste after digestion.

Fats

Fats are another source of energy for the body. They are considered to be a more compact source of energy because they have 2¼ times the number of calories as the other two energy sources—carbohydrates and proteins. Some vitamins require fats to carry them to the part of the body where they are needed. Although fat is necessary in the body, too much fat results in obesity and serious diseases, such as heart problems and high blood pressure. Fats are present in differing amounts in most foods. Foods that are known to be high in fat content include cheese, poultry skin, and avocados. Some foods that we are accustomed to require fats to prepare them. Baked goods, such as cakes and cookies, salad dressing, and fried foods, acquire fats through preparation.

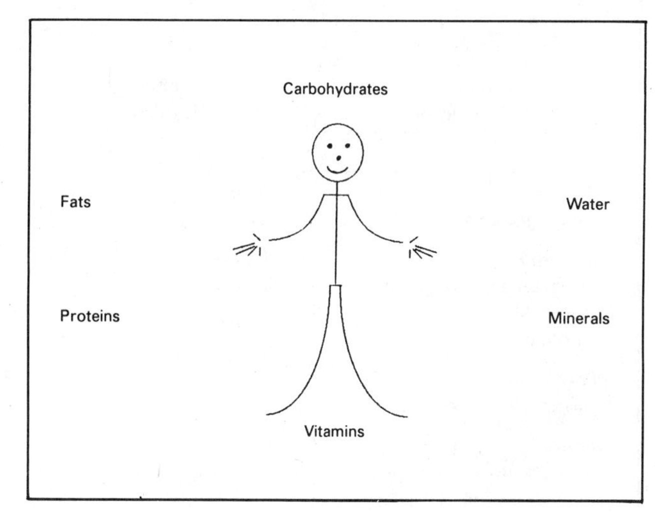

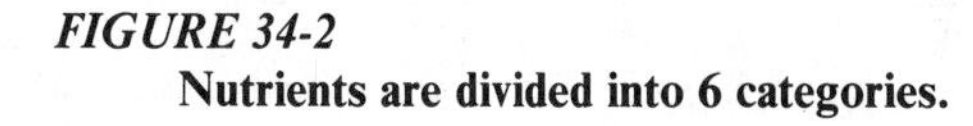

FIGURE 34-2
Nutrients are divided into 6 categories.

Proteins The body needs food with proteins to build and rebuild its cells. Hair, skin, teeth, and bones are all parts of your body that require protein. Proteins are in a continuous cycle of building up and breaking down. Approximately 3 to 5% of your protein is rebuilt each day. Beans, peanut butter, meats, eggs and cheese are high-protein foods.

Vitamins Vitamins are also essential to the functions of the body. Some vitamins are dissolved in your body fats and are stored in the body. Fat-soluble vitamins are not required in the diet each day, since they can be stored in the body. These include vitamins A, D, E, and K. Nine other vitamins—vitamin C and eight B vitamins—are water soluble and must be replenished daily. Specific vitamins have specific jobs in the body, and some foods are known to be richer in specific vitamins (Figure 34-3).

Minerals There are more than 20 minerals that are needed by the body. The amounts needed

FUNCTIONS AND SOURCES OF VITAMINS

Vitamin A	Vitamin B	Vitamin C	Vitamin D	Vitamin E	Vitamin K
FUNCTIONS					
vision bones skin healing wound	using protein, carbohydrates & fats to keep eyes, skin & mouth healthy brain nervous system	wound healing blood vessels bones teeth other tissues works with minerals	needed for using calcium & phosphorus bones teeth	preserve cell tissue	blood clotting
SOURCES					
yellow, orange & green vegetables	whole grain & enriched cereals, breads, meats, beans	citrus fruits, melons, berries, leafy green vegetables, broccoli, cabbage, spinach	fatty fish, liver, eggs, butter, added to most milk	vegetable oils, whole grain cereals	leafy green vegetables, peas, cauliflower, whole grains

FIGURE 34-3
Vitamins that we need and foods that provide them

FUNCTIONS

Bone Development	Fluid Regulation	Materials for Cells	Trigger Other Reactions
SOURCES			
calcium milk products *magnesium* nuts, seeds, dark green vegetables, whole grain products *phosphorus* no specific food group *fluorine* some seafood, some plants, added to drinking water	*sodium* salt *potassium* bananas *chlorine* salt	*iron* meats, liver, beans, green leafy vegetables, grains works with vitamin C *iodine* iodized salt added to salt	*zinc* whole grain breads & cereals, beans, meats, shellfish, eggs *copper* fish & meats, nuts, raisin, oils & grains

FIGURE 34-4
Minerals: some functions and sources

may be small, but they are required nonetheless. The 20 minerals are divided into four major groups. Some minerals are part of bones, others regulate bodily functions, some are needed to make special materials for cells, and others trigger chemical reactions in the body (Figure 34-4).

Water The human body is more than 50% water. Water carries nutrients to cells, removes waste, and maintains the body's proper temperature. Fluid foods, such as milk and juice, obviously supply the body's need for water. However, foods such as meat and bread also contain water.

Food Groups that Meet Needs

Each food is different in the types of nutrients it contains and ultimately provides to the body. Foods are divided into five major food groups, which represent the nutritional needs of the body. The five food groups are: fruits and vegetables; breads and cereals; milk and cheese; meat, fish, poultry, and beans; and fats and sweets. The fifth group contains items that are often present in foods that belong to the first four groups. You should generally not seek foods in this category for nutritional benefit. Often fats and sweets are used to prepare foods in the other groups, and they show up in adequate amounts in most people's diets in the United States. The names of the groups suggest some of the typical foods that they include.

Eating from each of the major food groups daily will ensure a well-balanced diet and provide the essential nutrients needed for growth and development. The number of portions consumed per day differs with each group. At different stages in life, requirements within each group may vary to some degree, but no food group should dominate or be eliminated from the diet (Figure 34-5). More information on nutrition is provided in Unit 26 of this text.

DAILY PORTIONS FROM FIVE FOOD GROUPS

Food Group	Portion
Vegetable and Fruit	Four servings each day. Such as: oranges, watermelon, potatoes, peas, spinach, etc. (small glass of juice equals one serving)
Breads and Cereals	Four servings each day. Such as: bread, rice, corn products, muffins, noodles, etc. (one piece of bread equals one serving)
Milk and Cheese	Portion varies with age: Teenagers should have 4 servings a day. Such as: cheese, yogurt, ice cream, milk, etc. (glass of milk equals one serving)
Meat, Fish, Poultry, Beans	Two servings per day. Such as: tunafish, peanut butter, meat, rice and beans, etc. (2-3 ounces is considered a serving—one piece of chicken is one serving)
Fats and Sweets	No portion suggested Found in foods such as butter, salad oil, potato chips, soft drinks, honey, etc.

FIGURE 34-5
The five food groups – portions and sources

Food Customs of Major World Populations

What is eaten to represent each of these food groups may vary around the world. This primarily reflects what is most readily available based on the climatic conditions discussed in the previous unit.

In hot and wet climates such as Southeast Asia, a great deal of rice is consumed. Likewise, in areas of the world where corn is plentiful, many food items contain corn in some form. For example, cornmeal may be used to make tortillas or pancakes and be mixed as a cereal in countries such as Mexico. Availability of food, and technology to prepare that food, has dictated eating habits over the years. Introducing dairy products to areas of the world where dairy cows are not raised presents an educational as well as a transportation and processing challenge.

Methods of Processing, Preserving, and Storing Foods

One of the oldest ways to preserve food for delayed use is fermenting and pickling. *Fermentation* is a chemical change that involves foaming as gas is released. Long ago, it was determined that some foods did not spoil when allowed to ferment naturally or by adding fermented liquids to the foods. Controlled fermentation is now used to produce cheese, wines, beers, vinegars, pickles, and sauerkraut (Figure 34-6).

Today, the primary objective of processing and preserving is to change raw commodities into sta-

FIGURE 34-6
Fermentation – one of the oldest forms of food preservation ***(Courtesy USDA)***

FIGURE 34-7
The cooling of milk begins on the farm. ***(Courtesy of Denmark High School, Agriculture Education Department, photo by Jim Jones)***

ble forms. The reality that most foods are not used immediately has allowed us to expect almost all foods at any time during the year and in any part of the world.

Slowing deterioration is the primary goal in food preservation. Tomatoes and cucumbers that will be sold raw are waxed to retard shriveling while they are in the grocery store. Apples are treated with a decay inhibitor. Table grapes are fumigated with sulfur dioxide to control mold. Similarly, silos where grains are stored are purged with 60% carbon dioxide to control insects.

Carbon dioxide inhibits the growth of bacteria. *Controlled atmosphere* (CA) is where oxygen and carbon dioxide are adjusted to preserve or enhance particular foods. Transporting foods in controlled atmosphere allows the transporting of cut lettuce without the edges turning brown.

There are many other processing and preservation techniques. These techniques slow deterioration processes and allow you to enjoy foods in a variety of forms around the year and around the globe.

Refrigeration is an important key to many processing and preservation techniques. *Refrigeration* is the process of chilling or keeping cool. Low temperatures reduce or stop processes that contribute to the deterioration of products. Refrigeration retards respiration, aging, ripening, textural and color change, moisture loss and shriveling, insect activity, and spoilage from bacteria, fungi, and yeasts. Refrigeration is costly and only used widely in well-developed countries.

When crops are harvested in the field, their temperatures are between 70 and 80°F. The goal of refrigeration is to quickly reduce that temperature to near 32°F. How quickly the food is cooled varies by the use of different techniques. Some vegetables are precooled in the field by cold air blast, hydrocooling, or vacuum. *Hydrocooling* means cooling with water. On the other hand, milk is cooled in a refrigerated tank on the farm (Figure 34-7).

The logical step after a product is cooled is to continue cooling it until it is frozen. When foods are kept at 0°F or lower, very little deterioration occurs. However, even frozen foods have storage limits. Fruits and vegetables should be consumed within 1 year after freezing. Meats should be consumed within 3 to 6 months. Vegetables that are to be frozen often require blanching prior to freezing. *Blanching* is the scalding of food for a brief time before freezing it. This process inactivates enzymes that cause undesirable changes when plant cells are frozen.

Another popular preservation technique is canning. *Canning* involves putting food in airtight containers. Food is sterilized during canning to kill all living organisms that could cause spoilage. Temperatures of 212 to 250°F are required to kill microorganisms that lead to spoilage. Metal cans are coated to reduce chemical reactions between the can and its contents. A two-year shelf life for canned food is considered normal.

Another popular way to process and preserve food is dehydration. *Dehydration* means lowering the moisture content to inhibit growth of microorganisms. Moisture can be removed by the sun, by indoor tunnel or cabinet dehydrators, or by freeze drying. *Freeze drying* is the newest method of dehydration. It involves the removal of moisture by rapid freezing at very low temperatures. When foods are dehydrated, they have a 2 to 10% moisture content. The shelf life of dehydrated foods is as long as 2 years. *Oxidative deterioration* is loss of quality due to a reaction with oxygen. This can occur when air reaches dried foods. Glass or metal containers are more airtight than plastic ones. Dried soup mixes, packaged salad dressing, spices, and dried fruits are foods that have been dehydrated. These foods are lighter in weight and lower in volume than are whole foods.

Another processing technique is dehydrofrozen. *Dehydrofrozen* involves precooking, evaporation of water, and freezing. Potatoes are being used in the development of this technique. They are precooked as cubes or slices, and water is evaporated to reduce the weight by 50%. The potatoes are then frozen.

Other factors that must be considered in the processing of foods include humidity and wrapping food products during storage and transportation. *Humidity* is the amount of moisture in the air. Humidity of 90 to 95% is required for high-moisture products such as meats and vegetables. Dry onions require only 75% humidity for optimum storage.

There are a variety of wrappings used in the processing and preservation of foods. Cardboard boxes, wood boxes, molded pulp trays to reduce bruising, and plastic wraps in a variety of thicknesses or plies all meet different needs. Newer retortable pouches provide protection from light, heat, moisture, and oxygen transfer, all in one wrapping. *Retortable pouches* are flexible packages consisting of two layers of film (or plastic) with a layer of foil between them. Although the cost is now high, the benefits of a 1- to 2-year shelf life are appealing.

The cost of each of the processes and techniques mentioned above is different and must be taken into account by processing plants. Where energy costs are lower, more sophisticated techniques will not cost the consumer as much as they would if energy costs were higher. Research is always looking at ways of preserving food more economically in relation to energy costs.

AGRI-PROFILE

CAREER AREAS: FOOD SCIENTIST/FOOD TECHNICIAN/NUTRITIONIST

Food scientists develop better ways to process, handle, and package foods. The peel practically falls off this grapefruit after vacuum infusion, a process developed by the United States Department of Agriculture at Winter Haven, Florida. *(Courtesy USDA Agricultural Research Service)*

Food science programs are available in high schools, colleges, technical institutes, and universities. Careers in food science may be in specialty areas such as meats, fruits, vegetables, baked goods, dairy products, wine, or other beverages.

Food companies and consumers rely on food scientists to develop new products to meet the ever-changing needs of a busy world. Convenience products such as instant and freeze-dried coffee, dried and vacuum-packed fruits and meats, processed chicken tenders, boneless rolled meat, yogurt, ice cream, pasteurized and homogenized milk, shelf-safe milk, low-calorie and low-fat food, space-age food, unrefrigerated foods, and many other products are the handiworks of food science and technology.

Food scientists work for food companies, universities, research centers, food chain stores, dairies, radio and television stations, and government agencies. Some become authors and publish journal and magazine articles, cook books, and recipe books.

Food Additives to Enhance Sales of Food Commodities The processing of some foods may reduce their natural nutritional value. To compensate for that loss, vitamins and minerals are added back into foods to restore their nutritional value. Bread, noodles, and rice have vitamins and minerals added to them before they are packaged. Vitamins A and D are added to fluid milk prior to its sale to the public.

A *food additive* is anything that is added to a food during processing and before it goes into a package. In addition to food additives that restore nutritional value, preservatives are also added to foods to extend their shelf life. *Shelf life* refers to the amount of time before spoilage begins. Food additives also enhance the color or appearance of foods. Other food additives reduce the cooking time of foods such as oatmeal.

Sugar is probably one of the most widely used food additives. It is found on the labels of many cereals and beverages. Food labels identify the contents of a food product. The order of ingredients on a food label indicates the proportions of each, in descending order (Figure 34-8).

The government is continually testing and evaluating the positive and negative effects of food additives. It has been determined that some food additives pose more harm than benefit; therefore, they have been banned from use.

Food Preparation Techniques Some foods can be eaten as they come from the package. Crackers and raisins do not require additional processing at home to guarantee safe eating. However, cooking is often required for other foods purchased in the grocery store. Raw meat should be cooked before it is eaten to ensure safety. How food is cooked is determined by the type of food and the appliances available.

There are two basic methods of cooking foods. The appliances used to accomplish these methods vary from household to household. *Dry-heat cooking* involves surrounding the food by dry air in the oven or under the broiler. This method is usually used for tender cuts of meat with little connective tissue, and for vegetables with a high-moisture content, such as potatoes. *Moist-heat cooking* involves surrounding the food by hot liquid or by steaming, braising, boiling, or stewing it. The warm moisture breaks down the connective tissues. Moist-heat cooking is a popular method used for less tender cuts of meat, and for vegetables with a low-moisture content.

READ LABELS TO KNOW WHAT YOU ARE PAYING FOR!

Ingredient Listing Ingredients are listed in order from the most to the least amount found in the product.

Grape Juice:

Grape juice, grape juice from concentrate, ascorbic acid (vitamin C). No artificial flavors or colors added.

This label tells you:

- mostly grape juice and juice concentrate
- vitamin C added

Grape Juice Drink:
(10% Grape Juice)

Water, high fructose corn syrup, sugar, grape juice concentrate, fumaric, citric and maltic acids (provide tartness), vitamin C, natural flavor, artificial color.

This label tells you:

- mostly added water, syrup, and sugar
- some grape juice
- vitamin C added, plus other things

Powdered Grape Drink:

Sugar, citric acid (provides tartness), natural and artificial flavor, artificial color, vitamin C.

This label tells you:

- mostly sugar
- no juice at all
- vitamin C added, plus other things

FIGURE 34-8
Products may look the same, but read the label.

Appliances have been designed to accommodate different types of food, as well as different time and energy demands. The goal of preparing good food in a short period of time, without expending excessive energy, has resulted in the development of several appliances that supplement the traditional gas or electric range and oven. Pressure cookers concentrate moisture by sealing it in, thus reducing cooking time. Crockpots and oven bags allow slow, moist cooking without the operator continually being near the cooking process.

Table-top *conventional ovens* are used to cook small amounts of foods by dry heat. This requires less energy than large traditional ovens. The *microwave oven* offers a more energy-efficient way to cook foods that require both dry and moist heat. Approximately 50% of the energy goes to the food in the microwave, whereas only 6 to 14% of the energy used on a conventional range actually goes to the food.

Convection ovens heat food with the forced movement of hot air. Ovens that combine convection and microwave functions are now on the market. Again, energy efficiency, as well as convenience for the food preparer, is a constant goal.

Dehydrators dry food; *smokers* preserve food by keeping smoke in contact with the food for prolonged periods. The drying or smoking of food for family use is not very popular. Often refrigeration is still required after foods are dried or smoked.

FOOD PRODUCTS FROM CROPS

These types of food products generally help meet the body's requirements for food in four out of the five food groups. Fruits, vegetables, breads, cereals, beans, and oils all come from crops.

Those who do not eat meat products can still meet the requirements for food from all five food groups. People with allergies to milk products can meet their nutritional needs for food from this group with substitutes produced primarily from soybeans (Figure 34-9).

FIGURE 34-9
Foods from crops ***(Courtesy of the American Soybean Association)***

FOOD PRODUCTS FROM ANIMALS

Animal products supply the needs for food from the milk and cheese group, as well as from the meats, fish, poultry, and proteins group (Figure 34-10).

Dairy Products

Milk is the main fluid product from dairy animals. It is used and consumed in a variety of ways. Approximately 40% of all milk is consumed in fluid form. Milk is used to make many products, including cheese, butter, frozen foods, dried whole milk, cottage cheese, and evaporated and condensed milk. The processing of milk in different ways results in these many different products (Figure 34-11).

The fat in milk is called *butterfat. Cream* is a component of milk that contains up to 40% butterfat. Butter, which is about 80% fat, is made from cream. Cream is made by concentrating the fat portion of the milk. This is accomplished by passing milk through a cream separator. Whipping cream contains about 40% fat, table cream 18 to 20% and "half-and-half" approximately 12%.

FIGURE 34-10
Foods from animals

FIGURE 34-11
Milk is the source of many products. ***(Courtesy of Denmark High School, Agriculture Education Department, photo by Ken Seering)***

Ice cream, ice milk, and sherbet are also dairy products. They account for most of the frozen desserts in the United States.

Nonfat dried milk is used both as human food and animal feed. It is frequently used as an ingredient in dairy and other food products.

Cheese is made by exposing milk to certain bacterial fermentation or by treating it with enzymes. Both methods are designed to coagulate some of the proteins found in milk. There are many types and varieties of cheese. *Cottage cheese* is made from skimmed milk.

Condensed and evaporated milk are canned milk products. *Condensed milk* and *evaporated milk* are both produced by removing large portions of water from the whole milk using a machine called a *vacuum pan.* Condensed milk is further treated by adding large amounts of sugar. The sugar content makes condensed milk an important ingredient for the baking and ice cream industries.

Meat Products Meat products, like milk, are enjoyed in a variety of forms. Since each animal species is different, the procedures for slaughtering and processing each is somewhat different. There are common by-products that result from the slaughter of animals, as well as the familiar meats that are important to American eating habits.

Steps in Slaughtering Beef In order for beef to be consumed, the beef animal must be slaughtered (Figure 34-12). *Slaughter* means to kill animals for market purposes. Each processing plant may have slightly different techniques, but the slaughtering of beef generally involves the following steps:

Step 1. Rendering Insensible *Rendering insensible* means making the animal unable to sense pain. There are several methods of accomplishing this that comply with the Humane Slaughter Act of 1958. Packers that do not comply with provisions of this act are unable to sell meat to the federal government. The approved methods must be rapid and effective. Methods used in rendering insensible include a single blow or gunshot, an electrical current, or use of carbon dioxide gas. Additionally, rendering insensible may be accomplished

FIGURE 34-12
Animals must be slaughtered if we are to benefit from animal products. ***(Courtesy of Omaha Livestock Market, Inc.)***

by following the ritual requirements of religious faiths.

Step 2. Shackling, Hoisting, Sticking, and Bleeding Once the animal is insensible to pain, it is shackled, hoisted, and stuck to permit bleeding. *Shackles* are mechanical devices that confine the legs and prevent movement; *hoist* means to raise into position. To stick means to cut a major artery to permit the blood to drain from an animal. A large artery is *stuck* for efficient *bleeding out*. The head of the animal is removed during or following the bleeding-out process.

Step 3. Skinning The removal of the *hide* or skin is the next step in the beef-slaughtering process. The hide is cut open at the median or midline of the belly of the animal, and hide pullers are used to remove the hide in one piece. The breast and rump bones are split at this time by sawing.

Step 4. Removing Viscera The term *viscera* refers to organs located in the cavity of the animal. Organs that are removed include the heart, liver, and intestines. The kidneys are not removed at this time. Plants that are regulated by the U.S. Department of Agriculture must have the carcass and viscera inspected at this stage of slaughtering. *Carcass* refers to the body meat of the animal, the part that is left after the offal has been removed. *Offal* means the nonmeat material that is converted to by-products. It includes the blood, head, shanks, tail, viscera, hide, and loose fat.

Step 5. Splitting Carcass and Removing Tail The next step in the slaughtering process is cutting the carcass through the center of the backbone and removing the tail.

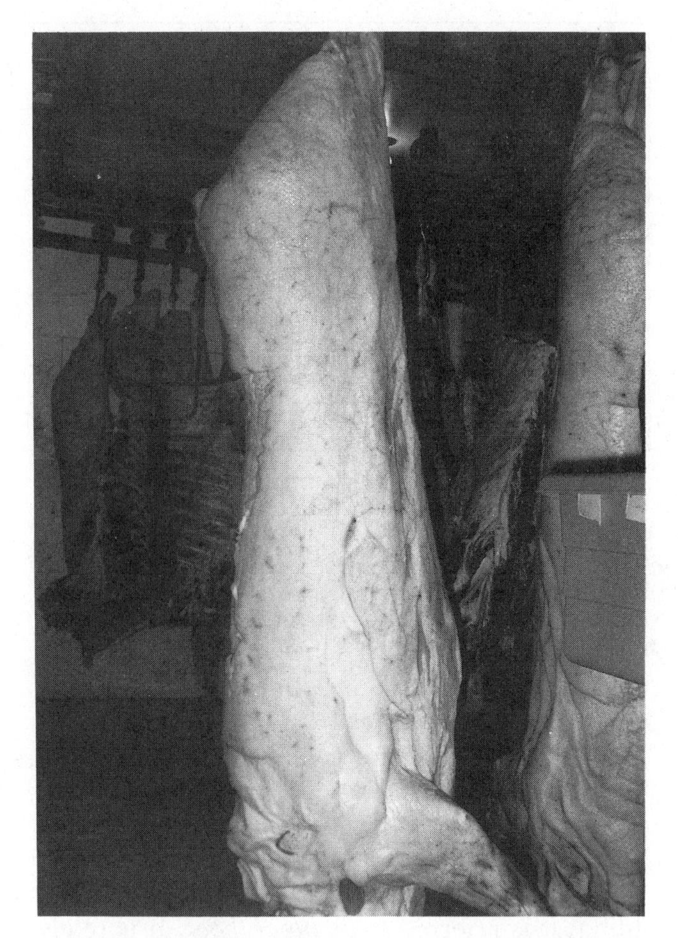

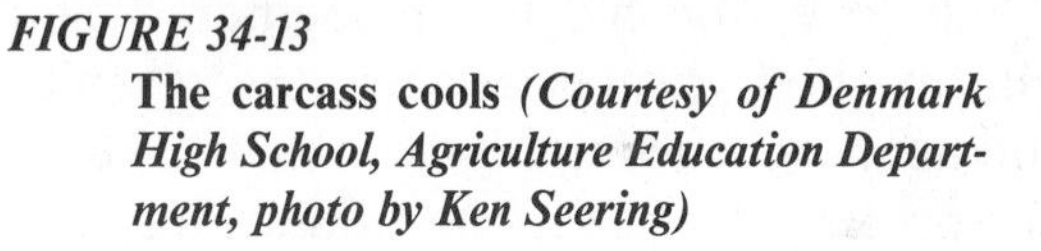

FIGURE 34-13
The carcass cools ***(Courtesy of Denmark High School, Agriculture Education Department, photo by Ken Seering)***

Step 6. Washing and Drying The *split carcass* (sides of the animal) is washed with warm water under pressure.

Step 7. Shrouding In order to provide a smooth appearance following cooling, high-quality carcasses are *shrouded*, or wrapped tightly with a cloth.

Step 8. Sending to the Cooler Sides are cooled for a minimum of 24 hours prior to ribbing and further processing (Figure 34-13). The meat is kept at 34°F until it is sold and consumed.

The Next Steps in Processing Meat Following slaughter, beef is aged or ripened. To *age or ripen* means to leave undisturbed for a period of time so that minor biological changes can take place while the beef cools. Fresh beef is not in its most tender state immediately after slaughter. While the beef is aging, evaporation (the loss of moisture) and discoloration are

TYPES OF AGING	TEMP.	HUMIDITY	TIME	TECHNOLOGY
Traditional	34-38°F	70–75% or 85–90%	1–6 weeks	cooler
Fast	70°F	85–90%	2 days	ultraviolet lights

NOTE: Vacuum packaging moisture-vaporproof film to protect meat—reduces weight loss and surface spoilage for 2-3 weeks. Travel time of 6-10 days allows sufficient aging.

FIGURE 34-14
Beef must age prior to further disassembly.

kept to a minimum. A fairly thick covering of fat on the carcass helps the aging process. There are three methods used to age beef—traditional aging, fast aging, and vacuum packaging. Time, temperature, and technology help define these three methods (Figure 34-14).

Beef carcasses are generally disposed of in three ways: as block beef; fabricated, boxed beef; and processed meats. Traditionally, meat is shipped in exposed halves, quarters, or wholesale cuts to be cut into retail cuts in supermarkets. In this condition, it is referred to as *block beef*. It is ready for sale "over the block" or counter. This traditional method creates concern over sanitation, shrinkage, spoilage, and discoloration. Therefore, more packers are using the disassembly process. The *disassembly process* means that the carcass is divided into smaller cuts, vacuum sealed, boxed, moved into storage, and shipped to retailers. This process is also known as *fabrication and boxing. Processed meats* are made from scraps of meat that are not in suitable form for sale over the block. Such meat has the bones removed and is sold as boneless cuts. It can also be canned, made into sausage, dried, or smoked.

Sheep The steps in slaughtering sheep are similar to those used with beef animals. However, due to differences in size and anatomical structure, some procedures are slightly different.

The sheep is rendered insensible and bled by inserting a double-edged knife into the neck just below the ear. The large blood vessel in the neck is severed. Next, the front feet are removed, the *pelt* or furlike covering is removed, and the hind feet and head are removed. The opening of the carcass and *evisceration* are similar to the procedures outlined for the slaughter of a beef animal. In view of the small size of a lamb, the forelegs are folded at the knees and are held in place by a skewer, after evisceration. Washing and cooling procedures are similar to those used in beef slaughter.

Hogs Again, the animal is rendered insensible, shackled, and hoisted prior to sticking. The hog is stuck just under the joint of the breast bone, severing the arteries and veins leading to the heart.

Vats of water at 150°F are used to scald the carcass for about 4 minutes. This process is required to loosen the hair and dry skin.

The hair of hogs is removed by mechanical scraping. A dehairing machine can remove the hair from about 500 hogs per hour (Figure 34-15).

Once the hair is removed, the hog is returned to overhead racks and the slaughtering process continues. The gam cords of the hind legs are exposed and gambrel sticks are inserted in the cords.

Seven procedures complete the slaughter and dressing procedures for hogs. The hog is washed and singed prior to the removal of the head. *Singe* means to burn lightly to remove hair. Next, the carcass is opened and eviscerated, prior to being split or halved with a cleaver or electric saw. The leaf fat is removed. *Leaf fat* is layers of fat inside the body cavity. Before the carcass is washed, the kidneys and facing hams are inspected. After washing, the carcass is sent to coolers at 34°F.

Unlike the fat from beef cattle, lard is considered a product along with the meat. *Lard* is the fat from hogs. It is used for a variety of cooking and baking products such as oleomargarine. Lard is often combined with other animal fats and with vegetable oils such as cottonseed, soybean, peanut, and coconut. Such mixes are used extensively for baking, cooking, frying, and other food preparations, and for commercial products.

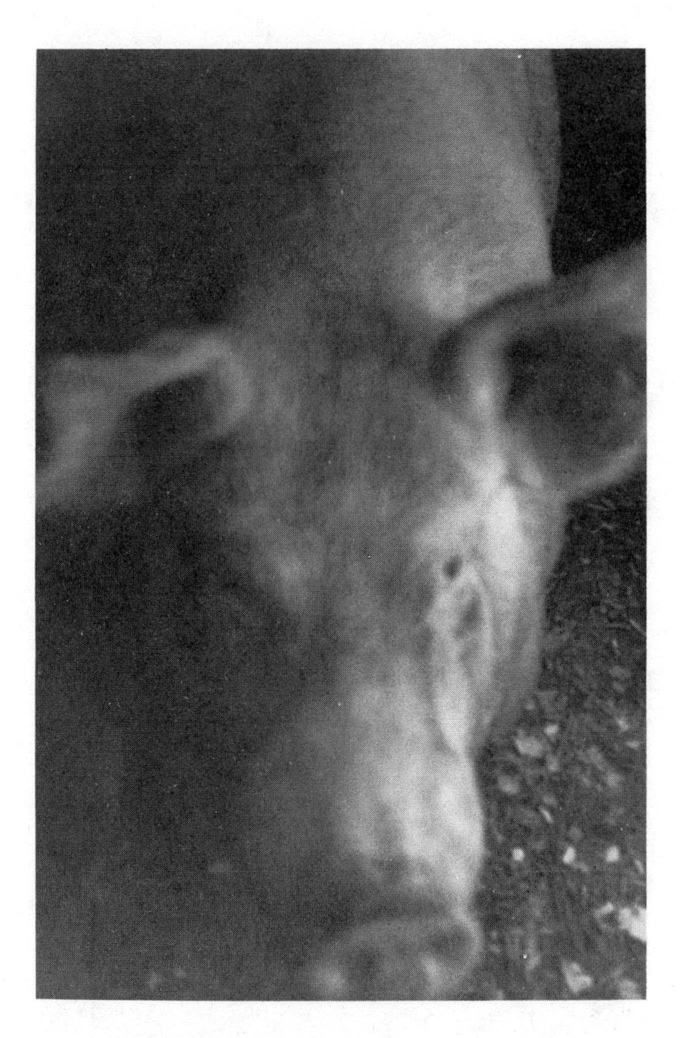

FIGURE 34-15
The hairs on the hog must be removed. ***(Courtesy of Denmark High School, Agriculture Education Department, photo by Jim Jones)***

Poultry The steps in slaughtering poultry are similar in many ways to those required for other animals. The feathers that cover a bird, like the hair that covers the hog, must be removed prior to evisceration.

The slaughtering process begins by shackling the bird on a conveyor belt. The jugular vein is slit, which begins the bleeding out. The *jugular vein* is the largest vein in the throat. Next, the bird is scalded prior to feather removal or picking. Singeing is required to remove the fine hairs that cover a bird under its feathers. Once the feathers and hair have been removed, the bird is washed and eviscerated. The giblets are then prepared. *Giblets* are the heart, liver, and gizzard of a bird. The bird is then cooled to 40°F, usually with ice.

Fish Once fish are caught or harvested, they too must be prepared for processing and consumption. The procedure depends upon the type of fish. Evisceration is usually done after the scales and head are removed. Washing and cooling follow.

Fish, like other foods, are consumed in a variety of forms. Whether a tuna is to be consumed whole or processed to be eaten later will determine whether the fish is left whole or is cut up. Some shellfish, like lobster, are kept alive until they are cooked. The heat used to cook the lobster kills it, and the processing continues. The cleaning may be done by the consumer after purchase in a restaurant.

Game Game slaughtering generally begins in the field. The game is typically killed by either gunshot or bow and arrow. The basic slaughtering procedures discussed above are followed for game. Most game is not slaughtered in processing plants. Like the differences between beef, sheep, swine, and poultry slaughter, procedures followed with game depend on the specific animal. Large animals such as deer, moose, and elk would be slaughtered in a fashion similar to cattle. Fowl such as geese, quail, and duck would be slaughtered following procedures similar to those for commercial poultry.

Kosher Slaughter *Kosher* means right and proper. Kosher slaughter is the major variation in the slaughter techniques discussed above. This type of slaughter is based on the religious ritual of the Jewish faith. Meat is slaughtered according to ancient Jewish rules or *Shehitah*. Shehitah requires that animals be killed by a rabbi or a specially trained representative called a *shohet or shochet*, which means slaughterer. The method of killing the animal, as well as the time by which the meat of the animal must be sold, are based on ritual that relates to concerns for sanitation. A special knife or *chalaf* is used to slit the neck of the animal for quick death and more efficient bleeding. The shohet inspects the carcass and places a cross inside a circle with the date of slaughter and the name of the inspector on the carcass. Meat must be consumed quickly. Neither packers nor retailers are permitted to hold kosher meat for more than 216 hours (9 days). Washing is required every 72 hours.

Major Cuts of Meat Once the meat is slaughtered, it is further prepared for use. Different areas of the animal are useful for different purposes. Cuts of beef, veal, lamb, and pork, with wholesale and retail terms, are shown in Figure 34-16.

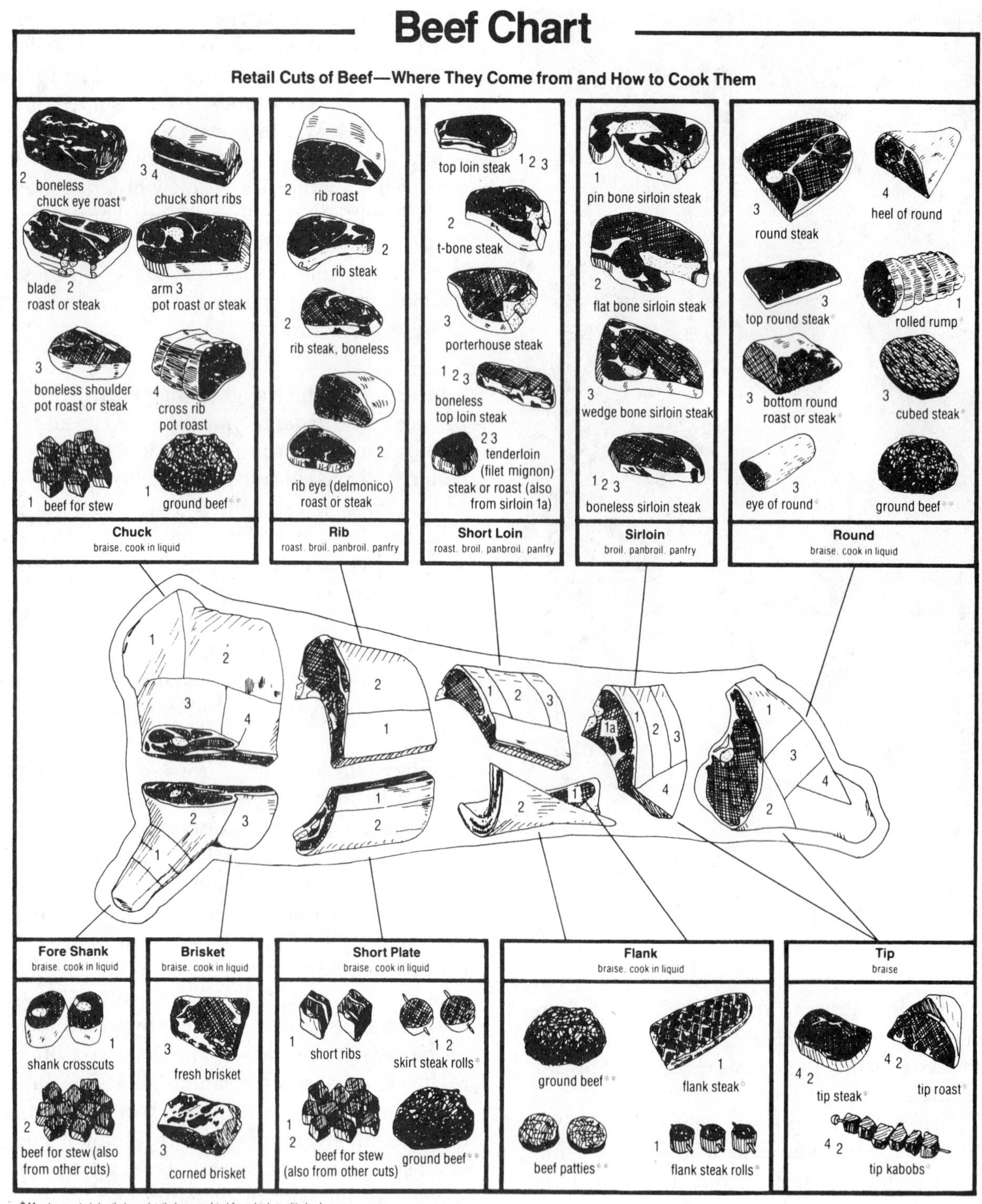

FIGURE 34-16

Where cuts of meat come from ***(Courtesy National Livestock and Meat Board)***

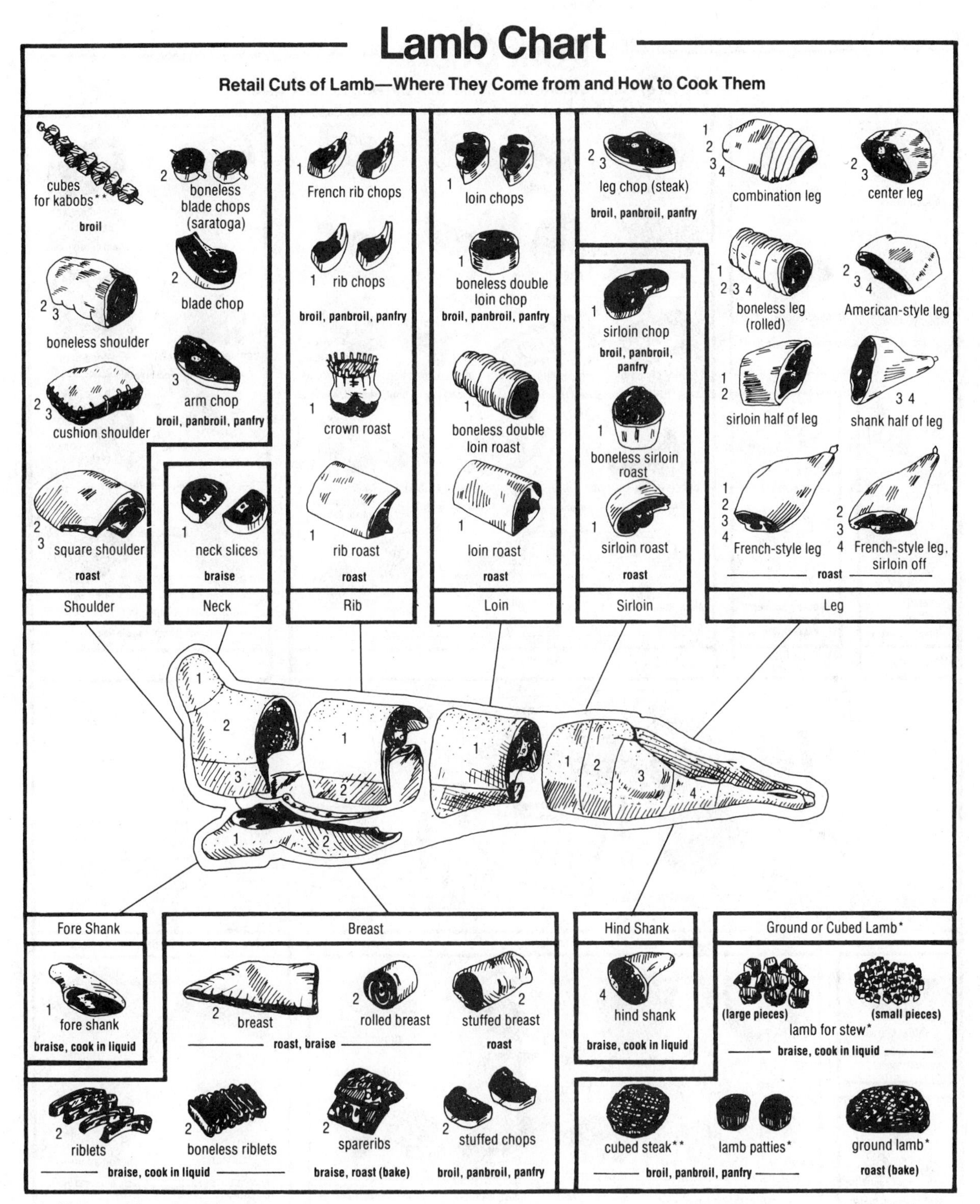

FIGURE 34-16
Where cuts of meat come from *(Courtesy National Livestock and Meat Board)* (continued)

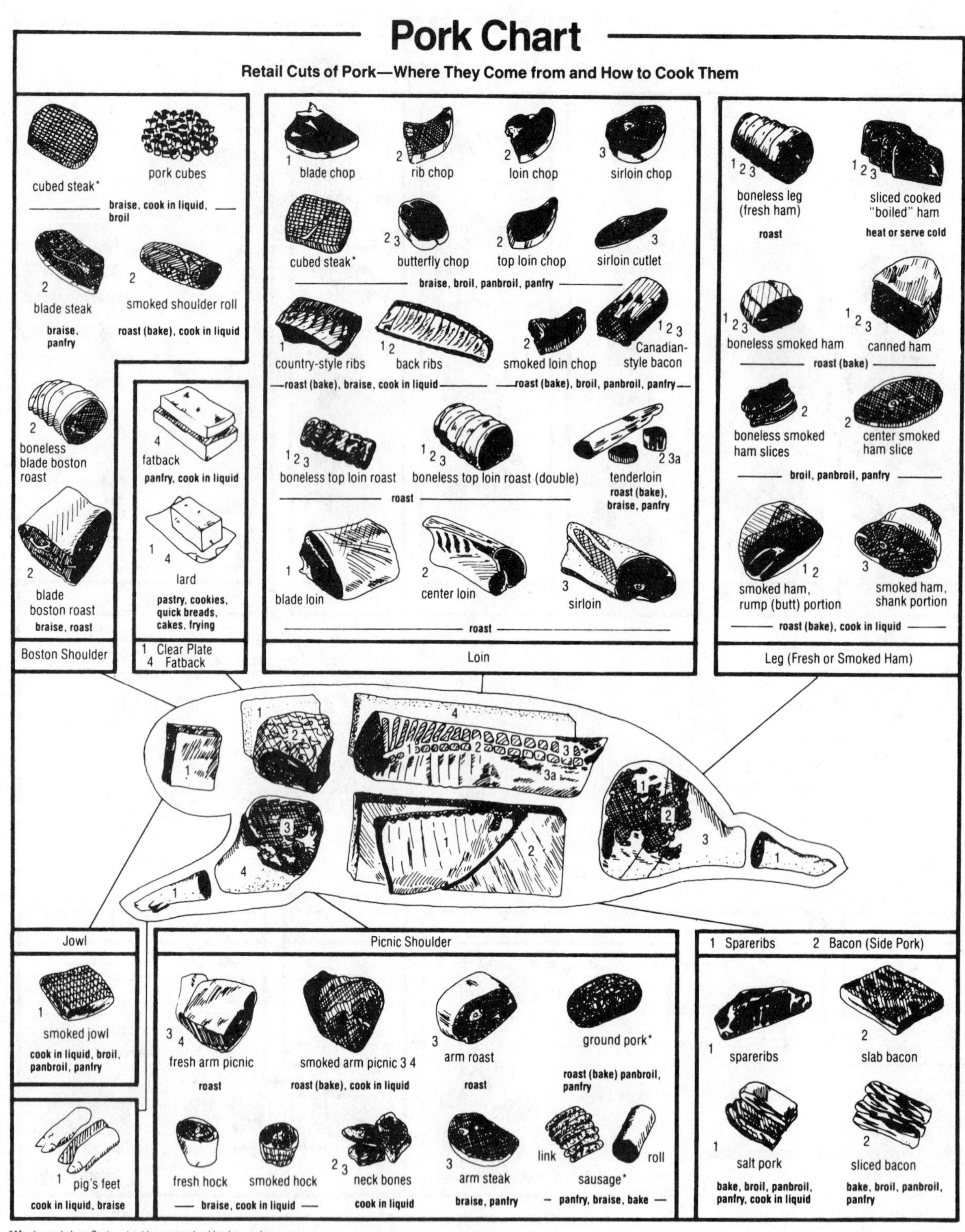

*May be made from Boston shoulder picnic shoulder, loin, or leg

FIGURE 34-16

Where cuts of meat come from *(Courtesy National Livestock and Meat Board)* (continued)

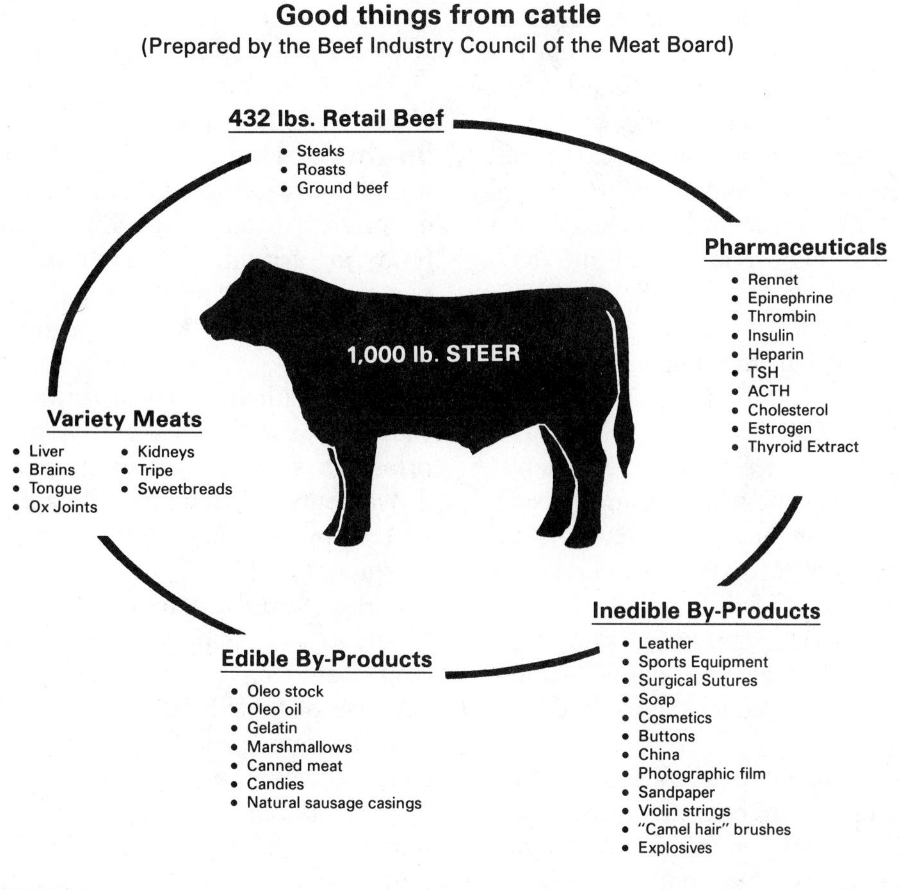

FIGURE 34-17
There are many useful by-products from animals. ***(Courtesy National Livestock and Meat Board)***

By-products—How Waste Products are Used Although most of the animal is consumed by humans, all of the animal is not edible. The *dressing percentage* is a term used to indicate the percentage or yield of hot carcass weight to the weight of the animal on foot. The offal is removed from the live animal to arrive at the dressing percentage. The formula is: hot carcass weight divided by the live weight times 100. The offal may be 40% of the live weight of the animal.

What happens to the offal accounts for many products that are used daily (Figure 34-17). By-products can generally be divided into 12 categories.

1. Hides—Leather from animal hides is used to make a variety of consumer products, such as shoes, harnesses, saddles, belting, clothing, sports equipment, hats, and gloves (Figure 34-18).
2. Fats—These are used to make products such as oleomargarine, soaps, animal feeds, lubricants, leather dressing, candles, and fertilizers.
3. Variety meats—The heart, liver, brains, kidneys, tongue, cheek meat, tail, feet, *sweetbreads* (thymus and pancreatic glands), and *tripe* (pickled *rumen* or stomach of cattle and sheep) are sold over the counter as variety or fancy meats.

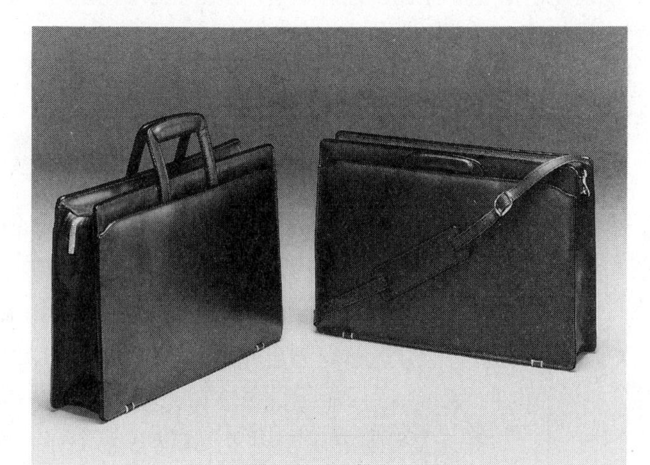

FIGURE 34-18
Leather products are all around us. ***(Courtesy Stebco Products)***

4. Hair—Brushes for artists are made from the fine hairs on the insides of the ears of cattle. Other hair from cattle and hogs is used for toothbrushes, paintbrushes, mattresses, upholstery for residential and commercial furniture, air filters, and baseball mitts.
5. Horns and hoofs—These items are used as a carving medium and are fashioned into decorative knife and umbrella handles, goblets, combs, and buttons.
6. Blood—The blood from animals is used in the refining of sugar, and also in making stock feeds and shoe polish.
7. Meat scraps and muscle tissue—After separation from the fat, meat scraps and muscle tissue are most often made into meat-meal or tankage. *Tankage* is the dried animal residue used as fertilizer and feed.
8. Bones—Some bones are put to the same uses mentioned above for horns and hoofs. In addition, bones are converted into stock feed, fertilizers, and glue.
9. Intestines and bladders—Sausage, lard, cheese, and snuff all use the intestines and bladders from cattle. Strings for musical instruments and tennis rackets are also made from these by-products.
10. Glands—The pharmaceutical industry relies heavily on animals glands for many drugs that are used today.
11. Collagen—*Collagen* is the chief constituent of the connective tissues. Glue and gelatin are made from collagen. These products, in various forms, are used in the furniture, photography, medical, and baking industries.
12. Contents of the stomach—Stomach contents of slaughtered animals are used primarily in the production of feed and fertilizer.

New Food Products on the Horizon The foods that we eat and how they reach us is an exciting area of agriscience. New foods, and new versions of familiar foods, are arriving on the market daily. Research is being conducted on improving the nutritional values of foods and keeping our food costs at a minimum.

The fruit industry uses an electronic fruit-shaped beeswax sensor to log the bumps and bruises sustained by fruit during shipment. Improvements in handling equipment are resulting in better apples. They also tend to reduce loss for producers.

Eggs are used in the preparation of other foods, as well as being eaten by themselves. Recently, eggs have been taken out of many diets because of their high cholesterol content, which leads to heart problems. Researchers are investigating ways of reducing the cholesterol levels in the egg without negating its nutritional value. Additionally, the poultry industry is looking at factors that make an animal fat and at ways to reduce the high-cholesterol component of poultry products.

Evaluation of different processing techniques with regard to cost, digestibility, nutrient retention, and flavor is constant. Nutrient loss in poultry has been found to be lower with freeze drying than with dehydration.

The conveniences that everyone enjoys will continue to be improved, as there continue to be new developments in the food science field of agriscience.

Student Activities

1. Define the "Terms To Know."
2. Keep a diary of what you eat for a week. Note which foods represent each of the five food groups.
3. With your teacher, arrange a visit to a slaughterhouse or meat department of a local grocery store to observe slaughter or meat processing.
4. Compare the end result when a food has been processed in a variety of ways, for example, fresh, canned, frozen, dehydrated, and freeze-dried potatoes.
5. Make a collage of items that exist because of animal by-products.
6. Using an outline of an animal, identify the major cuts of meat and where they come from on the animal.

Self-Evaluation

A. MULTIPLE CHOICE

1. Which of the following is NOT a carbohydrate?

   a. sugar
   b. starch
   c. fiber
   d. meat

2. The number of minerals needed by the body is approximately

   a. 2.
   b. 12.
   c. 20.
   d. 200.

3. The suggested number of daily portions from the meat, fish, poultry, eggs, and bean group is

   a. two, 2–3 oz. servings.
   b. four, 2–3 oz. servings.
   c. six, 2–3 oz. servings.
   d. eight, 2–3 oz. servings.

4. The butterfat content of table cream is approximately

   a. 10 to 12%.
   b. 18 to 20%.
   c. 30 to 32%.
   d. 38 to 40%.

5. To which milk product is sugar added during processing?

   a. evaporated milk
   b. condensed milk
   c. skimmed milk
   d. dried milk

6. Which mineral is added to most drinking water in the United States to assist in tooth development?

   a. zinc
   b. iron
   c. fluorine
   d. chlorine

7. Iron is important in the diet for the development of

   a. bones.
   b. vision.
   c. blood.
   d. hair.

8. Traditional aging of meat takes approximately

   a. 1 to 6 hours.
   b. 1 to 6 days.
   c. 1 to 6 weeks.
   d. 1 to 6 months.

B. MATCHING

_______ 1. Canning
_______ 2. Shrouding
_______ 3. Dehydration
_______ 4. Viscera
_______ 5. Sweetbreads
_______ 6. Food additive

a. Thymus and pancreatic glands
b. Internal organs
c. Added prior to packaging
d. Stored in airtight containers
e. Reduce moisture content
f. Wrap in cloth

C. COMPLETION

1. Vitamins ______, ______, ______, and ______ are fat soluble.
2. Minerals that assist in fluid regulation are ______, ______, and ______.

3. Meat from animals that have been slaughtered following Jewish ritual is called ______.

4. The method of precooking food, removing moisture, and then freezing is known as ______.

SECTION 10

Communications and Management in Agriscience

UNIT 35 Planning Agribusinesses

OBJECTIVE To define management, determine how decisions are made, and describe economic principles that affect management.

Competencies to be Developed

After studying this unit, you will be able to

- ☐ define management.
- ☐ describe the importance of management.
- ☐ describe kinds of agriscience management decisions.
- ☐ list eight steps in decision making.
- ☐ describe the economic principles of supply and demand, diminishing returns, comparative advantage, and resource substitutions.
- ☐ use capital and credit wisely in business management.

TERMS TO KNOW

Agribusiness management
Enterprise
Capital
Price
Supply
Credit
Demand
Diminishing returns
Comparative advantage
Resource substitution
Long-term loans
Intermediate loans
Capital investment
Short-term loans
Federal land banks
Production Credit Associations (PCA)
Commodity Credit Corporation (CCC)
Farmers Home Administration (FHA)
Small Business Administration (SBA)
Simple-interest loan
Promissory note
Discount loan
Add-on-loan

MANAGEMENT Agribusiness management is the human element that carries out a plan to meet goals and objectives in an agriscience business or *enterprise* generally referred to as an agribusiness. Management decides the types of business or production activities included in the agribusiness.

According to Dun and Bradstreet, 88% of all businesses fail because of poor management. Poor management of an agribusiness frequently results from no set objectives or goals. Management is usually considered to be good when maximum profits are achieved from the available resources. Generally, good management has established goals and objectives as guidelines.

Influences on Agribusiness Management Agribusiness management is usually influenced by the several members who make up the board of directors. Land, labor, and capital, being in limited supply, also influence management decisions. For instance, limited *capital*, which is money or property, may prevent the agribusiness manager from buying a new piece of equipment that would make the operation more efficient. Similarly, limited land will influence the agriscience manager's selection of enterprises. For example, a cow-calf beef producer will need more land than a feedlot operation will. In addition, limited labor will influence the manager's selection of enterprises. For instance, it will prevent planting crops with high labor requirements.

Estimating a Manager's Performance The performance or ability of the agribusiness manager can be estimated in several ways. Dollar income is the measurement most often used. According to a study conducted in Ohio, the managerial ability of an individual can be determined by the manager's economic orientation, decisiveness, ability to look at alternatives, and extent of social activities (Figure 35-1).

Agribusiness is constantly changing, and it is expected to change even faster in the future. As agribusiness becomes more complex, errors in management will be more costly. Therefore, the need for trained people in agribusiness management is greatly increasing.

1. Income goal (Economic orientation)
2. Willingness or tendency to make decisions (Decisiveness)
3. Manager's ability to recognize alternatives and opportunities
4. The extent of social activities or distractions from business goals

FIGURE 35-1

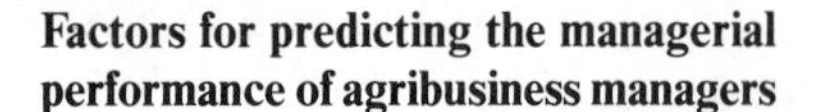

Factors for predicting the managerial performance of agribusiness managers

1. ORGANIZATIONAL DECISIONS

- Should I rent more land or should I borrow money and purchase land?
- Should the business be operated as a partnership or corporation?
- What lender is the best source of borrowed capital?

2. OPERATIONAL DECISIONS

- Should cattle be sold this week or next?
- Should I use high magnesium lime or regular lime?
- When should I start planting corn? Soybeans? Small grain?
- When should I change the photoperiod for my poinsettias?

FIGURE 35-2
Kinds of agribusiness decisions

CHARACTERISTICS OF DECISIONS IN AGRIBUSINESS

Although there are various types of decisions in agribusiness, decisions of managers are usually organizational or operational (Figure 35-2). Both types of decisions affect the success of the business. The degree of influence on the business is determined in part by the characteristic of the decision. The following discussion covers four characteristics of a decision.

Importance The importance of a decision may be determined by measuring the potential loss or gain of the decision involved. For example, the selection of an agribusiness enterprise or business venture is likely to be more important than the selection of a brand of a given commodity. The manager must spend a considerable amount of time selecting a type of business, since this decision cannot be changed easily once it is made. On the other hand, if a certain brand of wire or species of plant being sold is not successful, a different brand or species can be added or substituted at little additional cost.

Frequency The frequency, or how often a decision is made, varies greatly. Determining the cost per day to rent a truck may be used as an example. The cost of renting the truck for only one day, for a one-time job, might not be an important decision. However, the cost of renting the truck every several weeks as an ongoing expense makes the decision more important.

Urgency It is important that certain decisions be made immediately. Some decisions can be delayed. The urgency of a decision depends on the cost of waiting. Two examples of decisions with different degrees of urgency are hiring a new employee and buying a new office radio. If it is the busy time of year, an additional person may be needed immediately and should be hired. If you delay, you may lose sales. However, the delay in buying an office radio should not affect the overall management and profit of the agribusiness.

Available Choices Some situations offer several choices. If several choices are available, the manager should delete the least likely course of action. The manager should take time to gather important information. Such information should be helpful in making the correct decision. If no choices are available, there is no decision to be made.

STEPS IN DECISION MAKING

Making decisions based on facts will increase your management ability. The agribusiness manager must determine the process to be used for making better decisions. The following eight steps should be helpful in developing a management process (Figure 35-3).

1. Start the management process with a situation or established goal.
2. Gather all available facts and information for accurate analysis of the problem.
3. Analyze the available resources. Human resources are labor, time, skills, and interest.

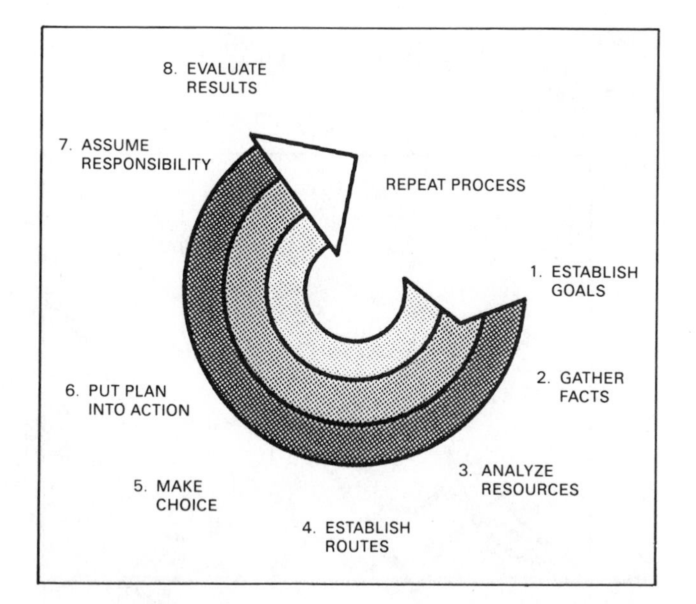

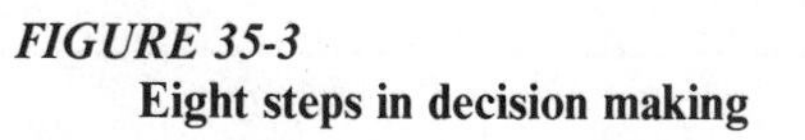
FIGURE 35-3
Eight steps in decision making

Material resources are land, equipment, and capital. Reevaluate your goal and adjust it if appropriate.

4. Determine the possible ways or established routes of accomplishing the goal or solving the problem.
5. Make a decision concerning the route(s) that will be taken.
6. Follow through with a plan of action.
7. Assume responsibility for implementing the plan.
8. Evaluate the results to determine whether the goals were accomplished. If not, could a better route have been selected?

After going through these eight steps, the manager should determine whether the process worked well. If not, where were the weak links? What changes should be made to improve the process? These and other questions will enable the manager to fine-tune the process until it fits his or her style, the personalities of the staff, and the structure of the organization.

PRINCIPLES OF ECONOMICS

Price *Price* is the amount received for an item that is sold. The price is determined by three price-making factors: (1) the supply of the item, (2) the demand for the item, and (3) the general price level. No one factor can be used to explain all price changes.

The general price level of a commodity is influenced by the supply or availability of the item, the demand for it, wars, depressions, and other factors. When supply and demand are in balance, a general price is established. An increase or decrease in either supply or demand may influence general prices.

Supply The quantity of a product that is available to buyers at a given time is called *supply*. A producer of vegetables has control over some factors that influence the vegetable supply. A big factor that the manager cannot control is the weather. If the rains do not come, the supply of vegetables will be reduced. The ability to buy items needed to produce the vegetables can also increase or reduce the supply. The ability to buy these items is influenced by their price. Credit may also influence supply. *Credit* is money loaned. It permits the agribusiness manager to buy items needed for production.

Demand *Demand* may be defined as the quantity of a product that buyers will purchase at varying prices at a given time. The quantity that a buyer is willing to purchase depends upon the quantity available and the price. Both the desire and the ability to purchase influence the law of demand. The population growth may also influence demand. As the population of the country grows, there will be increased demand for food (Figure 35-4).

DIMINISHING RETURNS

The term *diminishing returns* is often used in a conversation on agriscience economics. It refers to the amount of profit generated by additional inputs. Most of the time, the term is not completely understood. An understanding of the law of diminishing returns can be extremely helpful to the agribusiness manager in decision making.

The principle of diminishing returns has two parts—physical returns and economic returns. An explanation of both physical and economic returns follows.

Diminishing Physical Returns

Satisfaction from eating, or additional yields from applying additional fertilizer to a crop, may be used to illustrate diminishing physical returns (Figure 35-5). Each bite of food represents an additional unit of input. Each bite helps satisfy a part of your hunger. The added

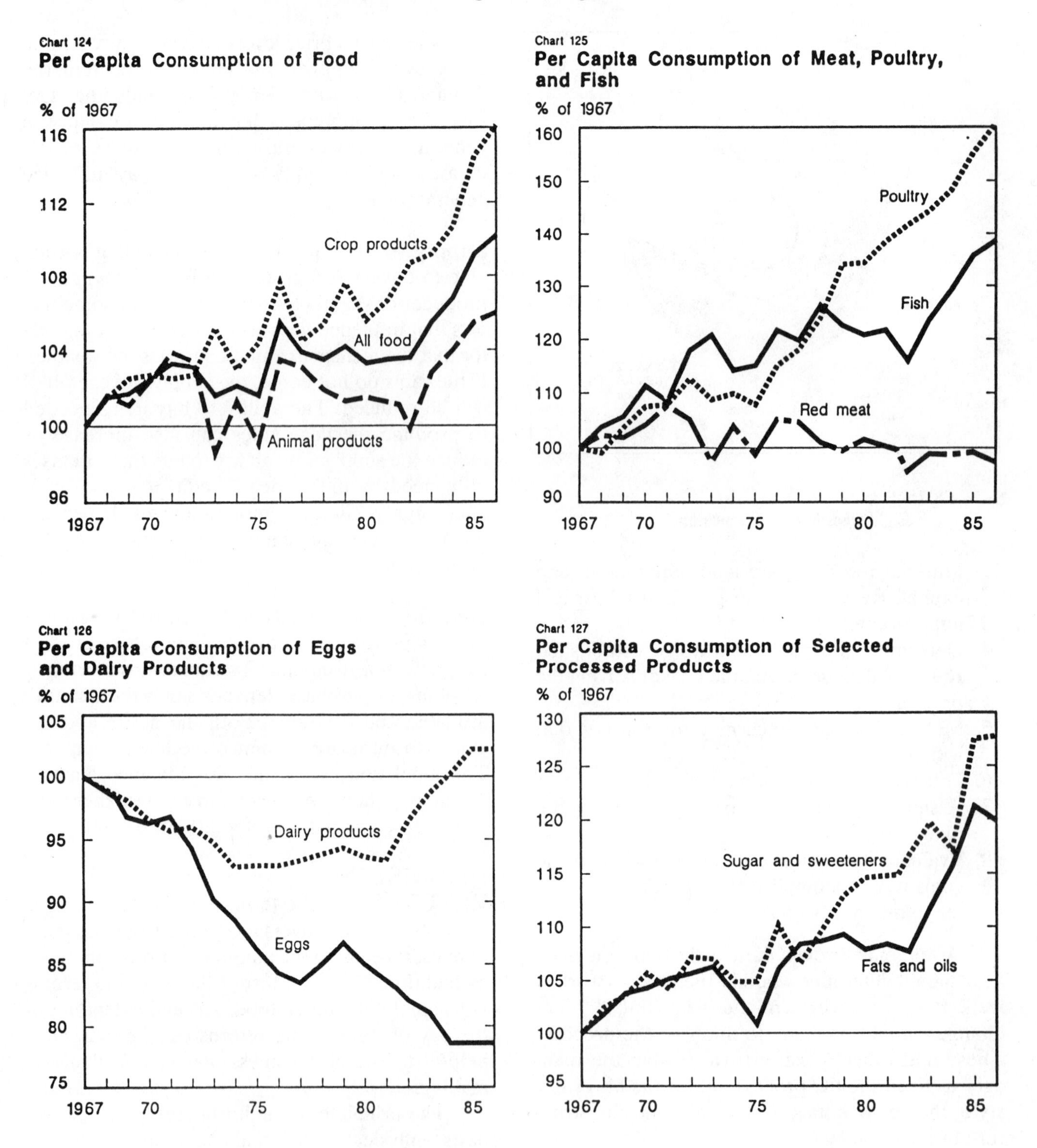

FIGURE 35-4
Per capita consumption of food *(Courtesy USDA)*

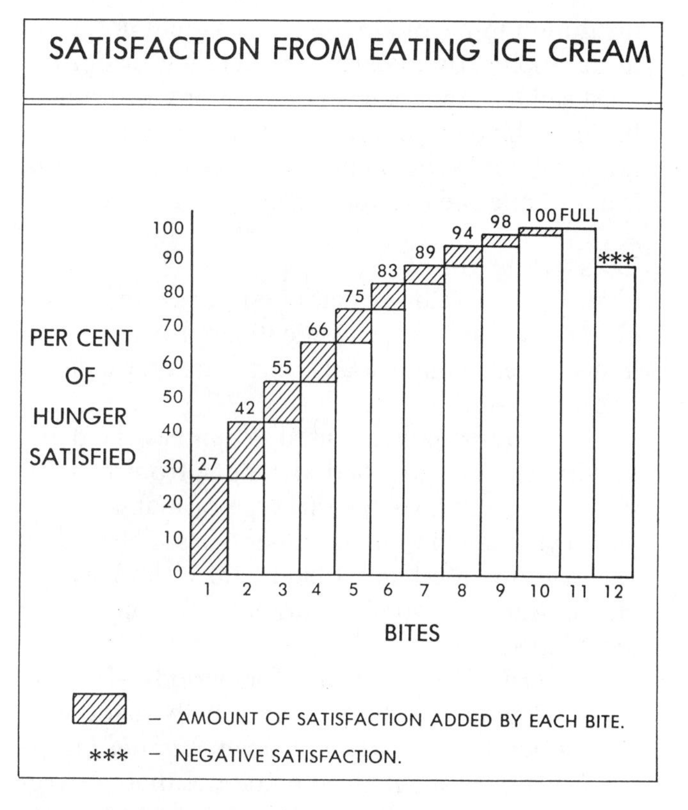

FIGURE 35-5
Diminishing returns

EFFECT OF NITROGEN ON CORN YIELDS*

Lbs. NITROGEN APPLIED	BUSHEL - YIELD
0	67
60	103
120	133
180	154
240	173
300	169

* Hypothetical and not based on research

FIGURE 35-6
Effect of nitrogen on corn yields

amount of satisfaction of your hunger diminishes as you eat more. This tendency is known as diminishing physical returns. At a certain point, the amount of hunger satisfied becomes negative with each bite taken. That means each additional bite takes away from the satisfaction gained by previous bites.

Diminishing Economic Returns An example of this is the addition of units of nitrogen fertilizer as inputs and the effect on corn yields as an output (Figure 35-6). It may be observed that adding the first unit of nitrogen produced the greatest increase in yields. The rate of increase in yield diminished as more units of input were added. The decision that the manager needs to make is how many units of nitrogen to apply to obtain the greatest profit from growing the corn. If the number is too low or too high, the highest potential income will not be reached.

COMPARATIVE ADVANTAGE

The United States has nine major farming areas that have developed over a period of years. These areas have developed because of changes in demand or other factors. Within each area, different commodities are produced. Each operation in an area has similar commodities and systems of production. The reason for the similarity in operations is comparative advantage found in following the program that has evolved.

Comparative advantage is the emphasis in the area where the most returns will be received. An understanding of comparative advantage will assist the agriscience manager in making management decisions.

Comparative advantage may be illustrated in several ways. A recent example has been the change in livestock production in New York State. At one time, most of the farms in New York raised sheep. Today, these farms are producing milk, because of the high demand and favorable prices for milk in New York. It has become more advantageous to produce milk than to produce lambs and wool.

RESOURCE SUBSTITUTION

A knowledge of the principle of resource substitution is essential to the agribusiness manager. *Resource substitution* means the use of a resource or item for another when the results are the same. It is often possible to substitute a less expensive item for a more expensive one.

For example, it may be possible to substitute barley for corn in making a cheaper dairy feed. This substitution can be done without affecting total production (Figure 35-7). Barley is frequently much less expensive than corn. Because barley and corn have slightly different feeding values for dairy cattle, barley must sell for $2.25 per cwt or less to be a better buy than corn selling for $2.50 per cwt.

The application of the principle of resource substitution is not limited to feeding livestock. This principle can be used in many management situations. How many possible applications of resource substitutions can you list?

AGRICULTURAL FINANCE

Importance and Uses of Credit

The management of finances is the most important single function of the agriscience manager. Possessing a great deal of technical knowledge and know-how is very important. However, this know-how will not make you a successful agriscience manager unless you are a competent money manager. Careful planning of finances is probably more important than planning other aspects of the agribusiness. Some very good agribusinesses have failed because of the manager's inability to manage money.

Credit is borrowed money. At one time, the agriscience manager who sought credit was considered to be a poor manager. Today this is no longer true. The manager approaching a lending institution to negotiate for credit should not do so in an apologetic manner. The lending institution must sell its commodity, money, in order to stay in business. The use of credit is a privilege that must not be abused. The reputation as a poor credit risk is a difficult one to overcome.

Classification of Credit

Credit is classified according to its period of use. The classifications are long-term loans, intermediate loans, and short-term loans (Figure 35-8).

Long-term loans are used to purchase land and buildings. The loan period ranges from 8 to 40 years. Interest rates for this type of loan are usually lower than they are for other types. These loans are made by the federal land banks, the Farmers Home Administration, insurance companies, and local banks.

Intermediate loans are for periods of 1 to 7 years. Capital investments are usually made with the money from this type of loan. *Capital investment* is money spent on commodities that are kept 6 months or longer. Examples are breeding stock, tractors, store equipment, and warehouse equipment. Production credit associations, the Farmers Home Administration, finance companies, and local banks are sources of this type of credit.

Short-term loans are for a period of 1 year or less. They are often referred to as production loans.

AN EXAMPLE OF RESOURCE SUBSTITUTION

A CONCENTRATE MIXTURE FOR DAIRY CATTLE:

	Ground corn	
substituted for corn -	Ground barley	1,130 lbs.
	Ground oats	500 lbs.
	Wheat bran	200 lbs.
	Linseed Meal	150 lbs.
	Salt	20 lbs.
		2,000 lbs.

Corn cost per cwt. $2.50

Barley cost per cwt. $2.05

FIGURE 35-7
An example of resource substitution.

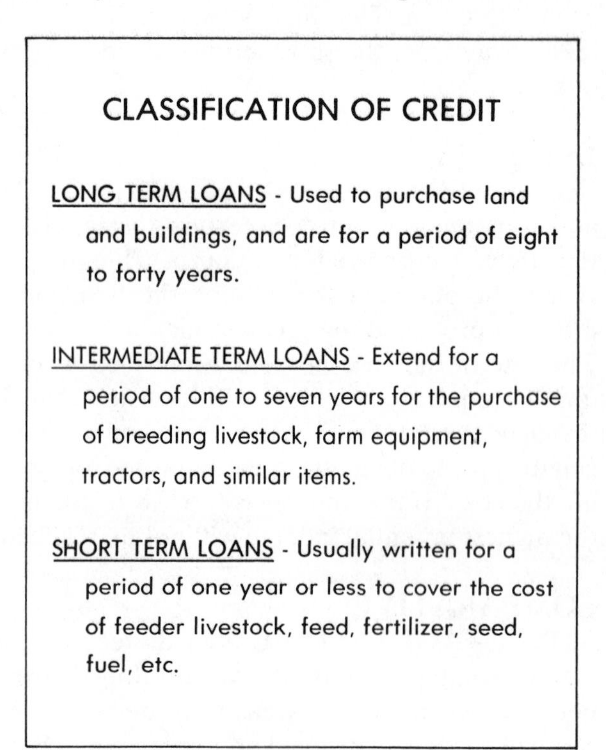
CLASSIFICATION OF CREDIT

LONG TERM LOANS - Used to purchase land and buildings, and are for a period of eight to forty years.

INTERMEDIATE TERM LOANS - Extend for a period of one to seven years for the purchase of breeding livestock, farm equipment, tractors, and similar items.

SHORT TERM LOANS - Usually written for a period of one year or less to cover the cost of feeder livestock, feed, fertilizer, seed, fuel, etc.

FIGURE 35-8
Classification of credit

Some managers are able to borrow short-term money without security. This is particularly true of managers who have a good reputation. Money borrowed without security is referred to as a signature loan. The borrower signs a promissory note that the loan will be paid on or before a specified date. Production credit associations and local banks are sources of short-term loans. It is not uncommon for local agribusinesses to permit their reliable customers to make purchases of this type on a short-term credit basis by simply signing a sales slip.

Types of Credit Two types of credit are productive credit and consumption credit (Figure 35-9). Productive credit is used to increase production or income. This is justifiable when the estimated increase in production will increase profits. This type of credit is used to purchase supplies, plants, flowers, livestock, land, equipment, storage facilities, seed, fertilizer, labor, and other materials.

Consumption credit is used to purchase consumable items used by the individual; it does not contribute to the business income. It is relatively easy for a family to abuse the use of consumption credit. This type of credit can also limit the amount of productive credit available in a family business.

A popular and growing form of consumer credit is credit card purchases. Most credit card purchases paid in 30 days or less are interest free.

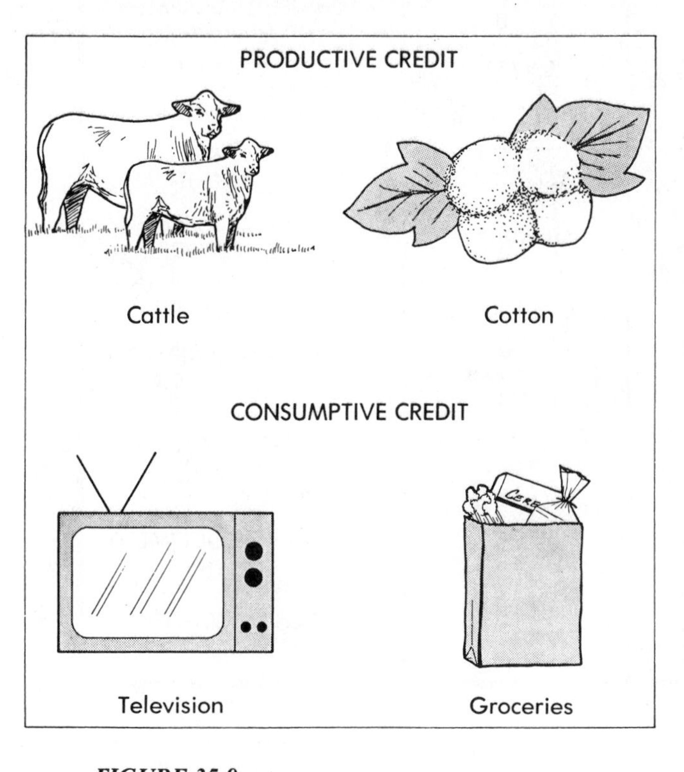

FIGURE 35-9
Kinds of credit

Federal statutes require firms charging interest on credit card purchases to publish their interest rates.

SOURCES OF CREDIT

There are many sources of credit available and differences in lending policies. Because of this, a knowledge and understanding of credit and credit practices should be helpful in

AGRI-PROFILE

CAREER AREAS: OWNER/MANAGER/ CONSULTANT

Agribusinesses constitute the backbone of many rural communities. Here a consultant utilizes a major computer information system to help plan future directions for this agribusiness. ***(Courtesy of the National FFA Organization)***

Agribusiness activities cut across the food and non-food spectrums of agriscience. It includes the supply side as well as the output side of production. It provides career opportunities in every sector wherever goods are bought or sold.

Agribusinesses sell products such as feed, seed, pesticides, fertilizer, tools, equipment, plants, or animals. Or, they may sell services such as animal care, crop spraying, or recreational fishing privileges. Careful planning is very important before starting an agribusiness, since many new businesses end in failure. Careful planning increases the chances of success.

Career opportunities in agribusiness planning include financial services such as banking, accounting services, management services, teaching, university extension service, marketing, market analysis, product specialization, product engineering, sales, ownership, and management. Agricultural economics, business management, accounting, finance, and personnel management are college programs that can lead to careers in agribusiness planning.

securing credit. Lenders differ in the interest rates they charge, the length of the loan, and the purposes for which money is loaned.

When selecting a lender, remember that the interest rate is not the most important factor in securing a loan. More important is the lender's willingness to extend a line of credit should an unexpected event occur. Another important factor is the lender's knowledge of the agribusiness.

Some agencies and institutions make only certain types of loans (Figure 35-10). Commercial banks are the most important source of credit. About 25% of the total agricultural debt is owed by banks. This percentage represents about 13.5% of the real estate loans and 45% of other loans. Commercial banks lead in loan volume because they make all types of loans.

Individuals are the next most important source of agribusiness credit. About 36% of agricultural real estate mortgages are held by individuals. Interest rates, tax deferments and reductions, and payment options are some of the reasons that individuals are willing to finance real estate.

Retail merchants and other agricultural businesses are included with individuals and others who supply non-real-estate debt. Their credit is usually in the form of an open account. Sometimes cash loans are made. The interest rates are usually higher from these sources than they are from commercial banks. The convenience of this type of credit is the reason for its popularity.

Federal land banks were created by Congress in 1916 to provide long-term credit for agriculture. These banks are an excellent source of credit because of the favorable interest rates.

Production Credit Associations (PCA) were established by an Act of Congress in 1933. The purpose was to provide favorable, short-term credit for agriculture. These funds are secured from the federal intermediate credit banks, who in turn secure their money from private lenders.

Life insurance companies have been excellent

SOURCES OF AGRICULTURAL CREDIT				
SOURCES	LENGTH OF LOAN	ANNUAL INTEREST RATE	PERCENT OF APPRAISAL LOAN VALUE	PURPOSE OF LOAN
COMMERCIAL BANKS & TRUST COMPANIES	6-12 MONTHS	10-13%	TO 100%	FARM PRODUCTION ITEMS
	2-3 YEARS	11½-12½ %	70-80%	MACHINERY, EQUIPMENT, AND LIVESTOCK
	10-20 YEARS	10%	60-75%	REAL ESTATE
FEDERAL LAND BANKS	20-35 YEARS	8½ % VARIABLE	UP TO 85%	REAL ESTATE
PRODUCTION CREDIT ASSOCIATION	1 YEAR	9% VARIABLE	TO 100%	PRODUCTION ITEMS
	3-7 YEARS	9%	VARIES	MACHINERY, EQUIPMENT, AND LIVESTOCK
FARMERS HOME ADMINISTRATION	40 YEARS	5%	TO 100%	REAL ESTATE
	7 YEARS	8¾ %	TO 100%	MACHINERY, EQUIPMENT, AND LIVESTOCK
INSURANCE COMPANIES	20-35 YEARS	VARIES	TO 75%	REAL ESTATE
INDIVIDUALS	15-30 YEARS	6-7%	TO 75%	REAL ESTATE
	2-5 YEARS	6-8%	TO 80%	MACHINERY, EQUIPMENT AND LIVESTOCK
EQUIPMENT MANUFACTURERS	1 MONTH-5 YEARS	17.5%	100%	MACHINERY AND EQUIPMENT

FIGURE 35-10
Sources of agribusiness credit

sources of credit for financing long-term real estate loans. Loans are made through brokers, correspondents, and company representatives. Insurance companies hold about 15% of agricultural real estate mortgages.

The *Commodity Credit Corporation (CCC)* was administered as an agency of the U. S. Department of Agriculture (USDA). The CCC is a government-owned corporation. Loans are made on eligible commodities such as grain, cotton, peanuts, and tobacco. Farmers use the commodities as security for the loan. Payment of the loan is made by purchasing the loan or delivering the commodity to the CCC.

Farmers Home Administration (FHA) was created by the government during the depression. The FHA's purpose was to assist tenant farmers in becoming land owners. It provides financing to farmers who are unable to secure credit from any other source. The advantages of borrowing money from the FHA are as follows:

1. A large percentage of the total cost of the property can be borrowed.
2. The repayment plan is based on the borrower's ability to repay.
3. Supervision of the loan and assistance with planning are provided.

The *Small Business Administration (SBA)* provides most of its loans for agribusinesses. In 1976 Congress passed legislation that permits the SBA to make agricultural production loans. These loans have a favorable interest rate.

COST OF CREDIT Interest is the major expense in borrowing money. In addition to interest, borrowers are also required to pay other fees. These fees include commissions, recording fees, title certification charges, an insurance fee, and service charges. There are several formulas used for calculating interest (Figure 35-11).

Simple Interest A *simple-interest loan* is made when the full amount of a loan is received by the borrower and is paid back with interest after a short period of time. The borrower generally signs a *promissory note* agreeing to the terms of the loan. When payments are made several times during the note period, interest is charged on the remaining balance of the principal.

Discount Loan The interest on a *discount loan* is subtracted from the principal at the time the loan is made. For example, the borrower would receive only $900 of a $1,000 loan if the interest was 10%. Since the full amount of the principal is not received, a higher interest rate is paid by the borrower.

Add-on Loan An *add-on loan* is the method used for calculating interest on consumer loans. The loan is repaid in installments of equal pay-

FORMULAS FOR CALCULATING INTEREST

SIMPLE INTEREST

Interest Rate = Principal X Rate X Time (I=P X R X T)

DISCOUNTED LOAN

$$\text{Interest Rate} = \frac{\text{Interest (dollar cost)}}{\text{Principal X Time}} \quad \left(R = \frac{I}{P \times T}\right)$$

ADD-ON LOAN

$$\text{Interest Rate} = \frac{2 \text{ X No. of Payments X Interest Charged in \$'s}}{\text{Beginning Principal X Years X (No. of Payments + 1)}}$$

A COMPARISON OF THE DIFFERENT METHODS OF CALCULATING INTEREST

A LOAN OF $600.00 FOR 12 MONTHS AT 10%. REPAID IN 12 EQUAL PAYMENTS.

Simple Interest

$600.00 X .10 X 1 = $60.00 = 10% True Interest

Discounted Loan

$$\frac{60.00}{540.00} \times 1 = .11 \text{ or } 11\% \text{ True Interest}$$

Add-On Loan (Installment Loan)

$600.00 X .10 = $60.00 660.00 = Total Payments

$$\frac{2 \times 12 \times 60}{600 \times 1 \times 13} = .184 \text{ or } 18.4\% \text{ True Interest}$$

FIGURE 35-11
Formulas for calculating credit

ments. Interest is added on at the beginning of the payments. For example, a $1,000 loan at 10% interest would be repaid in monthly installments of $91.67 ($1,000 principal + $100 interest = $1,100/12 = $91.67).

PRINCIPLES OF BORROWING

Most agribusinesses must borrow money to expand their operations and increase income. The decision to expand is not easily reached. A careful examination of alternatives available must be made. Then the best alternative for a particular situation must be chosen.

Seeking a Loan

When seeking a loan, the agriscience manager must be prepared to fully explain the benefits and risks to potential lenders. Many loan applications are not approved because the borrower does not present the details of the agribusiness. Other loan applications may not be approved because the borrower does not effectively present the advantages of expanding. To be effective, the presentation must be organized and the details of the operation should be put in writing. Lenders are concerned about the benefits and risks provided by a loan. Lenders are also concerned about a logical presentation of facts and the desires concerning specific agreements in the loan contract.

Information that should be included in the presentation includes (1) agribusiness plan; (2) business records (income statements, expense records, net worth statement, and financial history); (3) terms of the loan; and (4) method of repayment.

A lender should investigate several sources of credit before selecting a lender. The final selection should be one that best meets the needs of the borrower.

Selecting a Lender

Factors that should be considered in selecting a lender (Figure 35-12) include the:

1. lending institution representative's knowledge of agribusiness problems and practices.
2. lending institution representative's experience in handling agricultural credit of a similar nature.
3. reputation of the lending institution.
4. loan policies (interest rate, repayment schedule, closing costs, penalty clause, optional prepayment clause, and policy regarding failure to meet payment because of circumstances beyond borrower's control).
5. date the loan would be advanced.
6. possibility of increasing the loan.
7. availability of credit for other purposes.

FACTORS TO CONSIDER IN SELECTING A LENDER

* Lending institution representative's knowledge of agricultural practices and problems.

* Lending institution representative's experience in handling agricultural credit of a similar nature.

* Reputation of lending institution.

* Loan policies (interest rate, repayment schedule, closing cost, penalty clause, optional repayment clause, policy regarding late payments).

* Date loan would be advanced.

* Possibility of increasing loan.

* Availability of credit for other purposes.

FIGURE 35-12
Factors to consider when selecting a lender

WAYS IN WHICH BORROWER CAN MINIMIZE RISK

* Use production loans to increase income

* Limit amount borrowed on new or unfamiliar enterprises

* Keep debt as low as possible while still maintaining efficiency

* Keep abreast of markets and trends

* Maintain proper debt-net worth ratio

* Maintain proper debt-income relationship

* Select lender on dependability and terms

* Have a definite repayment schedule

* Be businesslike, fair, and frank

* Reduce risk by carrying adequate insurance

FIGURE 35-13
Ways in which a borrower can minimize risk

SOUND AND WISE USE OF CREDIT There is a chance that the borrower will fail to meet the financial obligations of a loan. In this case, the borrower may lose part or all that is owned. Because of this, the borrower takes a much greater risk than does the lender. The borrower, however, can minimize risk by applying certain rules (Figure 35-13) as follows:

1. Production loans should be used to increase income.
2. The borrower should limit the amount borrowed on new or unfamiliar business ventures.
3. The borrower should keep debt as low as possible while still maintaining efficiency.
4. The borrower should keep abreast of markets and trends.
5. A debt-to-net-worth ratio of 1:1 or less should be maintained.
6. A proper debt-to-income relationship should be maintained. The income must be greater than the principal and interest payments.
7. Dependability and the terms of the loan should be considered when selecting a lender.
8. The borrower should have a definite repayment plan and schedule.
9. The borrower should be businesslike, fair, and frank.
10. The borrower should have adequate insurance to reduce the lender's risk. Property, liability, and life insurance provide partial protection against risk.

The use of management in agriscience is extensive. Management is a vital part of the business and commerce associated with supplies, services, production, processing, distribution, and selling of plant, animal, and natural resource commodities. Careful planning, and the application of sound agribusiness principles and procedures, are necessary for success in many agriscience careers.

Student Activities

1. Define the "Terms To Know."
2. Explain how the eight steps in decision making can be used in planning an FFA or other school money-raising activity.
3. Explain how the principal of resource substitution may be used in making an agricultural mechanics project.
4. Discuss with a banker the procedures used in applying for an agribusiness loan.
5. Arrange to have a banker discuss agribusiness capital and credit in a class presentation.
6. Conduct an FFA or other organization fund-raising activity using business-type planning and management strategies.

Self-Evaluation

A. MULTIPLE CHOICE

1. Most businesses fail because of

   a. death of the manager.
   b. lack of capital.
   c. labor problems.
   d. poor management.

2. Management is considered good if

   a. labor is adequate at all times.
   b. maximum profits are achieved.
   c. the business survives for 5 years.
   d. the business survives for 10 years.

3. The importance of a decision may be measured by the

a. availability of the manager.
b. inventory.
c. potential for gain or loss.
d. time of year.

4. The amount of profits generated by additional inputs is known as

a. capital.
b. diminishing returns.
c. margin.
d. profit.

5. Using a resource in the place of another is known as

a. bailing out.
b. integration.
c. resource management.
d. resource substitution.

6. Loans for 8 to 10 years are called

a. amortized loans.
b. intermediate loans.
c. long-term loans.
d. short-term loans.

7. The most important source of credit is

a. commercial banks.
b. individuals.
c. insurance companies.
d. land banks.

8. The formula for calculating simple interest is

a. $I = P \times R \times T$.
b. $I = P/R \times T$.
c. $I = R/P \times T$.
d. $I = P \times T$.

9. A debt-to-net-worth ratio should not exceed

a. 5:1.
b. 3:1.
c. 2:1.
d. 1:1.

10. The type of insurance that a borrower should have is

a. liability.
b. life.
c. property.
d. all of the above.

B. MATCHING

______	**1.** Capital	**a.**	Money loaned
______	**2.** Price	**b.**	Quantity desired
______	**3.** Supply	**c.**	One year or less
______	**4.** Credit	**d.**	Quantity available
______	**5.** Demand	**e.**	1 to 7 years
______	**6.** Short-term loan	**f.**	Money or property
______	**7.** Intermediate loan	**g.**	Pay interest in the beginning
______	**8.** Generated income	**h.**	Consumptive credit
______	**9.** Used for the individual	**i.**	Amount received
______	**10.** Add-on loan	**j.**	Productive credit

C. COMPLETION–List the eight steps in decision making, in their correct order.

UNIT 36 Entrepreneurship in Agriscience

OBJECTIVE To define the boundaries of entrepreneurship and decide if it is the course for you.

Competencies to Be Developed

After studying this unit, you will be able to

- ☐ define and describe entrepreneurship.
- ☐ describe steps in planning a business venture.
- ☐ state five basic functions performed in the operation of a small business.
- ☐ select a product or service for a personal or group enterprise.
- ☐ determine the basic functions performed by all small business managers.
- ☐ analyze the outcome of a business venture.
- ☐ use small business financial records.
- ☐ analyze the benefits of self-employment versus other types of employment.

TERMS TO KNOW

Entrepreneur
Entrepreneurship
Inventor
Buying function
Selling function
Promoting function
Distribution function
Financing function
Short-term plans
Long-term plans
Actuating
Balance sheet
Assets
Liabilities
Net worth
Profit and loss statement
Inventory report

MATERIALS LIST

variety of publications with pictures
materials to make a collage or bulletin board

Entrepreneurship in agriscience provides extensive career possibilities at all levels of endeavor. It provides opportunities for the inventor, risk taker, profit seeker, owner, manager, and employee. The products and services of entrepreneurs include fresh food, processing plants, equipment sales, fishing, guide services, veterinarian work, forestry work, farming, ranching, teaching, research, mechanics, environmental efforts, law, insurance, real estate, finance, and many others.

THE ENTREPRENEUR The *entrepreneur* is the person who organizes a business, trade, or improves an idea. The word is taken from the French word, entreprendre, which means to undertake. *Entrepreneurship* is the process of planning and organizing a small-business venture. It also involves managing people and resources to create, develop, and implement solutions to problems to meet people's needs. The *inventor* is the person responsible for devising something new or making an improvement to an existing idea or product.

The entrepreneur can be the inventor as well as the small-business manager. However, in many cases, they are different people, each with distinct talents. The entrepreneur functions as the liaison between the inventor and the manager or management team. The entrepreneur brings these two groups together for the purpose of getting the invention to the persons it will serve.

It is the entrepreneur who visualizes the venture strategy and is willing to take the risk to get the venture off the ground. Inventors or business managers are not entrepreneurs unless they organize the venture. An understanding of the differences as well as the similarities among these terms is important to comprehending the role of entrepreneurship as a career option.

FIGURE 36-1
Entrepreneurship provides opportunities and many different jobs in agriscience.

ENTREPRENEURSHIP Various types of business enterprises are part of the entrepreneurship system. The element of individual, partners, or corporate ownership means private control as opposed to government ownership control. The profits (or losses) from entrepreneurship go to the owners. However, owners hire managers and other employees; therefore, entrepreneurships provide jobs for everyone.

A sales project conducted by a school organization such as the FFA is a type of entrepreneurship. However, it differs from a typical small-business venture in the following ways:

1. The sales project should be a valuable learning experience as well as a money-making activity.
2. The sales project is usually of a limited duration. It is completed in a short period of time. A small business typically is designed to operate indefinitely.
3. The sales project involves the voluntary participation of the class. The small business must hire and pay its employees.
4. The sales project usually involves very little risk of financial loss to the individual. The small-business owner (entrepreneur) could lose his or her investment.
5. The sales project may need to be approved by the school administration, but usually does not have to be licensed by the government, pay taxes, or file employee reports.

With these distinctions in mind, you can plan, organize, and implement a sales project and gain an insight into the world of entrepreneurship at the same time.

Operating Businesses Five basic functions are performed in the operation of both a small business and a sales project. These functions represent the basic steps in moving a product or service from the supplier to the consumer. They are listed below.

Buying Function The *buying function* involves selecting a product or service to be marketed or sold for a profit. The selection of a product or service is based on thorough marketing research, which determines consumer (customer) needs and wants. It also determines who, if anyone, is already promoting the product or service.

Selling Function The *selling function* includes studying the product or service to determine the reasons that customers will want and need the product or service. This function also involves developing suitable customer approaches. Planning sales presentations, determining methods of overcoming objections, and planning for the close of the sale are all part of this function.

Promoting Function The *promoting function* involves developing a plan to identify ways to make potential customers aware of the product or service to be offered. Examples of promotional activities include newspaper, TV, radio, and outdoor advertising.

Distribution Function The *distribution function* involves physically organizing and delivering the selected product or service. For products, this includes receiving, storing, and distributing the merchandise. Many of the same activities are involved with services, particularly distribution to the customers.

Financing Function The *financing function* includes obtaining capital for the initial inventory, recording sales, maintaining inventory, computing profit or loss, and reporting the results of the venture. A venture is an activity with some risk involved.

SELECTING A PRODUCT OR SERVICE

The first function in the process of establishing a new business or developing a fund-raising project is the selection of a product or service (buying function). This involves analyzing potential products/services in terms of the potential demand (number of customers). The analysis of potential products or services should answer the following questions:

1. Who are the potential purchasers of the proposed products or services? Customers can often be viewed in groups that have similar interests.
2. On what basis does each group make decisions on products or services? Typical responses to this question are price, quality, service, and other.
3. What are the competing products or services? How do they compare to your products or services relative to price, quality, and service?
4. What is the total estimated demand for your product or service?
5. How much profit would you make on the sale of this product or service? (Multiply the number of units sold by your markup per unit and subtract any expenses you incur in the purchase, sale, and delivery of the product or service.)

AGRI-PROFILE

CAREER AREAS: OWNER/MANAGER/ ASSISTANT MANAGER/MANAGEMENT TRAINEE

The counter is the focal point of most agribusinesses. Here businessperson meets customer and the transaction is completed, usually with the help of a computer. *(Courtesy of the National FFA Organization)*

Agribusiness management is seen by many as an ideal career. It has the advantage of continuing in the work of the family where one gains experience as they grow up. Or, one may start a business and work in a locality where there may not be jobs in specialized areas.

Many high school and college students are well established in an agribusiness before they finish their education. Some popular agribusinesses include lawn services, logging, lumber business, greenhouse or nursery operations, machinery repair, hunting and trapping, agricultural supplies, home and garden centers, florist shops, retail flower sales, livestock sales, farming, and ranching.

Preparation for careers in agribusiness management includes both formal education and on-the-job training. High school agribusiness and agriscience programs provide excellent training for business. Such programs should include classroom, laboratory, supervised agriscience experience, and leadership development. Advanced agribusiness programs may be taken at a technical school, college, or university to obtain appropriate training in economics, finance, and management.

ORGANIZATION AND MANAGEMENT

The role of a venture's organization and management is to make things happen so the venture can achieve its goals. To accomplish this, managers work mainly with data and people. Managing involves getting all the parts of the business—including personnel, marketing strategies, finances, and records—to function together to achieve the venture's goals.

No two managers have jobs that are the same. The jobs are shaped by the type of venture and the personality of the individual manager. Most managers perform the same functions. They are:

Planning work
Organizing people and resources for work
Actuating work
Controlling and evaluating work

Planning

When managers make a plan, they set objectives or goals. They recommend policies to achieve the goals, and they develop procedures, methods, or programs to implement those policies. Plans must be constantly reviewed and updated. No matter how thorough, the plans themselves do not guarantee success. In planning, managers must make short-term and long-term plans. *Short-term plans* are accomplished in several days or weeks. *Long-term plans* are accomplished over several months or years.

Organizing

After a plan is made, the work must be organized. Managers must identify the people needed to carry out the plan. The equipment to be used must also be identified.

Actuating

Actuating simply means putting the plan into action. Managers need to inform employees of the plan. All persons involved must understand their roles. Actuation also includes motivating people to work efficiently and effectively and to want to get the job done.

Controlling and Evaluating

Managers must carefully control implementation of the plan once the work begins. The quality and quantity of all results must be evaluated. If the results are satisfactory, work can continue. If problems arise, changes must be made and alternative plans may have to be developed. Managers must make adjustments in personnel, equipment, policies, or procedures whenever they are needed.

In a very small venture, all management functions may be carried out by the same person. As a venture grows and its goals and objectives become more involved, additional managers may be required. At such time, organizational lines of responsibility must be developed and job descriptions written to identify the functions that are performed by each manager.

SMALL-BUSINESS FINANCIAL RECORDS

Financial records are invaluable tools to the entrepreneur and manager. Good financial management will allow the manager to maintain control of the business venture and to increase profits or reduce losses.

Financial records can reveal which items are selling and which are not. These records can also show the extent of success of each person on the sales force. Financial records should indicate the amount of inventory on hand and how much has been sold. How much profit is made and the venture's total value at any time can also be computed with financial records.

The small-business owner or entrepreneur needs and uses more detailed financial records than are needed to maintain the typical sales project for a club or organization. Items such as buildings, fixtures and equipment, credit, debt, and return on investments are examples of financial records maintained by small businesses. However, balance sheets, profit and loss statements, and inventory and sales journals may be very useful for a group project, as well as for a small business. Some typical financial records are discussed in the following paragraphs.

Balance Sheet

A *balance sheet* is a photograph of the business at a given time. It shows the assets, liabilities, and owner's investment on a particular date. The equation for the balance sheet is: assets = liabilities + net worth. *Assets* are anything the venture owns, including cash on hand, equipment, and inventory. *Liabilities* are both current and long-term debts. *Net worth* is the owner's investment in the business including profits as they occur. A sample balance sheet is shown in (Figure 36-2).

Profit and Loss Statement

The *profit and loss statement* projects costs and other expenses against sales and revenue over a period of time. The five basic sections in a profit and loss statement are (Figure 36-3):

Balance Sheet: A photograph of the business at a given time, showing its assets, liabilities, and the owner's investment on a particular date.

a. The equation for the balance sheet is:

Assets = Liabilities + net worth

- Assets included anything the venture owns, including cash on hand, equipment and inventory.
- Liabilities include both current and long-term debts.
- Net worth is the owner(s) investment in the business including profits as they occur.

b. A sample of a balance sheet:

Balance Sheet

Date: ________

ASSETS		LIABILITIES	
Current: Cash	_____	Current: Notes Payable	_____
Merchandise	_____	Accounts Payable	_____
Accounts Receivable	_____		
Fixed: Land	_____	Non-Current Liabilities: Debts more than one year maturity	_____
Building	_____		
Machinery	_____	TOTAL LIABILITIES:	_____
Equipment	_____		
		OWNERSHIP OR NET WORTH	
Other:	_____	Proprietorship Net Worth	_____
		or	
		Partnership Net Worth	
		or	
		Corporate Net Worth	
TOTAL ASSETS:	_____	TOTAL LIABILITIES AND NET WORTH:	_____

FIGURE 36-2
Elements of a balance sheet

1. Total sales
2. Cost of goods sold
3. Gross profit
4. Expenses
5. Net profit

Inventory Report The *inventory report* includes how many units of each product were received, how many were sold, each item's cost, total sales, and profit. It can be used to make decisions such as reducing the price of slow items, reordering other items, and learning which items are yielding the best profit (Figure 36-4).

SELF-EMPLOYMENT VERSUS OTHER FORMS OF EMPLOYMENT

The decision to continue working for someone else or to open a business is a difficult one to make. A way to help make that decision is to look at the advantages and disadvantages of both working for someone else and working for yourself.

The Employee The advantages of being an employee over working for someone else center on security. The salaried employee has no personal financial risk or responsibility to the company for

The profit and loss statement projects costs and other expenses against sales and revenue over a period of time. The five basic sections to a profit and loss statement are:

a. total sales
b. costs of goods sold
c. gross profit
d. expenses
e. net profit

Profit and Loss Statement

For period ending ________________

TOTAL SALES	$75,000
COST OF GOODS SOLD	
Beginning Inventory	55,000
(Plus) Purchases	10,000
(Less) Ending Inventory	15,000
TOTAL COST OF GOODS SOLD	50,000
GROSS PROFIT	25,000
EXPENSES	
Salaries	18,000
Payroll Taxes	2,500
Rent	4,500
Advertising	1,000
TOTAL EXPENSES	26,000
NET PROFIT (LOSS) (BEFORE TAXES)	(1,000)

FIGURE 36-3
Sample profit and loss statement

which he or she works. Employees generally put in regular hours. If they work additional hours, they may be paid overtime for those hours. In addition, employees are guaranteed vacation time and fringe benefits such as life and health insurance. In many cases, they are offered a retirement plan. An employee can count on a somewhat stable life-style, with a fairly accurate idea of what the income will be from year to year. Additionally, the employee may move up the career ladder within the company.

However, there are some disadvantages to working for someone else. The company does not have any financial responsibility to the employee should there be a recession resulting in a cutback in personnel. Some companies have a predetermined salary scale, so the employee could remain at a particular salary level until the right combination of years or experience is met. This is also true of promotions that could be based on years accumulated rather than merit. The work pattern could become fairly routine. If no positions open within the company, the employee has to wait until the position opens or look for a position with another company. The company management controls these decisions, not the employee.

Self-Employment and Employer One of the major reasons cited for opening a business is that it gives the owner control over his or her own destiny. The owner has

Item No.	Description	Beginning Inventory	Items Sold	Ending Inventory	Total Cost Beg./Inv.	Total Cost End/Inv.	Total Sales	Total Profit
000	Tick Tack Tote	45	45	0	33.75	.00	45.00	11.25
001	Brown Ashtray	32	32	0	25.60	.00	48.00	22.40
002	Fish Ashtray	20	20	0	15.00	.00	30.00	15.00
003	Lion Ashtray	6	6	0	4.80	.00	9.00	4.20
004	Guitar Ashtray	8	8	0	6.40	.00	12.00	5.60
005	Asst. Animal Ashtray	34	34	0	25.50	.00	51.00	25.50
006	Elephant Ashtray	10	10	0	5.00	.00	15.00	10.00
007	Screwdriver Set	22	22	0	11.00	.00	33.00	22.00
008	Brush & Shoehorn	36	36	0	32.40	.00	81.00	48.60
010	Fully Auto. Umbrella	9	9	0	20.70	.00	31.50	10.80
011	Auto. Umbrella	8	8	0	16.00	.00	24.00	8.00
012	Dad's No. 1 Keychain	200	190	10	90.00	4.50	190.00	104.50
013	Grandpa Keychain	100	79	31	45.00	13.95	79.00	47.95
014	Dad's Pad	48	30	18	28.80	10.80	45.00	27.00
015	Mini Screwdriver Set	75	75	0	60.00	.00	150.00	90.00
016	Dad's Plaque	72	72	0	28.80	.00	108.00	79.20
017	Tic Tac Toe	36	36	0	21.60	.00	36.00	14.40
018	Dad's Pen	1	1	0	.80	.00	1.50	.70

Inventory Report: The inventory report identifies how many units of each product were received, how many were sold, each item's cost, total sales, and profit. It can be used to make decisions such as reducing the price of slow items, reordering other items and learning which items are yielding the best profit.

FIGURE 36-4
Sample inventory report

the opportunity to set personal goals and recruit a team to help carry out those goals. Successfully meeting those goals results in a sense of achievement.

Owning the business and being responsible for the decisions related to the business gives the entrepreneur a sense of independence. The owner's ability to make money is not restricted to a particular level.

The disadvantages of owning a business are not as obvious as the advantages. However, the required capital outlay could jeopardize the family savings or even the family home. Also, the number of hours required to run the business will mean a definite commitment from the owner as well as the family members involved. Should the company have difficulty, the responsibility for both the management and the financial problems rests with the owner.

Contributions of Small Businesses Entrepreneurs have been credited with being the cornerstones of the American enterprise system. Furthermore, many see entrepreneurs as the self-renewing agents of the economic environment in the United States. In recent years, numerous socialistic countries of the world, such as China, Russia, and those in eastern Europe, have encouraged entrepreneurship in those countries that have opposed capitalism for decades.

The role of entrepreneurs in the United States is highlighted by the following:

1. Most businesses in the United States (95%) are classified as small by the Small Business Administration.
2. New businesses are formed at a rapid rate (600,000 per year).
3. Small businesses generate almost half (48%) of the U.S. gross national product (GNP).
4. Small business employs almost half (48%) of all U.S. workers.
5. Small business created 60% of the new jobs in our economy in the last two decades.
6. Small business produces 2½ times as many innovations (products, services, techniques) as do larger businesses.

As the American economy continues toward an emphasis on services, the role and importance of small business and entrepreneurship should increase. This should occur because small businesses are especially dominant in the service sector of the economy.

New businesses can be the center of innovation because they are not generally tied to existing ways of doing things. They have a sense of energy and vitality that comes from the entrepreneurial spirit.

The contribution of small business to the American private-enterprise system is important. It is likely to remain so in the future.

Student Activities

1. Define the "Terms To Know."
2. Develop a collage on the bulletin board illustrating entrepreneurship opportunities in agriscience.
3. Make a table showing the advantages and disadvantages of being (1) an employee, and (2) an entrepreneur. The following format is suggested:

Being an Employee	**Being Self-Employed**
Advantages	*Advantages*
1.	1.
2.	2.
etc.	etc.
Disadvantages	*Disadvantages*
1.	1.
2.	2.
etc.	etc.

4. Participate in a group sales project in the FFA, some other school group, 4-H, or some other community group. Encourage the group to operate the project to help all participants obtain useful business skills.

5. Organize a cooperative business within your class or FFA for buying or marketing a product of interest to the group.
6. Become familiar with the record book and recordkeeping system, provided for members of your local FFA chapter or 4-H club, for keeping records on individual projects.
7. Become an entrepreneur—own and operate your own business venture.

Self-Evaluation

A. MULTIPLE CHOICE

1. The selection of a product or service to be sold for profit is called the
   a. buying function.
   b. distribution function.
   c. promoting function.
   d. selling function.
2. Determining reasons that customers may wish to buy a product or service is called the
   a. buying function.
   b. distribution function.
   c. promoting function.
   d. selling function.
3. Development of a plan to identify ways to make potential customers aware of a product or service is called the
   a. buying function.
   b. distribution function.
   c. promoting function.
   d. selling function.
4. When analyzing competing products or services, which is NOT a factor?
   a. price
   b. quality
   c. service
   d. supply
5. Which is not a function of managers?
   a. actuating
   b. controlling and evaluating
   c. organizing people and resources
   d. setting policy
6. A disadvantage of being an employee is
   a. overtime pay.
   b. regular hours.
   c. little or no control over future job.
   d. fringe benefits.
7. A major advantage of owning a business is
   a. control over your destiny.
   b. financial responsibility.
   c. the relationship of business income and family finances.
   d. a shorter work week.

B. MATCHING

______	**1.** Entrepreneur	**a.** Business venture
______	**2.** Entrepreneurship	**b.** Organizer; risk taker
______	**3.** Balance sheet	**c.** Expenses versus revenues
______	**4.** Profit and loss statement	**d.** One who devises
______	**5.** Inventor	**e.** Photograph of a business

Glossary

Abiotic — non-living.
Abortion — loss of a fetus before it is viable.
Accent — distinctive feature or quality.
Accent color — attention-getting color.
Acid — pH of less than 6.9.
Acidity — sourness.
Active ingredient — a component which achieves one or more purposes of the mixture.
Actuating — putting a plan into action.
Acute toxicity — a measurement of the immediate effects of a single exposure to a chemical.
Add-on-loan — the method used for calculating interest on consumer loans. The loan is repaid in installments of equal payments. Interest is added-on at the beginning of the payments.
Adjourn — a motion used to close a meeting.
Adventitious roots — root other than the primary root or a branch of a primary root.
Aeration — the mixing of air with water or soil to improve the oxygen supply of plants and other organisms.
Aeroponics — the plant roots hang in the air and are misted regularly with a nutrient solution.
Aerosol — a can with contents under pressure.
Age or ripen — leave undisturbed for a period of time.
Aggregate culture — a material such as sand, gravel, or marbles supports the plant roots.
Aggregates — soil units containing mostly clay, silt, and sand particles held together by a gel-type substance formed by organic matter.
Agribusiness — commercial firms that have developed with or stems out of agriculture.
Agribusiness management — the human element that carries out a plan to meet goals and objectives in an agriscience business.
Agricultural — business, employment or trade in agriculture, agribusiness or renewable natural resources.
Agricultural economics — management of agricultural resources, including farms and agribusinesses.
Agricultural education — teaching and program management in agriculture.
Agricultural engineering — application of engineering principles in agricultural settings.
Agriculture — activities concerned with the production of plants and animals, and the related supplies, services, mechanics, products, processing, and marketing.
Agriculture/agribusiness and renewable natural resources — broad range of activities in agriculture.
Agriscience — all jobs relating in some way to plants, animals, and renewable natural resources. Also, the application of scientific principles and new technologies to agriculture.
Agriscience mechanics — design, operation, maintenance, service, selling, and use of power units, machinery, equipment, structures, and utilities in agriscience.
Agriscience processing, products, and distribution — industry which hauls, grades, processes, packages, and markets commodities from production sources.
Agriscience professions — professional jobs dealing with agriscience situations.
Agriscience supplies and services — businesses that sell supplies and agencies that provide services for people in agriscience.
Agronomy — science of soils and field crops.
Air — colorless, odorless, and tasteless mixture of gases.
Air layering — plant propagation by girdling a plant stem, wrapping with sphagnum peat and protecting with plastic.
Alkaline — pH of more that 7.1
Alkalinity — sweetness.
Alluvial — soils transported by streams.
Amend — a type of motion used to add to, subtract from, or strike out words in a main motion.
Amendment — in addition to; change in.
Ammonia/Nitrite/Nitrate — the chemical components generated during biological breakdown of animal wastes.
Amphibian — organisms that complete part of their life cycle in water and part on land.
Anemia — low red blood cell count often caused by a deficiency of iron, copper, or pyridoxine.
Angiosperm — a plant with its seeds enclosed in a pod or seed case.
Angora — wool from wool-producing rabbits.
Animal science technology — use of modern principles and practices in animal growth and management.
Animal sciences — animal growth, care and management.
Anions — ions that are negatively charged.
Annual weed — a weed which completes its life cycle within one year.
Anorexia — the result of too little nutrition.
Anther — portion of the male part that contains the pollen.

Antibiotic — substance used to help prevent or control certain diseases of animals.

Apiary — area where bee hives are kept.

Appressorium — the swollen tip of a fungal hypha which allows the fungus to attach itself to a plant.

Aquaculture — raising of finfish, shellfish and other aquatic animals under controlled conditions. Also, the management of the aquatic environment for production of plants and animals.

Aquaculturist — trained professional involved in the production of aquatic plants and animals.

Aquifer — water-bearing rock formation.

Area of cell division — area of the root tip where new cells are formed.

Area of cell elongation — portion of the new root where the cells start to become specialized and begin their function.

Area of cell maturation — area where cells mature.

Arid — an area deficient in rainfall; dry.

Arthropod — eight-legged animals such as spiders and mites.

Artificial insemination — the placing of sperm cells in contact with female reproductive cells by a method other than natural mating.

Asbestos — heat- and friction-resistant material.

Asexual — propagation utilizing a part or parts of one parent plant.

Assets — anything the business owns.

Asymmetrical — not equal on both sides of center.

Auction markets — markets where products are sold by public bidding.

Auctioneer — person who conducts the sales at auction markets.

Axillary bud — bud that occurs in the axil of the leaf.

B & B — an abbreviation for balled and burlapped.

B horizon — soil below the A horizon or topsoil and generally referred to as subsoil.

Bacteria — one-celled microscopic plants.

Balance — state of quality and calm between items.

Balance sheet — statement of assets, liabilities, and owner's investment.

Balled and burlapped — plants that have been dug in the field, wrapped in burlap, and tied with a heavy cord.

Balling gun — a device used to place a pill in an animal's throat.

Band application — placing fertilizer about 2″ to one side and slightly below the seed.

Bantam — miniature chicken.

Barbed wire — wire with sharp points used to discourage livestock from touching fences.

Bare rooted — plants that are dug and the soil is shaken or washed from the roots.

Barrow — castrated male of the swine family.

Basic color — background color.

Bays — natural open water areas along coastlines where freshwater and seawater mix.

Bedrock — the area below horizon C consisting of larger soil particles.

Beef — meat from cattle.

BelRus — superior baking potato bred to grow well in the Northeast.

Beltsville Small White — breed of turkey that weighs only 8 to 12 pounds at maturity.

Biennial weed — a weed that will live for two years.

Binomial — having two names.

Biochemistry — chemistry as it applies to living matter.

Biological control — pest control that uses natural control agents.

Biology — basic science of the plant and the animal kingdoms.

Biotechnology — management of the characteristics transmitted from one generation to another.

Biotic diseases — diseases caused by living organisms.

Bit — metal mouthpiece used to control the horse.

Blade — the upper portion of the turfgrass leaf.

Blanching — the brief scalding of food prior to freezing.

Bleeding out — draining blood from an animal.

Blemish — in horses, any abnormality that does not affect the use of the horse.

Block beef — meat sold over the counter to consumers.

Boar — male of the swine family.

Board foot — a unit of measurement for lumber that equals 1″ by 12″ by 12″.

Bogs — water-logged areas.

Bone — the main component of the skeletal system.

Bone marrow — material inside of bones that makes blood cells.

Border plant — a planting that is used to separate some part of the landscape from another. It might be used as a fence or a windbreak.

Brackish water — waters influenced by tide and river flow with intermediate salinity of 3–22%.

Bracts — modified leaf that is often brightly colored and showy.

Bran — skin or covering of a wheat kernel.

Breed — a group of animals having similar physical characteristics that are passed on to their offspring.

Broadcast planter — scatters seed in a random pattern.

Broadcasting — scattering seeds rather than sowing them in rows.

Broiler — young chicken grown for meat.
Buck — male of the goat or rabbit family.
Bulbs — short underground stem surrounded by many overlapping, fleshy leaves.
Bull — male of the cattle family.
Bunch-type grass — a turfgrass which grows in clumps.
Business meeting — a gathering of people working together to make decisions.
Butterfat — fat found in milk.
Buying function — selection of a product or service to be marketed or sold for a profit.
By-product — a secondary product left from the production of a primary commodity.
C horizon — soil below the B horizon; it is important for storing and releasing water to the upper layers of the soil.
Calf — young member of the cattle family.
Calve — to give birth in cattle.
Calyx — group of sepals of a flower.
Cane cutting — stems are cane like and cuttings are cut into sections that have one or two eyes or nodes.
Canning — storing food in airtight containers.
Cannula — blunt needle.
Canopy — the top of the plant that has the framework and leaves.
Canter — in horses, a fast 3-beat gait.
Capability class — soil classification indicating the most intensive, but safe land use.
Capability subclass — soil group within a class designated by a small letter.
Capability unit — soil group within a subclass.
Capillary water — water held by soil particles and available for plant use.
Capital — money or property.
Capital investment — money spent on commodities that are kept six months or longer.
Capon — castrated male chicken used for meat.
Carbohydrate — starches and sugars that provide energy in the diet.
Carbon dioxide — by-product of respiration.
Carbon monoxide — colorless, odorless and highly poisonous gas; carbon dioxide and water combine to make plant food and release oxygen.
Carcass — body of meat after the animal has been eviscerated.
Carcinogen — a chemical capable of producing a tumor.
Carrying capacity — the number of animals that a pasture will provide feed for.
Cash crop — a crop grown for cash sale.
Castration — removal of the male sex organs.
Cations — ions that are positively charged.
Cattle — members of the bovine family over one year old.
Causal agent — an organism which produces a disease.
Cell — a unit of protoplasmic material with a nucleus and walls.
Cellulose — woody fiber parts that make up plant cell walls.
Central nervous system — the brain and the spinal cord.
Cereal crop — grasses grown for their edible seeds.
Chalaf — a special knife used to slit the throat of animals in kosher slaughter.
Cheese — milk that is exposed to bacterial fermentation.
Chemical control — the use of pesticides for pest control.
Chemistry — science dealing with the characteristics of elements or simple substances.
Chevon — meat from goats.
Chick — newborn chicken or pheasant.
Chlorofluorocarbons — group of compounds consisting of chlorine, fluorine, carbon and hydrogen used as aerosol propellants and refrigeration gas.
Chlorophyll — green pigment in leaves.
Chloroplast — membrane-bound body inside a cell containing chlorophyll pigment; necessary for photosynthesis.
Chlorosis — yellowing of the leaf.
Chromosome — the rod-like carrier for genes.
Chronic toxicity — a measurement of the effect of a chemical over a long period of time and under lower exposure doses.
Circulation — activity in a plantscaped area.
Circulatory system — the system that provides food and oxygen to the cells of the body and filters waste materials from the body.
Clean culture — any practice that removes breeding or over-wintering sites of a pest.
Clearcut — removal of all marketable trees from an area.
Climate — the weather conditions of a specific region.
Climatic conditions — temperature, temperature range, and precipitation.
Clods — a lump or mass of earth.
Clone — an exact duplication of the parent.
Closebreeding — mating of father to daughter, mother to son, or brother to sister.
Coach horse — type of horse developed to pull stage coaches.
Coarse texture — a turfgrass with a wide leaf blade.

Coccidiosis — disease of poultry which costs growers nearly $300 million a year.
Cock — adult male chicken or pheasant.
Cockerel — young male chicken or pheasant.
Collagen — chief component of connective tissue.
Collar — light green or white banded area on the outside of a leaf blade.
Colluvial — soils deposited by gravity.
Color breed — breed of horses based on color.
Colostrum — first milk produced by mammals; high in antibodies.
Colt — young male horse or pony.
Comb — wax material on and in which bees store honey.
Combine — machine which cuts and threshes grain in the field.
Commensalism — one type of wildlife living in, on, or with another without either harming or helping it.
Commission — fee for selling a product.
Commodity Credit Corporation — lends money for production of farm commodities.
Common name — name given to a pesticide by a recognized committee on pesticide nomenclature.
Comparative advantage — putting the emphasis in the area where the most returns will be received.
Competition — two types of wildlife eating the same source of food.
Complete fertilizer — a fertilizer having nitrogen, phosphorus and potassium.
Compost — mixture of partially decayed organic matter.
Compound — a chemical substance that is composed of more than one element.
Compound leaf — two or more leaves arising from the same part of the stem.
Concentrate — feeds high in total digestible nutrients and low in fiber.
Condensed milk — milk that has had water removed and sugar added.
Condominium — apartment building or unit in which the apartments are individually owned.
Conifer — evergreen trees with needle-like leaves.
Conservation tillage — techniques of soil preparation, planting and cultivation that disturbs the soil the least and leaves the maximum amount of plant residue on the surface.
Consumers — people who use a product.
Contact herbicide — a herbicide that will not move or translocate within the plant.
Contagious — diseases that can be spread by contact.
Container grown — plants that are grown in pots or other type of container and are shipped in the container.
Contaminate — to add material that will change the purity or usefulness of a substance.
Continuous flow systems — the nutrient solution flows constantly over the plant roots.
Contour — level line around a hill.
Contour practice — operations such as plowing, discing, planting, cultivating and harvesting across the slope and on the level.
Controlled atmosphere — adjusting oxygen and carbon dioxide where food is stored and while it is being transported.
Convection oven — an oven which heats food by circulating hot air.
Conventional tillage — land is plowed, turning over all crop residues.
Cool season turfgrass — turfgrass adapted to the northern regions of the United States which grow best at 60 to 75°F.
Cooperative Extension Service — an educational agency of the United States Department of Agriculture and an arm of land grant state universities.
Cooperatives — groups of producers who join together to market a commodity.
Corms — short, flattened underground stem surrounded by scaly leaves.
Corn picker — machine that removes ears of corn from stalks.
Corolla — collectively all of the petals of the flower.
Cottage cheese — a product made of skimmed milk.
Cotton gin — machine that removes cotton seed from cotton fiber.
Courage — willingness to proceed under difficult conditions.
Course-textured (sandy) soil — loose and single grained soil.
Cover crop — close-growing crop planted to protect the soil and prevent erosion.
Cow — female of the cattle family that has given birth.
Cream — milk containing 40% butterfat.
Credit — money borrowed.
Creep feed — feed provided to young animals to supplement their mother's milk.
Crop rotation — planting of different crops in a given field every year or every several years.
Crop science — use of modern principles in growing and managing crops.
Crossbreed — animal with parents of two different breeds.

Crown — an unelongated stem of major meristematic tissue of turfgrass.
Crumbs — aggregates.
Crustaceans — aquatic organisms with exoskeleton that molt during growth.
Cull — remove from the herd.
Cultivar — a group of plants with a particular species that has been cultivated and is distinguished by one or more characteristics; through sexual or asexual propagation it will keep these characteristics.
Cultivation — the act of preparing and working soil.
Cultural control — pest control that adapts farming practices to better control pests.
Cuticle — top-most layer of the leaf; waxy protective covering of the leaf.
Cutting — vegetative part removed from the parent plant and managed so it will regenerate itself.
Dam — mother.
Deciduous — plants that lose their leaves every year.
Decompose — decay into soil-building materials and nutrients.
Deficiency — less available than needed for optimum growth.
Deficiency diseases — conditions resulting from improper levels or balances of nutrients.
Defoliate — to strip a plant of its leaves.
Dehydration — removing moisture with heat.
Dehydrator — a device for drying food.
Dehydrofrozen — removing moisture after partial cooking and then freezing.
Delta — land created when soil is deposited by water at the mouth of a river.
Demand — the amount of a product wanted at a specific time and price.
Dermatitis — a skin disorder caused by the deficiency of riboflavin, niacin, or biotin.
Development limitations — limitations imposed by resources or use in a plantscape.
Dicotyledon (dicot) — plant with seed leaves.
Digestive system — system that provides food for the body of the animal and for all of its systems.
Diminishing returns — point where each additional unit of input decreases the output or returns.
Dipping — the process of treating animals for external parasites by walking or swimming them through a medicated bath.
Direct sales — refers to the selling of animals directly to processors by the producer.
Disassembly process — dividing the carcass into smaller cuts.
Discount loan — is a loan where interest is subtracted from the principal at the time the loan is made.
Disease triangle — the term applied to the relationship of the host, pathogen, and the environment in disease development.
Diseases — infective agents that result in lowered health in living things.
Disinfectants — materials that destroy infective agents such as bacteria and viruses.
Dissolved oxygen — oxygen that is dissolved in water.
Distribution function — physical organization and delivery of product or service.
Distributor — person or business storing food for transport to regional markets.
DNA (deoxyribonucleic acid) — the substance that serves as the coding mechanism for heredity.
Dock — remove or shorten tails of certain animals.
Doe — female of the goat or rabbit family.
Domestic — of or pertaining to the household.
Domesticate — the act of taming an animal for use of man.
Dominant — gene that expresses itself to the exclusion of other genes.
Donkey — member of the horse family with long ears and a short, erect mane.
Dormant — resting stage, no active growth.
Double-eye cutting — used when the plant has leaves that are opposite.
Draft — animals used for work.
Draft horse — type of horse bred for work.
Drake — male duck.
Drenching — a process of administering drugs orally to animals.
Dressing percentage — the proportion of the live weight of an animal and the eight of the carcass prior to cooling.
Drift — movement of a pesticide through the air to non-target sites.
Drill planter — plants seed in narrow rows at a high population.
Drip line — the edge of the tree where the branches stop.
Drone — male bee.
Drupe — a stone fruit.
Dry heat cooking — surrounding food with dry air while cooking.
Dry matter — material left after all water is removed from a feed material.
Dry period — period of time when a cow is not producing milk.
Duckling — young duck.

Dwarf — tree whose rootstock limits growth to 10 feet or less.
Ear notch — system of permanently marking animals by cutting notches from their ears.
Ease of working — refers to the difficulty in cutting, shaping, nailing, and finishing wood.
Economic threshold — the level of pest damage to justify the cost of a control measure.
Edible — fit to eat or consume by mouth.
Egg — female reproductive cells.
Ejaculate — the amount of semen produced at one time by a male.
Element — a uniform substance which can not be further decomposed by ordinary means.
Embryo — fertilized egg.
Embryo transfer — a process that removes fertilized eggs from a female and places them in another female who carries them until birth.
Endocrine or hormone system — a group of ductless glands that release hormones into the body.
Endosperm — interior of a wheat kernel which will become wheat flour.
Enterprise — commercial establishment to generated profits.
Enthusiasm — energy to do a job and inspiration to encourage others.
Entomology — science of insect life.
Entomophagous insects — are insects that feed on other insects.
Entrepreneur — person organizing a business, trade, or entertainment.
Entrepreneurship — process of planning and organizing a business.
Environment — space and mass around us.
Epidermis — surface layer on the lower and upper side of the leaf.
Equitation — the art of riding on horseback.
Eradicant fungicide — a fungicide applied after disease infection has occurred.
Eradication — complete control or removal of a pest from a given area.
Erosion — wearing away.
Estrogen — hormone that regulates the heat period.
Estrus — heat period or time when female is receptive to the male.
Estuaries — ecological systems including bays, streams and tidal areas influenced by brackish water.
Evaporated milk — milk that has had water removed.
Evaporation — changing from a liquid to a vapor or gas.
Evergreen — plants that do not lose their leaves on a yearly basis.
Evisceration — removal of the viscera.
Ewe — female of the sheep family.
Executive meeting — meeting of the officers to conduct business of the organization between regular meetings.
Exoskeleton — the external body wall of an insect.
Experience — anything and everything observed, done or lived through.
Extemporaneous — speech delivered with little or no preparation.
Extravaginal growth — turfgrass growth in which growth originates from an axillary bud on the crown.
Fabricated, boxed beef — vacuum sealing smaller cuts of meats prior to shipment.
Famine — widespread starvation.
Farmers' Home Administration — assists farmers to become land owners.
Farrier — a person who shoes horses.
Farrow — giving birth in the swine family.
Farrowing crate — special cage or pen used for swine to give birth in.
Fat — nutrients that have 2.25 times as much energy as carbohydrates.
Feces — solid body waste.
Federal Land Banks — provides long-term credit for agriculture.
Feed — animal food.
Feed additive — a non-nutritive substance added to feed to improve growth, increase feed efficiency, or to maintain health.
Feeder calves — calves intended to be fed to heavier weights before slaughter.
Feedlots — areas in which large numbers of animals are grown for food.
Feedstuff — any edible material used for animal feed.
Fermentation — a chemical change that results in gas release.
Fertility — amount and type of nutrients in the soil.
Fertilizer — material that supplies nutrients for plants.
Fertilizer analysis — is the percentage of nutrients by weight in the fertilizer.
Fertilizer grade — percentages of primary nutrients in fertilizer.
Fetus — embryo from the attachment to the uterine wall until birth.
FFA — an intra-curricular organization for students enrolled in agriscience programs.
Fibrous root — one of the two major root systems, consisting of many fine hair-like roots.
Filament — structure that supports the anther.

Filly — young female horse or pony.
Financing function — obtaining capital for a business.
Fine texture — a turfgrass with narrow leaves.
Fine textured (clay) soils — usually forms very hard lumps or clods when dry; plastic when wet.
Finish — fat covering on a market animal.
Flock — group of birds or sheep.
Floriculture — production and distribution of cut flowers, potted plants, greenery, and flowering herbaceous plants.
Flower — reproductive part of the plant.
Flukes — very small flat worms that are parasites.
Foal — newborn horse or pony. Also, to give birth in horses.
Foliage — stems and leaves.
Foliar sprays — liquid fertilizer sprayed directly to the leaves of plants.
Food — material needed by the body to sustain life.
Food additive — anything added to food prior to packaging.
Food chain — interdependence of plants or animals on each other for food.
Food industry — production, processing, storage, preparation, and distribution of food.
Forage — crop plants grown for their vegetative growth and fed to animals.
Forest — large group of trees and shrubs.
Forester — a person who studies and manages forests.
Forestry — industry which grows, manages, and harvests trees for lumber, poles, posts, panels and many other commodities.
Formal — equal in number, size, or texture.
Formulation — the physical properties of the pesticide and its inert ingredients.
4-H — network of clubs directed by Cooperative Extension Service personnel to enhance personal development and provide skill development in many areas.
Free choice — all of the feed that an animal wants to eat.
Free water — water which drains out of soil after it has been wetted.
Freemartin — sexually imperfect female calf (usually sterile) born twin to a male in cattle.
Freeze drying — removing moisture with cold.
Fresh water — water that flows from the land to oceans and contains little or no salt.
Fructose — simple fruit sugar.
Fruit — mature ovary; seed.
Fry — small, newly-hatched fish.
Fungal endophytes — microscopic plants growing within a plant.
Fungi — are plants which lack chlorophyll.
Fungicide — a material used to destroy fungi or protect plants against their attack.
Furrows — grooves made in the soil.
Gait — way of moving.
Galactose — simple milk sugar.
Gallop — fast 3-beat gait.
Gamete — a reproductive cell.
Gander — male goose.
Gelding — castrated male of the horse family.
Gene — a unit of hereditary material located on a chromosome.
General Use Pesticide — a pesticide which poses a minimal amount of risk when applied according to label directions.
Genes — components of cells which determine the individual characteristics of living things.
Genetic control — pest control using resistant varieties of crops.
Genetic engineering — the process of transferring genes from one individual to another individual or organism.
Geneticist — a person who studies genetics.
Genetics — the study of heredity.
Genotype — what the genes look like.
Genus — taxonomic category between family and species.
Genus — (plural of genera) is defined as a closely related and definable group of plants comprising one or more species.
Germ — new wheat plant inside the kernel.
Germinate — a seed sprouting or starting to grow.
Gestation — length of pregnancy.
Giblets — heart, liver, and gizzard of poultry.
Gills — organ of aquatic animal that absorbs oxygen from the water.
Gilt — female of the swine family that has not given birth.
Ginning — the process of removing the seeds from cotton.
Girl Scouts and Boy Scouts — organization which provide opportunities for leadership development and skill development.
Glacial — soils deposited by ice.
Glucose — simple sugar and the building blocks for other nutrients.
Goiter — enlarged thyroid gland caused by a deficiency of iodine.
Goose — female of the goose family.
Gosling — young goose.
Gourmet — sensitive and discriminating taste in food preferences.

Grade — animal that is not purebred or of unknown ancestry.

Grader — person who inspects the food for freshness, size, and quality.

Grades — quality standards.

Gradient — a measurable change in an amount over time or distance.

Grading up — crossing a purebred male with a grade female.

Grafting — joining two parts of a plant together so that they will grow as one.

Grain crop — crop grown for its edible seeds.

Grass waterway — strip of grass growing in the low area of a field where water can gather and cause erosion.

Gravitational water — water that drains out of soil after it has been wetted.

Green manure crop — crop grown to be plowed under to provide organic matter to the soil.

Green revolution — process where many countries became self-sufficient in food production.

Green roughage — plant materials with high moisture content such as pastures and root plants.

Group planting — trees or shrubs are planted together so as to point out some special feature or to provide privacy or a small garden area.

Guard cells — cells that surround the stoma.

Gully erosion — removal of soil to form relatively narrow and deep trenches known as gullies.

Habitat — area or type of environment in which an organism or biological population normally lives.

Hand — unit of measurement for horses equal to 4 inches.

Hand mating — system of breeding where the male and female are kept apart except during mating.

Hardness — wood's resistance to compression.

Hardwood — wood from deciduous trees.

Hardy — the ability of a plant to survive and grow in a given environment.

Harvester — person responsible for taking products from plants in the field.

Harvesting — taking a product from the plant where it was grown or produced.

Hatchery — place where eggs are hatched.

Hay — forages that have been cut and dried to a low level of moisture, used for animal feed.

Haylage — silage made from forages dried to 40 to 55 percent moisture.

Heel cutting — a shield-shaped cut made about halfway through the wood around the leaf and the axial bud.

Heifer — female of the cattle family that has not given birth.

Hen — adult female chicken, duck, turkey, and pheasant.

Herb — plant kept for aroma, medicine, or seasoning.

Herbaceous — a plant whose stem does not turn woody; it is more or less soft and succulent, and not winter hardy.

Herbicide — a substance for killing weeds.

Heredity — the passing on of traits or characteristics from parents to offspring.

Heterozygous — pairs of genes that are different.

Hide — skin of an animal.

High technology — use of electronics and ultramodern equipment to perform tasks and control machinery and processes.

Hinny — cross between a stallion and a jennet.

Hive — place or box in which bees are kept.

Hog — swine more than four months old.

Hoist — lift into position.

Homozygous — pairs of genes that are alike.

Honey — a thick, sweet substance made by bees from the nectar of flowers.

Horizon — layer.

Horizon A — soil located near the surface that is made up of desirable proportions of mineral and organic matter.

Horizon O — natural layer of organic matter on the surface of soils.

Hormones — chemicals that regulate many of the activities of the body.

Horse — member of the horse family 14.2 or more hands tall.

Horsemanship — the art of riding and knowing the needs of the horse.

Horticulture — the science of producing, processing, and marketing fruits, vegetables, and ornamental plants.

Host animal — a species of animal in or on which diseases or parasites can live.

Hothouse lamb — lamb less than three months old.

Humidity — moisture in the air.

Hutch — a rabbit cage or house.

Hybrid — plant or animal offspring from crossing of two different species or varieties.

Hybrid vigor — off-spring of greater strength and potential resulting when two different breeds or varieties are crossed.

Hydrocarbons — organic compounds containing hydrogen and carbon.

Hydrocooling — removal of heat by immersing in cold water.

Hydroponics — the practice of growing plants without soil.

Hygroscopic water — water that is held too tightly by soil particles for plant roots to absorb.
Hyphae — the thread-like vegetative structure of fungi.
Ice-minus — bacteria genetically altered to retard frost formation on plant leaves.
Imbibition — the absorption of water into the cell causing it to swell.
Immune — not affected by.
Impatiens — popular, easy-to-grow, summer flowering plant.
Imperfect flower — flower that is missing one or more of the following parts; stamen, pistil, petals, or sepals.
Implant — to place a substance under the skin to improve growth of animals.
Improvement by selection — picking the best plants or animals for producing the next generation.
Improvement projects — projects that improve beauty, convenience, safety, value, or efficiency learned outside of the regularly scheduled classroom or laboratory classes.
In vitro — fertilization of eggs outside of the body.
Inbreeding — mating of animals that are related.
Incomplete dominance — neither gene expresses itself to the exclusion of the other.
Incorporated — mixed into the soil by spading, plowing, or discing.
Induction — a specific set of conditions must occur to cause flower development.
Inert — inactive.
Inflorescence — are the flowers and ultimately the seed area of the plant.
Informal — when speaking of landscaping it refers to planting areas that contain elements that are not equal in number, size, or texture.
Infusion — the process for treating udder problems through the teat canal.
Inheritability — the capacity to be passed down from parent to offspring.
Injection — the process of administering drugs by needle and syringe.
Inorganic compound — a compound that does not contain carbon.
Insect — a six-legged animal with three body segments.
Insecticide — a material used to kill insects or protect against their attack.
Insemination — the placement of semen in the female reproductive tract.
Instar — the stage of the insect during the period between molts.
Integrated pest management — pest control program based on multiple-control practices.
Integrity — honesty.
Intermediate loans — loans with payment periods that range from one to seven years.
Internode — area in between two nodes.
Intradermal — between layers of skin.
Intramuscular — in a muscle.
Intraperitoneal — in the abdominal cavity.
Intraruminal — in the rumen.
Intravaginal growth — a type of growth in which shoots develop within the lower leaf sheath at a crown's axillary bud.
Intravenous — in a vein.
Inventor — person devising something new or improving an existing idea or product.
Inventory report — units received and sold, unit cost, total sales, and profit.
Involuntary muscles — muscles that operate in the body without control by the will of the animal.
Ions — atom or a group of atoms that have an electrical charge.
IPM — integrated pest management.
Irrigation — addition of water to plants to supplement that provided by rain or snow.
Jack — male donkey or mule.
Jennet — female donkey or mule.
Jog — in horses, a slow, smooth 2-beat diagonal gait.
Jugular vein — the largest vein in the throat.
Jungle fowl — wild ancestor of the chicken.
Katahdin — popular potato variety of the 1930s.
Key pest — a pest that occurs on a regular basis for a given crop.
Kid — young goat.
Knife application — injection of fertilizer into soil in gaseous form.
Knowledge — familiarity, awareness, and understanding.
Kosher — a Hebrew word meaning right and proper.
Lactation — milk production.
Lactation period — period of time when mammals are producing milk.
Lactose — compound milk sugar.
Lacustrine — soils deposited by lakes.
Lamb — member of the sheep family less than one year old; also meat from young sheep.
Lambing — act of giving birth in sheep.
Lard — rendered pork fat.
Larvae — mobile aquatic organisms that become fixed and grow into non-mobile adults.
Lay on the table — a motion used to stop discussion on a motion until the next meeting. The way to table a motion is to say, "I move to table the motion."

Layer — chicken developed for the purpose of laying eggs.
Layering — a practice inducing formation of roots for new plants while maintaining the health and vigor of the parent plant.
LC_{50} — lethal concentration of a pesticide in the air required to kill 50% of the test population.
LD_{50} — lethal dose of a pesticide required to kill 50% of a test population.
Leached — certain elements have been washed out of the soil.
Lead — is to show the way by going in advance or to guide the action or opinion of.
Leadership — is defined as the capacity or ability to lead.
Leaf — plant part consisting of a stipule, petiole, and blade.
Leaf fat — loose fat on hogs.
Leaf mold — partially decomposed plant leaves.
Legume — plant in which certain nitrogen-fixing bacteria utilize nitrogen gas from the air and convert it to nitrates that the plant can use as food.
Lenticels — pores in the stem that allow the passage of gasses in and out of the plant.
Lethal — genetic characteristic that causes an animal to be born dead or to die shortly after birth.
Liabilities — current and long-term debts.
Light horse — type of horse developed for riding.
Light intensity — brightness of light.
Ligule — is located on the inside of the leaf and is a membranous or hairy structure.
Lime — material which reduces the acid content of soil and supplies nutrients such as calcium and magnesium to improve plant growth.
Linebreeding — mating of individuals with a common ancestor.
Linen — fabric produced from fibers in flax plants.
Linseed oil — oil produced from flax seed.
Litter — group of young born at the same time with the same parents.
Loamy — a granular soil containing a good balance of sand, silt and clay.
Loess — soils deposited by wind.
Long-term loans — loans with payment periods that range for 8 to 40 years.
Long-term plans — plans accomplished over months or years.
Lope — in horses, a very slow canter.
Loyalty — reliable support for an individual, group, or cause.
Lumber — boards cut from trees.
Lymph glands — glands that secrete disease-fighting materials for the body.
Macronutrients — elements used in relatively large quantities.
Main motion — a basic motion used to present a proposal for the first time. The way to state it is to get recognized and then say, "I move . . ."
Malting — process of preparing grain for the production of beer and alcohol.
Maltose — compound malt sugar.
Mammal — milk-producing animal.
Mammary system — milk production system in the female.
Manage — is to use people, resources, and processes to reach a goal.
Mange — crusty skin condition caused by mites.
Mare — adult female horse or pony.
Margins — edge of the leaf.
Mastitis — infection of the milk-secreting glands of cattle, goats, and other milk production animals.
Maturity — the state or quality of being fully grown.
Mechanical control — pest control that affects the pest's environment or the pest itself.
Media — material that provides plants with nourishment and support through their root systems.
Medium — surrounding environment in which something functions and thrives.
Medium-textured (loamy) soil — a relatively even mixture of sand, silt, and clay.
Meiosis — cell division that results in the formation of gametes.
Meristematic tissue — plant tissue responsible for plant growth.
Mesophyll — tissue of the leaf where photosynthesis occurs.
Metamorphosis — the change in growth stages of an insect.
Microbe — living organism that requires the aid of a microscope to be seen.
Micronutrients — elements used in very small quantities.
Microorganisms — tiny plants or animals that may contribute to food spoilage.
Microwave oven — an oven which heats food by using an electromagnetic wave.
Middlemen — people who handle an agricultural product between the farm and the consumer.
Migratory labor — workers who move to places where harvesting is occurring.
Milk — white or yellowish liquid produced by the mammary glands for the purpose of feeding young.
Milk replacer — dried dairy or vegetable products that are mixed in warm water and fed to calves in place of milk.

Milking machine — machine that milks cows and goats.
Mineral — element essential for normal body functions.
Mineral matter — non-living items such as rocks.
Minimum tillage — soil is worked only enough so that seed will germinate.
Minutes — is the name of the official written record of a business meeting.
Misuse statement — pesticide label information which reminds the user to follow the label directions when using the products.
Mitosis — simple cell division for growth.
Mode of action — how a pesticide affects its target.
Mohair — hair produced by Angora goats.
Moist heat cooking — surrounding the food with liquid while cooking.
Moldboard plow — plow with a curved bottom that will turn prairie soils.
Molt — softening and cracking of the exoskeleton so that crustaceans may escape and grow a larger exoskeleton.
Monitor — to check or observe pest numbers and damage.
Monocotyledon (monocot) — plant with one seed leaf.
Monogastric — animal with a single compartment stomach.
Mosaic — a virus disease of plants where a leaf shows a symptom of light and dark green mottling of the foliage.
Mulch — material placed on soil to break the fall of raindrops (preventing erosion), prevent weeds from growing, or improve the appearance of the area.
Mule — cross between a jack and a mare.
Muscular system — the lean meat of the animal.
Mutation — change in genes.
Mutton — meat from mature sheep.
Mutualism — two types of wildlife living together for the mutual benefit of both.
Mycelium — a collection of fungal hyphae.
Natural fisheries — existing breeding groups of wild fishes, i.e. oceans, continental shelves, lakes and bays.
Natural service — male mates directly with the female.
Nematode — a type of roundworm.
Net contents — the amount of pesticide in the container.
Net worth — owner's investment in the business.
Neutral — neither acid nor alkaline.
Nitrates — a form of nitrogen used by plants.
Nitrogen-gas laser — device used to determine wavelengths given off by plants.
Nitrous oxides — compounds containing nitrogen and oxygen; they make up about 5% of the pollutants in automobile exhaust.
No-till — seed is planted directly into the residue of the previous crop, without exposing the soil surface.
Node — portion of the stem that is swollen or slightly enlarged that gives rise to buds.
Nomenclature — a systematic method of naming plants or animals.
Non-selective herbicide — a chemical that kills all plants it comes in contact with.
Noncontagious — diseases that cannot be spread to other animals.
Noxious weed — plant that is prohibited by state law.
Nurse crop — a crop used to protect another crop until it can get established.
Nursery — a place where young trees, shrubs, and other plants are grown.
Nut — a fruit or seed contained within a removable outer cover
Nutriculture — water culture.
Nutrients — substances necessary for the functioning of an organism.
Nutrition — the process where all body parts receive materials needed for function, growth, and renewal.
Obesity — the result of too much or improper types of food being eaten.
Offal — materials thrown away during slaughter including the head, blood, viscera, loose fat, etc.
Oil meal — by-product of the production of vegetable oil.
Oilseed crop — crop produced for the oil content of their seeds.
Olericulturalist — one who studies the cultivation of vegetables.
Olericulture — the cultivation of vegetables.
On-the-job training — experience obtained while working in an actual job setting.
Order of business — refers to the items and sequence of activities conducted at a meeting.
Organic compound — a compound that contains carbon.
Organic food — food that has been grown without the use of certain chemical pesticides.
Organic matter — dead plant and animal tissue that originates from living sources such as plants, animals, insects, and microbes.
Ornamental plant — used to improve the appearance of a structure or an area.

Ornamentals — science of fruits, vegetables, and ornamental plants.
Osmosis — movement of materials through a semipermeable membrane.
Osteomalacia — weak or brittle bones in mature animals caused by a deficiency of calcium, phosphorus, or vitamin D.
Outcrossing — mating of unrelated animal families within one breed.
Ovary — female organ that produces eggs or female sex cells; also, that portion of the flower that contains the ovules or seeds.
Over grazing — damage to plants or soil due to animals eating too much of the plants at one time or reducing the plant's ability to hold soil or recover after grazing.
Overripe — beyond maturity.
Overseeding — seeding a second crop into one that is already growing.
Ovulation — process of releasing mature eggs from the ovary.
Ovule — unfertilized seed.
Oxidative deterioration — decay resulting from exposure to air.
Ozone — compound which exists in limited quantities about 15 miles above the earth's surface.
Pace — 2-beat side to side gait.
Packer — person or firm responsible for preparing commodities for shipment.
Palisade cells — elongated, vertical cells that give the leaf strength and manufactures food.
Parakeratosis — a skin disease similar in appearance to mange caused by a deficiency of zinc in the diet.
Parasite — organism which lives in or on another organism with no benefit to the host.
Parasitism — one type of wildlife living and feeding on another without killing it.
Parent material — the horizon of unconsolidated material from which soil develops.
Parliamentary procedure — is a system of guidelines or rules for making group decisions in business meetings.
Parturition — birth.
Pasture — forages that are harvested by the livestock itself.
Pasture mating — male allowed to roam freely with the herd and mate at random.
Pathogen — organisms which produce disease.
Peat moss — a type of organic matter made from sphagnum moss.
Pedigree — record of ancestry.
Pelt — skin of an animal with the hair attached.
Pencil shrink — the estimated amount of loss that an animal will sustain during the trip to market.
Percolation — movement of water through the soil.
Perennial weed — a weed that lives for more than two years.
Perfect flower — flower containing all of the parts; stamen, pistil, petals, and sepals.
Peripheral nervous system — system that controls the functions of the body tissues including the organs.
Perlite — natural volcanic glass material having water-holding capabilities.
Permeable — permitting movement.
Pest — any organism which adversely affects man's activities.
Pest resurgence — ability of a pest population to recover.
Pesticide — chemical used to control pests.
Pesticide resistance — the ability of a pest to tolerate a lethal level of a pesticide.
Petals — brightly colored, sometimes fragrant portion of the flower.
Petiole — stem of the leaf.
pH — measurement of acidity or alkalinity from 1 to 14.
Phenotype — physical appearance of an individual.
Phloem — cells that carry manufactured food to areas of the plant where it is stored or used.
Phloem tissue — conductive tissue responsible for transport of carbohydrates.
Photodecomposition — chemical breakdown caused by exposure to light.
Photosynthesis — process where chlorophyll in green plants enable those plants to utilize light.
Photosynthesis — use of light to manufacture sugar from carbon dioxide and water in plants.
Physiology — study of the functions and vital processes of living creatures and their organs.
Pig — swine less than four months of age.
Piglet — newborn of the swine family.
Pistil — female part of the flowers consisting of stigma, style, ovary, and ovule.
Plan — to think through.
Plant disease — any abnormal plant growth.
Plant fertilization — addition of elements to a plant-growing environment.
Plant Hardiness Zone map — a map developed by the U.S. Department of Agriculture, dividing the country into 10 zones based on average winter temperatures.
Plant nutrition — provision of elements to plants.
Playability level — is the suitability for the intended recreational or sporting use of the turf.

Plugging — establishment of turf by using small pieces of existing turfgrass.
Plywood — construction material made of thin layers of wood glued together.
Point of order — a procedure used to object to some item in or about the meeting that is not being done properly. The procedure to use is to say, "Point of order!" The presiding officer should then recognize the member by saying, "State your point."
Polled — hornless, either naturally or otherwise.
Pollen — small male sperm or grains that are necessary for fertilization in the flower.
Pollination — transfer of pollen from anther to stigma.
Polluted — containing harmful chemicals or organisms.
Pome — fleshy fruits with embedded core and seeds.
Pomologist — a fruit grower or scientist.
Pomology — the science and practices of fruit growing.
Pony — member of the horse family less than 14.2 hands tall.
Pork — meat from swine.
Port — town having a harbor for ships to take on cargo.
Postemergence herbicide — a herbicide applied after the weed or crop is present.
Potable — drinkable, that is, free of harmful chemicals and organisms.
Poult — young turkey.
Poultry — group name for domesticated birds.
PPM — parts per million.
PPT — parts per thousand.
Precipitate — a chemical action that results in the dropout of solids in a solution.
Precipitation — formation of rain and snow.
Precooling — rapid removal of heat before storage or shipment.
Predation — a way of life where one type of wildlife eats another type.
Predator — an animal that feeds on a small or weaker animal.
Preemergence herbicide — a herbicide applied prior to weed or crop germination.
Presiding officer — is a president, vice president, or chairperson who is designated to lead a business meeting.
Prey — animal eaten by another animal.
Price — amount received for an item that is sold.
Primary nutrients — in agriculture: nitrogen, phosphorous, and potash.
Processed meats — meat that is not suitable for sale over the counter.
Processing — turning raw agricultural products into consumable food.
Processor — person or business cleaning, separating, handling, and preparing a product before it is sold to the distributor.
Producer — person who grows the crop.
Production agriscience — farming and ranching.
Production Credit Association — provides short-term credit for agriculture.
Production or productive project — enterprise conducted for wages or profit.
Profession — occupation requiring an education: especially law, medicine, teaching, or the ministry.
Profile — a cross-section view of soil.
Profit and loss statement — projection of costs and expenses against sales and revenue over time.
Progeny — offspring.
Progesterone — hormone that prevents estrus during pregnancy and causes development of the mammary system.
Program — total plans, activities, experiences, and records.
Project — a series of activities related to a single objective or enterprise.
Promissory note — note signed by borrower agreeing to the terms of the loan.
Promoting function — plan to make potential customers aware of the product or service.
Propagated — the increase or multiplication of animals and plants by sexual or asexual methods.
Propagation — process of increasing the numbers of a species or perpetuating a species.
Protectant fungicide — a fungicide applied prior to disease infection.
Protein — nutrient made up of amino acids and essential for maintenance, growth, and reproduction.
Protozoa — microscopic one-celled animals that are parasites of animals.
Pruning — the removal of dead, broken, unwanted, diseased, or insect-infested wood.
Pullet — young female chicken.
Pulpwood — wood used for making fiber for paper and other products.
Purebred — animal with both parents of the same breed and with registration papers.
Purify — removal of all foreign material.
Putting green — golf playing area with turfgrass mowed very short.
Quality grade — grade based on amount and distribution of finish on an animal.

Quarantine — isolation of pest-infested material.
Queen — only fertile, egg laying female bee in the hive.
Rack — in horses, a fast 4-beat gait.
Radioactive material — material that is emitting radiation.
Radon — colorless, radioactive gas formed by disintegration of radium.
Ram — male member of the sheep family.
Ratio — proportion.
Ration — the amount of feed fed in one day.
Real world experience — an activity conducted in the daily routine of our society.
Reaper — machine which cuts grain.
Recessive — gene that expresses itself only in the absence of a dominant gene.
Rectum — the last organ in the digestive tract.
Recuperative potential — is the ability of the plant to recover after being damaged.
Refer — a motion used to refer some other motion to a committee or person for finding more information and or taking action on the motion on behalf of the members. The way to state a referral is to say, "I move to refer this motion to . . ."
Refrigeration — keeping cool.
Registration papers — records proving that an animal is purebred.
Reins — leather or rope lines attached to the bit for control of the horse.
Relative humidity — relationship between the actual amount of water vapor in the air to the greatest amount possible, at a given temperature.
Render — cook and press the oil from.
Rendering insensible — making the animal unable to feel pain.
Renewable natural resources — resources provided by nature that can replace themselves.
Reproduction — the making of a new plant or animal.
Residual soils — parent materials formed in place.
Resource substitution — the use of a resource or item for another when the results are the same.
Resources inventory — summary of the resources that may be available for conducting a SAEP.
Respiration — a process by which living cells take in oxygen and give off carbon dioxide.
Respiratory system — system that provides oxygen to the blood of the animal.
Restricted use pesticide — a pesticide which poses a greater risk to man and the environment than one that is not so labeled.
Retail marketing — the selling of a product directly to consumers.
Retailer — person or store that sells directly to the consumer.
Retortable pouches — packages with foil between two layers of plastic.
Rhizome — horizontal underground stems.
Rickets — deformity of bones in young animals caused by a deficiency of calcium, phosphorus, or vitamin D.
Roaster — old chicken used for meat.
Rollover — rapid change in water quality.
Rooster — adult male chicken or pheasant.
Root crop — crop grown for the thick, fleshy storage root that it produces.
Root cutting — section of a root cut and used for propagation purposes.
Root pruning — systematic cutting of the roots by hand or machine to encourage the roots to develop within the root ball range.
Rootbound — roots restricted by a container.
Rooting hormone — chemical used to stimulate root formation on a cutting.
Rootstock — root system and stem of a plant on which another plant is grafted.
Roughage — grass, hay, or silage and other feeds high in fiber and low in TDN.
Round worms — slender worms that are tapered on both ends.
Row crop planter — plants seed in precise rows and with even spacing within the row.
Rumen — multi-compartment stomach of cattle and sheep.
Ruminant — animal that has a stomach with more than one compartment.
Saddle — padded leather seat for the rider of a horse.
SAEP — supervised agricultural experience program.
Safe water — water that is free of harmful chemicals and disease-causing organisms.
Salinity — a measure of total dissolved mineral solids in water.
Salmonids — any of the family of soft-finned fish such as salmon or trout that have the last vertebrae upturned.
Saltwater marsh — the ecological systems of plants and animals influenced by tidal waters with salinity of 15–34 PPT.
Saturated — water added until all the spaces or pores are filled.
Scale — size of items.
Scarify — to soak, keep moist, or mechanically scrape a seed coat to aid germination.
Scurvy — vitamin C deficiency disease.

Seasonal — pertaining to a particular season of the year.
Secondary host — a plant or animal which carries a disease or parasite during part of the life cycle.
Secretary — a person elected or appointed to take notes and prepare minutes of the meeting.
Seed blend — combination of different cultivars of the same species.
Seed certification — program administered by state governments to insure seed quality.
Seed culm — stem which supports the inflorescence of the plant.
Seed legume — crop that is nitrogen-fixing and produces edible seeds.
Seed mixture — combination of two or more species.
Seed pieces — vegetative means of propagating sugarcane.
Seedling — young plant grown from seed.
Seining — physical netting of fish in open water.
Selective breeding — selection of parents to get desirable characteristics in the offspring.
Selective herbicide — a chemical that kills or affects certain types of plants.
Selling function — market research, sales plans, and sales closures.
Semen — the sperm cells and accompanying fluids.
Semi-arid — an area partially deficient in rainfall; dry.
Semi-dwarf — tree whose rootstock limits growth to 15 feet or less.
Seminal roots — roots formed at the time of seed germination to anchor the seed in soil.
Semi-permeable membrane — membrane which permits a solution to move through it.
Sepals — small green leaf-like structures found at the base of the flower.
Sequence — related or continuous series.
Sewage system — receives and treats human waste.
Sex-linked — genes carried on chromosomes that determine sex.
Sexual reproduction — union of an egg and sperm to produce a seed or fertilized egg.
Shackles — mechanical devices that restrict movement.
Shear — remove wool from sheep.
Sheath — lower portion of the turfgrass leaf.
Sheet erosion — removal of soil from broad areas of the land.
Shehitah — ancient rules of the Jewish faith that describe the slaughter ritual.
Shelf life — time between packaging and spoilage.
Shellfish — any aquatic animal having a shell or shell-like exoskeleton.
Shepherd — one who cares for sheep.
Shod — wearing shoes.
Shohet or shochet — a trained person who can supervise kosher slaughter.
Short-term loans — loans with a payment period of one year or less.
Short-term plans — plans accomplished in days or weeks.
Shrinkage — changes in dimensions of wood as it reacts to changes in humidity and temperature.
Shroud — wrap tightly with a cloth.
Shrubs — woody perennial plants that normally grow, low, produce many stems, or shoots from the base, and do not reach more than 15 feet high.
Side dressing — placing fertilizer in bands about 8″ from the sides of growing plants.
Signal word — the required word on the label which denotes the relative toxicity of the product.
Silage — feed resulting from the storage and fermentation of green crops in the absence of air.
Silo — air-tight storage facility for silage.
Silviculture — the scientific management of forests.
Simple interest — interest calculated when the full amount of a loan is received by the borrower and paid back with interest after a short period of time.
Simple layering — stem is bent to the ground, held in place, and covered with soil.
Simulate — to look like or act like.
Singe — burn lightly.
Single eye — cutting made with the node of a plant with alternate leaves.
Sire — father.
Skeletal system — bones joined together by cartilage and ligaments.
Slaughter — the killing of animals for market.
Slaughter cattle — cattle intended for slaughter.
Small Business Administration — provides loans to agribusinesses.
Smoker — device used to add smoke flavor and taste to food.
Sodding — removal of a portion of the soil and turfgrass plant for vegetative establishment purposes.
Softwood — wood from conifers.
Soil — top layer of the earth's surface suitable for the growth of plant life.
Soil science — study of the properties and management of soil to grow plants.
Sow — female of the swine family that has given birth.
Spat — the young shellfish just after they attach to underwater structures.

Spawn — egg-laying process of fish.
Species — the basic unit in the classification system whose members have similar structure, common ancestors, and maintain their characteristics; subgroup of genus.
Specimen plant — plant that is used as a single plant to highlight it or some other special feature of the landscape.
Sperm cells — male reproductive units.
Sphagnum — pale and ashy mosses used to condition soil.
Split carcass — halves of the animal after it is killed.
Split vein cutting — cutting made by slitting a large leaf on the veins before placing in a rooting medium.
Spoiled — chemical changes that reduce nutritional value, or render food unfit to eat.
Spongy layer — lower, irregular layer in the leaf that allows the veins, or vascular bundle, to extend into the leaf.
Sprigging — planting a section of a rhizome or stolon (sprig) in the soil.
Spring lamb — lamb between three and seven months old.
Sprinklers — devices that spray water into the air to water crops; its effect is much like rain.
Square foot — a unit of area measuring one foot on each side of a square.
Stag — male hog or cattle castrated after reaching sexual maturity.
Stallion — adult male horse or pony.
Stamen — male part of the flower which contains the pollen, anther, and filament.
Starch — major energy source in livestock feeds.
Starter solutions — dilute mixtures of single or complete fertilizers used when plants are transplanted.
Steer — castrated male member of the cattle family.
Stem — supports the leaves and conducts the flow of water and nutrients.
Stem tip cutting — cutting taken from the end of a stem or branch, normally including the terminal bud.
Sterility — inability to reproduce.
Stiff lamb disease — muscle degeneration in lambs caused by a deficiency of vitamin E.
Stigma — part of the pistil that receives the pollen.
Stimulant crop — crop that stimulates the senses of users.
Stipule — a small appendage that is located at the base of the petiole on some species of plants.
Stirrup — attached to the saddle by the stirrup leather, this device is a footrest used when riding horses.
Stolon — a stem which grows above ground.
Stolonizing — type of vegetative establishment in which sprigs are broadcast onto the soil surface.
Stoma — small openings, usually on the lower side of the leaf, that controls movement of gases.
Strip cropping — alternating strips of row crops with strips of close-growing crops.
Structure — refers to the tendency of soil particles to cluster together and function as soil units.
Stuck — cutting of a large blood vessel.
Style — enlarged terminal part of the pistil.
Subcutaneous — under the skin.
Subsurface irrigation — water is supplied to the crop from underground pipes.
Successive — following one right after the other.
Succulent — in horticulture: thick, fleshy leaves or stems that store moisture.
Sucrose — compound cane sugar.
Sugar crop — crop grown for the sugar content of its stems or roots.
Sulfur — pale-yellow element occurring widely in nature.
Super — box in which bees store excess honey.
Supervised — looked after and directed.
Supervised agricultural experience program — SAEP; all supervised agriscience experiences.
Supplement — extra nutrient added to animal feed.
Supplementary agriscience skills — agriscience skills learned anywhere except in the instructional settings at school or in production enterprises or improvement projects in the SAEP.
Supply — the quantity of a product that is available to buyers at a given time.
Surface irrigation — water flows over the soil surface to the crop.
Susceptible — subject to some influence.
Swarm — group of bees with a queen that leaves an overcrowded hive to find a new home.
Sweetbreads — thymus and pancreatic glands of animals.
Symmetrical — in landscaping, a planting that is equal in number, size, and texture on both sides.
Symptom — the visible change to the host caused by a pest.
Synthetic nutrients — nutrients manufactured chemically.
Syringe — an instrument used to give injections of medicine or to draw body fluids from animals.
Syringing — a light application of water to a turfgrass.

Systemic herbicide — a herbicide which is absorbed by the roots of the plant and translocated throughout the plant.
Tack — equipment or gear for horses.
Tact — skill of encouraging others in positive ways.
TAN — measurement of total ammonial nitrogen.
Tankage — dried animal residue after slaughter.
Taproot — large main root of the system; usually has little or no branch roots.
Targeted pest — an identified pest which, if introduced, poses a major economic threat.
Tattoo — means of marking animals for identification.
Taxonomy — systematic classification of plants and animals.
TDN — total digestible nutrients, the measure of digestibility of feed.
Teats — the appendages of an udder.
Technology — application of science to an industrial or commercial objective; also, the equipment, and expertise to cultivate, harvest, store, process, and transport crops for consumption.
Terminal — growing at the end of a branch or stem.
Terminal bud — bud at the end of a twig or branch.
Terminal market — a stockyard that acts as a place to hold animals until they are sold to another party.
Terrace — soil or wall structure built across the slope to capture water and move it safely to areas where it will not cause erosion.
Terrestrial — land organism.
Testes — organs that produce sperm cells.
Testostestrone — male sex hormone.
Tetraethyl lead — colorless, poisonous, and oily liquid.
Texture — visual or surface quality.
Thatch — building of organic matter on the soil and around turfgrass plants.
Thermo — classification of plants according to growing season required.
Threshing — removal of grain from the plant.
Tidewater — water that flows up the mouth of a river as the ocean tide rises and comes in.
Tillable — soil that is workable with tools and equipment.
Tillers — new shoots of a grass plant that develop at the axillary bud of the crown and form within the lower leaf sheath.
Tip layering — tip of a shoot placed in media and covered.
Tissue culture — plant reproduction using very small, actively growing plant parts under sterile conditions and medium.
Tofu — food made by boiling and crushing soybeans and letting it coagulate into curds.
Tom — male turkey.
Topdressing — a procedure where fertilizer is broadcast lightly over close-growing plants.
Topsoil — desirable proportions of plant nutrients, chemicals, and living organisms located near the surface which support good plant growth.
Townhouse — one of a row of houses connected by common side walls.
Toxicity — a measurement of how poisonous a chemical is.
Toxin — a poisonous substance, causing injury to animals or plants.
Tractor — source of power for belt-driven machines as well as for pulling.
Trade name — the manufacturer's name for its pesticide product.
Transpiration — process by which a plant loses water vapor.
Transplants — plants grown from seed in special structures.
Trap crop — a susceptible crop planted to attract a pest into a localized area.
Tree — woody plant that produces a main trunk and has a more or less distinct and elevated head (a height of 15 feet or more).
Tree spade — a piece of equipment, operated by hydraulics, that will dig a tree in a matter of a few minutes to a very specific size ball.
Tripe — the pickled rumen of cattle and sheep.
Trot — in horses, fast 2-beat diagonal gait.
Trucker — person transporting commodities from farm to consumer.
Tuber crop — crop grown for its fleshy, underground storage stem.
Tubers — specialized food-storage stem that grows underground.
Turbidity — measure of suspended solids.
Turf — the turfgrass plant and soil immediately below it; also, grass used for decorative or soil-holding purposes.
Turfgrass — grasses that are mowed frequently for a short and even appearance.
Turgor — swollen or stiff condition as a result of being filled with liquid.
Two-year-old — cattle that are two or more years old.
Udder — the milk secreting glands of the animal.
Underripe — as applied to vegetation, any that is not mature.
Universal solvent — material which dissolves or otherwise changes most other materials.

Unselfishness — placing the desires and welfare of others above oneself.
Unsoundness — in horses, any abnormality that affects the use of the horse.
Urea — synthetic source of nitrogen made from air, water, and carbon.
Urinary system — system that removes waste materials from the blood.
Urine — liquid body waste.
USDA — United States Department of Agriculture.
Utility-type grass — a turfgrass adapted to lower maintenance levels.
Vaccination — the injection of an agent into an animal to prevent disease.
Vacuum pan — a device used to remove water from milk.
Vapor drift — movement of pesticide vapors due to chemical volatilization of the product.
Variegated — a leaf having streaks, marks, or patches of color.
Variety — a subdivision of a species, it has various inheritable characteristics of form and structure that are continued through both sexual and asexual propagation.
Vascular bundle — food, water, and nutrient-carrying tissue extending from roots to tips of plants.
Veal — young calves or the meat from young calves.
Vector — a living organism which transmits a disease.
Vegetable — any herbaceous plant whose fruit, seeds, roots, tubers, bulbs, stems, leaves, or flower parts are used as food.
Vegetative — plant parts such as stems, leaves, roots, and buds, but not flowers, fruits, and seeds.
Vegetative spreading — reproduction in plants other than by seed.
Veneer — thin sheet of wood used in paneling and furniture.
Vermiculite — mineral matter used for starting plant seeds and cuttings.
Vertebrates — animals with backbones.
Vertical integration — several steps in the production, marketing, and processing of animals are joined together.
Veterinarian — animal doctor.
Viable — alive and capable of functioning.
Vice — bad habit in horses.
Virgin — a forest that has never been harvested.
Virus — pathogenic entities consisting of nucleic acid and a protein sheath.
Viscera — internal organs of an animal including heart, liver, and intestines.
Vitamin — complex chemicals essential for normal body functions.
Volatilization — process in which a liquid or solid changes to the gaseous phase.
Voluntary muscles — muscles that can be controlled by animals to do things such as walk and eat food.
Walk — in horses, a slow 4-beat gait.
Warm-blooded animal — animal with the ability to regulate its body temperature.
Warm season turfgrass — turfgrass adapted to the southern region of the United States which grow best at 80 to 95°F.
Warp — the tendency of wood to bend permanently due to moisture change.
Water — clear, colorless, tasteless and near odorless liquid.
Water culture — growing of plants with roots in a nutrient solution.
Water cycle — movement of water surface, to atmosphere, to surface.
Water hardness — measured as the amount of Calcium (ppm) in the solution.
Water quality — measurement of factors that affect the utilization of water.
Water resources — all aspects of water conservation and management.
Water table — level below which soil is saturated or filled with water.
Waterfowl — ducks and geese.
Waterlogged — soaked or saturated with water.
Wean — to accustom an animal to take food other than by nursing.
Weed — a plant that is not wanted.
Wether — castrated male member of the sheep family.
Wetlands — a lowland area often associated with ponds or creeks that is saturated with freshwater.
White muscle disease — muscle degeneration in calves caused by a deficiency of vitamin E.
Wholesale marketing — the marketing of a product through a middleman.
Wholesaler — person who sells to the retailer.
Wildlife — animals that are adapted to live in a natural environment without the help of humans.
Wood grain — hard and soft patterns in wood caused by growth of the tree as it adds successive annual growth rings.
Woodlot — small, privately owned forest.
Woody — hard, stiff, dark-colored growth of plants; they are winter hardy.

Worker — female bee that does the work in the hive.

X-Gal — compound that causes ice-minus bacteria to turn blue.

Xylem — vessels of the vascular bundle that carry the water and nutrients from roots to leaves.

Xylem tissue — conductive tissue responsible for transport of water and nutrients within the plant.

Yardage fee — a fee for caring for animals until they are sold.

Yearling — cattle between one and two years of age.

Yield grade — grade based on the amount of lean meat in relation to the amount of fat and bone in cattle and sheep.

Zygote — fertilized egg.

Index